# COMBINATORIAL AND ALGORITHMIC MATHEMATICS

## From Foundation to Optimization

## First Edition

**BAHA M. ALZALG**
Professor of Mathematics
The University of Jordan

kindle
direct
publishing

Published by Baha Alzalg 2022 I Kindle Direct Publishing, Washington, United States 2022/9.
Published simultaneously in Jordan.

For general information on our other products and services please contact Baha Alzalg
http://sites.ju.edu.jo/sites/alzalg/pages/contact.aspx or email b.alzalg@ju.edu.jo

The publisher also publishes this book in a variety of electronic formats. Some content that appears in print,
however, may not be available in electronic format.

*Library of Congress Cataloging-in-Publication Data: May 7, 2022.*
A certificate was issued under the seal of the Copyright Office in accordance with title 17, *United States Code*, attesting
that registration has been made for this work. The following information has been taken from the certificate and made a
part of the Copyright Office records.
Registration Number: TXu 2-285-355.
Effective Date of Registration: October 14, 2021.
Registration Decision Date: November 04, 2021.
Title of Work: Combinatorial and Algorithmic Mathematics.
Author/Owner: Baha Mahmoud Nahar Alzalg.
Author Created: Novel.
Domiciled in: The United States of America.


Combinatorial and Algorithmic Mathematics:
From Foundation to Optimization / Baha Alzalg.
   Includes bibliographical references and index.
   1. Logic.  2. Combinatorics  3. Algorithms.  4. Optimization
 Propositional Logic—Predicate Logic—Set-Theoretic Structures—Analytic Structures—Algebraic Structure—Graphs—
Recurrences—Counting—Algorithmic Analysis—Array Algorithms—Numeric Algorithms—Combinatorial Algorithms—
Linear Programming—Second-Order Cone Programming—Semidefinite Programming. Alzalg, Baha.

ISBN 9-798-353-92572-9
DOI 10.5281/zenodo.7110553

Printed in the United States of America.
8  3  5  3  9  2  5  7  2

# Contents

## PART III ALGORITHMS

## PART IV OPTIMIZATION

# PREFACE

While books of high quality have been proliferated in combinatorics, algorithms, and optimization during recent years, there has been to date no single reference work accessible to the students, covering all the major aspects of three fields starting from their foundations. We wrote this book to provide a concrete and readable text for the traditional courses in combinatorial and algorithmic mathematics with optimization that mathematics, computer science, and engineering students take following the first course in calculus. This book developed from a series of lecture notes for courses at the Ohio State University and the University of Jordan. The text is intended primarily for use in undergraduate courses in discrete structures, combinatorics, algorithms, and optimization. It was written for all students and scientists in all disciplines in which algorithms are used. The book is crowned with modern optimization methodologies. Without the optimization part, the book can be used as a textbook in a one- or two-term undergraduate course in combinatorial and algorithmic mathematics. The optimization part can be used in a one-term high-level undergraduate course, or a low-to medium-level graduate course.

The book is divided into four major parts, and each part is divided into three chapters. Part I is devoted to studying mathematical foundations. This includes mathematical logic and basic structures. This part is divided into three chapters (Chapters 1 − 3). In Chapter 1, we study the propositional logic and the predicate logic. In Chapter 2, we study basic set-theoretic structures such as sets, relations and functions. In Chapter 3, we study basic analytic and algebraic structures such as sequences, series, subspaces, convex structures, and polyhedra.

Part II is devoted to studying combinatorial structures. Discrete mathematics is the study of countable structures. Combinatorics is that area of discrete mathematics which studies how to count these objects using various representations. Part II, which is divided into three chapters (Chapters 4 – 6), studies recursion techniques, counting methods, permutations, combinations, arrangements of objects and sets, and graphs. Specifically, Chapter 4 introduces graph basics and properties, Chapter 5 presents some recurrence solving techniques, and Chapter 6 introduces some counting principles, permutations, and combinations.

Part III is devoted to studying algorithmic mathematics, which is a branch of mathematics that deals with the design and analysis of algorithms. Analyzing algorithms allows us to determine and express their efficiency. This part is divided into three chapters (Chapters 7 – 9). In Chapter 7, we discuss the asymptotic notations which are one of the important tools in algorithmic analysis. Then we dive more into determining the computational complexity of various algorithms. In Chapter 8, we present and analyze standard integer, array and numeric algorithms. In Chapter 9, we present elementary combinatorial algorithms.

Part IV is devoted to studying linear and conic optimization problems, and is divided into three chapters (Chapters 10 – 12). Chapter 10 introduces linear optimization and studies its duality and geometry. In this chapter, we also study simplex and non-simplex algorithms for linear optimization. Second-order cone programming is linear programming over vectors belonging to second-order cones. Semidefinite programming is linear programming over positive semidefinite matrices. One of the chief attractions of these conic optimization problems is their diverse applications, many in engineering. In Chapter 11, we introduce second-order cone optimization applications and algorithms. In Chapter 12, we introduce semidefinite optimization applications (especially combinatorial applications) and algorithms.

Solutions to all chapter exercises in this edition are provided in Appendix A. The references [AU94, CLRS01, CS21, MKB86, HV16, Jos89, MR04, Ros02, SH71, SEA14, KVL63] are good sources for information relative to the above topics. The reader can visit the official book website at: sites.ju.edu.jo/sites/alzalg/pages/camfobook.aspx If you find an error not listed in the errata list, which is available on the book website, please do let us know about it. Comments, criticisms, errors or suggestions should be directed to the author at the email address: b.alzalg@ju.edu.jo

*Acknowledgment*  The author would like to express his thanks to a number of individuals. The author is grateful to his colleagues: Rephael Wenger from The Ohio State University and Fuad Kittaneh from The University of Jordan, for reading parts of the book and providing some constructive comments. Some of the author's students at The Ohio State University have contributed by reading parts of the manuscript, pointing out misprints, and working on the exercises: Yiqing Li, Sarthak Mohanty, Kishore Prakash Sailaja, Luke Evers, Ron Chen, and Sery Gunawardena. The author's Ph.D. student at The University of Jordan, Asma Gafour, has also read some chapters from the optimization part and pointed out some misprints. The author is also grateful for very fruitful discussions with Kenneth Supowit, Nickalaus Painter, and Doreen Close. Great people and wonderful experience.

The author is grateful to the Department of Mathematics at the University of Jordan for giving him sabbatical and unpaid leaves he has used to work on this book. The author also thanks the Department of Computer Science and Engineering at The Ohio State University for hosting him as a visiting associate professor during the academic years of 2019/20 to 2021/22.

And of course, I would like to thank my wife, Ayat Ababneh, for her patience and belief in making this book a reality. She gave me support and help, and discussed with me some of its ideas. She also supported the family during much of this work.

I am also grateful to my father: Mahmoud N. Alzalg, my mother: Aysheh H. Alzoubi, my brother: Lewa Alzaleq, my kids: Heba, Rami and Alma Alzalg, and my sisters for their blessings and inspirations. Lastly, Alhamdulillah, I praise and thank Allah for giving me the strength and ability to work on this book.

B. M. ALZALG

*Columbus, Ohio*

**Core Outline**

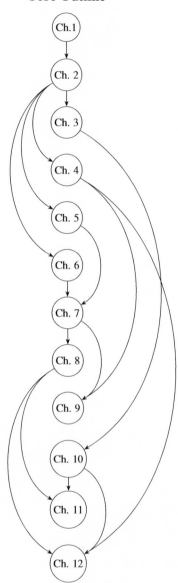

**Part I**

---

# FOUNDATIONS

# CHAPTER 1

# MATHEMATICAL LOGIC

## Contents

© 2022 by Baha Alzalg | Kindle Direct Publishing, Washington, United States 2022/9
B. Alzalg, *Combinatorial and Algorithmic Mathematics: From Foundation to Optimization*,
DOI 10.5281/zenodo.7110742

The precise definition of "logic" is quite broad and literally hundreds of logics have been studied by philosophers, mathematicians, and computer scientists. When most people say "logic", they mean either: propositional logic, or predicate logic. The propositional logic is the classical one, in which there are two possible truth values (i.e., true and false). The predicate logic extends propositional logic with the ability to explicitly speak about objects and their properties.

In this chapter, we first study the propositional logic which is the simplest, and most abstract logic we can study. After that we study the predicate logic. We start by introducing the notion of a proposition.

Throughout this chapter (and the entire book), $\mathbb{N} = \{1, 2, 3, \ldots\}$ denotes the set of all natural numbers, $\mathbb{Z} = \{0, \pm 1, \pm 2, \pm 3, \ldots\}$ denotes the set of all integers, and $\mathbb{Q} = \{a/b : a, b \in \mathbb{Z} \text{ and } b \neq 0\}$ denotes the set of all rational numbers. For example, 2 and $3/4$ are rational numbers, but $\sqrt{2}$ and $\pi$ are irrational numbers. We also let $\mathbb{R}$ denote the set of all real (rational and irrational) numbers.

## 1.1. Propositions

In this section, we define a proposition and introduce two different types of propositions with examples. To begin with, we have following definition.

> **Definition 1.1** *A proposition is a statement that can be either true or false.*

Note that the proposition must be true or false, and it cannot be both. We give the following example for more illustration.

**Example 1.1** Identify each of the following statements as a proposition or not, and add any necessary comments.

(*a*) 1+2 =3.

(*b*) 2+4 =4.

(*c*) Hillary Clinton is a former president of the United States.

(*d*) Bill Clinton is a former president of the United States.

(*e*) Be careful.

(*f*) Is Abraham Lincoln the greatest president that the United States has ever had?

(*g*) x+y =3.

**Solution** (*a*) "1+2 =3" is a proposition (a true proposition).

(*b*) "2+4 =4" is a proposition (a false proposition).

(*c*) "Hillary Clinton is a former president of the United States" is a proposition (a false proposition).

(*d*) "Bill Clinton is a former president of the United States" is a proposition (a true proposition).

(*e*) "Be careful" is not a proposition.

(*f*) "Is Abraham Lincoln the greatest president that the United States has ever had?" is not a proposition.

(*g*) "x+y =3" is not a proposition. In fact, the values of the variables have not been assigned. So, we do not know the values of x and y, and hence it neither true or false.

■

Sometimes a sentence does not provide enough information to determine whether it is true or false, so it is not a proposition. An example is the following: "Your answer to Question 13 is incorrect". The sentence does not tell us who we are talking about. If we identify the person, say "Adam's answer to Question 13 is incorrect", then the sentence becomes a proposition.

Note that, for a given statement, *'being not able to decide whether it is true or false (due to the lack in information)'* is different from *'being not able to know how to verify whether it is true or false'*. Consider the statement: "Every even integer greater than 2 can be written as the sum of two primes"[1]. This statement is the famous Goldbach's conjecture[2], which dates back to 1742. Nobody has ever proved or disproved this claim, so we do not know whether it is true or false, even though computational data suggest it is true. Nevertheless, it is a proposition because it is either true or false but not both. It is impossible for this sentence to be true sometimes, and false at other times. With the advancement of mathematics, someone may be able to either prove or disprove it in the future.

There are some sentences that cannot be determined to be either true or false. For example, the sentence: "This statement is false". This is not a proposition because we cannot decide whether it is true or false. In fact, if the sentence: "This statement is false" is true, then by its meaning it is false. On the other hand, if the sentence: "This statement is false" is false, then by its meaning it is true. Therefore, the sentence: "This statement is false" can have neither true nor false for its truth value. This type of sentence is referred to as paradoxes.[3] The study of a paradox has played a fundamental role in the development of modern mathematical logic.

Note that the sentence "This statement is false", which is self contradicting, is different from the sentence "This statement is true", which is self consistent on either choice. Nevertheless, neither is a proposition. The later type of sentence is referred to as an anti-paradox.[4] The question that arises now is: Why the sentence "This

---

[1] Prime numbers are those that are only divisible by one and themselves. For example, 7 is a prime number.
[2] Christian Goldbach (1690–1764) was a German mathematician who also studied law. He is remembered today for Goldbach's conjecture, which is one of the oldest and best-known unsolved problems in number theory and all of mathematics.
[3] A paradox is a logically self-contradictory statement that is not a proposition.
[4] An anti-paradox is a self-supporting, self-validating statement that is not a proposition.

statement is true" is not a proposition? This question is left as an exercise for the reader.

We now define atomic (or primitive) propositions. Intuitively, these are the set of smallest propositions.

> **Definition 1.2** *An atomic proposition is one whose truth or falsity does not depend on truth or falsity of any other proposition.*

According to Definition 1.2, we find that all the propositions in items $(a) - (d)$ of Example 1.1 are atomic.

> **Definition 1.3** *A compound proposition is a proposition that involves the assembly of multiple atomic propositions.*

The following connectives allow us to build up compound propositions:

$$\text{AND } (\wedge), \quad \text{OR } (\vee), \quad \text{NOT } (\neg), \quad \text{IF-THEN } (\rightarrow), \quad \text{IFF } (\leftrightarrow).$$

**Example 1.2**   The following proposition is compound.

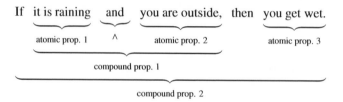

Note that

- compound proposition 1 = (atomic proposition 1) $\wedge$ (atomic proposition 2),

- compound proposition 2 = (compound proposition 1) $\rightarrow$ (atomic proposition 3).

We have the following remark.

> **Remark 1.1** *Sentences in natural languages such as English can often be ambiguous and words can have different meanings on the context in which they are used.*

To illustrate Remark 1.1, we consider the following sentences:

- The sentence *"You can download Whats App or Skype to ring friends" could mean that you can only download one of the applications, or it could mean that you can download just one or download both.*

- *The sentence* "I smelled a chlorine-like odor and felt ill" *implies that the odor of chlorine made you sick, but the sentence* "I am majoring in CS and minoring Math" *does not imply that majoring in CS caused you to minor in Math.*

- *The sentence "I went to Chicago and took a plane" could mean that you took a plane to travel to Chicago, or it could mean that you went to Chicago and then took a plane from Chicago to another destination, such as Las Vegas.*

In mathematics and computer science, it is important to avoid ambiguity and for sentences to have a precise meaning. This is why people have invented artificial languages such as Java.

**Notations** Rather than writing out propositions in full, we will abbreviate them by using propositional variables: P, Q, R, ..., etc.

**Example 1.3 (Example 1.2 revisited)** In Example 1.2, let P be the proposition "it is raining". Q be the proposition "you are outside", and R be the proposition "you get wet". Then we can abbreviate the compound proposition

"If it is raining and you are outside, then you get wet" as $P \wedge Q \rightarrow R$.
$\qquad\quad$ P $\qquad\quad$ ∧ $\qquad\quad$ Q $\qquad\qquad$ R

In the next section, we will learn more about the connectives AND, OR, NOT, IF-THEN, and IFF.

## 1.2. Logical operators

In this section, we study the following logical operators or connectives: Negation, conjunction, disjunction, exclusive disjunction, implication, and double implication.

Before starting with the logical operators, we introduce the truth tables which help us understand how such operators work by calculating all of the possible return values of atomic propositions.

> **Definition 1.4** *A truth table is a mathematical table used to show the truth or falsity of a compound proposition depending on the truth or falsity of the atomic propositions from which it is constructed.*

Examples of truth tables will be seen very frequently throughout this chapter.

### Negation, conjunction and disjunction

This part is devoted to introducing the logical operators: Negation (NOT), conjunction (AND), disjunction (OR), and exclusive disjunction (XOR).

**Negation** "NOT" transforms a proposition into its opposite truth value via a negation.

**Definition 1.5** *If P is an arbitrary proposition, then the negation of P is written ¬P (NOT P) which is true when P is false and is false when P is true.*

The truth table for "¬" is shown below.

| P | ¬P |
|---|----|
| T | F  |
| F | T  |

**Example 1.4** If P is the proposition "it is raining", then ¬P is the proposition "it is not raining".

**Remark 1.2** *Negation does not mean "opposite". For instance, if x is a real number and ¬P is the proposition "x is not positive", then you cannot conclude that ¬P is the proposition "x is negative" because x could be 0, which is neither positive nor negative.*

**Double negation**   In real numbers, two negative signs cancel each other out. Similarly, in propositional logic, two negations also cancel each other out. The double negation of P is ¬(¬P) or ¬¬P. The truth table for "¬¬" is shown below.

| P | ¬P | ¬¬P |
|---|----|-----|
| T | F  | T   |
| F | T  | F   |

**Logical equivalence**   Logical equivalence is a type of relationship between two propositional formulas in propositional logic.

**Definition 1.6** *When the truth values for two propositional formulas P and Q are the same, the propositional formulas are called logically equivalent, and this is denoted as $P \equiv Q$ or $P \Longleftrightarrow Q$.*

For instance, from the truth table for "¬¬", we have $P \equiv \neg\neg P$. This equivalence is called the double negation law.

The negation connective is unary as it only takes one argument. The upcoming connectives are binary as they take two arguments.

**Conjunction**   "AND" connects two or more propositions via a conjunction.

**Definition 1.7** *If P and Q are arbitrary propositions, then the conjunction of P and Q is written $P \wedge Q$ (P AND Q) which is true when both P and Q are true.*

The truth table for "∧" is shown below.

| P | Q | P ∧ Q |
|---|---|-------|
| T | T | T |
| T | F | F |
| F | T | F |
| F | F | F |

**Example 1.5** If P is the proposition "it is raining" and Q is the proposition "I have an umbrella", then P∧Q is the proposition "it is raining and I have an umbrella".

*Disjunction*   "OR" connects two or more propositions via a disjunction.

> **Definition 1.8** *If P and Q are arbitrary propositions, then the disjunction of P and Q is written P∨Q (P OR Q) which is true when either P is true, or Q is true, or both P and Q are true.*

The truth table for "∨" is shown below.

| P | Q | P ∨ Q |
|---|---|-------|
| T | T | T |
| T | F | T |
| F | T | T |
| F | F | F |

**Example 1.6** If P is the proposition "I get married" and Q is the proposition "I live alone", then P∨Q is the proposition "I get married or I live alone".

*Exclusive disjunction*   In English, the word "or" can be used in two different ways: inclusively or exclusively. For example, let us say that: a computer science course might require that a student be able to program in either Java or C++ before enrolling in the course. In this case, "or" is used inclusively: a student who can program in both Java and C++ would be eligible to take the course. On the other hand, for another example, let us say that: I get married or stay single. In this case, "or" is used exclusively: you can either get married or stay single, but not both.

> **Definition 1.9** *If P and Q are arbitrary propositions, then the exclusive-disjunction of P and Q is written P ⊕ Q (P XOR Q) which is true when exactly one of P and Q is true and false otherwise.*

The truth table for "⊕" is shown below.

| P | Q | P ⊕ Q |
|---|---|---|
| T | T | F |
| T | F | T |
| F | T | T |
| F | F | F |

## Implication and double implication

In this part, we study the following logical operators: implication (IF-THEN) and double implication (IFF).

***Implication*** "IF-THEN" connects two or more propositions via an implication, thus forming a conditional proposition.

> **Definition 1.10** *Let P and Q be arbitrary propositions. The implication P implies Q (also-called the conditional of P and Q) is written P→Q (IF P THEN Q), which is false when P is true and Q is false, and true otherwise. Here, P is called the hypothesis and Q is called the conclusion.*

**Example 1.7** If P is the proposition "you live in Russia" and Q is the proposition "you live in the coldest country in the world", then P→Q is the proposition "If you live in Russia, then you live in the coldest country in the world". ∎

We have five different ways to read P → Q, as is seen in the following remark.

> **Remark 1.3** *Let P and Q be two atomic propositions. The following statements have the same meaning as the conditional statement "If P then Q".*
>
> (*a*)  *Q if P.*                    (*c*)  *Q is necessary for P.*
>
> (*b*)  *P is sufficient for Q.*     (*d*)  *P only if Q.*

**Example 1.8** Consider the implication:

- *If a person is president of the United States, then s/he is at least 35 years old.*

According to Remark 1.3, the above implication can be restated in the following equivalent ways:

- *A person is at least 35 years old if s/he is president of the United States.*

- *Being president of the United States is sufficient for being at least 35 years old.*

- *Being at least 35 years old is necessary for being president of the United States.*

- *A person is president of the United States only if s/he is at least 35 years old.*

■

The truth table for "→" is shown below.

| P | Q | P → Q |
|---|---|-------|
| T | T | T |
| T | F | F |
| F | T | T |
| F | F | T |

Therefore, the implication is true if we are on the lines (tt), (ft), or (ff) only, and it is false if we are on the line (tf). We now justify the truth table for "→". Suppose that Sara is a math teacher and that she told her students in a class before the first midterm exam the statement that "If everyone in the class gets an A on the midterm, then I will make cookies for the class". Note that this statement says nothing about what will happen if not everyone gets an A. So, if a student did not get an A, Sara is free to make cookies or not and she will have told the truth in either case. Thus, the only case where Sara did not tell the truth (i.e., the implication is false) is the case that if everyone got A's in the midterm (i.e., the hypothesis is true) but Sara did not make cookies for the class (i.e., the conclusion is false). This is the case that made the (tf)-line in the implication truth table.

**Example 1.9** Let $\mathbb{R}$ denote the set of real numbers and $x \in \mathbb{R}$.[5] Decide whether each of the following implications is true or false. Justify your answer.

(a) If $x \geq 10$, then $x \geq 0$.

(b) If $x \geq 0$, then $x \geq 10$.

(c) If $x^2 \geq 0$, then $x = 42$.

(d) If $x^2 < 0$, then $x = 42$.

(e) If the Riemann hypothesis is true, then $x^2 \geq 0$.[6]

**Solution** The very first look tells us that the proposition in item (a) is true, and those in items (b) and (c) are false, but we might be uncertain about the propositions in items (d) and (e). Since all these propositions have the P → Q form, we shall analyze each item according to the truth table for "→". As mentioned earlier, the conditional statement P → Q is true if we are on the lines (tt), (ft), or (ff) only, and it is false if we are on the line (tf).

---

[5]To express the fact that an object $x$ is a member of a set $A$, we write $x \in A$ (see Section 2.2 for definitions).
[6]The Riemann hypothesis states that all non trivial zeros of a mathematical function, the Riemann zeta function, have a real part equal to 0.5. This hypothesis is still a mysterious unsolved problem of mathematics.

(*a*) Since the hypothesis is $x \geq 10$ and the conclusion is $x \geq 0$, we are

$$\text{on the line:} \begin{cases} \text{(ff),} & \text{if } x < 0; \\ \text{(ft),} & \text{if } 0 \leq x < 10; \\ \text{(tt),} & \text{if } 10 \leq x. \end{cases}$$

So, we cannot be on the (tf)-line for any $x \in \mathbb{R}$. Thus, the proposition is true.

(*b*) Since the hypothesis is $x \geq 0$ and the conclusion is $x \geq 10$, we are

$$\text{on the line:} \begin{cases} \text{(ff),} & \text{if } x < 0; \\ \text{(tf),} & \text{if } 0 \leq x < 10; \\ \text{(tt),} & \text{if } 10 \leq x. \end{cases}$$

So, there is $x \in \mathbb{R}$ that makes us on the (tf)-line. Thus, the proposition is false.

(*c*) Take $x = 2$, then $x^2 = 4 \geq 0$, but $x \neq 42$. So, we are on the (tf)-line. Hence, the proposition is false.

(*d*) P is always false, so we cannot be on the (tf)-line. Hence, the proposition is true.

(*e*) The Riemann hypothesis is an unsolved conjecture in mathematics. That is, no one knows if it is true or false at this time. But this does not prevent us from answering the question for this item. In fact, as Q is always true, we cannot be on the (tf)-line. Hence, the proposition is true.

■

■ **Example 1.10** Use a truth table to show that $P \rightarrow Q \equiv \neg P \vee Q$. This equivalence is called the implication law.

**Solution** We prove that two propositional formulas are logically equivalent by showing that their truth values are the same. A truth table that contains the truth values for $P \rightarrow Q$ and $\neg P \vee Q$ is shown below and ends up the proof.

| P | Q | $P \rightarrow Q$ | $\neg P$ | $\neg P \vee Q$ |
|---|---|---|---|---|
| T | T | T | F | T |
| T | F | F | F | F |
| F | T | T | T | T |
| F | F | T | T | T |

■

***Why implications are important in mathematics?*** Implications are important in mathematics because many mathematical theorems can be restated in the IF-THEN form, which enables us to prove them. See, for instance, the theorem that is stated and proved in the following example.

**Example 1.11** Prove the result of the following theorem.

*Theorem: The sum of two even integers is even.*

Solution   To prove the theorem, we restate it in IF-THEN form:

*Theorem Restatement: If x and y are both even integers, then x + y is even.*

**Proof** If $x$ and $y$ are even, then each of $x$ and $y$ is the product of 2 and an integer. That is,

$$x = 2n \text{ and } y = 2m, \text{ for some } n, m \in \mathbb{Z},$$

where $\mathbb{Z}$ denotes the set of integers. Then $x + y = 2n + 2m = 2(n + m)$. Since the sum of any two integers is an integer, this means that $x + y$ is the product of 2 and an integer, and hence is even. The proof is complete. ∎

***Contrapositive of an implication***   The contrapositive of the implication $P \rightarrow Q$ is the implication $\neg Q \rightarrow \neg P$.

**Example 1.12** The contrapositive of the proposition "If today is Monday, then Adam has a class today" is "If Adam do not have a class today, then today is not Monday". ∎

The truth table for the contrapositive is given below.

| P | Q | $\neg$ Q | $\neg$ P | $\neg Q \rightarrow \neg P$ |
|---|---|---|---|---|
| T | T | F | F | T |
| T | F | T | F | F |
| F | T | F | T | T |
| F | F | T | T | T |

From the truth table for contrapositive and that for the implication, we conclude that $P \rightarrow Q \equiv \neg Q \rightarrow \neg P$. This is called the law of contrapositive.

The contrapositive is important and useful in mathematics for two reasons:

- Once a theorem in IF-THEN form is proven, there is no need to prove the contrapositive of the theorem. A proof by contrapositive, or a proof by contraposition, is a rule of inference used in proofs, where one infers a conditional statement from its contrapositive.

- If we are trying to prove a theorem in IF-THEN form, sometimes it is easier to prove its contrapositive.

Now, we give an example supporting the first reason.

**Example 1.13** The Pythagorean theorem states that in any right triangle, the square of the length of the hypotenuse (the side opposite the right angle) is to equal the sum of the squares of the lengths of the legs of the right triangle. The proof of this theorem is beyond the scope of our discussion.

In IF-THEN form, the Pythagorean theorem can be rewritten as: "If a triangle is right with hypotenuse c and legs a and b, then $a^2 = b^2 + c^2$", or equivalently: "If a triangle is right-angled, then its three sides satisfy a Pythagorean triple", where a Pythagorean triple is a set of positive integers, $a, b$ and $c$, that fits the rule $a^2 = b^2 + c^2$.

Since Pythagorean theorem was proven by Pythagoras, its contrapositive: "If triangle sides do not satisfy a Pythagorean triple, then the triangle is not right-angled" is obtained for free. ∎

We also give an example supporting the second reason.

**Example 1.14** Let $x$ be a positive integer. Prove the result of the following theorem.

*Theorem: If $x^2$ is odd, then x is odd.*

**Solution** The implication in the theorem statement is equivalent to its contrapositive:

*If x is even, then $x^2$ is even,*

which is easier to prove. Let $x$ be an even integer, then $x$ can be written in the form $x = 2k$, for some positive integer $k$. It follows that $x^2 = (2k)^2 = 2(2k^2)$. Thus, $x^2$ is even. The proof is complete. ∎

Note that a proof by contraposition is different than the so-called direct proof. In a direct proof, we write a sequence of statements which are either evident or evidently follow from previous statements, and whose last statement is the desired conclusion (the one to be proved). For instance, in Example 1.14, when we proved the statement that "*If x is even, then $x^2$ is even*" we used a direct proof.

***Converse of an implication*** The converse of the implication P → Q is the implication Q → P. The truth table for the converse is given below.

| P | Q | P → Q | Q → P |
|---|---|-------|-------|
| T | T | T | T |
| T | F | F | T |
| F | T | T | F |
| F | F | T | T |

From the above truth table, it is clear that Q → P ≢ P → Q.

**Example 1.15** The converse of the proposition "If today is Monday, then Adam has a class today" is "If Adam has a class today, then today is Monday". Suppose that all Adam's classes are on Mondays, Wednesdays and Fridays though, then the original implication is true while its converse is false. This illustrates why we found that P → Q and Q → P are not logically equivalent. ∎

***Inverse of an implication***    The inverse of the implication $P \to Q$ is the implication $\neg P \to \neg Q$. The truth table for the inverse is given below.

| P | Q | $\neg P$ | $\neg Q$ | $\neg P \to \neg Q$ |
|---|---|---|---|---|
| T | T | F | F | T |
| T | F | F | T | T |
| F | T | T | F | F |
| F | F | T | T | T |

It is clear that a conditional statement is not logically equivalent its inverse. That is, $P \to Q \not\equiv \neg P \to \neg Q$. It is also clear, from the truth table for the inverse and that for converse, that the inverse and converse of any conditional statement are logically equivalent. That is, $\neg P \to \neg Q \equiv Q \to P$.

**Example 1.16**  The inverse of the proposition "If today is Monday, then Adam has a class today" is "If today is not Monday, then Adam does not have a class today".

We now give the last logical operator, which is the double implication (IFF, which is read "if and only if").

***Double implication***    The following definition defines a double implication as the combination of an implication and its converse.

> **Definition 1.11**  *Let P and Q be arbitrary propositions. The double implication (also-called the biconditional) of P and Q is written $P \leftrightarrow Q$ (P IFF Q), which is true precisely when either P and Q are both true or P and Q are both false.*

**Example 1.17**  If P is the proposition "you live in Russia" and Q is the proposition "you live in the coldest country in the world", then $P \leftrightarrow Q$ is the proposition "You live in Russia iff you live in the coldest country in the world".

The truth table for "$\leftrightarrow$" is given below.

| P | Q | $P \leftrightarrow Q$ |
|---|---|---|
| T | T | T |
| T | F | F |
| F | T | F |
| F | F | T |

Note that $P \leftrightarrow Q$ means that P is both necessary and sufficient for Q. One of the chapter exercises asks to prove that $P \leftrightarrow Q \equiv (P \to Q) \wedge (Q \to P)$.

To prove an "if and only if" statement, you have to prove two directions. To disprove an "if and only if" statement, you only have to disprove one of the two

| $P$ | $Q$ | $R$ | $\neg R$ | $P \wedge Q$ | $(P \wedge Q) \leftrightarrow \neg R$ |
|-----|-----|-----|----------|--------------|----------------------------------------|
| T | T | T | F | T | F |
| T | T | F | T | T | T |
| T | F | T | F | F | T |
| T | F | F | T | F | F |
| F | T | T | F | F | T |
| F | T | F | T | F | F |
| F | F | T | F | F | T |
| F | F | F | T | F | F |

Table 1.1: A truth table for $(P \wedge Q) \leftrightarrow \neg R$.

directions. For instance, letting $x$ be an integer, to prove that $x^2 = 0$ iff $x = 0$, we have to prove two the directions: First, we prove that if $x^2 = 0$ then $x = 0$. And second, we prove that if $x = 0$ then $x^2 = 0$ (the proofs are trivial). To disprove $x^2 = 4$ iff $x = 2$, we only disprove the direction: If $x^2 = 4$ then $x = 2$ (the disproof is trivial: take $x = -2$).

🔲 **Example 1.18** Construct a truth table for the compound proposition $(P \wedge Q) \leftrightarrow \neg R$.

**Solution**   Since we have three propositional variables, $P, Q$ and $R$, the number of rows in our truth table is eight rows (see Remark 1.4). The truth table for $(P \wedge Q) \leftrightarrow \neg R$ is given in Table 1.1. ▪

We end this section with the following remark.

> **Remark 1.4** *The number of rows in a truth table for a propositional formula with n variables is $2^n$ rows.*

For instance, the numbers of rows in the truth tables for the propositional formulas $\neg P$, $P \wedge Q$ and $(P \wedge Q) \leftrightarrow \neg R$ are $2^1, 2^2$ and $2^3$ rows, respectively.

## 1.3. Propositional formulas

In this section, we study special propositional formulas, which are tautologies, contradictions and contingencies. Then we study how to negate compound propositions and show how to derive propositional formulas. Before we dive into these, we illustrate the order in which the logical operators will be applied.

*Order of logical operations*   In arithmetic, multiplication has higher precedence than addition. Hence, the value of the expression 4+5×2 is not 18, but 14. Similarly, in

| Operation(s) | Operator(s) | Precedence |
|---|---|---|
| Parentheses | ( ) | 1 |
| Negation (NOT) | ¬ | 2 |
| Conjunction (AND) | ∧ | 3 |
| Disjunction (OR, XOR) | ∨, ⊕ | 4 |
| Implication (IF-THEN) | → | 5 |
| Double implication (IFF) | ↔ | 6 |

Table 1.2: The precedence order of logical operations.

propositional logic, logical operators have operator precedence the same as arithmetic operators. We preserve the precedence order specified in Table 1.2.

**Example 1.19** Determine the value of the propositional formula $P \vee Q \wedge R$ the propositional variables $P, Q$ and $R$ have values of false, true and false, respectively.

**Solution** From Table 1.2, the operator $\wedge$ has higher precedence than the operator $\vee$. So $P \vee Q \wedge R = P \vee (Q \wedge R)$. Plugging in the values of the variables, we have $F \vee (T \wedge F)$, which has the value of false. ∎

Note that, for longer expressions, we recommend using parentheses to group expressions and control which ones are evaluated first. For example, the propositional formula $P \vee Q \wedge R \to P \vee R$ is written as $(P \vee (Q \wedge R)) \to (P \vee R)$, and the propositional formula $\neg P \vee Q \leftrightarrow R \to S$ is written as $((\neg P) \vee Q) \leftrightarrow (R \to S)$. Note also that if we have two logical operators with the same precedence, then their associativity is from left to right. For example, $P \oplus Q \vee R$ is written as $(P \oplus Q) \vee R$, and $P \to Q \to R$ is written as $(P \to Q) \to R$.

### Tautologies, contradictions and contingencies

Tautologies, contradictions and contingencies are propositional formulas of particular interest in propositional logic. We have the following definition.

**Definition 1.12** *A propositional formula is called*

(a) *a tautology if it is always true,*

(b) *a contradiction if it is always false,*

(c) *a contingency if it is neither a tautology nor a contradiction.*

Let P be an arbitrary proposition. From Table 1.3, it is seen that $P \vee \neg P$ is a tautology, $P \wedge \neg P$ is a contradiction, and $P \to (P \wedge \neg P)$ is a contingency.

**Example 1.20** Use a truth table to determine if each of the following propositional formulas is a tautology, contradiction or contingency.

| P | ¬P | P ∨ ¬P | P ∧ ¬P | P → (P ∧ ¬P) |
|---|----|--------|--------|--------------|
| T | F  | T      | F      | F            |
| F | T  | T      | F      | T            |

Table 1.3: A tautology, contradiction and contingency.

| P | Q | P ∧ Q | P ∨ Q | (P ∧ Q) → P | (P ∨ Q) → P |
|---|---|-------|-------|-------------|-------------|
| T | T | T     | T     | T           | T           |
| T | F | F     | T     | T           | T           |
| F | T | F     | T     | T           | F           |
| F | F | F     | F     | T           | T           |

Table 1.4: A truth table for the propositional formulas of Example 1.20.

(a)  (P ∧ Q) → P.

(b)  (P ∨ Q) → P.

(c)  (P ∨ Q) → (¬P ∨ Q).

**Solution**   From Table 1.4, it is seen that (P ∧ Q) → P is a tautology and (P ∨ Q) → P is a contingency. This answers items (a) and (b). Item (c) is left as an exercise for the reader.   ∎

Last, but not least, it is worth mentioning that Definition 1.12 suggests an alternative definition to the notion of logical equivalence: Two propositions P and Q are logically equivalent if P ↔ Q is a tautology.

A proof by contradiction is a common technique of proving mathematical statements. In a proof by contradiction, to prove that a statement $P$ is true, we begin by assuming that $P$ false, and show that this assumption leads to a contradiction, then this contradiction tells us that our original assumption is false, and hence the statement $P$ is true. The following example illustrates how this proof technique is used.

**Example 1.21** Prove that $\sqrt{2}$ is an irrational number.

**Solution**   Suppose, in the contrary, that $\sqrt{2}$ is rational. It follows that there are two integers, say $a$ and $b$, such that $\sqrt{2} = a/b$. Without loss of generality, we can assume that $a$ and $b$ have no common factors (otherwise, we can reduce the fraction and write it in its simplest form). Multiplying both sides of the equation $\sqrt{2} = a/b$ by $b$ and squaring, we get $a^2 = 2b^2$. This means that $a^2$ is even. From the theorem stated in Example 1.14, $a$ itself must be even. Thus, $a = 2m$ for some integer $m$. It follows that

$$2b^2 = a^2 = (2m)^2 = 4m^2,$$

| $P$ | $Q$ | $\neg P$ | $\neg Q$ | $P \wedge Q$ | $\neg(P \wedge Q)$ | $\neg P \vee \neg Q$ |
|-----|-----|----------|----------|--------------|---------------------|----------------------|
| T   | T   | F        | F        | T            | F                   | F                    |
| T   | F   | F        | T        | F            | T                   | T                    |
| F   | T   | T        | F        | F            | T                   | T                    |
| F   | F   | T        | T        | F            | T                   | T                    |

Table 1.5: A truth table verifying the first DeMorgan's law.

and hence, after dividing by 2, we have $b^2 = 2m^2$. This means that $b^2$. By the same theorem stated in Example 1.14, $b$ itself must be even. We have shown that both $a$ and $b$ are even, and hence they are both multiples of 2. This contradicts the fact that $a$ and $b$ have no common factors. This contradiction tells us that our original assumption is false, and hence $\sqrt{2}$ must not be rational. Thus, $\sqrt{2}$ is irrational. The proof is complete. ∎

## Negating compound propositions

In this part, we study how to negate negations, conjunctions, disjunctions and implications.

***Negating negations***   When we say "Results of the numerical experiment are not inconclusive" means "Results of the numerical experiment are conclusive". So, if you negate a negation they effectively cancel each other out. This leads us to the double negation law $\neg\neg P \equiv P$, which was previously stated in this chapter.

***Negating conjunctions***   Suppose your roommate made the statement "I will get an A in Software I and an A in Foundations I". If your roommate's prediction is not correct, then s/he did not get an A in Software I or did not get an A in Foundations I. This leads us to the first DeMorgan's law:

$$\neg(P \wedge Q) \equiv \neg P \vee \neg Q.$$

Table 1.5 verifies the first DeMorgan's law.

***Negating disjunctions***   Suppose your roommate made the statement "I will get an A in Software I or an A in Foundations I". If your roommate's prediction is not correct, then s/he did not get an A in Software I and did not get an A in Foundations I. This leads us to the second DeMorgan's law:

$$\neg(P \vee Q) \equiv \neg P \wedge \neg Q.$$

Constructing a truth table to verify the second DeMorgan's law is left as an exercise for the reader.

Note that DeMorgan's laws can be extended to expressions with more than one conjunction and disjunction as follows:

$$\neg(P \wedge Q \wedge R) \equiv \neg P \vee \neg Q \vee \neg R,$$
$$\neg(P \vee Q \vee R) \equiv \neg P \wedge \neg Q \wedge \neg R.$$

**Example 1.22** Negate and simplify the following two compound propositions.

(*a*) $(P \vee Q) \wedge \neg R.$                    (*b*) $(P \vee Q) \vee (R \wedge \neg P).$

Solution   (*a*)   $\neg((P \vee Q) \wedge \neg R)$   $\equiv$   $\neg(P \vee Q) \vee \neg\neg R$      (DeMorgan's law)

$\equiv$   $\neg(P \vee Q) \vee R$      (Double negation law)

$\equiv$   $(\neg P \wedge \neg Q) \vee R.$      (DeMorgan's law).

(*b*) This item is left as an exercise for the reader.

**Negating implications**   Suppose your (rich) roommate who is stingy made the request that "If you go to the supermarket, buy him organic food with your money". Then you realized that it is time to teach him an unforgettable lesson for having these miserly attitudes and asking you this. What should you do? Is the correct answer by not go to supermarket and not buy him organic food? No. In fact, the best way is "You go to the supermarket and you do not buy him organic food". This leads us to the law

$$\neg(P \rightarrow Q) \equiv P \wedge \neg Q, \tag{1.1}$$

which we call the implication negation law. The logical equivalence in (1.1) immediately follows by noting that

$\neg(P \rightarrow Q)$   $\equiv$   $\neg(\neg P \vee Q)$      (Implication law)

$\equiv$   $(\neg\neg P) \wedge \neg Q$      (DeMorgan's law)

$\equiv$   $P \wedge \neg Q.$      (Double negation law)

We can negate and simplify conditional statements either by directly applying the implication negation law, such as

$$\neg((P \wedge Q) \rightarrow R) \equiv P \wedge Q \wedge \neg R, \tag{1.2}$$

or by applying earlier logical equivalences, such as

$\neg(P \rightarrow (Q \vee R))$   $\equiv$   $P \wedge \neg(Q \vee R)$      (Implication negation law)

$\equiv$   $P \wedge \neg Q \wedge \neg R.$      (DeMorgan's law)

$(1.3)$

In Exercise 1.13, we learn how to negate exclusive disjunctions and double implications.

**Universal sets**   We have seen that we can transform any propositional formula into an equivalent formula in DNF that uses only operators from the set $\{\wedge, \vee, \neg\}$.

> **Definition 1.13** *A set of operators, such as* $\{\wedge, \vee, \neg\}$, *that can be used to express any proposition is called a universal set.*

Note that the set of operators $\{\vee, \neg\}$ is universal. To see this, it is noted the proposition $P \vee Q$ can be expressed using just $\neg$ and $\wedge$. In fact, using double negation and DeMorgan's laws, we have

$$P \vee Q \equiv \neg\neg(P \vee Q) \equiv \neg(\neg P \wedge \neg Q).$$

So, we can for instance write $P \vee (\neg Q \wedge R) \equiv \neg[\neg P \wedge \neg(\neg Q \wedge R)]$.

Note also that the set of operators $\{\wedge, \neg\}$ is universal, while $\{\vee, \wedge\}$ is not.

## Modeling using propositional logic

Propositional logic modeling is the process of abstracting a propositional logic problem from the real world. We have the following example.

### Example 1.23

(*a*)  Model the following statement using propositional logic tools.

"If it is raining and Paul does not have an umbrella, then he will get wet".

(*b*)  Simplify the contrapositive of the propositional formula obtained in item (*a*), then translate it back to English.

Solution   (*a*)  Let $P, Q$ and $R$ be three propositional variables defined below.

$$\underbrace{\text{If it is raining}}_{P} \text{ and } \underbrace{\text{Paul does not have an umbrella}}_{\neg Q}, \text{ then } \underbrace{\text{he will get wet}}_{R}.$$

(Here, $Q$ is the proposition "Paul has an umbrella"). Therefore, the statement can be symbolized as $(P \wedge \neg Q) \rightarrow R$.

(*b*)  The contrapositive of the propositional formula $(P \wedge \neg Q) \rightarrow R$ is the propositional formula $\neg R \rightarrow \neg(P \wedge \neg Q)$. Using the DeMorgan's and double negation laws, this formula can be simplified to $\neg R \rightarrow (\neg P \vee Q)$. Translating back to English, we get $\underbrace{\text{If Paul did not get wet}}_{\neg R}, \text{then } \underbrace{\text{it was not raining}}_{\neg P} \text{ or } \underbrace{\text{he had an umbrella}}_{Q}$.

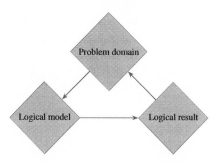

Figure 1.1: Conceptual relationships among problem domain, modeling and solution strategies.

The general scheme for solving a logical word problem is as follows. We construct a propositional logical model from a logical word problem statement in a problem domain. The model is then solved to obtain a logical conclusion, which can be finally translated back into the problem domain. See Figure 1.1.

We have the following examples.

**Example 1.24**  Negate the following statements.

(*a*)  If Sara gets an A in Foundations I and COVID-19 disappears, then she will travel to San Francisco.

(*b*)  If COVID-19 disappears, then Adam will travel to San Francisco or Los Angeles.

Solution   (*a*)  This statement can be symbolized as the propositional formula $(P \wedge Q) \to R$, where $P$ is "Sara gets an A in Foundations I", $Q$ is "COVID-19 disappears" and $R$ is "Sara will travel to San Francisco". According the equivalence (1.2), the negation is $P \wedge Q \wedge \neg R$. Translating back to English, we get the statement "Sara got an A in Foundations I, COVID-19 disappeared and she did not travel to San Francisco", which is the negation of the original statement.

(*b*)  Let P be the proposition "COVID-19 disappears", Q be the proposition "Adam will travel to San Francisco" and R be the proposition "Adam will travel to Los Angeles". The given statement can then be represented as the propositional formula $P \to (Q \vee R)$, which, using the logical equivalence (1.3), has the negation $P \wedge \neg Q \wedge \neg R$. Translating back to English, we get the statement "COVID-19 disappeared and Adam did not travel to San Francisco nor to Los Angeles", which is the negation of the original statement.

| $L$ | $M$ | $N$ | $\neg L$ | $N \vee M$ | $M \rightarrow L$ | $\neg L \wedge (N \vee M) \wedge (M \rightarrow L)$ | (1.4) |
|---|---|---|---|---|---|---|---|
| T | T | T | F | T | T | F | T |
| T | T | F | F | T | T | F | T |
| T | F | T | F | T | T | F | T |
| T | F | F | F | F | T | F | T |
| F | T | T | T | T | F | F | T |
| F | T | F | T | T | F | F | T |
| F | F | T | T | T | T | T | T |
| F | F | F | T | F | T | F | T |

Table 1.6: A truth table verifying that the conditional statement in (1.4) is a tautology.

**Example 1.25** Consider the following propositional logic word problem:

Problem statement: 
$\begin{cases} \text{Laila is not a Lebanese.} \\ \text{Nora is a New Zealander or Maria is a Macedonian.} \\ \text{If Maria is a Macedonian then Laila is a Lebanese.} \end{cases}$

Can you conclude that Nora is a New Zealander? Justify your reasoning and conclusion by linking them to the propositional logic.

**Solution** Let $L$ be "Laila is a Lebanese", $M$ be "Maria is a Macedonian", and $N$ be "Nora is a New Zealander", Then the problem statement can be formulated in the following propositional logical model.

$$\text{Problem model:} \begin{cases} \neg L, \\ N \vee M, \\ M \rightarrow L. \end{cases}$$

The desired conclusion is an affirmative answer to the following question.

Is the propositional formula $[(\neg L) \wedge (N \vee M) \wedge (M \rightarrow L)] \longrightarrow N$ a tautology? (1.4)

Otherwise, we cannot conclude that Nora is indeed a New Zealander. It is seen from Table 1.6 that Nora is a New Zealander. ∎

***Propositional formulas in computer programs*** Propositions and logical connectives arise all the time in computer programs. For example, consider the code snippet given[7] in Algorithm 1.1.

---

[7]"If-statement" written in the code is formally introduced in Section 7.1.

---

**Algorithm 1.1:** A code snippet

---

1: **if** $(x > 0 \parallel (x \leq 0 \mathbin{\&\&} y > 100))$ **then**

2: | ⋮

3: | (further instructions)

4: **end**

---

In Algorithm 1.1, $\parallel$ denotes the logical OR operator and $\&\&$ denotes the logical AND operator. The condition in the "if-statement" can be simplified. Letting $P$ be "x > 0" and $Q$ be "y > 100", then the condition can be symbolized as $P \vee (\neg P \wedge Q)$. In Example 1.26(*a*) (see also Example 1.27(*a*)), we will prove that $P \vee (\neg P \wedge Q) \equiv P \vee Q$. This means that we can simplify the code snippet without changing the program's behavior. See Algorithm 1.2. Simplifying expressions in software can increase the speed of your program.

---

**Algorithm 1.2:** The code snippet in Algorithm 1.1 revisited

---

1: **if** (x > 0 $\parallel$ y > 100) **then**

2: | ⋮

3: | (further instructions)

4: **end**

---

### Deriving logical equivalences

We can derive (and prove) logical equivalences either by constructing corresponding truth tables or by applying earlier logical equivalences.

***Using truth tables to prove logical equivalences***   If the truth values of two propositional formulas are identical under all possible interpretations [8], then they are logically equivalent. We have the following example.

▨ **Example 1.26**   Prove or disprove each of the following logical equivalences.[9]

(*a*)  $P \vee (\neg P \wedge Q) \equiv P \vee Q$.        (*b*)  $P \vee (Q \oplus R) \equiv (P \vee Q) \oplus (P \vee R)$.

**Solution**   From the last two columns of Table 1.7, it is seen that the truth values of the propositional formulas $P \vee (\neg P \wedge Q)$ and $P \vee Q$ are identical under all interpretations, hence they are logically equivalent. This proves the desired result in item (a). Item (b) is left for the reader an exercise.      ▨

---

[8] An interpretation assigns a truth value to each propositional variable.

[9] "Prove or disprove" means that either you choose to give a proof that the given statement is correct or you give a counterexample (assignments for the variables $P, Q, \ldots$) that shows the given statement is incorrect.

| $P$ | $Q$ | $\neg P$ | $\neg P \wedge Q$ | $P \vee (\neg P \wedge Q)$ | $P \vee Q$ |
|-----|-----|----------|-------------------|-----------------------------|------------|
| T | T | F | F | T | T |
| T | F | F | F | T | T |
| F | T | T | T | T | T |
| F | F | T | F | F | F |

Table 1.7: A truth table verifying $P \vee (\neg P \wedge Q) \equiv P \vee Q$.

**_Applying earlier equivalences to prove logical equivalences_**    Recall that a direct proof is a sequence of statements which are either givens or deductions from previous statements, and whose last statement is the conclusion to be proved. In particular, we use a sequence of logical equivalences to prove the logical equivalence of two compound propositions. In propositional logic, many common logical equivalences exist and are often listed as laws or properties. In Table 1.8, we include a list of the most common such logical equivalences. Item (*a*) of Example 1.27 uses Table 1.8 to re-verify the logical equivalence in item (*a*) of Example 1.26.

**Example 1.27**    Use logical equivalences to show that:

(*a*)  $P \vee (\neg P \wedge Q) \equiv P \vee Q$.

(*b*)  $(P \wedge Q) \to P$ is a tautology.

(*c*)  $[P \wedge (P \to Q)] \wedge \neg Q$ is a contradiction.

**Solution**    We use direct proofs. The following sequence of logical equivalences proves the logical equivalence of $P \vee (\neg P \wedge Q)$ and $P \vee Q$.

$$
\begin{aligned}
P \vee (\neg P \wedge Q) &\equiv (P \vee \neg P) \wedge (P \vee Q) && \text{(Distributive law)} \\
&\equiv T \wedge (P \vee Q) && \text{(Tautology law)} \\
&\equiv P \vee Q. && \text{(Identity law)}
\end{aligned}
$$

This proves the desired result in item (a). For item (b), we prove that $(P \wedge Q) \to P$ is a tautology by showing that $(P \wedge Q) \to P$ and $T$ are logically equivalent, which is seen from the sequence of equivalences.

$$
\begin{aligned}
(P \wedge Q) \to P &\equiv \neg(P \wedge Q) \vee P && \text{(Implication law)} \\
&\equiv (\neg P \vee \neg Q) \vee P && \text{(DeMorgan's law)} \\
&\equiv \neg P \vee (\neg Q \vee P) && \text{(Associative law)} \\
&\equiv \neg P \vee (P \vee \neg Q) && \text{(Commutative law)} \\
&\equiv (\neg P \vee P) \vee \neg Q && \text{(Associative law)} \\
&\equiv T \vee \neg Q \equiv T. && \text{(Tautology and domination laws)}
\end{aligned}
$$

This proves the desired result in item (b). Item (c) is left for the reader an exercise. ∎

| Name | Formula(s) |
|---|---|
| Identity laws | $P \wedge T \equiv P$ |
| | $P \vee F \equiv P$ |
| Domination laws | $P \wedge F \equiv F$ |
| | $P \vee T \equiv T$ |
| Idempotent laws | $P \vee P \equiv P$ |
| | $P \wedge P \equiv P$ |
| Tautology law | $P \vee \neg P \equiv T$ |
| Contradiction law | $P \wedge \neg P \equiv F$ |
| Double negation law | $P \equiv \neg(\neg P)$ |
| DeMorgan's laws | $\neg(P \wedge Q) \equiv \neg P \vee \neg Q$ |
| | $\neg(P \vee Q) \equiv \neg P \wedge \neg Q$ |
| Contrapositive law | $P \rightarrow Q \equiv \neg Q \rightarrow \neg P$ |
| Implication law | $P \rightarrow Q \equiv \neg P \vee Q$ |
| Implication negation law | $\neg(P \rightarrow Q) \equiv P \wedge \neg Q$ |
| Double implication law | $P \leftrightarrow Q \equiv (P \rightarrow Q) \wedge (Q \rightarrow P)$ |
| Commutative laws | $P \vee Q \equiv Q \vee P$ |
| | $P \wedge Q \equiv Q \wedge P$ |
| | $P \oplus Q \equiv Q \oplus P$ |
| Associative laws | $(P \vee Q) \vee R \equiv P \vee (Q \vee R)$ |
| | $(P \wedge Q) \wedge R \equiv P \wedge (Q \wedge R)$ |
| | $(P \oplus Q) \oplus R \equiv P \oplus (Q \oplus R)$ |
| Distributive laws | $P \vee (Q \wedge R) \equiv (P \vee Q) \wedge (P \vee R)$ |
| | $P \wedge (Q \vee R) \equiv (P \wedge Q) \vee (P \wedge R)$ |
| | $P \wedge (Q \oplus R) \equiv (P \wedge Q) \oplus (P \wedge R)$ |
| Exclusive distributive laws | $P \oplus Q \equiv (P \vee Q) \wedge \neg(P \wedge Q)$ |
| | $P \oplus Q \equiv (P \wedge \neg Q) \vee (\neg P \wedge Q)$ |

Table 1.8: A list of the most common logical equivalences.

# 1.4. Logical normal forms

From mathematics classes, we know that $f(x) = x^2$ defines a function that squares its one input, and $g(x, y) = 2x + y$ defines a function that doubles its first input and adds the second input. The domain and codomain of such functions are real numbers.

We can also define functions whose domain and codomain are $\{T, F\}$. Such functions are called Boolean functions.[10] For example, the truth table in Table 1.9 defines a Boolean function $f(\cdot, \cdot, \cdot)$ on three variables. In this section, we learn how to derive a propositional formula for the Boolean function $f(\cdot, \cdot, \cdot)$ represented by the truth table in Table 1.9. In particular, the Boolean disjunctive and conjunctive normal forms are given.

## Disjunctive normal forms

To present the logical normal forms, we first have some definitions.

> **Definition 1.14** *A literal is an atomic proposition such as $P, Q$ or $R$, or the negation of an atomic proposition such as $\neg P, \neg Q$ or $\neg R$.*

> **Definition 1.15** *A conjunctive clause is the conjunction of one or more literals.*

For example, $P, \neg Q \wedge R$ and $\neg P \wedge Q \wedge R \wedge \neg P$ are all conjunctive clauses.

| $P$ | $Q$ | $R$ | $f(P, Q, R)$ |
|-----|-----|-----|--------------|
| T | T | T | F |
| T | T | F | T |
| T | F | T | F |
| T | F | F | T |
| F | T | T | F |
| F | T | F | F |
| F | F | T | T |
| F | F | F | F |

Table 1.9: A truth table defining a Boolean function $f(\cdot, \cdot, \cdot)$ on three variables.

---

[10] Named after George Boole, a 19th century logician.

> **Definition 1.16** *A proposition is in disjunctive normal form (DNF for short) if it is the disjunction of one or more conjunctive clauses.*

Informally, a proposition is in DNF if it is an ORing of AND clauses. For example, the propositions

$$\neg P, \; P \wedge \neg Q, \; \neg Q \vee R \text{ and } (P \wedge \neg Q) \vee \neg P \vee (P \wedge Q \wedge R)$$

are all in DNF. In contrast, the propositions

$$\neg(P \vee Q) \text{ and } (P \wedge \neg Q) \vee (P \vee Q \vee R)$$

are not in DNF.

When we derive explicit propositional formulas for Boolean functions, such as $f(\cdot, \cdot, \cdot)$ represented by the truth table in Table 1.9, we will see that the obtained formulas are in logical normal forms such as DNF. The two-step procedure in the following workflow, followed by two examples, will teach us to create such propositional formulas.

> **Workflow 1.1** *We create a propositional formula for a Boolean function defined by a truth table by following two steps:*
>
> (i) *Create a connective clause for each row where the value of the function is true.*
>
> (ii) *Create a disjunction of the conjunctive clauses obtained in (i).*

**Example 1.28** Create a propositional formula using logical operators that is equivalent to the function $f$ defined in the truth table given in Table 1.9.

**Solution**   First, by creating a connective clause for each row where the value of the function is true, we get

| $P$ | $Q$ | $R$ | $f(P, Q, R)$ | Conjunctive clauses |
|-----|-----|-----|--------------|---------------------|
| T | T | T | F | |
| T | T | F | T $\Longrightarrow$ | $P \wedge Q \wedge \neg R$ |
| T | F | T | F | |
| T | F | F | T $\Longrightarrow$ | $P \wedge \neg Q \wedge \neg R$ |
| F | T | T | F | |
| F | T | F | F | |
| F | F | T | T $\Longrightarrow$ | $\neg P \wedge \neg Q \wedge R$ |
| F | F | F | F | |

Note that the clause $P \wedge Q \wedge \neg R$, for instance, evaluates to false for all other values of $P, Q$ and $R$ since it is a conjunction.

Next, by creating a disjunction of the conjunctive clauses obtained above, we obtain the formula

$$f(P, Q, R) = (P \wedge Q \wedge \neg R) \vee (P \wedge \neg Q \wedge \neg R) \vee (\neg P \wedge \neg Q \wedge R),$$

which is in DNF.

**Example 1.29** Write a truth table for each of the following propositions and use it to find an equivalent proposition in DNF.

(a) $\neg(P \rightarrow Q)$. 　　　　　　　　　　(b) $(P \wedge Q) \vee \neg R$.

**Solution** (a) A truth table for the proposition $\neg(P \rightarrow Q)$ is given below. It is also seen from the table that $\neg(P \rightarrow Q) \equiv P \wedge \neg Q$ which is the equivalent proposition in DNF.

| $P$ | $Q$ | $P \rightarrow Q$ | $\neg(P \rightarrow Q)$ | Conjunctive clauses |
|-----|-----|-----|-----|-----|
| T | T | T | F | |
| T | F | F | T $\Longrightarrow$ | $P \wedge \neg Q$ |
| F | T | T | F | |
| F | F | T | F | |

(b) A truth table for the proposition $(P \wedge Q) \vee \neg R$ is given below.

| $P$ | $Q$ | $R$ | $P \wedge Q$ | $\neg R$ | $(P \wedge Q) \vee \neg R$ | Conjunctive clauses |
|-----|-----|-----|-----|-----|-----|-----|
| T | T | T | T | F | T $\Longrightarrow$ | $P \wedge Q \wedge R$ |
| T | T | F | T | T | T $\Longrightarrow$ | $P \wedge Q \wedge \neg R$ |
| T | F | T | F | F | F | |
| T | F | F | F | T | T $\Longrightarrow$ | $P \wedge \neg Q \wedge \neg R$ |
| F | T | T | F | F | F | |
| F | T | F | F | T | T $\Longrightarrow$ | $\neg P \wedge Q \wedge \neg R$ |
| F | F | T | F | F | F | |
| F | F | F | F | T | T $\Longrightarrow$ | $\neg P \wedge \neg Q \wedge \neg R$ |

It is seen that

$$(P \wedge Q) \vee \neg R \;\equiv\; (P \wedge Q \wedge R) \vee (P \wedge Q \wedge \neg R) \vee (P \wedge \neg Q \wedge \neg R)$$
$$\vee (\neg P \wedge Q \wedge \neg R) \vee (\neg P \wedge \neg Q \wedge \neg R),$$

which is in DNF.

The three-step procedure in the following workflow, followed by an example, will learn us to convert any proposition into DNF.

> **Workflow 1.2**  *We transform a proposition into DNF using a sequence of logical equivalences by following three steps:*
>
> (i) *Convert any symbols that are not $\vee$, $\wedge$ or $\neg$ using logical equivalence laws. For instance, $P \to Q \equiv \neg P \vee Q$.*
>
> (ii) *Use the DeMorgan's laws and the double negation law, as needed, so that any negation symbols are directly on the variables.*
>
> (iii) *Use the distributive, associative and commutative laws, if needed.*

**Example 1.30** Transform the proposition $(P \to (Q \to R)) \vee \neg(P \vee \neg(R \vee S))$ into DNF.

**Solution**  The following sequence of logical equivalences transforms the given proposition into DNF.

$$
\begin{aligned}
&(P \to (Q \to R)) \vee \neg(P \vee \neg(R \vee S)) \\
\equiv\ & (\neg P \vee (Q \to R)) \vee \neg(P \vee \neg(R \vee S)) && \text{(Implication law)} \\
\equiv\ & (\neg P \vee (\neg Q \vee R)) \vee \neg(P \vee \neg(R \vee S)) && \text{(Implication law)} \\
\equiv\ & (\neg P \vee (\neg Q \vee R)) \vee (\neg P \wedge \neg\neg(R \vee S)) && \text{(DeMorgan's law)} \\
\equiv\ & (\neg P \vee (\neg Q \vee R)) \vee (\neg P \wedge (R \vee S)) && \text{(Double negation law)} \\
\equiv\ & (\neg P \vee (\neg Q \vee R)) \vee ((\neg P \wedge R) \vee (\neg P \wedge S)) && \text{(Distributive law)} \\
\equiv\ & (\neg P) \vee (\neg Q) \vee R \vee (\neg P \wedge R) \vee (\neg P \wedge S).
\end{aligned}
$$

## Conjunctive normal forms

The conjunctive normal form and the disjunctive normal form can be viewed as the dual of each other. We have this definition.

> **Definition 1.17**  *A proposition is in conjunctive normal form (CNF for short) if it is the conjunction of one or more disjunctive clauses, where each clause is a disjunction of one or more literals.*

Informally, a proposition is in CNF if it is an ANDing of OR clauses. For example, the following propositions are all in CNF:

$$\neg P,\ \neg Q \vee R,\ P \wedge \neg Q \text{ and } (P \vee \neg Q) \wedge \neg P \wedge (P \vee Q \vee R).$$

In contrast, the following propositions are not in CNF:

$$\neg(P \vee Q) \text{ and } (P \vee \neg Q) \wedge (P \vee Q \wedge R).$$

The three-step procedure in the following workflow, followed by an example, will learn us to convert a proposition into CNF.

---

**Workflow 1.3** *We transform a propositional formula into CNF by following three steps:*

(*i*) *Create a truth table for the negation of the proposition.*

(*ii*) *Use the truth table from (i) to create an equivalent proposition in DNF.*

(*iii*) *Negate the proposition from (ii), and apply DeMorgan's law and the double negation law to convert the proposition to CNF.*

---

**Example 1.31** Convert the proposition $(P \wedge Q) \vee \neg R$ into CNF.

**Solution** A truth table for the negation of the proposition $(P \wedge Q) \vee \neg R$ is given below.

| $P$ | $Q$ | $R$ | $P \wedge Q$ | $\neg R$ | $(P \wedge Q) \vee \neg R$ | $\neg[(P \wedge Q) \vee \neg R]$ | Conjun. cls. |
|---|---|---|---|---|---|---|---|
| T | T | T | T | F | T | F | |
| T | T | F | T | T | T | F | |
| T | F | T | F | F | F | T $\Longrightarrow$ | $P \wedge \neg Q \wedge R$ |
| T | F | F | F | T | T | F | |
| F | T | T | F | F | F | T $\Longrightarrow$ | $\neg P \wedge Q \wedge R$ |
| F | T | F | F | T | T | F | |
| F | F | T | F | F | F | T $\Longrightarrow$ | $\neg P \wedge \neg Q \wedge R$ |
| F | F | F | F | T | T | F | |

This shows that

$$\neg[(P \wedge Q) \vee \neg R] \equiv (P \wedge \neg Q \wedge R) \vee (\neg P \wedge Q \wedge R) \vee (\neg P \wedge \neg Q \wedge R). \quad (1.5)$$

which is in DNF. Then

$$
\begin{aligned}
(P \wedge Q) \vee \neg R &\equiv \neg[\neg[(P \wedge Q) \vee \neg R]] && \text{(DNL)} \\
&\equiv \neg[(P \wedge \neg Q \wedge R) \vee (\neg P \wedge Q \wedge R) \vee (\neg P \wedge \neg Q \wedge R)] && \text{By (1.5)} \\
&\equiv \neg(P \wedge \neg Q \wedge R) \wedge \neg(\neg P \wedge Q \wedge R) \wedge \neg(\neg P \wedge \neg Q \wedge R) && \text{(DL)} \\
&\equiv (\neg P \vee \neg\neg Q \vee \neg R) \wedge (\neg\neg P \vee \neg Q \vee \neg R) \wedge (\neg\neg P \vee \neg\neg Q \vee \neg R) && \text{(DL)} \\
&\equiv (\neg P \vee Q \vee \neg R) \wedge (P \vee \neg Q \vee \neg R) \wedge (P \vee Q \vee \neg R) && \text{(DNL)}
\end{aligned}
$$

where DNL and DL stand for the double negation law and DeMorgan's law, respectively. ∎

# 1.5. The Boolean satisfiability problem

The Boolean satisfiability is the problem of determining if a given propositional formula is satisfiable or not.

> **Definition 1.18** *A propositional formula is said to be satisfiable if it evaluates to true for some values of its variables. If a propositional formula is not satisfiable, it is called unsatisfiable. An unsatisfiable formula evaluates to false on any values of its variables.*

For example, the proposition $(P \lor Q) \to \neg P$ is satisfiable because it evaluates to true when $P$ is false and Q is true. Note that a satisfiable proposition is either a tautology or contingency, and that an unsatisfiable proposition is a contradiction.

The DNF satisfiability problem is the problem of determining whether or not there is any variable assignment that makes a DNF formula true. We have the following example.

**Example 1.32**   The following proposition

$$(\neg P \land Q) \lor (P \land \neg Q \land R) \lor (\neg P \land \neg R)$$

which is in DNF, is satisfiable. To see this, note that the first clause evaluates to true when $P$ is false and Q is true. Since the proposition is a disjunction of clauses, it is not necessary to evaluate the rest of the clauses. ∎

Example 1.32 leads us to the following remark.

> **Remark 1.5** *A propositional formula in DNF is satisfiable if and only if at least one of its conjunctive clauses is satisfiable. A conjunctive clause is satisfiable if and only if for every atomic proposition P, the clause does not contain both P and ¬P as literals.*

From Remark 1.5, it can be seen that the DNF satisfiability is "easy". In fact, we evaluate one conjunction at a time, if at least one conjunction is not a contradiction, the formula is satisfiable.

The CNF satisfiability problem is the problem of determining whether or not there is some variable assignment that makes a DNF formula true. We have the following example.

**Example 1.33**   The following proposition

$$(P \lor Q) \land (\neg Q \lor R \lor \neg S) \land (\neg P \lor S)$$

which is in CNF, is satisfiable. To see this, note that all clauses evaluate to true when $P$ is true, Q is false, $R$ is false, and $S$ is true. Since the proposition is a conjunction of clauses, it is necessary to evaluate all clauses. ∎

If the number of propositional variables is very small in a CNF formula, we can determine whether the formula is satisfiable or not is by using a truth table. We will leave proving (by constructing a truth table) whether the following proposition is satisfiable or not as an exercise for the reader.

$$(P \vee Q \vee R) \wedge (\neg P \vee \neg Q) \wedge (\neg P \vee \neg R) \wedge (\neg R \vee \neg Q).$$

In general, the CNF satisfiablity problem is not as easy as this looks in Example 1.33 or in the above exercise.

> **Definition 1.19** *A propositional formula is in k-CNF if it is the AND of clauses of ORs of exactly k variables or their negations.*

For example, the propositional formula $(P \vee Q) \wedge (\neg Q \vee \neg R) \wedge (\neg P \vee R)$ is in 2-CNF, and the propositional formula $(P \vee Q \vee R) \wedge (\neg Q \vee \neg R \vee \neg S) \wedge (\neg P \vee R \vee S)$ is in 3-CNF.

In fact, the 2-CNF satisfiability problem is a version of the satisfiability problem that can be formulated and solved using the so-called depth-first search method, which is the topic to be covered in Sections 9.4 and 9.5. However, it is worth mentioning that 3-CNF satisfiability is a "hard" problem. Further discussion about this can be found in any textbook on computational complexity theory.

# 1.6. Predicates and quantifiers

In the second part of this chapter, we study the predicate logic which extends propositional logic by adding predicates and allowing the presence of quantifiers. The predicate logic is also known as first-order logic. First-order logic quantifies only variables that range over individuals. Second-order logic and third-order logic admit quantifications over sets and sets of sets, respectively. In general, higher-order logic quantifies over sets that are nested deeply as necessary.

### Predicates and quantified statements

*Predicates*   A predicate can be thought of as a function whose codomain is {T, F}. As a matter of examples, the following functions are predicates.

- $P(x) = $ "$x > 4$" is a predicate on one variable. Note that, for instance, $P(5)$ is true because $5 > 4$, but $P(3)$ is false because $3 \not> 4$.

- $Q(x, y) = $ "$x \geq y$" is a predicate on two variables. Note that, for instance, $Q(5, 4)$ is true while $Q(4, 5)$ is false.

The above two examples illustrate the concept of a predicate. A formal definition looks like:

> **Definition 1.20** *A predicate is a statement whose truth value depends on the value of one or more variables.*

***Quantifiers***    Defining quantifiers leads us to define the predicate logic.

> **Definition 1.21** *A quantifier is an expression, such as "every" and "there exists", that indicates the scope of a term to which it is attached.*

Many theorems, conjectures and definitions in mathematics use quantifiers. For example,

- "Every prime number[11] greater than 2 is odd".

- "There exists an even prime number".

- "For any integer value of $n$ greater than 2, there are no three positive integers $a, b$, and $c$ that satisfy the equation $a^n + b^n = c^n$".[12]

There are two fundamental kinds of quantification in the predicate logic:

($i$)  The universal quantification: The quantifier that we symbolize by "$\forall$" and read as "every" or "for all" is called the universal quantifier.

($ii$)  The existential quantification: The quantifier that we symbolize by "$\exists$" and read as "there is" or "there exists" is called the existential quantifier.

The following example helps us to always remember that the upside-down "A" stands for "All" and the backward "E" stands for " Exists". Recall that $\mathbb{N} = \{1, 2, 3, \ldots\}$ denotes the set of all natural numbers.

**Example 1.34**    If $P(x) = "x^2 > 4"$, then

($a$)  "$\exists x \in \mathbb{N}, P(x)$" is a proposition that means "there is a natural number $x$ such that $x^2 > 4$". This is a true predicate logic proposition.

($b$)  "$\forall x \in \mathbb{N}, P(x)$" is a proposition that means "for every natural number $x$ we have $x^2 > 4$". This is a false predicate logic proposition. (To see its falsity, take $x = 1$.)

In the two predicate logic propositions given in Example 1.34, the set $\mathbb{N}$ is called the universe of discourse. We have the following definition.

---

[11] Recall that a prime number is a natural number which is greater than 1 and can be divided only by 1 and the number itself.

[12] This is called the Fermat's last theorem (also known as Fermat's conjecture) and was first conjectured by Pierre de Fermat around 1637.

| Quantifiers | Universal quantifier | Existential quantifier |
|---|---|---|
| Formulas | $\forall x \in D, P(x)$ | $\exists x \in D, P(x)$ |
| When true? | If $P(x)$ is true for every $x$ | If there is an $x$ s.t. $P(x)$ is true |
| When false? | If there is an $x$ s.t. $P(x)$ is false | If $P(x)$ is false for every $x$ |

Table 1.10: The truth and falsity of quantifiers.

**Definition 1.22** *The universe of discourse, which is also-called the domain of discourse (or simply the domain) and is denoted by D, is the set of the possible values of the variable(s) in the predicate.*

When we use quantifiers, it is important to state the universe of discourse. In Calculus courses, unless specified otherwise, the domain of a function is usually the set of real numbers $\mathbb{R}$. In propositional logic, the default domain is everything: All numbers, people, computers, cats, etc.

Let $P(x)$ be a predicate. The simplest form of a quantified formula is as follows.

$$\underbrace{\left\{ \begin{array}{c} \forall \\ \exists \end{array} \right\}}_{\text{a quantifier}} \quad \underbrace{x \in D,}_{\substack{\text{a variable} \\ \text{in a domain}}} \quad \underbrace{P(x).}_{\text{a predicate}}$$

More specifically,

- we write "$\forall x \in D, P(x)$" to say that a predicate, $P(x)$, is true for all values of $x$ in some set $D$,

- we write "$\exists x \in D, P(x)$" to say that a predicate, $P(x)$, is true for at least one value of $x$ in some set $D$.

For example, the proposition "$\forall n \in \mathbb{N}, n < n^2$", which means that "Every natural number is less than its own square", is false, while the proposition "$\exists n \in \mathbb{N}, n = n^2$", which means that "Some natural number is equal to its own square", is true because $n = 1$ is a witness. Table $1.10^{13}$ shows when the truth and falsity of quantifiers are asserted.

Some propositions have multiple quantifiers and/or multiple predicates. In this part we study this concretely.

***Multiple quantifiers*** More than one quantifier can be used in a proposition. Let $P(x, y)$ be a predicate on two variables. The simplest forms of a quantified formula

---

[13] In Table 1.10, "s.t." refers to "such that".

with multiple quantifiers are as follows.

$$\forall x \in D, \forall y \in D, P(x, y),$$
$$\exists x \in D, \exists y \in D, P(x, y).$$

We have the following examples.

**Example 1.35** Let $P(x, y) = "x + y = 5"$. Then:

(*a*) The proposition "There exist two natural numbers whose sum is 5", or equivalently the more detailed proposition "There is a natural number $x$ and a number natural $y$ such that $x + y = 5$", can be symbolized as $\exists x \in \mathbb{N}, \exists y \in \mathbb{N}, P(x, y)$. This is a true proposition.

(*b*) The proposition "The sum of every pair of natural numbers is 5", or equivalently the more detailed proposition "For any two natural numbers $x$ and $y$ we have $x + y = 5$", can be symbolized as $\forall x \in \mathbb{N}, \forall y \in \mathbb{N}, P(x, y)$. This is a false proposition. To see its falsity, take the pair of numbers 1 (to be $x$) and 2 (to be $y$).

**Example 1.36** Let $P(x, y) = "x + y = 2"$ and $Q(x, y) = "xy \geq 0"$. Decide whether each of the following propositions is true or false. Justify your answer.

(*a*) $\exists x \in \mathbb{N}, \exists y \in \mathbb{N}, P(x, y)$.      (*c*) $\forall x \in \mathbb{N}, \forall y \in \mathbb{N}, Q(x, y)$.

(*b*) $\exists x \in \mathbb{N}, \exists y \in \mathbb{N}, P(x, y) \wedge x \neq y$.      (*d*) $\forall x \in \mathbb{Z}, \forall y \in \mathbb{Z}, Q(x, y)$.

**Solution**  The proposition in item (*a*) is true because $x$ and $y$ can both be 1. For item (*b*), note that "$\neq$" has higher precedence than "$\wedge$". The only assignment for $x$ and $y$ that makes the proposition in item (*b*) true is $x = y = 1$. Thus, the proposition in item (*b*) is false. Items (*c*) and (*d*) are left as an exercise for the reader.

***Multiple predicates***  Some propositions have more than one predicate. Let $P(x)$ and $Q(x)$ be two predicates, the simplest forms of a quantified formula with multiple predicates are as follows.

$$\forall x \in D, P(x) \circ Q(x),$$
$$\exists x \in D, P(x) \circ Q(x),$$

where "$\circ$" is any binary logical operator such as $\wedge, \vee, \oplus, \rightarrow$ and $\leftrightarrow$. For example, the proposition "Some natural number is equal to its own square and is equal to its own cube". This can be symbolized as $\exists n \in \mathbb{N}, n = n^2 \wedge n = n^3$.

Note that "=" has higher precedence than "$\wedge$". So, $\exists n \in \mathbb{N}, n = n^2 \wedge n = n^3$ means $\exists n \in \mathbb{N}, (n = n^2) \wedge (n = n^3)$. In general, mathematical symbols such as addition, subtraction, multiplication, division, equality and inequality have higher precedence than logical operators such as negation, conjunction, disjunction, implication and double implication.

When a quantifier is used on the variable $x$ we say that this occurrence of $x$ is bound. When the occurrence of a variable is not bound by a quantifier or set to a particular value, the variable is said to be free. The part of a logical expression to which a quantifier is applied is the scope of the quantifier. So, a variable is free if it is outside the scope of all quantifiers. For example, if we consider the statement $(\forall x \in D, P(x)) \wedge Q(x)$, the $x$ in $P(x)$ is bound by the universal quantifier, while the $x$ in $Q(x)$ is free. The scope of the universal quantifier is only $P(x)$. In fact, the statement $(\forall x \in D, P(x)) \wedge Q(x)$ is not a proposition since there is a free variable.

The quantifiers "$\forall$" and "$\exists$" have higher precedence than logical operators. For example, $\forall x \in D, P(x) \wedge Q(x)$ means $(\forall x \in D, P(x)) \wedge Q(x)$, and it does not mean $\forall x \in D, (P(x) \wedge Q(x))$.

Some of the questions that arise now are the following.

- Is $\forall x \in D, (P(x) \wedge Q(x))$ logically equivalent to $\forall x \in D, P(x) \wedge \forall x \in D, Q(x)$?

- Is $\forall x \in D, (P(x) \vee Q(x))$ logically equivalent to $\forall x \in D, P(x) \vee \forall x \in D, Q(x)$?

Before answering these questions, we need the following definition.

---

**Definition 1.23** *Two predicate logic propositions S and T are logically equivalent if they have the same truth value regardless of the interpretation.* [a]

---

[a]That is regardless of the meaning that is attributed to each propositional function, and regardless of the domain of discourse.

---

Proving that the proposition $\forall x \in D, (P(x) \wedge Q(x))$ is logically equivalent to $\forall x \in D, P(x) \wedge \forall x \in D, Q(x)$ is left as an exercise for the reader. To disprove that $\forall x \in D, (P(x) \vee Q(x))$ is logically equivalent to $\forall x \in D, P(x) \vee \forall x \in D, Q(x)$, we give the following counterexample to the assertion that they have the same truth value for all possible interpretations. Let $D$ denote the set of people in the world, $P(x)$ ="$x$ is male", and $Q(x)$ ="$x$ is female". Then $\forall x \in D, (P(x) \vee Q(x))$ is true, while $\forall x \in D, P(x) \vee \forall x \in D, Q(x)$ is false. In fact, the first proposition means that every person in the world is a male or a female, while the second one means that every person in the world is a male or every person in the world is a female. Note that the formula $\forall x \in D, P(x) \vee \forall x \in D, Q(x)$ can also be written as $\forall x \in D, P(x) \vee \forall y \in D, Q(y)$ or as $\forall x, y \in D, (P(x) \vee Q(y))$.

Consider also the following two assertions.

- $\exists x \in D, (P(x) \wedge Q(x))$ is logically equivalent to $\exists x \in D, P(x) \wedge \exists x \in D, Q(x)$.

- $\exists x \in D, (P(x) \vee Q(x))$ is logically equivalent to $\exists x \in D, P(x) \vee \exists x \in D, Q(x)$.

The first assertion is false, and a counterexample is $D = \mathbb{Z}$, $P(x) = $ "$x < 0$", and $Q(x) = $ "$x \geq 0$". Stating whether the second assertion is true or false and providing a sentence or two of justification are left as an exercise for the reader.

***Mixing quantifiers***    Some propositions have two different kinds of quantifiers. Let $P(x, y)$ be a predicate on two variables. The simplest forms of a quantified formula with mixed quantifiers are as follows.

$$\forall x \in D, \exists y \in D, P(x, y),$$
$$\exists x \in D, \forall y \in D, P(x, y). \tag{1.6}$$

When quantifiers of the same kind are used in a proposition, the order of the quantified variables does not matter. For example, let $P(x, y)$ be a predicate on two variables, the following two propositions have always the same meaning regardless of the predicate $P(x, y)$.

$$\exists x \in D, \exists y \in D, P(x, y),$$
$$\exists y \in D, \exists x \in D, P(x, y).$$

When both universal and existential quantifiers are used in the same proposition, the order can matter. For example, the first proposition in (1.6) and the proposition

$$\exists y \in D, \forall x \in D, P(x, y)$$

can have different meanings. Also, the second proposition in (1.6) and the proposition

$$\forall y \in D, \exists x \in D, P(x, y)$$

can have different meanings. For more illustration, we give the following example.

**Example 1.37**  Let $P(x, y) = $ "$x + y = 0$". Then:

(*a*)  The proposition $\forall x \in \mathbb{R}, \exists y \in \mathbb{R}, P(x, y)$ means that for any real number $x$ there exists a real number $y$ such that $x + y = 0$. This is clearly a true proposition because $y = -x$ is a witness.

(*b*)  The proposition $\exists x \in \mathbb{R}, \forall y \in \mathbb{R}, P(x, y)$ means that there is a real number $x$ such that for every real number $y$ we have $x + y = 0$. This is clearly a false proposition because there does not exist a real number whose sum with every real number is equal to 0.

Some propositions have two or more domains of discourse, one for each kind of quantifiers, as we will see in the following example.

**Example 1.38**  Symbolize Goldbach's conjecture which states that "Every even integer greater than 2 is the sum of two primes".

**Solution**  Goldbach's conjecture can be restated as "For every even integer $n$ greater than 2, there exist primes $p$ and $q$ such that $n = p + q$".

Let Evens be the set of even integers greater than 2 and Primes be the set of prime numbers. Then the Goldbach's conjecture can be symbolized as

$$\forall n \in \text{Evens}, \exists p \in \text{Primes}, \exists q \in \text{Primes}, n = p + q.$$

■

In this part, we study how to negate quantified statements.

The negation of a universal statement is an existential statement with the predicate negated. That is,

$$\neg(\forall x, P(x)) \equiv \exists x, \neg P(x). \tag{1.7}$$

Often, a proposition such as $\neg(\forall x, P(x))$ is simply written as $\neg \forall x, P(x)$. As a direct application, we have the following example.

**Example 1.39** The proposition "Every prime number is odd" is false because 2 is a prime number that is not odd. So its negation, which is the proposition "There exists a prime number that is not odd", is true.

Symbolically, to see why the second proposition is the negation of the first one, let $\forall x$ be "For every prime number $x$" and $P(x)$ be "$x$ is an odd number". The negation of the proposition "Every prime number is odd" is expressed as

$$\neg \forall x, P(x) \text{ or equivalently using } (1.7) \ \exists x, \neg P(x),$$

which is translated back to English as "There exists a prime number that is not odd".

■

The negation of an existential statement is a universal statement with the predicate negated. That is,

$$\neg(\exists x, P(x)) \equiv \forall x, \neg P(x). \tag{1.8}$$

Often, a proposition such as $\neg(\exists x, P(x))$ is simply written as $\neg \exists x, P(x)$. As a direct application, we have the following example.

**Example 1.40** The proposition "There is a negative integer whose square is negative" is clearly false. So its negation, which is the proposition "The square of every negative integer is nonnegative", is true.

Symbolically, to see why the second proposition is the negation of the first one, let $\exists x$ be "There is a negative integer $x$" and $P(x)$ be "The square of $x$ is negative". The negation of the proposition "Every prime number is odd" is expressed as

$$\neg \exists x, P(x) \text{ or equivalently using } (1.8) \ \forall x, \neg P(x),$$

which is translated back to English as "The square of every negative integer is nonnegative".

■

**Negating propositions with multiple quantifiers** The procedure of negating quantified statements can be extended to propositions with more than one quantifier. For example, let $P(x, y, z)$ be a predicate on three variables. The negation of the proposition $\forall x \in D, \exists y \in D, \forall z \in D, P(x, y, z)$ is created using the following sequence of logical equivalences.

$$
\begin{aligned}
\neg\,(\forall x \in D, \exists y \in D, \forall z \in D, P(x, y, z)) &\equiv\ \exists x \in D, \neg\,(\exists y \in D, \forall z \in D, P(x, y, z)) \\
&\equiv\ \exists x \in D, \forall y \in D, \neg\,(\forall z \in D, P(x, y, z)) \\
&\equiv\ \exists x \in D, \forall y \in D, \exists z \in D, \neg P(x, y, z).
\end{aligned}
$$

We also give the following example.

**Example 1.41** The proposition $\forall x \in \mathbb{R}, \exists y \in \mathbb{R}, xy = 1$ means that "Every real number has a multiplicative inverse". This is false because picking $x = 0$, then for any $y \in \mathbb{R}, xy \neq 1$. The negation of the proposition $\forall x \in \mathbb{R}, \exists y \in \mathbb{R}, xy = 1$ can be obtained as follows.

$$
\begin{aligned}
\neg\,(\forall x \in \mathbb{R}, \exists y \in \mathbb{R}, xy = 1) &\equiv\ \exists x \in \mathbb{R}, \neg\,(\exists y \in \mathbb{R}, xy = 1) \\
&\equiv\ \exists x \in \mathbb{R}, \forall y \in \mathbb{R}, \neg\,(xy = 1) \\
&\equiv\ \exists x \in \mathbb{R}, \forall y \in \mathbb{R}, xy \neq 1.
\end{aligned}
$$

■

***Expressing propositions using only one type of quantifiers***   Sometimes, for convenience, we need to express propositions using only one type of quantifiers.

In fact, every universal statement can be expressed as an existential statement as follows.

$$\forall x \in D, P(x) \equiv \neg\neg\forall x \in D, P(x) \equiv \neg\exists x \in D, \neg P(x). \tag{1.9}$$

As a direct application, we have the following example.

**Example 1.42** The proposition "Every US senator is at least 30 years old" is the same as the proposition "There does not exist a US senator who is not at least 30 years old".

Symbolically, to see why the first proposition is equivalent to the second one, let $\forall x$ be "Every US senator $x$" and $P(x)$ be "$x$ is at least 30 years old". The proposition "Every US senator is at least 30 years old" is expressed as

$$\forall x \in D, P(x) \ \text{ or equivalently using } (1.9) \ \neg\exists x \in D, \neg P(x),$$

which is translated back to English as "There does not exist a US senator who is not at least 30 years old". ■

Similarly, every existential statement an be expressed as a universal statement as follows.

$$\exists x \in D, P(x) \equiv \neg\neg\exists x \in D, P(x) \equiv \neg\forall x \in D, \neg P(x). \tag{1.10}$$

As a direct application, we have the following example.

■  **Example 1.43** The proposition "There exists a white cat" is the same as the proposition "Not every cat has color different than white".

Symbolically, to see why the first proposition is equivalent to the second one, let $\exists x$ be "There exists cat $x$" and $P(x)$ be "$x$ has the color white". The proposition "There exists a white cat" is expressed as

$$\exists x \in D, P(x) \text{ or equivalently using } (1.10) \ \neg \forall x \in D, \neg P(x),$$

which is translated back to English as "Not every cat has color different than white".

■

Note that the above procedure can be extended to propositions with more than one quantifier. For example, let $P(x, y, z)$ be a predicate on three variables, then

$$\forall x \in D, \exists y \in D, \forall z \in D, P(x,y,z) \ \equiv \ \forall x \in D, \neg \neg \exists y \in D, \forall z \in D, P(x,y,z)$$
$$\equiv \ \forall x \in D, \neg \forall y \in D, \neg \forall z \in D, P(x,y,z).$$

## 1.7. Symbolizing statements of the form "All P are Q"

In this section, we study modeling of the statements of the form "All P are Q" based on the predicate logic. Perhaps the best way to understand how to symbolize this form of predicate logic propositions is to give illustrative examples.

The proverb saying "All that glitters is not gold" means that not everything that is shiny and superficially attractive is valuable, i.e., appearances can be deceiving. On the contrary, Led Zeppelin[14] in the song "Stairway to heaven" said "All that glitters is gold". In the following example, we express the statement 'All that glitters is gold" and its negation using quantifiers.

■  **Example 1.44**

(*a*) Symbolize the statement "All that glitters is gold".

(*b*) Symbolize the statement "All that glitters is gold" using only an existential quantifier.

(*c*) Symbolize the negation of the statement "All that glitters is gold" and translate it back to English.

(*d*) Symbolize the statement "Nothing that glitters is gold".

Solution. We let $D$ be the set of all elements, whether they are gold or not, whether they glitter or not. We also define the two predicates:

$$GLT(x) = \text{"}x \text{ glitters"} \text{ and } GLD(x) = \text{"}x \text{ is gold"}.$$

[14]Led Zeppelin were an English rock band formed in London in 1968.

(*a*) The statement "All that glitters is gold" means that if "If $x$ glitters, then $x$ is gold". This is an implication, so the correct way to express it as

$$\forall x \in D, (GLT(x) \longrightarrow GLD(x)). \qquad (1.11)$$

By the implication law, formulation in $(1.11)$ is equivalent to the following formulation.

$$\forall x \in D, (\neg GLT(x) \vee GLD(x)). \qquad (1.12)$$

It is worth mentioning that a common error is to symbolize the statement "All that glitters is gold" as $\forall x \in D, (GLT(x) \wedge GLD(x))$, which is not correct because it means that everything in the universe both glitters and is gold. Another error that some people make is to symbolize "All that glitters is gold" as $\forall x \in GLT(x), GLD(x)$, which is not correct because "$\in$" is a set operator but $GLT(x)$ is a predicate, not a set.

(*b*) From $(1.12)$, it is seen that the statement "All that glitters is gold" is symbolized as $\forall x \in D, (\neg GLT(x) \vee GLD(x))$. The following sequence of logical equivalences expresses this formula using only an existential quantifier.

$$
\begin{aligned}
\forall x \in D, (\neg GLT(x) \vee GLD(x)) \;\equiv\; & \neg\neg\forall x \in D, (\neg GLT(x) \vee GLD(x)) \\
\equiv\; & \neg \exists x \in D, \neg(\neg GLT(x) \vee GLD(x)) \\
\equiv\; & \neg \exists x \in D, (\neg\neg GLT(x) \wedge \neg GLD(x)) \\
\equiv\; & \neg \exists x \in D, (GLT(x) \wedge \neg GLD(x)).
\end{aligned}
$$

Here, the first equivalence follows from the double negation law, the second equivalence follows from $(1.7)$, and the last two follow from the DeMorgan's law and double negation law.

(*c*) From $(1.11)$, the negation of the statement "All that glitters is gold" is symbolized as $\neg \forall x \in D, (\neg GLT(x) \vee GLD(x))$. Using the rules of propositional logic, we have

$$
\begin{aligned}
\neg \forall x \in D, (\neg GLT(x) \vee GLD(x)) \;\equiv\; & \exists x \in D, \neg(\neg GLT(x) \vee GLD(x)) \\
\equiv\; & \exists x \in D, (\neg\neg GLT(x) \wedge \neg GLD(x)) \\
\equiv\; & \exists x \in D, (GLT(x) \wedge \neg GLD(x)),
\end{aligned}
$$

which is translated back to English as "There is something in the universe that glitters but is not gold".

(*d*) Using the predicates defined above, the statement "Nothing that glitters is gold" is symbolized as $\forall x \in D, (GLT(x) \rightarrow \neg GLD(x))$.

■

**Example 1.45**

(*a*)  Symbolize the statement "There are no orange cats" after writing it in the form "All P are Q".

(*b*)  Can you obtain the same formula by symbolizing after negating the negation of the given statement? Justify your answer.

**Solution.** We let $D$ be the set of all creatures (or species), whether they are cats or not, whether they are orange-colored or not. We also define the two predicates:

$$CAT(x) = ``x \text{ is a cat''} \quad \text{and} \quad ORG(x) = ``x \text{ is orange''}.$$

(*a*)  The statement "There are no orange cats" is equivalent to the implication "If something is a cat, then is not orange". So, a correct way to symbolize the given statement is $\forall x \in D, (CAT(x) \to \neg ORG(x))$.

(*b*)  We can obtain the same formula by negating the negation of the given statement. In fact, another way to get the same formula is to symbolize the negation of the statement "There is an orange cat" as follows.

$$
\begin{aligned}
\neg \exists x \in D, (CAT(x) \wedge ORG(x)) \quad &\equiv \quad \forall x \in D, \neg(CAT(x) \wedge ORG(x)) \\
&\equiv \quad \forall x \in D, (\neg CAT(x) \vee \neg ORG(x)) \\
&\equiv \quad \forall x \in D, (\neg\neg CAT(x) \to \neg ORG(x)) \\
&\equiv \quad \forall x \in D, (CAT(x) \to \neg ORG(x)).
\end{aligned}
$$

**Example 1.46**  Symbolize the following statements using the predicate logic.

(*a*)  "There is at most one piano".          (*c*)  "There are at most two pianos".

(*b*)  "There are at least two pianos".       (*d*)  "There are exactly two pianos".

**Solution.** We let $D$ be the set of all musical instruments, whether they are pianos or not. We also define the predicate: $P(x) = ``x \text{ is a piano''}$.

(*a*)  The statement "There are at least two pianos" is equivalent to the implication "If there are two musical instruments that are pianos, then they must be the same instrument". This can be symbolized as

$$\forall x \in D, \forall y \in D, (P(x) \wedge P(y) \longrightarrow x = y),$$

or equivalently, using the implication and DeMorgan's laws, as

$$\forall x \in D, \forall y \in D, (\neg P(x) \vee \neg P(y) \vee x = y).$$

(*b*) An incorrect way to symbolize the statement "There are at least two pianos" is the following. (Why is this incorrect?)

$$\exists x \in D, \exists y \in D, (P(x) \wedge P(y)).$$

To obtain the correct formula, note that the statement "There are at least two pianos" is the negation of the statement "There is at most one piano". Using item (*a*) and the rules of propositional logic, we have

$$\neg \forall x \in D, \forall y \in D, (\neg P(x) \vee \neg P(y) \vee x = y)$$
$$\equiv \exists x \in D, \neg \forall y \in D, (\neg P(x) \vee \neg P(y) \vee x = y)$$
$$\equiv \exists x \in D, \exists y \in D, \neg(\neg P(x) \vee \neg P(y) \vee x = y)$$
$$\equiv \exists x \in D, \exists y \in D, (P(x) \wedge P(y) \wedge x \neq y).$$

Thus, the statement "There are at least two pianos" is symbolized as

$$\exists x \in D, \exists y \in D, (P(x) \wedge P(y) \wedge x \neq y).$$

Note that the obtained formula explicitly says that $x$ and $y$ are not the same instrument. In the earlier incorrect formula, $x$ and $y$ could refer to the same instrument.

(*c*) The statement "There are at most two pianos" is symbolized as

$$\forall x \in D, \forall y \in D, \forall z \in D, ((P(x) \wedge P(y) \wedge P(z)) \longrightarrow (x = y) \vee (x = z) \vee (y = z)).$$

(*d*) The statement "There are exactly two pianos" means that "There are at least two pianos and there are at most two pianos". From items (*b*) and (*c*), this can be symbolized as

$$[\exists x \in D, \exists y \in D, (P(x) \wedge P(y) \wedge x \neq y)] \wedge$$
$$[\forall x \in D, \forall y \in D, \forall z \in D, (P(x) \wedge P(y) \wedge P(z) \rightarrow (x = y) \vee (x = z) \vee (y = z))].$$

An alternative way to symbolize the statement "There are exactly two pianos" is the following.

$$\exists x \in D, \exists y \in D, \left((x \neq y) \wedge \left[\forall z \in D, \left(P(z) \longleftrightarrow [(x = z) \vee (y = z)]\right)\right]\right).$$

∎

Finally, it is worth mentioning that sometimes statements of the form "All P are Q" are expressed using quantifiers and predicates with more than one variable. For example, to symbolize the statement "Every mail message larger than two megabytes will be compressed", we define the predicates $ML(x, k) =$ "mail message $x$ is larger than $k$ megabyte" and $MC(x) =$ "mail message $x$ will be compressed". Then the given statement is symbolized as follows.

$$\forall x \in D, \big(ML(x, 2) \longrightarrow MC(x)\big).$$

Here, the domain of discourse $D$ is the set of all mail messages.

# EXERCISES

**1.1**    Choose the correct answer for each of the following multiple-choice questions/items.

(*a*) True or False: The statement "$x + 2 = 7$ if $x = 2$" is a proposition.

    (*i*) True.                    (*ii*) False.

(*b*) Let $P$ be the sentence "$Q$ is false", $Q$ be the sentence "$R$ is false", and $R$ be the sentence "$P$ is false". Then $P$ is

    (*i*) a proposition.

    (*ii*) not a proposition, but a paradox.

    (*iii*) not a proposition, but an anti-paradox.

    (*iv*) not a proposition, neither a paradox nor an anti-paradox.

(*c*) Let $P$ represent a true statement, and let $Q$ and $R$ represent false statements. The truth value of the compound statement $\neg(\neg P \wedge \neg Q) \vee (\neg R \vee \neg P)$.

    (*i*) True.                    (*ii*) False.

(*d*) Let $P$ be the statement "Kids are happy" and $Q$ be the statement "Parents are happy". If we translate the compound proposition $\neg(P \vee \neg Q)$ into words, we get

    (*i*) It is not the case that kids are happy or parents are not happy.

    (*ii*) Kids are not happy and parents are not happy.

    (*iii*) It is not the case that kids are happy and parents are not happy.

    (*iv*) Kids are not happy or parents are not happy.

(*e*) True or False: If $6 < 1$, then $11 < 4$.

    (*i*) True.                    (*ii*) False.

(*f*) A sufficient condition that a triangle T be a right triangle is that its three sides satisfy a Pythagorean triple. An equivalent statement is

(*i*) If T is a right triangle then its three sides satisfy a Pythagorean triple.

(*ii*) If the three sides of a triangle T satisfy a Pythagorean triple, then T is a right triangle.

(*iii*) If the three sides of a triangle T do not satisfy a Pythagorean triple, then T is not a right triangle.

(*iv*) T is a right triangle only if its three sides satisfy a Pythagorean triple.

(*g*) Consider the following argument:

*If I am thirsty, then I will drink a glass of water.*

*I am not thirsty.*

*I will not drink a glass of water.*

Then this argument is

(*i*) valid by the implication.

(*ii*) invalid by the fallacy of the inverse.

(*iii*) valid by the contrapositive.

(*iv*) invalid by the fallacy of the converse.

(*h*) The total number of rows in the truth table for the compound proposition $(P \lor Q) \land (R \lor \neg S) \to T$ is

(*i*) 25.          (*ii*) 15.          (*iii*) 23.          (*iv*) 32.

(*i*) Consider the statement, "If $n$ is divisible by 12, then $n$ is divisible by 2 and by 3 and by 4". This statement is equivalent to the statement

(*i*) If $n$ is not divisible by 12, then $n$ is divisible by 2 or divisible by 3 or divisible by 4.

(*ii*) If $n$ is not divisible by 12, then $n$ is not divisible by 2 or not divisible by 3 or not divisible by 4.

(*iii*) If $n$ is divisible by 2 and divisible by 3 and divisible by 4, then $n$ is divisible by 12.

(*iv*) If $n$ is not divisible by 2 or not divisible by 3 or not divisible by 4, then $n$ is not divisible by 12.

(*j*) Consider the statement "Given that people who are in need of refuge and consolation are apt to do odd things, it is clear that people who are apt to do odd things are in need of refuge and consolation." This statement, of the form $(P \to Q) \to (Q \to P)$, is logically equivalent to

  (*i*) People who are in need of refuge and consolation are not apt to do odd things.

  (*ii*) People are apt to do odd things if and only if they are in need of refuge and consolation.

  (*iii*) People who are apt to do odd things are in need of refuge and consolation.

  (*iv*) People who are in need of refuge and consolation are apt to do odd things.

(*k*) Let M = "Adam is a Math major", P = "Adam is a Physics major", A = "Adam's wife is an Astronomy major", K = "Adam's wife has read Al-Khwarizmi's algebra", and H = "Adam's wife has read Al-Haytham's optics". One of the following propositions expresses the statement "Adam is a Physics major and a Math major, but his wife is an Astronomy major who has not read both Al-Haytham's optics and Al-Khwarizmi's algebra", which is

  (*i*) $P \wedge M \wedge (A \vee (\neg K \vee \neg H))$.     (*iii*) $P \wedge M \wedge A \wedge (\neg K \vee \neg H)$.

  (*ii*) $P \wedge M \wedge A \wedge (\neg K \wedge \neg H)$.     (*iv*) $P \wedge M \wedge (A \vee (\neg K \wedge \neg H))$.

(*l*) The negation of the proposition "If 3 is positive, then -3 is negative" is the proposition

  (*i*) "If 3 is positive, then -3 is also positive".

  (*ii*) "If 3 is negative, then -3 is positive".

  (*iii*) "3 is positive and -3 is positive".

  (*iv*) "3 is positive and -3 is nonnegative".

(*m*) One of the following propositions is not logically equivalent to the proposition $(P \wedge Q) \vee (\neg P \wedge Q) \vee (P \wedge \neg Q)$, which is

  (*i*) $\neg P \wedge \neg Q$.         (*iii*) $Q \vee (P \wedge \neg Q)$.

  (*ii*) $P \vee Q$.              (*iv*) $P \vee (Q \wedge \neg P)$.

(*n*) The contradiction law is an example of

  (*i*) a tautology.           (*iii*) a contingency.

  (*ii*) a contradiction.        (*iv*) none of the above.

(*o*) One of the following compound propositions is in both DNF and CNF, which is

  (*i*) $(P \lor Q) \land \neg R$.      (*iii*) $P \land Q \land \neg R$.

  (*ii*) $(P \land Q) \lor \neg R$.     (*iv*) None of the above.

(*p*) True or False: A propositional formula in CNF is satisfiable if and only if at least one of its disjunctive clauses is satisfiable.

  (*i*) True.                    (*ii*) False.

**1.2**    Which of the following sentences are propositions? What are the truth values of those that are propositions? (Here, in items (*c*)-(*f*), $x$ and $y$ are any real numbers and $\theta$ is any angle in the standard position).

(*a*) $2 + 3 = 5$.

(*b*) $2 + 3$.

(*c*) $x + 2 = 11$.

(*d*) $x + y = y + x$.

(*e*) $\cos^2 \theta = 1$.

(*f*) $\cos^2 \theta + \sin^2 \theta = 1$.

(*g*) Which of the following sentences are propositions?

(*h*) John F. Kennedy is the 35th president of the United States.

(*i*) Do not be in close contact with a sick person with COVID-19.

(*j*) Repeat your answer to item (*l*).

(*k*) Your answer to item (*l*) is incorrect.

(*l*) The truth value of the statement in item (*h*) is true.

(*m*) The truth value of the statement in item (*i*) is true.

(*n*) The truth value of the statement in this item is false.

**1.3**    What is the negation of each of the following propositions?

(*a*) Today is Thursday.

(*b*) There is no pollution in New Jersey.

(*c*) $2 + 1 = 3$.

(*d*) Sara's first answer to item (*l*) in Exercise 1.1 was incorrect.

(*e*) The summer in Santiago is not hot.

(*f*) The summer in Santiago is hot but bearable.

(*g*) The summer in Santiago is not hot or it is not humid.

(*h*) If the sun is shining in Santiago's sky, then I will go to the nearest beach.

(*i*) If the sun is shining in Santiago's sky, then I will go to the nearest beach and do a little physical exercise.

**1.4** Let $P$ and $Q$ be propositions

$P$ : I bought a lottery ticket this week.

$Q$ : I won the million dollar jackpot on Friday.

Express each of the following propositions as an English sentence.

(*a*) $\neg P$.  (*b*) $P \vee Q$.  (*c*) $P \wedge Q$.  (*d*) $\neg P \wedge \neg Q$.

**1.5** Let $P$ and $Q$ be propositions

$P$ : It is below freezing.

$Q$ : It is snowing.

Write the following propositions using $P$ and $Q$ and logical connectives.

(*a*) It is below freezing and snowing.

(*b*) It is below freezing but not snowing.

(*c*) It is not below freezing and it is not snowing.

(*d*) It is either snowing or below freezing (or both).

**1.6** Consider the implication: "A passing score on the final exam is required in order to receive a passing grade for the course". Restate this statement as five implications, all have the same meaning, and each using one of the five equivalent forms listed in Remark 1.3.

**1.7** The following implications are mainly about the twin primes conjecture.[15] Decide whether each of the these implications is true or false. Justify your answer.

(*a*) If the twin primes conjecture is unsolved, then the twin primes conjecture is conjectured by a monkey.

(*b*) If the twin primes conjecture is true, then the Pythagoras' theorem is true.

(*c*) If the twin primes conjecture is false, then the Pythagoras' theorem is true.

(*d*) If the twin primes conjecture is either proven or disproven, then the twin primes conjecture is solved.

(*e*) If the twin primes conjecture is true, then the twin primes conjecture is true.

---

[15]Twin primes are primes that are two steps apart from each other on the number line, such as 3 and 5, 5 and 7, 29 and 31, and so on. The twin primes conjecture (also known as the Polignac's conjecture) states that there are infinitely many twin primes. This conjecture is still a mysterious unsolved problem in the field of Number Theory.

($f$) If the twin primes conjecture is either true or false, then the twin primes conjecture is solved.

($g$) If the twin primes conjecture is either true or false, then the twin primes conjecture is unsolved.

**1.8**   Prove the result of the following theorem. Start by restating the theorem in IF-THEN form.

*Theorem I: The sum of two odd integers is even.*

**1.9**   Write the contrapositive, converse, and inverse of the proposition: "If the Sun is shrunk to the size of your head, then the Earth will be the size of the pupil of your eye".

**1.10**   Give, if possible, an example of a true conditional statement for which

($a$) the converse is true.          ($b$) the contrapositive is false.

**1.11**   Let $x$ be a positive integer. Prove the results of the following two theorems. Use the low of contrapositive.

*Theorem I: If $x^2$ is even, then $x$ is even.*
*Theorem II: $x^2$ is odd iff $x$ is odd.*

**1.12**   Use truth tables to prove that $P \leftrightarrow Q \equiv (P \rightarrow Q) \land (Q \rightarrow P)$.

**1.13**   Negate and simplify the following two compound propositions.

($a$) $P \leftrightarrow Q$. (Hint: Use the logical equivalence $P \leftrightarrow Q \equiv (P \rightarrow Q) \land (Q \rightarrow P)$).

($b$) $P \oplus Q$. (Hint: Use the logical equivalence $P \oplus Q \equiv (P \lor Q) \land \neg(P \land Q)$).

**1.14**   Construct truth tables for the following compound propositions.

($a$) $P \land \neg P$.                            ($h$) $(P \oplus Q) \land (P \oplus \neg Q)$.

($b$) $(P \lor \neg Q) \rightarrow Q$.                ($i$) $P \rightarrow \neg Q$.

($c$) $(P \lor Q) \rightarrow (P \land Q)$.           ($j$) $\neg P \leftrightarrow Q$.

($d$) $(P \rightarrow Q) \leftrightarrow (\neg Q \rightarrow \neg P)$.      ($k$) $(P \rightarrow Q) \lor (\neg P \rightarrow Q)$.

($e$) $P \oplus P$.                              ($l$) $(P \leftrightarrow Q) \lor (\neg P \leftrightarrow Q)$.

($f$) $P \oplus \neg Q$.                       ($m$) $(P \land Q) \lor R$.

($g$) $\neg P \oplus \neg Q$.                   ($n$) $(P \land Q) \land R$.

(*o*) $(P \vee Q) \vee R$.

(*r*) $(P \rightarrow Q) \vee (\neg P \rightarrow R)$.

(*p*) $(P \wedge Q) \vee \neg R$.

(*s*) $(P \rightarrow Q) \wedge (\neg P \rightarrow R)$.

(*q*) $P \rightarrow (\neg Q \vee R)$.

(*t*) $(P \leftrightarrow Q) \vee (\neg Q \leftrightarrow R)$.

**1.15**    Consider the following propositional logic word problem: Either Lillian is forceful or she is creative. If Lillian is forceful, then she will be a good executive. It is not possible that Lillian is both efficient and creative. If she is not efficient, then either she is forceful or she will be a good executive. Can you conclude that Lillian will be a good executive? Your reasoning and conclusion should be justified by linking them to the propositional logic.

**1.16**    List the names of the logical equivalence laws from Table 1.8 that are used to prove Item (*c*) of Example 1.27.

**1.17**    Determine if each of the following implications is a tautology, contradiction or contingency. Justify your answer without using truth tables.

(*a*) $[\neg P \wedge (P \vee Q)] \rightarrow Q$.

(*c*) $[P \wedge (P \rightarrow Q)] \rightarrow Q$.

(*b*) $[(P \rightarrow Q) \wedge (Q \rightarrow R)] \rightarrow (P \rightarrow R)$.

(*d*) $[(P \vee Q) \wedge (P \rightarrow R) \wedge (Q \rightarrow R)] \rightarrow R$.

**1.18**    Without using truth tables, prove that the following propositions are logically equivalent.

(*a*) $P \leftrightarrow Q$.

(*b*) $(P \wedge Q) \vee (\neg P \wedge \neg Q)$.

(*c*) $\neg (P \oplus Q)$.

Hint: You can show that the propositions (*a*), (*b*) and (*c*) are logically equivalent by showing that the propositions in (*a*) and (*b*) are logically equivalent and that those in (*b*) and (*c*) are logically equivalent.

**1.19**    Construct a *disjunctive normal form* having the following truth table.

| P | Q | R | Statement |
|---|---|---|-----------|
| T | T | T | F |
| T | T | F | T |
| T | F | T | T |
| T | F | F | F |
| F | T | T | F |
| F | T | F | T |
| F | F | T | F |
| F | F | F | F |

**1.20**   For each of the following statements, find an equivalent statement in conjunctive normal form.

(*a*)  $\neg(A \lor B)$.                (*b*)  $\neg(A \land B)$.                (*c*)  $A \lor (B \land C)$.

**1.21**   Give an example of a proposition in DNF with three variables and four distinct clauses that is unsatisfiable.

**1.22**   Is the following propositional formula statisfiable? Explain.

$$P \land Q \land (\neg R \lor S \lor \neg T).$$

**1.23**   Choose the correct answer for each of the following multiple-choice questions/items.

(*a*)  The propositions in items (*c*) and (*d*) of Example 1.36 are

    (*i*)  both true.                    (*iii*)  true and false, respectively.

    (*ii*)  both false.                    (*iv*)  false and true, respectively.

(*b*)  Expressing the proposition $\exists x \in D, \forall y \in D, \forall z \in D, P(x, y, z)$ using only the existential quantifier, we get

    (*i*)  $\exists x \in D, \neg \exists y \in D, \neg \exists z \in D, \neg P(x, y, z)$.

    (*ii*)  $\exists x \in D, \neg \exists y \in D, \neg \exists z \in D, P(x, y, z)$.

    (*iii*)  $\exists x \in D, \exists y \in D, \neg \exists z \in D, \neg P(x, y, z)$.

    (*iv*)  $\exists x \in D, \neg \exists y \in D, \exists z \in D, \neg P(x, y, z)$.

(*c*)  Let $V$ be the set of graph vertices and $C$ be the set of all colors (see Sections 4.1 and 4.4). Let also edge$(u, v)$ mean that "vertex $v$ is adjacent to vertex $u$", and color$(v, x)$ mean that "vertex $v$ has color $x$". Which one of the following is the correct formulation for the statement "Adjacent vertices do not have the same color"?

    (*i*)  $\forall v \in V, \forall u \in V, \forall x \in C, (\text{edge}(v, u) \land \text{color}(v, x) \land \neg \text{color}(u, x))$.

    (*ii*)  $\forall v \in V, \forall u \in V, \forall x \in C, (\text{edge}(v, u) \land \text{color}(v, x) \rightarrow \neg \text{color}(u, x))$.

    (*iii*)  $\forall v \in V, \forall u \in V, \forall x \in C, (\text{edge}(v, u) \land \text{color}(v, x) \lor \neg \text{color}(u, x))$.

    (*iv*)  $\forall v \in V, \forall u \in V, \forall x \in C, (\text{edge}(v, u) \land \text{color}(v, x) \lor \text{color}(u, x))$.

(*d*)  Let $F$ be the set of all (direct or indirect) flights, and define the predicate

$$DF(u, x, y) = \text{"There is a direct flight } u \text{ from a city } x \text{ to a city } y\text{"}.$$

One of the following statements is a correct translation for the quantified statement $\neg \exists x \in F, DF(x, \text{Columbus}, \text{Rochester})$, which is

(*i*) There is no flight between Columbus and Rochester.

(*ii*) There is no flight from Columbus to Rochester.

(*iii*) There is no direct flight between Columbus and Rochester.

(*iv*) All flights from Columbus to Rochester are indirect.

(*e*) The negation of "Nothing that glitters is gold" is

(*i*) "All that glitters is gold".

(*ii*) "There is something that glitters that is gold".

(*iii*) "It is not all that glitters that is gold".

(*iv*) "All that glitters that is not gold".

(*f*) Let $S$ be the set of all shapes, and define the three predicates $FS(x) =$ "$x$ is a footprint shape", $C(x) =$ "$x$ is circular-shaped", and $E(x) =$ "$x$ is elliptical-shaped". One of the following is the correct formulation for the statement "Each footprint shape is either circular or elliptical, but not both", which is

(*i*) $\forall x \in S, (F(x) \longrightarrow [C(x) \vee E(x)])$.

(*ii*) $\forall x \in S, (F(x) \longrightarrow [C(x) \rightarrow \neg E(x)])$.

(*iii*) $\forall x \in S, (F(x) \longrightarrow [C(x) \leftrightarrow \neg E(x)])$.

(*iv*) $\forall x \in S, (F(x) \longrightarrow [C(x) \oplus \neg E(x)])$.

**1.24**    Let the domain of discourse consist of the non-zero integers, i.e., $D = \mathbb{Z} - \{0\}$. Give the truth value of each of the following quantified statements. Find a value of $x$ which supports your conclusion.

(*a*) $\forall x \in D, x < 2x$.

(*b*) $\exists x \in D, x + x = x - x$.

(*c*) $\exists x \in D, x^2 = 2x$.

(*d*) $\forall x \in D, \dfrac{x}{2x} < x$.

(*e*) $\exists x \in D, \dfrac{x}{x} = x$.

**1.25**    Consider the following two propositions.

$$\exists x \in \mathbb{N}, \forall y \in \mathbb{N}, P(x, y), \tag{1.13}$$

$$\forall x \in \mathbb{N}, \exists y \in \mathbb{N}, P(x, y). \tag{1.14}$$

Using the above two propositions, give a predicate $P(x, y)$ that makes (or explain why it is not possible):

(*a*) Propositions (1.13) and (1.14) are both true.

(*b*) Propositions (1.13) and (1.14) are both false.

(*c*) Proposition (1.13) is true and Proposition (1.14) is false.

(*d*) Proposition (1.13) is false and Proposition (1.14) is true.

**1.26**    Negate the false proposition $\exists x \in \mathbb{R}, \forall y \in \mathbb{R}, x + y = 0$.

**1.27**    Symbolize the following statements using the predicate logic. Use the four predicates $A(x) = $ "$x$ is an architect", $C(x) = $ "$x$ is a constructor", $I(x) = $ "$x$ is invited", and $L(x) = $ "$x$ is late".

(*a*) There is at least a person who is on time.

(*b*) There is at least an invited person who is neither a constructor nor an architect.

(*c*) All architects and constructors invited to the party are late.

**1.28**    Let $D$ be the universe of all living things, and define the three predicates: Child($x$) = " $x$ is a child", Mime($y$) = "$y$ is a mime", and Like($x, y$) = "$x$ likes y". Symbolize the negation of the quantified statement "Some children do not like mimes".

**1.29**    Symbolize the following statements using the predicate logic.

(*a*) For each owner, there is at least a house with this owner.

(*b*) Adjacent houses do not have the same owner.

(*c*) Each house has at most one owner.

**1.30**    Symbolize the following statements using the predicate logic.

(*a*) Friends of friends are friends.

(*b*) Friendless people smoke.

(*c*) Smoking causes cancer.

(*d*) If two people are friends, either both smoke or neither does.

# CHAPTER 2

# SET-THEORETIC STRUCTURES

## Contents

© 2022 by Baha Alzalg | Kindle Direct Publishing, Washington, United States 2022/9
B. Alzalg, *Combinatorial and Algorithmic Mathematics: From Foundation to Optimization*,
DOI 10.5281/zenodo.7110890

Much of mathematical discrete structures are written in terms of sets. In this chapter, we introduce basic concepts of set theory. After introducing the mathematical induction as a method for proving mathematical statements, we study set-theoretic structures such as sets, relations, and functions.

## 2.1. Induction

Mathematical induction is a powerful and elegant technique for proving certain types of mathematical statements.

As explained in Figure 2.1, there are two steps involved in knocking over a row of dominoes. These steps are the same as the steps in a proof by induction. See Figure 2.1 which shows the idea of induction.

We have an infinite number of claims that we wish to prove: Claim (1), Claim (2), Claim (3), ..., Claim ($n$), .... We have the following workflow principle.

---

**Workflow 2.1** *If we can perform the following two steps, then we are assured that all these claims are true:*

(*i*)  *Prove that Claim (1) is true.*

(*ii*)  *Prove that, for every natural number k, if Claim (k) is true, then Claim (k + 1) is true.*

---

In Workflow 2.1, (i) represents the knocking over the first domino, and (ii) shows that if the $k^{th}$ domino is knocked over, then it follows that the $(k + 1)^{st}$ domino will be knocked over.

There are two steps involved in knocking over a row of dominoes: First, you have to knock over the first domino. You have to be sure that when domino $k$ falls, it knocks over domino $k + 1$. These steps are the same as the steps in a proof by induction.

Figure 2.1: Falling dominoes.

To apply the principle of mathematical induction to sets of natural numbers, let $S$ be a set of natural numbers (in other words, $S$ is a subset of $\mathbb{N}$) such that

   (*i*)  $1 \in S$, and                      (*ii*)  for each $k \in \mathbb{N}, k \in S \to k + 1 \in S$,

then $S = \mathbb{N}$. The intuitive justification is as follows: By (*i*), we know that $1 \in S$. Now apply (*ii*) with $k = 1$, we have $2 = 1 + 1 \in S$. Now, apply (*ii*) with $k = 2$, we get $3 = 2 + 1 \in S$. And so forth.

***Principle of induction for predicates***    We have the following principle.

**Principle 2.1** *Let $P(n)$ be a predicate whose domain is $\mathbb{N}$ such that*

   (*i*)  *$P(1)$ is true, and*

  (*ii*)  *if $P(k)$ is true, then $P(k + 1)$ is true, for all $k \in \mathbb{N}$.*

*If (i) and (ii) are true, then $P(n)$ is true for all $n \in \mathbb{N}$.*

In Principle 2.1, we call (*i*) the base case (basis step), and (*ii*) the inductive step, which consists the inductive hypothesis and the inductive conclusion.

   **Example 2.1** Use mathematical induction to prove that the sum of the first $n$ positive integers is $n(n + 1)/2$.

**Solution**    By induction, let $P(n)$ be

$$"1 + 2 + 3 + \cdots + n = \frac{n(n + 1)}{2}" . \tag{2.1}$$

for $n \in \mathbb{N}$. First, we look at the base case for $n = 1$ to see whether or not $P(1)$ is true. Substituting $n = 1$ for both sides of (2.1), we get $1 = 1(1 + 1)/2 = 1$.

Next, we look at the inductive step for $n = k$ to see whether or not $P(k + 1)$ is true if the inductive hypothesis $P(k)$

$$"1 + 2 + 3 + \cdots + k = \frac{k(k + 1)}{2}" \tag{2.2}$$

is true. Using the inductive hypothesis (2.2), we have

$$
\begin{aligned}
1 + 2 + 3 + \cdots + k + (k + 1) &= \frac{k(k + 1)}{2} + (k + 1) \\
&= \frac{k(k + 1)}{2} + \frac{2(k + 1)}{2} \\
&= \frac{k(k + 1) + 2(k + 1)}{2} = \frac{(k + 1)(k + 2)}{2}.
\end{aligned}
$$

Thus, if $P(k)$ is true, then $P(k + 1)$ is true for any $k$ greater than or equal to 1 which proves the inductive conclusion.

Therefore, by induction, for each $n \in \mathbb{N}$, $P(n)$ is true. In other words, the equality in (2.1) holds. The proof is complete.       ■

---

**Corollary 2.1** *As one of the most known series, we have[a]*

$$\sum_{k=1}^{n} k = \frac{1}{2}n(n+1),$$

*which is well known in mathematics as the arithmetic series.*

---

[a]The symbol $\sum$ is the capital Greek letter "sigma". We write $\sum_{k=1}^{n} k$ (read "the sum of $k$ from $k$ equals 1 to $k$ equals $n$") to indicate the sum $1 + 2 + \cdots + n$.

---

**Example 2.2** Use mathematical induction to prove that the sum of the first $n$ positive odd integers is $n^2$.

**Solution**   By induction on $n$, let $P(n)$ be

$$"1 + 3 + 5 + \cdots + (2n - 1) = n^2" \tag{2.3}$$

for $n \in \mathbb{N}$. First, we look at the base case for $n = 1$ to see whether or not $P(1)$ is true. Substituting $n = 1$ for both sides of (2.3), we get $2n - 1 = 2(1) - 1 = 1$.

Now, we look at the inductive step for $n = k$ to see whether or not $P(k + 1)$ is true if the inductive hypothesis $P(k)$

$$"1 + 3 + 5 + \cdots + (2k - 1) = k^2" \tag{2.4}$$

is true. Using the inductive hypothesis (2.4), we have

$$
\begin{aligned}
1 + 3 + 5 + \cdots + (2k - 1) + (2(k + 1) - 1) &= k^2 + (2(k + 1) - 1) \\
&= k^2 + (2k + 1) = (k + 1)^2.
\end{aligned}
$$

Thus, if $P(k)$ is true, then $P(k + 1)$ is true for any $k$ greater than or equal to 1 which proves the inductive conclusion.

Therefore, by induction, for each $n \in \mathbb{N}$, $P(n)$ is true. In other words, the equality in (2.3) holds. This completes the proof. ∎

**Induction proves recursion**   A recurrence is a well-defined mathematical function written in terms of itself.

For example, it can be shown (see Exercise 2.5), that the following recurrence

$$T(1) = 1, \quad T(n) = 3T(n - 1) + 4, \quad n = 1, 2, 3, \ldots, \tag{2.5}$$

describes the function

$$T(n) = 3^n - 2, \quad n = 0, 1, 2, \ldots. \tag{2.6}$$

Recurrences will be studied extensively in Chapter 5. In this part, we learn briefly how to use induction to verify solutions of recurrence formulas.

The formula (2.6) is called the solution of the recurrence relation (2.5) and can be obtained by using the method of mathematical induction. A simpler example is the following.

**Example 2.3** Use the induction method to show that the recurrence

$$T(n) = \begin{cases} 1, & \text{if } n = 0; \\ T(n-1) + 1, & \text{if } n = 1, 2, 3, \dots, \end{cases} \tag{2.7}$$

has the solution

$$T(n) = n + 1, \quad n = 0, 1, 2, \dots.$$

**Solution** By induction on $n$. First, the base case for $n = 0$ is seen to be true as $T(0) = 1 = 0 + 1$, where the first equality follows from (2.7). Now, assume that the inductive hypothesis holds for $n = k$, i.e., $T(k) = k + 1$. Then

$$T(k + 1) = T(k) + 1 = (k + 1) + 1 = k + 2.$$

Therefore, if $T(k) = k + 1$, then $T(k + 1) = k + 2$ for any integer $k \geq 0$. This proves the inductive conclusion. Thus, by induction on $n$, the equality in (2.7) holds. ∎

## 2.2. Sets

The concept of a set is used throughout mathematics, and its formal definition closely matches our intuitive understanding of the word.

> **Definition 2.1** *A set is an unordered collection of distinct objects.*

We can build sets containing any objects that we like, but usually we consider sets whose elements have some property in common.

> **Definition 2.2** *The objects comprising a set are called its elements or members.*

Note that sets do not contain duplicates and that the order of elements in a set is not significant. A finite set is a set that has a finite number of elements. An infinite set is a set with an infinite number of elements. By convention, a set can be defined by enumerating its components in curly brackets. We have the following example.

**Example 2.4** The following are examples of sets.

- Vowels = {a, e, i, o, u}.

- Weekend = {Saturday, Sunday}.

- Days = {1, 2, 3, ..., 365}.

- $\mathbb{N} = \{1, 2, 3, \ldots\}$.

- $\mathbb{Z} = \{\ldots, -3, -2, -1, 0, 1, 2, 3, \ldots\}$.

- Truth-Values = {true, false}.

Note that the above sets are finite, except $\mathbb{N}$ and $\mathbb{Z}$ which are infinite sets.     ■

***Set membership***    To express the fact that an object $x$ is a member of a set $A$, we write $x \in A$. Note that an expression with this form is either true or false; the thing is either a member or not. For instance, in Example 2.4, the statements u ∈ Vowels, $10 \in \mathbb{N}, 21 \in$ Days are true, but the statement $-21 \in$ Days is false.

Note that an expression such as $\{1, 2, 3\} \in \{4, 5, 6\}$ is not allowed, but expressions such as $\{1, 2, 3\} \in \{\{6, 7, 8\}, \{1, 2, 3\}\}$ and $\{1, 2, 3\} \in \{\{6, 2, 1\}\}$ are allowed. (Why? And which one is true?).

***Cardinality of sets***    The cardinality of a set $A$, denoted by $|A|$, is the number of elements in it. For instance, in Example 2.4, |Vowels| = 4 and |Days| = 365. A finite set is a set that has a finite number of elements. Therefore, in Example 2.4, Vowels and Days are finite sets, but $\mathbb{N}$ and $\mathbb{Z}$ are infinite sets.

***Set equality***    Two arbitrary sets $A$ and $B$ are equal, denoted by $A = B$, if $A$ and $B$ contain precisely the same members. We emphasize that the order in which values occur in the set is not significant. For instance, in Example 2.4, the equality $\{1, 2, 3\} = \{2, 3, 1\}$ is true, but the equality Days = $\mathbb{N}$ is false.

***Subsets and proper subsets***    Let $A$ and $B$ be arbitrary sets of the same type. Then $A$ is said to be a subset of $B$, denoted by $A \subseteq B$, if every member of $A$ is also a member of $B$. For example, the expressions $\{1, 2, 3\} \subseteq \mathbb{N}, \mathbb{N} \subseteq \mathbb{Z}$ and $\{1, 2, 3\} \subseteq \{1, 2, 3\}$ are true, but the expressions $\{1, 2, 3, 4\} \subseteq \{1, 2, 3\}$ and $\mathbb{Z} \subseteq \mathbb{N}$ are false. Note that equal sets are subsets of each other. That is, $A = B \leftrightarrow (A \subseteq B) \wedge (B \subseteq A)$.

A set $A$ is said to be a proper subset of a set $B$, denoted by $A \subset B$, if $A \subseteq B$ and $A \neq B$. For instance, in Example 2.4, the expression {Saturday} ⊂ Weekend is true, but the expression {Monday} ⊂ Weekend is false.

***The empty set***    The empty set, denoted by { } or $\emptyset$, is a special set that has the property of having no members. Note that nothing is a member of the empty set. That is, if $x$ is any element, then the expression $x \in \emptyset$ must be false. Therefore $|\emptyset| = 0$, while $|\{\emptyset\}| = 1$ for instance. Note also that the empty set is a subset of every set. That is, if $A$ is an arbitrary set, then $\emptyset \subseteq A$. This includes of course $\emptyset \subseteq \emptyset$.

***The powerset operator***    Let $A$ be an arbitrary set. The set of all subsets of $A$ is called the powerset of $A$ and is denoted as $\mathcal{P}(A)$. So, $B \in \mathcal{P}(A) \longleftrightarrow B \subseteq A$.

For example, if $A = \{1\}, B = \{1, 2\}$ and $C = \{a, b, c\}$, then $\mathcal{P}(A) = \{\emptyset, \{1\}\}, \mathcal{P}(B) = \{\emptyset, \{1\}, \{2\}, \{1, 2\}\}$, and $\mathcal{P}(C) = \{\emptyset, \{a\}, \{b\}, \{c\}, \{a, b\}, \{a, c\}, \{b, c\}, \{a, b, c\}\}$. Note that $\mathcal{P}(A) = 2, |\mathcal{P}(B)| = 4 = 2^2$ and $|\mathcal{P}(C)| = 8 = 2^3$. In general, for any finite set $A$, one can prove that $|\mathcal{P}(A)| = 2^{|A|}$. Proving this is left as an exercise for the reader (see Exercise 2.4). See also Example 6.5.

***Manipulating sets***     Venn diagrams use overlapping circles or other shapes to illustrate the logical relationships between two or more sets. See Figure 2.2 which shows Venn diagrams for the most common combined sets obtained from two sets $A$ and $B$, such as the union of $A$ and $B$, denoted by $A \cup B$, the intersection of $A$ and $B$, denoted by $A \cap B$, the complement of $A$, denoted by $A'$, the difference of $A$ and $B$, denoted by $A - B$, and the symmetric difference of $A$ and $B$, denoted by $A \triangle B$.

    Table 2.1 shows a list of most common laws of algebra of sets.

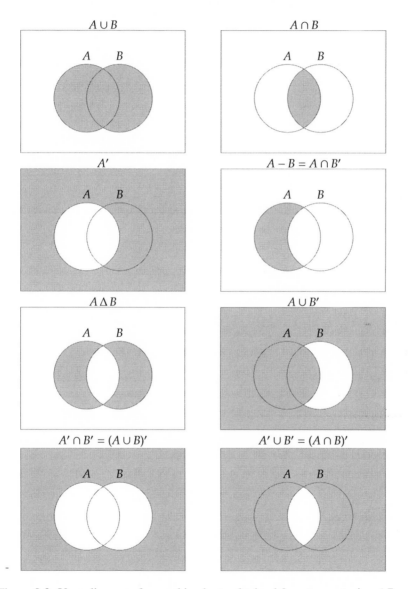

Figure 2.2: Venn diagrams for combined sets obtained from two sets $A$ and $B$.

For example,

- $\{a, e, i\} \cup \{o, u\} = \{a, e, i, o, u\}$,
- $\{a, e\} \cup \{e, i\} = \{a, e, i\}$,
- $\emptyset \cup \{a, e\} = \{a, e\}$,
- $\{a, e, i\} \cap \{o, u\} = \emptyset$,

- $\mathbb{N} \cap \mathbb{Z} = \mathbb{N}$,
- $\mathbb{N} \cup \mathbb{Z} = \mathbb{Z}$,
- $\{a, e, i, o, u\} - \{o\} = \{a, e, i, u\}$,
- $\emptyset - \{a, e, i, u\} = \emptyset$.

The Cartesian product of two sets $A$ and $B$, denoted $A \times B$, is the set of all possible ordered pairs where the elements of $A$ are first and the elements of $B$ are second. For example, $\{a, b, c\} \times \{1, 2\} = \{(a, 1), (a, 2), (b, 1), (b, 2), (c, 1), (c, 2)\}$.

**Sets defined by predicates**    The set of all elements $x$ in a domain $D$ such that a predicate $P(x)$ is true is denoted as $\{x \in D : P(x)\}$ and is called the set-builder notation. For example, if $P(x)$ is

$$\text{"}x \text{ is an even integer between 1 and 11"}$$

then the set $\{x \in \mathbb{Z} : P(x)\} = \{2, 4, 6, 8, 10\}$. For another example, the set $\{n \in \mathbb{N} : (n \geq 1) \wedge (n < 4)\}$ is the set $\{1, 2, 3\}$ in the set-builder notation. We also have the following example.

**Example 2.5**  Redefine each of the following sets using the set-builder notation.

(a) $A = \{-1, \sqrt{2}\}$.    (b) $B = (-3, 5]$.    (c) $\emptyset$.    (d) $\mathbb{R}$.

**Solution**    It is easy to see that the given sets can be redefined as follows.

(a) $A = \{x \in \mathbb{R} : (x + 1)(x - \sqrt{2}) = 0\}$.    (c) $\emptyset = \{x \in \mathbb{Z} : x + \sqrt{2} = 0\}$.

(b) $B = \{x \in \mathbb{R} : (-3 < x) \wedge (x \leq 5)\}$.    (d) $\mathbb{R} = \{x \in \mathbb{R} : x^2 + 1 > 0\}$.

Let $S$ be a space that contains the sets $A$ and $B$. We can write the union, intersection, complement, difference, and Cartesian product introduced above in set-builder notation as follows.

$$\begin{aligned}
A \cup B &= \{x \in S : x \in A \ \vee \ x \in B\}, \\
A \cap B &= \{x \in S : x \in A \ \wedge \ x \in B\}, \\
A' &= \{x \in S : x \notin A\}, \\
A - B &= \{x \in S : x \in A \ \wedge \ x \notin B\}, \\
A \triangle B &= \{x \in S : x \in A - B \ \vee \ x \in B - A\}, \\
A \times B &= \{(a, b) : a \in A \text{ and } b \in B\}.
\end{aligned}$$

| Name | Formula(s) |
|------|-----------|
| Identity laws | $A \cup A = A$ |
| | $A \cap A = A$ |
| Domination laws | $A \cap \emptyset = \emptyset$ |
| | $A \cup S = S$ |
| Idempotent laws | $A \cap A = A$ |
| | $A \cup A = A$ |
| Tautology law | $A \cup A' = S$ |
| Contradiction law | $A \cap A' = \emptyset$ |
| Double negation law | $A = (A')'$ |
| DeMorgan's laws | $(A \cup B)' = A' \cap B'$ |
| | $(A \cap B)' = A' \cup B'$ |
| | $A - (B \cup C) = (A - B) \cap (A - C)$ |
| | $A - (B \cap C) = (A - B) \cup (A - C)$ |
| Contrapositive law | $(A \subseteq B) \equiv (B' \subseteq A')$ |
| Commutative laws | $A \cup B = B \cup A$ |
| | $A \cap B = B \cap A$ |
| Associative laws | $(A \cup B) \cup C = A \cup (B \cup C)$ |
| | $(A \cap B) \cap C = A \cap (B \cap C)$ |
| Distributive laws | $(A \cup B) \cap C = (A \cap B) \cup (A \cap C)$ |
| | $(A \cap B) \cup C = (A \cup B) \cap (A \cup C)$ |
| More laws | $(A - B) \cap B = \emptyset$ |
| | $A - B) \cup B = A \cup B$ |
| | $A - (B \cap C) = (A - B) \cup (A - C)$ |
| | $A - (B \cup C) = (A - B) \cap (A - C)$ |
| | $A \cap (B - C) = (A \cap B) - (A \cap C)$ |
| | $A \cap (B \triangle C) = (A \cap B) \triangle (A \cap C)$ |
| | $(A - B) \cup (B - A) = (A \cup B) - (A \cap B)$ |

Table 2.1: A list of the most common laws of algebra of sets. Here, $A, B$ and $C$ are finite sets in a space $S$.

## 2.3. Relations

The concept of a relation is important in mathematics and computer science. For example, relations are used in the definition of functions which will be introduced in Section 2.5. Relations can be also used in the definition of directed graphs which will be introduced in Section 4.5.

Given a set of objects, we may want to say that certain pairs of objects are related in some way. For example, we may say that two students are related if they are attending the same university or working for your company, or if they are from the same hometown. Mathematically speaking, we have the following definition.

---

**Definition 2.3** *A binary relation* $\mathcal{R}$ *on two sets A and B is a subset of the Cartesian product* $A \times B$. *The notation* $a \, \mathcal{R} \, b$ *is read a is* $\mathcal{R}$-*related (or simply related) to b and it means that* $(a, b) \in \mathcal{R}$. *If* $(a, b) \notin \mathcal{R}$, *we write* $a \, \cancel{\mathcal{R}} \, b$. *A relation from A to A is called a relation on A.*

---

There are several ways to represent a relation in a usable form. For example, we can list the ordered pairs inside set brackets, construct its corresponding table, or plot its corresponding rectangular coordinate graph. We can also describe a relation with an expression such as an equality or inequality. We have the following example.

### Example 2.6

Let $A = \{0, 1, 2, 3, 4\}$ and $B = \{0, 1, 2, 3\}$. We represent a binary relation $\mathcal{R}$ from the set $A$ to the set $B$ in three different ways. We first describe $\mathcal{R}$ by explicitly listing its ordered pairs inside set brackets:

$$\mathcal{R} = \{(1, 1), (2, 0), (3, 3), (4, 2)\}.$$

In table form, $\mathcal{R}$ can be represented by the table shown to the right. In graphical representation, $\mathcal{R}$ can be represented by the graph in the figure shown to the right.

| 1 | 2 | 3 | 4 |
|---|---|---|---|
| 1 | 0 | 3 | 2 |

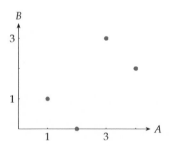

In general, there are many different relations from a set $A$ to a set $B$, because every subset of $A \times B$ is a relation from $A$ to $B$. This includes the empty set $\emptyset$, which is called the empty relation from $A$ to $B$, and includes the set $A \times B$, which is called the full relation (or universal relation) from $A$ to $B$.

**Definition 2.4** *The domain of the binary relation $\mathcal{R}$ from $A$ to $B$ is the set*

$$Dom(\mathcal{R}) = \{x \in A : \text{ there exists } y \in B \text{ such that } x\,\mathcal{R}\,y\}.$$

*The range of the relation $\mathcal{R}$ is the set*

$$Rng(\mathcal{R}) = \{y \in B : \text{ there exists } x \in A \text{ such that } x\,\mathcal{R}\,y\}.$$

**Example 2.7** Define the binary relations

$$\begin{aligned}
\mathcal{R} &= \left\{(x_1, x_2) \in \mathbb{R} \times \mathbb{R} : x_1^2 + x_2^2 \leq 1\right\}, \\
\mathcal{T} &= \left\{(x_1, x_2) \in \mathbb{R} \times \mathbb{R} : \max\{|x_1|, |x_2|\} \leq 1\right\}.
\end{aligned}$$

The graphs of $\mathcal{R}$ and $\mathcal{T}$ are shown below.

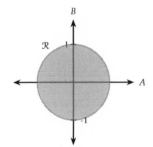

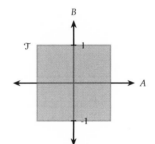

$Dom(\mathcal{R}) = Rng(\mathcal{R}) = [-1, 1].$ $\qquad$ $Dom(\mathcal{T}) = Rng(\mathcal{T}) = [-1, 1].$

**Definition 2.5** *An n-ary relation on sets $A_1, A_2, \ldots, A_n$ is a subset of the Cartesian product $A_1 \times A_2 \times \cdots \times A_n$.*

As an example, the following 3-dimensional unit sphere

$$\mathcal{S} = \left\{(x_1, x_2, x_3) \in \mathbb{R}^3 : \sum_{i=1}^{3} x_i^2 \leq 1\right\}$$

is a ternary relation on $\mathbb{R}^3 = \mathbb{R} \times \mathbb{R} \times \mathbb{R}$.

## Equivalence relations

Relations that possess the three properties in the following definition are particularly valuable.

> **Definition 2.6** *Let A be a set and $\mathcal{R}$ be a relation on A. Then*
>
>    (i) *$\mathcal{R}$ is said to be reflexive if for all $x \in A$, $x\,\mathcal{R}\,x$, and irreflexive otherwise.*
>
>   (ii) *$\mathcal{R}$ is said to be symmetric if for all $x, y \in A$ if $x\,\mathcal{R}\,y$, then $y\,\mathcal{R}\,x$, and asymmetric otherwise.*
>
>  (iii) *$\mathcal{R}$ is said to be transitive if for all $x, y, z \in A$ if $x\,\mathcal{R}\,y$ and $y\,\mathcal{R}\,z$, then $x\,\mathcal{R}\,z$.*

In light of Definition 2.6 and using 1.7, if $\mathcal{R}$ is a relation on a set $A$, then

- $\mathcal{R}$ is not reflexive if there is some $x \in A$ such that $x\,\not\mathcal{R}\,x$.

- $\mathcal{R}$ is not symmetric if there are some $x, y \in A$ such that $x\,\mathcal{R}\,y$, and $y\,\not\mathcal{R}\,x$.

- $\mathcal{R}$ is not transitive such that there are some $x, y, z \in A$ such that $x\,\mathcal{R}\,y$ and $y\,\mathcal{R}\,z$ but $x\,\not\mathcal{R}\,z$.

**Example 2.8**  Let $A = \{1, 2, 3\}$, and define the following relations on $A$:

$$
\begin{aligned}
\mathcal{R} &= \{(1,1),(2,2),(1,2),(2,1)\}, \\
\mathcal{S} &= \{(1,1),(2,2),(3,3),(1,2),(2,1),(1,3)\}, \\
\mathcal{T} &= \{(1,1),(2,2),(3,3),(1,2),(2,1),(1,3),(3,1)\}, \\
\mathcal{U} &= \{(1,1),(2,2),(3,3),(1,2),(2,1)\}, \\
\mathcal{V} &= \{(1,1),(2,2),(3,3),(1,2),(2,1),(1,3),(3,1),(2,3),(3,2)\}.
\end{aligned}
$$

Based on Definition 2.6, it can be seen that

- $\mathcal{R}$ is not reflexive because $3\,\not\mathcal{R}\,3$.

- $\mathcal{S}$ is reflexive, but not symmetric because $1\,\mathcal{S}\,3$ and $3\,\not\mathcal{S}\,1$.

- $\mathcal{T}$ is reflexive and symmetric, but not transitive because $2\,\mathcal{T}\,1$ and $1\,\mathcal{T}\,3$ but $2\,\not\mathcal{T}\,3$.

- $\mathcal{U}$ and $\mathcal{V}$ are reflexive, symmetric, and transitive on $A$. The next definition calls $\mathcal{U}$ and $\mathcal{V}$ equivalence relations.

■

> **Definition 2.7** *A relation on a set A is called an equivalence relation on A if it is reflexive on A, symmetric, and transitive.*

| Relations on $\mathbb{N}$: | "=" | "$\leq$" | "$<$" |
|---|---|---|---|
| Is the relation reflexive? | Yes | Yes | No |
| Is the relation symmetric? | Yes | No | No |
| Is the relation transitive? | Yes | Yes | Yes |
| Is it an equivalence relation? | Yes | No | No |

Table 2.2: Some relations on the natural numbers.

**Example 2.9** Among the three relations "=", "$\leq$" and "$<$" on $\mathbb{N}$, "=" is the only equivalence relation on $\mathbb{N}$. See Table 2.2.  ∎

An equivalence class is a set of elements that are considered to be similar or equivalent. We have the following definition.

**Definition 2.8** *Let $\mathcal{R}$ be an equivalence relation on a set $A$. For $x \in A$, the equivalence class of $x$ determined by $\mathcal{R}$ is the set*

$$x/\mathcal{R} = \{y \in A : x\,\mathcal{R}\,y\}.$$

*This is read "the class of $x$ modulo $\mathcal{R}$" or "$x$ mod $\mathcal{R}$". When $\mathcal{R}$ is known from the context, equivalence class $x/\mathcal{R}$ is also denoted by $[x]$.*
*The set of all equivalence classes is called $A$ module $\mathcal{R}$ and is denoted*

$$A/\mathcal{R} = \{x/\mathcal{R} = [x] : x \in A\}.$$

The following are examples of equivalence classes.

**Example 2.10** In Example 2.8, we have seen that $\mathcal{U} = \{(1,1),(2,2),(3,3),(1,2),(2,1)\}$ is an equivalence relation on $A = \{1,2,3\}$. It is also seen that

$$1/\mathcal{U} = 2/\mathcal{U} = \{1,2\} \text{ and } 3/\mathcal{U} = \{3\}.$$

Thus $A/\mathcal{U} = \{\{1,2\},\{3\}\}$.
In Example 2.8, we have also seen that $\mathcal{V}$ is an equivalence relation on $A$. Finding $1/\mathcal{V}, 2/\mathcal{V}, 3/\mathcal{V}$ and $A/\mathcal{V}$ is left as an exercise for the reader.  ∎

**Example 2.11** The relation "$\diamond$" on $\mathbb{R}$ defined as $x \diamond y$ iff $x^2 = y^2$ is an equivalence relation on $\mathbb{R}$. In this example, we have $[1] = \{1,-1\}$, $[-2] = \{-2,2\}$, and $[0] = \{0\}$. In fact, for any $x \in \mathbb{R}$, we have $[x] = \{x,-x\}$. Thus

$$\mathbb{R}/\diamond = \{0,\{\pm 1\},\{\pm 2\},\{\pm 3\},\ldots\}.$$

∎

**Example 2.12 (Congruence classes)** Given an integer $n > 1$, called a modulus, two integers $x$ and $y$ are said to be congruent modulo $n$, denoted by $x \equiv y \pmod{n}$, if $n$ is a divisor of their difference (i.e., if there is an integer $k$ such that $a - b = nk$). For example, $76 \equiv 52 \pmod{12}$ because $76 - 52 = 24$ which is a multiple of 12.

Congruence modulo $n$ is a relation called a congruence relation. This congruence relation is an equivalence relation. This can be seen as follows:

- Reflexivity: $x \equiv x \pmod{n}$.

- Symmetry: $x \equiv y \pmod{n}$ if $y \equiv x \pmod{n}$ for all $x, y$, and $n$.

- Transitivity: If $x \equiv y \pmod{n}$ and $y \equiv z \pmod{n}$, then $x \equiv z \pmod{n}$.

So, congruence modulo $n$ is an equivalence relation, and the equivalence class of the integer $a$, denoted by $[a]_n$, is the set

$$[a]_n = \{\ldots, a - 2n, a - n, a, a + n, a + 2n, \ldots\}.$$

When the modulus $n$ is known from the context that residue is also denoted $[a]$. This set, consisting of all the integers congruent to a modulo $n$, is called the congruence class, residue class, or simply residue of the integer a modulo $n$.

## Ordering relations

In Table 2.2, we found that the relation "$\leq$" on $\mathbb{N}$ (or on any of the number systems $\mathbb{Z}$ or $\mathbb{R}$) is not an equivalence relation because it is not symmetric. However, the reflexively and transitivity properties of "$\leq$" relation can be used in some applications such as the Big-Oh notation of algorithms which will be studied in Chapter 7. The following definition introduces a property of relations that bridges the gap between symmetric and asymmetric relations.

**Definition 2.9** *A relation $\mathcal{R}$ on a set $A$ is called antisymmetric if, for all $x, y \in A$, if $x \mathcal{R} y$ and $y \mathcal{R} x$, then $x = y$.*

For example, the relation "$\leq$" on $\mathbb{R}$ has the property that if $x \leq y$ and $y \leq x$, then $x = y$. Therefore, this relation is antisymmetric. Other examples of antisymmetric relations will be presented after the next definition.

Note that if a relation $\mathcal{R}$ is antisymmetric, then $x \mathcal{R} y$ and $x \neq y$ implies that $y \not\mathcal{R} x$. Note also that a relation may be antisymmetric and not symmetric, symmetric and not antisymmetric, both, or neither. Antisymmetry is one of the prerequisites for a partial ordering on a set $A$. We have the following definition.

**Definition 2.10** *A relation $\mathcal{R}$ on a set $A$ is called a partial order (or partial ordering) for $A$ if $\mathcal{R}$ is reflexive on $A$, antisymmetric and transitive. A set $A$ with partial order R is called a partially ordered set, or poset.*

Besides the above example on the relation "$\leq$" which is a partial order for $\mathbb{R}$, below are two other examples of partial orders. The first example shows than the relation "divides" on the set $\mathbb{N}$ of natural numbers is a partial order. Let $\mathcal{D}$ be the relation 'divides' on $\mathbb{N}$, then $1 \mathcal{D} 7$, $7 \mathcal{D} 7$, $7 \mathcal{D} 35$ and $13 \mathcal{D} 78$.

**Example 2.13** The relation "divides" on $\mathbb{N}$, is reflexive on $\mathbb{N}$ because every natural number divides itself. Note that if $a$ divides $b$ and $b$ divides $c$, then $a$ divides $c$. This implies that the relation "divides" is transitive. Note also that if $a$ divides $b$ and $b$ divides $a$, then $a = b$, which implies that this relation is antisymmetric. Thus, the relation "divides" is a partial order for $\mathbb{N}$. ∎

**Example 2.14** Let $A$ be a set. The set inclusion relation "$\subseteq$" on the powerset of $A$ is reflexive on $\mathcal{P}(A)$ and transitive. Note that if $A, B \in \mathcal{P}(A)$ and $A \subseteq B$ and $B \subseteq A$, then $A = B$, which means that this relation is antisymmetric. Thus, the relation "$\subseteq$" is a partial order on the powerset $\mathcal{P}(A)$ for any set $A$. ∎

Let $\mathcal{R}$ be a relation on a set $A$. Two elements of $A$ are called comparable if they are related by $\mathcal{R}$. For example, in the relation "divides" on $\mathbb{N}$, 7 and 35 are comparable but 10 and 31 are not comparable. Comparability is one of the prerequisites for a total ordering on a set $A$. We have the following definition.

> **Definition 2.11** *A partial ordering $\mathcal{R}$ on a set $A$ is called a linear order (or total order) for $A$ if $\mathcal{R}$ if for any two elements $x$ ad $y$ of $A$, either $x \mathcal{R} y$ or $y \mathcal{R} x$, i.e., $x$ and $y$ are comparable in $\mathcal{R}$. A set $A$ with total order $R$ is called a totally ordered set, linearly ordered set, or loset.*

For example, the relation "$\leq$" is a total order on the set $\mathbb{N}$ of natural numbers, but the "divides" relation is not a total order on $\mathbb{N}$. The total ordering will be used in Chapter 9 to define the topological ordering of graphs.

## 2.4. Partitions

Partitioning can be used to organize many things around us. The months of a year, for example, are partitioned in several ways: By the four seasons, by the month-lengths, by social/cultural events, etc; see Figure 2.3. A partition of a set is a grouping of its elements into nonempty subsets, in such a way that every element is included in exactly one subset. Formally, we have the following definition.

> **Definition 2.12** *Let $A$ be a nonempty set and $\mathcal{A}$ be a set of subsets of $A$. Then $\mathcal{A}$ is called a partition of $A$ if the following conditions hold:*
>
> *(i) If $X \in \mathcal{A}$, then $X \neq \emptyset$.*
>
> *(ii) If $X \in \mathcal{A}$ and $Y \in \mathcal{A}$, then $X = Y$ or $X \cap Y = \emptyset$.*
>
> *(iii) $\cup_{X \in \mathcal{A}} X = A$.*

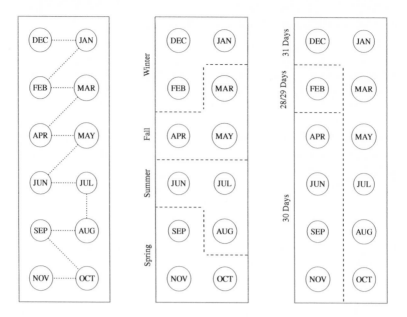

Figure 2.3: The year's months (left), and their partitions by seasons (middle) and by month-lengths (right).

The following are examples of partitions.

**Example 2.15** Here are four different partitions of $\mathbb{N}$:

- $\{\{1\}, \{2\}, \{3\}, \ldots\}$.

- $\{A_0, A_1, A_2, A_3\}$, where $A_i = \{4k + i : k \in \mathbb{N}\}$ for $i = 0, 1, 2, 3$.

- $\{E, O\}$, where $E$ (respectively, $O$) is the set of even (respectively, odd) numbers.

- $\{\mathbb{N}\}$.

Note that the first and fourth partitions in the above example are the extremes in terms of the number of elements. In fact, for any nonempty set $A$, $\{\{x\} : x \in A\}$ and $\{A\}$ are always partitions of $A$.

**Example 2.16** Here are three different partitions of the set $\mathbb{R}$ of real numbers:

- $\{[n, n + 1) : n \in \mathbb{Z}\}$.

- $\left\{ \left[ \frac{2n-1}{2}, \frac{2n+1}{2} \right) : n \in \mathbb{Z} \right\}$.

- $\{\mathbb{Q}, \mathbb{Q}'\}$, where $\mathbb{Q}$ (resp., $\mathbb{Q}'$) is the set of rational (resp., irrational) numbers.

A basic theorem of equivalence classes is the following.

> **Theorem 2.1** *The equivalence classes of any equivalence relation $\mathcal{R}$ in a set $A$ from a partition of $A$, and any partition of $A$ determines an equivalence relation on $A$ for which the sets in the partition are the equivalence classes.*

**Proof** For the first part of the proof, we must show that the equivalence classes of $\mathcal{R}$ are nonempty, pairwise-disjoint sets whose union is $A$. Because $\mathcal{R}$ is reflexive, $x \in [x]$, and so the equivalence classes are nonempty; moreover, since every element $x \in A$ belongs to the equivalence class $[x]$, the union of the equivalence classes is $A$. It remains to show that the equivalence classes are pairwise disjoint, that is, if two equivalence classes $[x]$ and $[y]$ have an element $a$ in common, then they are in fact the same set. Suppose that $x \mathcal{R} a$ and $y \mathcal{R} a$. By symmetry, $a \mathcal{R} y$, and by transitivity, $x \mathcal{R} y$. Thus, for any arbitrary element $b \in [x]$, we have $b \mathcal{R} x$ and, by transitivity, $b \mathcal{R} y$, and thus $[x] \subseteq [y]$. Similarly, $[y] \subseteq [x]$, and thus $[x] = [y]$.

For the second part of the proof, let $\mathcal{A} = \{A_i\}$ be a partition of $A$, and define $\mathcal{R} = \{(x, y) : \text{ there exists } i \text{ such that } x \in A_i \text{ and } y \in A_i\}$. We claim that $\mathcal{R}$ is an equivalence relation on $A$. Reflexivity holds, since $x \in A_i$ implies $x \mathcal{R} y$. Symmetry holds, because if $x \mathcal{R} y$, then $x$ and $y$ are in the same set $A_i$, and hence $y \mathcal{R} x$. If $x \mathcal{R} y$ and $y \mathcal{R} z$, then all three elements are in the same set $A_i$, and thus $x \mathcal{R} z$ and transitivity holds. To see that the sets in the partition are the equivalence classes of $\mathcal{R}$, observe that if $x \in A_i$, then $a \in [x]$ implies $a \in A_i$, and $a \in A_i$ implies $a \in [x]$. The proof is complete. ∎

**Example 2.17** The set $\mathcal{B} = \{B_0, B_1, B_2\}$ is a partition of $\mathbb{Z}$, where

$$B_0 = \{3k : k \in \mathbb{Z}\}, \ B_1 = \{3k + 1 : k \in \mathbb{Z}\}, \ \text{and} \ B_2 = \{3k + 2 : k \in \mathbb{Z}\}.$$

The integers $x$ and $y$ are in the same set $B_i$ iff $x = 3n + i$ and $y = 3m + i$ for some integers $n$ and $m$ or, in other words, iff $x - y$ is a multiple of 3. Thus, the equivalence relation associated with the partition $\mathcal{B}$ is the relation of congruence modulo 3 and each $B_i$ is the residue class of $i$ modulo 3, for $i = 0, 1, 2$. ∎

## 2.5. Functions

Functions are extremely important in mathematics and computer science. For example, but not limited to: In Section 3.1, sequences will be defined using functions. In Sections 5.1–5.4, solutions of recurrences will be expressed as functions. In Chapter 7, functions are used to represent how long it takes an a computer program to solve problems of a given size.

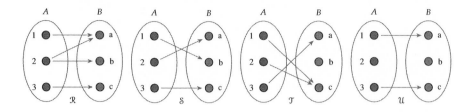

Figure 2.4: Different relations from $A$ to $B$, some of them are functions from $A$ to $B$.

The concept of a function is very old, but it is relatively recently that it has become standard to define a function as a relation with special properties. We have the following definition.

---

**Definition 2.13** *A function (or mapping) from a set $A$ to a set $B$ is a relation $f$ from $A$ to $B$ such that if $(x, y) \in f$ and $(x, z) \in f$, then $y = z$. We write $f : A \to B$ and this is read "$f$ is a function from $A$ to $B$", or "$f$ maps $A$ to $B$". The set $A$ is called the domain of $f$, and $B$ is called the codomain of $f$. In the case where $A = B$, we say that $f$ is a function on $A$.*

---

Note that, in Definition 2.13, no restriction is placed on the sets $A$ and $B$. They may be sets of numbers, ordered pairs, functions, or even sets of sets of functions.

**Example 2.18** The following sets are relations from the set $A = \{1, 2, 3\}$ to the set $B = \{a, b, c\}$.

$$\begin{array}{ll} \mathcal{R} = \{(1, a), (2, a), (2, b), (3, c)\}, & \mathcal{T} = \{(1, c), (2, c), (3, a)\}, \\ \mathcal{S} = \{(1, b), (2, a), (3, c)\}, & \mathcal{U} = \{(1, a), (3, c)\}. \end{array}$$

The relation $\mathcal{R}$ is not a function from $A$ to $B$ because $(2, a)$ and $(2, b)$ are distinct ordered pairs with the same first coordinates. See Figure 2.4. The relations $\mathcal{S}$ and $\mathcal{T}$ are functions from $A$ to $B$. The relation $\mathcal{U}$ is a function from $\{1, 3\}$ to $B$ because the domain of $\mathcal{U}$ is not $A$, but the set $\{1, 3\}$. ∎

The vertical line test is a graphical method for testing whether the graph of a relation represents a graph of a function. The test states that if every vertical line intersects the graph of a relation at most once, then the relation is a function.

**Example 2.19** In Figure 2.5, we show two graphs. The graph shown on the left-hand side of Figure 2.5 is for the relation $\mathcal{R}_1 = \{(x, y) \in \mathbb{R} \times \mathbb{R} : x^2 + y^2 = 1\}$ with domain $[-1, 1]$. Since vertical segments intersect the graph of $\mathcal{R}_1$ in two different point, $\mathcal{R}_1$ is not a function from $[-1, 1]$ to $\mathbb{R}$. The graph shown on the right-hand side of Figure 2.5 is for the relation $\mathcal{R}_2 = \{(x, y) \in [-\pi, \pi] \times \mathbb{R} : y = \sin x\}$ with domain $[-\pi, \pi]$. Since every vertical segment intersects the graph of $\mathcal{R}_2$ at at most one point, $\mathcal{R}_2$ is a function from $[-\pi, \pi]$ to $\mathbb{R}$. ∎

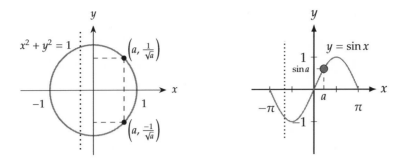

Figure 2.5: The graphs of the circle $x^2 + y^2 = 1$ and the function $y = \sin x$.

The function $y = \sin x$ whose graph is shown on the right-hand side of Figure 2.5 is the standard trigonometric sine function. There are different kinds of common functions, such as polynomials, trigonometric, exponential, logarithmic, etc. This can be found in any calculus textbook. In particular, a polynomial of one variable is a function that has the form

$$P(x) = a_n x^n + a_{n-1} x^{n-1} + \cdots + a_2 x^2 + a_1 x + a_0$$

where $n$ is a nonnegative integer and the constant numbers $a_0, a_1, a_2, \ldots, a_n$ are the coefficients.

**Definition 2.14** *Let a and b be elements of sets A and B, respectively. If f is a function from A to B and $f(a) = b$, we say that b is the image of a under f and a is a preimage of b under f. The range, or image, of f is the set of all elements of A.*

Figure 2.6 visually illustrates the difference between the terms range and codomain. When we define a function we specify its domain, its codomain, and the mapping of elements of the domain to elements in the codomain.

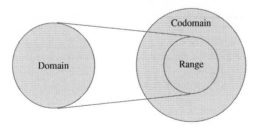

Figure 2.6: Range versus codomain.

Define the function $f : \mathbb{R} \to \mathbb{R}$ by $f(x) = 1 + \sin x$. It can be seen easily by looking at the graph of $f(x)$ which is shown to the right that:
- The domain of $f$ is $\mathbb{R}$.
- The range of $f$ is $[0, 2]$.
- The codomain of $f$ is $\mathbb{R}$.
Since $f(a) = 1 + \sin a$, the value $1 + \sin a$ is the image of $a$ under $f$, and the value $a$ is a preimage of $1 + \sin a$ under $f$.

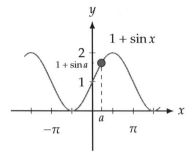

Two functions are equal when they have the same domain, and map each element of their common domain to the same image. A function is called real-valued if its codomain is $\mathbb{R}$, and it is called integer-valued if its codomain is $\mathbb{Z}$. Figure 2.7 shows an example of an integer-valued function.

Another common example of integer-valued functions is the floor function. This function, also known the greatest integer function, has domain $\mathbb{R}$ and codomain $\mathbb{Z}$. It maps each real number $x$ to the greatest integer $n$ such that $n \leq x$. We use the notation $\lfloor x \rfloor$ for this function. Specifically, $\lfloor \pi \rfloor = 3$ and $\lfloor -\pi \rfloor = -4$. The ceiling function is also an example of integer-valued functions. This function, also known the smallest integer function, has domain $\mathbb{R}$ and codomain $\mathbb{Z}$. It maps each real number $x$ to the smallest integer $n$ such that $n \geq x$. We use the notation $\lceil x \rceil$ for this function. Specifically, $\lceil \pi \rceil = 4$ and $\lceil -\pi \rceil = -3$.

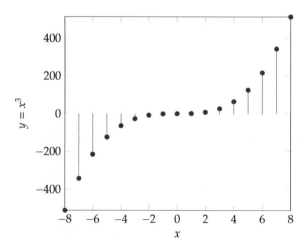

Figure 2.7: The graph of the integer-valued cubic function $y = x^3$ with domain $[-8, 8] \cap \mathbb{Z}$.

***Surjections***    A function is said to be a surjection or an onto function if its range and codomain are the same. For example, the two functions

$$f : \mathbb{R} \to \mathbb{R}, \quad \text{where} \quad f(x) = |x|,$$
$$g : \mathbb{R} \to \mathbb{R}_+, \quad \text{where} \quad g(x) = |x|,$$

are equal. Here $\mathbb{R}_+$ denotes the set of nonnegative real numbers. The functions $f$ and $g$ have the same range, which is $\mathbb{R}_+$, but $f$ maps to $\mathbb{R}$ while $g$ maps to $\mathbb{R}_+$. Therefore, $f$ is not onto while $g$ is onto. We formally have the following definition.

> **Definition 2.15** *A function $f : A \to B$ is said to be onto, or a surjection, if every element in B has a preimage, i.e., for every element $b \in B$ there is an element $a \in A$ with $f(a) = b$.*

**Example 2.20** To show that the function $f : \mathbb{R} \to \mathbb{R}$, where $f(x) = 2x + 1$, is onto, we must show that for every $t \in \mathbb{R}$, there exists $s \in \mathbb{R}$ such that $f(s) = t$. Let $t \in \mathbb{R}$, and choose $s = (t - 1)/2$. Then $f(s) = 2s + 1 = 2\frac{t-1}{2} + 1 = t$. Thus, $f$ is onto.

In light of Definition 2.15, to show that a function $f : A \to B$ is not a surjection, we need to find a particular $b \in B$ such that $f(a) \neq b$ for all $a \in A$. We have the following example.

**Example 2.21** To show that the function $g : \mathbb{R} \to \mathbb{R}$, where $g(x) = x^2$, is not onto, we must find a value $t \in \mathbb{R}$ that has no preimage in $\mathbb{R}$. Let $t = -1$. Since $x^2 \geq 0$ for every $x \in \mathbb{R}$, there is no $x \in \mathbb{R}$ such that $g(x) = -1$. Thus $g$ is not onto.

***Injections***    A function $f$ is said to be an injection or a one-to-one function if no two elements in the domain of $f$ with equal images. For example, the two functions

$$f : \mathbb{R} \to \mathbb{R}, \quad \text{where} \quad f(x) = |x|,$$
$$g : \mathbb{R}_+ \to \mathbb{R}, \quad \text{where} \quad g(x) = |x|,$$

are different. The functions $f$ and $g$ have the same range, which is $\mathbb{R}_+$, but the domain of $f$ is $\mathbb{R}$ while the domain of $g$ is $\mathbb{R}_+$. Note that $f(1) = f(-1) = 1$, which means that there are two different numbers $-1, 1 \in \mathbb{R}$ with equal images under $f$. In contrast, no two different numbers in $\mathbb{R}_+$ with equal images under $g$. Therefore, $f$ is not one-to-one while $g$ is one-to-one. We formally have the following definition.

> **Definition 2.16** *A function $f : A \to B$ is said to be one-to-one, or an injection, if $f(a) = f(b)$ implies that $a = b$ for all $a$ and $b$ in the domain of $f$.*

**Example 2.22** Determine whether each of the following functions is one-to-one on its domain. Justify your answer.

(i) $f(x) = x^3 - 8.$       (ii) $g(x) = \dfrac{x}{x-8}.$       (iii) $h(x) = x^4 + 3.$

**Solution**     (i) Clearly, the domain of $f$ is $\mathbb{R}$. Let $a, b \in \mathbb{R}$, then

$$f(a) = f(b) \implies a^3 - 8 = b^3 - 8 \implies a^3 = b^3.$$

It follows from this that $a = b$. Thus, $f$ is one-to-one.

(ii) The domain of $g$ is $\mathbb{R} - \{8\}$. Let $a, b \in \mathbb{R} - \{8\}$, then

$$g(a) = g(b) \implies \frac{a}{a-8} = \frac{b}{b-8} \implies ab - 8a = ab - 8b \implies 8a = 8b.$$

It follows from this that $a = b$. Thus, $g$ is one-to-one.

(iii) Clearly, the domain of $h$ is $\mathbb{R}$. Let $a, b \in \mathbb{R}$, then

$$h(a) = h(b) \implies a^4 + 3 = b^4 + 3 \implies a^4 = b^4.$$

It does not follow from this that $a = b$. In fact, this failed "proof" suggests a way to find real numbers with equal images. Indeed, $h(1) = 4 = h(-1)$ while $1 \neq -1$. This shows that $h$ is not one-to-one.     ∎

In light of Definition 2.16, to show that a function $f : A \rightarrow B$ is not an injection, we need to find particular elements $a, b \in A$ such that $a \neq b$ and $f(a) = f(b)$. We have the following example.

**Example 2.23** Which one of the following functions is not an injection?

(i) $f_1(x) = \sin x$ for $x \in [0, \pi]$.       (iii) $f_3(x) = x^2 + 1$ for $x \geq 0$.

(ii) $f_2(x) = |x + 1|$ for $x \geq 0$.       (iv) $f_4(x) = \cos x$ for $x \in [-\pi, 0]$.

**Solution**   The correct answer is (i). In fact, the function $f_1(x) = \sin x$, $x \in [0, \pi]$, is not an injection because

$$f_1\left(\frac{\pi}{3}\right) = f_1\left(\frac{2\pi}{3}\right) = \frac{\sqrt{3}}{2} \quad \text{while} \quad \frac{\pi}{3} \neq \frac{2\pi}{3}.$$

∎

The horizontal line test is a graphical method for testing whether the graph of a function represents a graph of an injection. The test states that if every horizontal line intersects the graph of a function at most once, then the function is an injection.

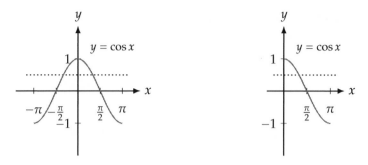

Figure 2.8: The graphs of the function $y = \cos x$ with two different domains.

![icon] **Example 2.24** In Figure 2.8, we show two graphs. The graph shown on the left-hand side of Figure 2.8 is for the trigonometric function $f(x) = \cos x$ with domain $[-\pi, \pi]$. Since vertical segments intersect this graph in two different point, $f(x)$ with domain $[-\pi, \pi]$ is not an injection. The graph shown on the right-hand side of Figure 2.8 is for the same function, $f(x) = \cos x$, but with domain $[0, \pi]$. Since every vertical segment intersects this graph in at most one point, $f(x)$ with domain $[0, \pi]$ is an injection. ◼

**Bijections** Bijective functions are essential to many areas of mathematics. For example, in Section 4.2, we will see that bijections arise in the definition of graph isomorphism.

> **Definition 2.17** *A function is said to be a one-to-one correspondence, or a bijection, if it is both injection and surjection.*

The function $f : \mathbb{R}_+ \to \mathbb{R}_+$ defined by $g(x) = |x|$ is an example of a bijection. See also Figure 2.9 which shows examples of different types of functions. Only the shown on the right-hand side of Figure 2.9 is bijection.

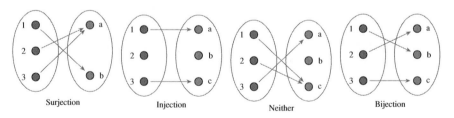

Figure 2.9: Examples of different types of functions.

# EXERCISES

**2.1**    Choose the correct answer for each of the following multiple-choice questions/items.

(*a*) One of the following statements is true about mathematical induction proofs, which is

   (*i*) Every mathematical induction proof is a two-part proof, and only the first part is necessary.

   (*ii*) Every mathematical induction proof is a two-part proof, and only the second part is necessary.

   (*iii*) Every mathematical induction proof is a two-part proof, and both parts are absolutely necessary.

   (*iv*) None of the above is true.

(*b*) If induction on $n$ is used to prove that

$$\sum_{i=1}^{n} i^4 = \frac{1}{30}n(n + 1)(2n + 1)(3n^2 + 3n - 1), \ \ \text{for } n \in \mathbb{N},$$

then the base case (basis step) is

   (*i*) $\sum_{i=1}^{n} 1^4 = \frac{1}{30}1(1 + 1)(2 + 1)(3 + 3 - 1), \ n \in \mathbb{N}.$

   (*ii*) $\sum_{i=1}^{n} i^1 = \frac{1}{30}1(1 + 1)(2 + 1)(3 + 3 - 1), \ n \in \mathbb{N}.$

   (*iii*) $\sum_{i=1}^{1} i^4 = \frac{1}{30}1(1 + 1)(2 + 1)(3 + 3 - 1).$

   (*iv*) none of the above.

(*c*) If induction on $n$ is used to prove that

$$\sum_{i=1}^{n} i^3 = \frac{1}{4}n^2(n + 1)^2, \ \ \text{for } n \in \mathbb{N},$$

then the inductive step shows that

   (*i*) $\sum_{i=1}^{k+1} i^3 = \frac{1}{4}(k + 1)^2(k + 2)^2.$

   (*ii*) $\sum_{i=1}^{k+1} i^3 = \frac{1}{4}(k + 1)^2(k + 2)^2 \longrightarrow \sum_{i=1}^{k} i^3 = \frac{1}{4}k^2(k + 1)^2.$

   (*iii*) $\sum_{i=1}^{k} i^3 = \frac{1}{4}k^2(k + 1)^2 \longrightarrow \sum_{i=1}^{k+1} i^3 = \frac{1}{4}k^2(k + 1)^2.$

   (*iv*) $\sum_{i=1}^{k} i^3 = \frac{1}{4}k^2(k + 1)^2 \longrightarrow \sum_{i=1}^{k+1} i^3 = \frac{1}{4}(k + 1)^2(k + 2)^2.$

(*d*) If we list all of the proper subsets of the set $\{\{\emptyset\}\}$, we get

(*i*) $\emptyset$.                                    (*iii*) $\{\emptyset\}$.

(*ii*) $\emptyset, \{\emptyset\}$.                    (*iv*) no proper subsets.

(*e*) Let $D$ be a domain and $X = \{x \in D : P(x)\}$. Only one of the following statements is false, which is

(*i*) $a \in X \longrightarrow P(a)$.                (*iii*) $\neg P(a) \longrightarrow a \notin X$.

(*ii*) $P(a) \longrightarrow a \in X$.                (*iv*) $\neg P(a) \longrightarrow a \notin D$.

(*f*) The binary relation $\{(a,a),(b,a),(b,b),(b,c),(b,d),(c,a),(c,b)\}$ on the set $\{a,b,c\}$ is

(*i*) reflective, symmetric and transitive.

(*ii*) irreflexive, symmetric and transitive.

(*iii*) irreflexive and antisymmetric.

(*iv*) neither reflective, nor irreflexive but transitive.

(*g*) The relation $x \mathcal{R} y$ if $|x| = |y|$ is

(*i*) transitive and symmetric.                  (*iii*) reflexive and asymmetric.

(*ii*) reflexive, symmetric and transitive.    (*iv*) irreflexive, antisymmetric and transitive.

(*h*) Let $A = \{1,2,3,4,5,6,7\}$. Which one of the following is not a partition of $A$?

(*i*) $\{\{1,2,5\},\{3,6\},\{4,7\}\}$.            (*iii*) $\{\{1,5,7\},\{3,4\},\{2,5,6\}\}$.

(*ii*) $\{\{1,2,5,7\},\{3\},\{4,6\}\}$.          (*iv*) $\{\{1,2,3,4,5,6,7\}\}$.

(*i*) Which one of the following relations is not a function?

(*i*) $\{(x,y) \in \mathbb{R} \times \mathbb{R} : y = |x|\}$.            (*iii*) $\{(x,y) \in \mathbb{R} \times \mathbb{R} : y = x^3\}$.

(*ii*) $\{(x,y) \in \mathbb{R} \times \mathbb{R} : x = |y|\}$.          (*iv*) $\{(x,y) \in \mathbb{R} \times \mathbb{R} : x^2 = y^3\}$.

(*j*) Which one of the following functions is not a bijection?

(*i*) $\{(x,y) \in \mathbb{R} \times \mathbb{R} : y = x\}$.            (*iii*) $\{(x,y) \in \mathbb{R}_+ \times \mathbb{R} : y = |x|\}$.

(*ii*) $\{(x,y) \in \mathbb{R} \times \mathbb{R} : y = x^3\}$.          (*iv*) $\{(x,y) \in \mathbb{R} \times \mathbb{R} : y = x^2\}$.

**2.2**    Archimedean principle states that for all natural numbers $n$ and $m$, there exists a natural number $s$ such that $m < sn$. This principle can be proven by induction on

$n$. To prove this for $n = 1$, choose $s$ to be $m + 1$, then $m < m + 1 = sn$. So the base step is true. Assuming the principle is true when $n = k$ for some $k \in \mathbb{N}$, complete the induction proof by establishing the inductive step.

**2.3**    Use mathematical induction to prove that, for all $n \in \mathbb{N}$, we have

$$\sum_{i=1}^{n} i^2 = \frac{n(n + 1)(2n + 1)}{6}.$$

**2.4**    Use mathematical induction to prove that the cardinality of the powerset of a finite set $A$ is equal to $2^n$ if the cardinality of $A$ is $n$. Show all the steps in the proof.

**2.5**    Use the induction method to show that the recurrence

$$T(n) = \begin{cases} 1, & \text{if } n = 1; \\ 3T(n - 1) + 4, & \text{if } n = 2, 3, 4, \ldots, \end{cases}$$

has the solution given in (2.5).

**2.6**    Let $A, B$ and $C$ be any three sets. Decide whether each of the following implications true or false? And if it is false, give an example that shows that.

(a) $(A \in B) \wedge (B \in C) \longrightarrow (A \in C)$.      (b) $A \cup C \subseteq B \cup C \longrightarrow A \subseteq B$.

**2.7**    Find the number of subsets of the given set. Justify your answer.

(a) $S = \{$fundamentals, discrete, structures, combinatorics, optimization$\}$.

(b) $T = \{n \in \mathbb{N} : n$ is an even number between 21 and 41$\}$.

**2.8**    Let $S = \{q, r, s, t, u, v, w, x, y, z\}$ be a space or universe of three sets $A, B$ and $C$, where $A = \{q, s, u, w, y\}, B = \{q, s, y, z\}$, and $C = \{v, w, x, y, z\}$. Identify the set $(A \cup C') \cap B'$ by listing its member in set braces.

**2.9**    Let $\mathcal{R}$ be the relation on a set of all people defined as $x \mathcal{R} y$ if and only if $x$ and $y$ have the same birthday (out of 365 possible birthdays). Show that $\mathcal{R}$ is an equivalence relation. How many equivalence classes does $\mathcal{R}$ have?

**2.10**    Show that the exponential function $f : \mathbb{R} \to [0, \infty)$ defined by $f(x) = e^x$ is a bijection. Is the exponential function $g : \mathbb{R} \to \mathbb{R}$ defined by $g(x) = e^x$ a bijection? Justify your answer.

# CHAPTER 3

# ANALYTIC AND ALGEBRAIC STRUCTURES

## Contents

© 2022 by Baha Alzalg | Kindle Direct Publishing, Washington, United States 2022/9
B. Alzalg, *Combinatorial and Algorithmic Mathematics: From Foundation to Optimization*,
DOI 10.5281/zenodo.7110896

This chapter introduces basic analytic and algebraic structures in four sections. The first two sections study sequences, summations and series, which are used extensively in solving recurrences in Chapter 5. The last two sections study subspaces, bases, convex sets, convex cones, convex hulls, and polyhedra, which are used extensively in solving linear programming in Chapter 10.

## 3.1. Sequences

A sequence of real numbers is a function $a : \mathbb{N} \to \mathbb{R}$. The values $a(1), a(2), \ldots$, which for simplicity are usually written as $a_1, a_2, \ldots$, are called the terms of the sequence. In particular, $a_n$ is called the $n$th term of $a$. The sequence is often written as

$$\{a_1, a_2, a_3, \ldots\}, \{a_n\}_{n=1}^{\infty}, \text{ or even more simply } \{a_n\}.$$

**Example 3.1**   The first five terms of the sequence $\{a_n\} = \{\frac{n}{n+2}\}$ are

$$a_1 = 1/3, a_2 = 2/4, a_3 = 3/5, a_4 = 4/6, \text{ and } a_5 = 5/7.$$

■

**Example 3.2**   It is not hard to see that the $n$th term of the sequence that has the five given terms $b_1 = -1/4, b_2 = 2/9, b_3 = -3/16, b_4 = 4/25$ and $b_5 = -5/36$ is

$$b_n = (-1)^n \frac{n}{(n+1)^2}.$$

■

A sequence can be infinite, as in Examples 3.1 and 3.2, or finite if it has a limited number of terms.

A sequences can also defined by a recurrence relation, which expresses each term as a combination of the previous terms. For example, the Fibonacci sequence[1], $f_0, f_1, f_2, \ldots$, is defined by the recurrence relation

$$f_n = f_{n-1} + f_{n-2}, \text{ for } n = 2, 3, 4, \ldots, \text{ where } f_0 = 0 \text{ and } f_1 = 1.$$

Using the recurrence formula we have

$$\begin{aligned}
f_2 &= f_1 + f_0 = 1 + 0 = 1, & f_5 &= f_4 + f_3 = 3 + 2 = 5, \\
f_3 &= f_2 + f_1 = 1 + 1 = 2, & f_6 &= f_5 + f_4 = 5 + 3 = 8, \\
f_4 &= f_3 + f_2 = 2 + 1 = 3, & f_7 &= f_6 + f_5 = 8 + 5 = 13.
\end{aligned}$$

---

[1] The Fibonacci sequence is named after the Italian mathematician who was born in 1170.

Solving recurrences is an important subject that will be studied exclusively in this chapter.

A sequence $\{a_n\}$ is called bounded above if there is a number $M$ such that $a_n \leq M$ for all $n$. Here, $M$ is called an upper bound for $\{a_n\}$. A sequence $\{a_n\}$ is called bounded below if there is a number $m$ such that $a_n \geq m$ for all $n$. Here, $m$ is called a lower bound for $\{a_n\}$. The sequence is called bounded if it is both bounded above and bounded below. For example, since $0 \leq \frac{1}{n} \leq 2$ for every $n$, the sequence $\{\frac{1}{n}\}$ is bounded below by 0 and is bounded above by 1.

Now, we study the limits of sequences. Because we are only concerned with limits involving infinity, we introduce the following definition.

---

**Definition 3.1 (Limit)** *Let $a_n : \mathbb{N} \to \mathbb{R}$ be a sequence and $L \in \mathbb{R}$. It is said that the limit of $a_n$, as $n$ approaches infinity, is $L$, and written $\lim_{n\to\infty} a_n = L$, if for every real number $\epsilon > 0$, there exists $n_0 \in \mathbb{N}$ such that $|a_n - L| < \epsilon$ whenever $n \geq n_0$.*

---

The following example illustrates how Definition 3.1 is applied.

**Example 3.3** Prove that $\lim_{n\to\infty} 1/n = 0$.

**Solution** Let $\epsilon > 0$ and choose an integer $n_0 > 1/\epsilon$. Then $1/n_0 < \epsilon$. Now, for any $n \geq n_0$, we have

$$0 < \frac{1}{n} < \frac{1}{n_0},$$

and so

$$\left|\frac{1}{n} - 0\right| = \left|\frac{1}{n}\right| = \frac{1}{n} \leq \frac{1}{n_0} < \epsilon.$$

By Definition 3.1, we conclude that $\lim_{n\to\infty} 1/n = 0$. ∎

**Example 3.4**

Note that the function $n^2$ becomes arbitrarily large as $n \to \infty$. See the discrete graph shown to the right. Therefore

$$\lim_{n\to\infty} n^2$$

does not exist.

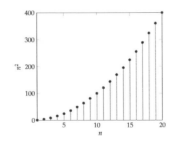

∎

A sequence that has a limit is said to be convergent. A sequence that has no limit is said to be divergent. For instance, the sequence $\{1/n\}$ in Example 3.3 is convergent to 0, and the sequence $\{n^2 + 4\}$ in Example 3.4 is divergent.

> **Theorem 3.1** *Every convergent sequence is bounded.*

**Proof** Assume that $a_n \to L$. Choose any positive number, say 1, and use it as $\epsilon$. Then we can see that there exists a positive integer $n_0$ such that

$$|a_n - L| \le 1 \quad \text{for all} \quad n \ge n_0.$$

Note that

$$|a_n| - |L| \le ||a_n| - |L|| \le |a_n - L|.$$

It follows that

$$|a_n| \le |L| + 1 \quad \text{for all} \quad n \ge n_0.$$

Consequently, we have

$$a_n \le \max\{|a_1|, |a_2|, \ldots, |a_{n_0-1}|, L + 1\} \quad \text{for all} \quad n.$$

This proves that $\{a_n\}$ is bounded, and hence proves the theorem.   ■

The truth of the following corollary follows from the contrapositive law (see Table 1.8) and Theorem 3.1.

> **Corollary 3.1** *Every unbounded sequence is divergent.*

For example, the sequences

$$\{a_n\} = n^2 \quad \text{and} \quad \{b_n\} = n\,e^n$$

are unbounded. Therefore, each of these sequences is divergent.

Using Definition 3.1, we can prove the following theorems.

> **Theorem 3.2** *Let $\lim_{n\to\infty} a_n = L$ and $\lim_{n\to\infty} b_n = M$, where $L$ and $M$ are finite real numbers. Then*
>
> (i) $\lim_{n\to\infty} k\,a_n = kL$ for any $k \in \mathbb{R}$.    (iii) $\lim_{n\to\infty}(a_n b_n) = LM$.
>
> (ii) $\lim_{n\to\infty}(a_n \pm b_n) = L \pm M$.    (iv) If $M \ne 0$, $\lim_{n\to\infty} \frac{a_n}{b_n} = \frac{L}{M}$.

> **Theorem 3.3** *Let $k$ be a constant. The following limits hold.*
>
> (i) $\lim_{n\to\infty} k = k$.    (iii) If $k > 0$, $\lim_{n\to\infty} \log_k n = \infty$.
>
> (ii) $\lim_{n\to\infty} n^k = \begin{cases} \infty, & \text{if } k > 0; \\ 0, & \text{if } k < 0. \end{cases}$    (iv) $\lim_{n\to\infty} k^n = \begin{cases} \infty, & \text{if } k > 1; \\ 0, & \text{if } 0 < k < 1. \end{cases}$

**Theorem 3.4** *Let $\{a_n\}$ be a sequence. If $\lim_{n\to\infty} a_n = \infty$, then $\lim_{n\to\infty} \frac{1}{a_n} = 0$.*

A well-known and very useful theorem for computing limits is the following.

**Theorem 3.5 (L'Hospital's rule at infinity)** *Let $f(n)$ and $g(n)$ be two real valued differentiable functions on $\mathbb{N}$. If*

$$\lim_{n\to\infty} f(n) = \lim_{n\to\infty} g(n) = 0 \quad or \quad \lim_{n\to\infty} f(n) = \lim_{n\to\infty} g(n) = \infty,$$

*then*

$$\lim_{n\to\infty} \frac{f(n)}{g(n)} = \lim_{n\to\infty} \frac{f'(n)}{g'(n)}.$$

This version of L'Hospital's rule is restricted to the limits where the variable $n$ approaches infinity since these are only limits of interest in our context.

**Example 3.5** Test the following sequences for convergence.

(i) $\{a_n\} = \left\{ \dfrac{\ln n}{n^2} \right\}.$     (ii) $\{b_n\} = \left\{ \dfrac{n^2}{e^n} \right\}.$     (iii) $\{c_n\} = \left\{ \dfrac{n^2}{e^n} \right\}.$

**Solution**    (i) Both numerator and denominator tend to $\infty$ as $x \to \infty$. Using L'Hospital's rule, we have

$$\lim_{n\to\infty} a_n = \lim_{n\to\infty} \frac{\ln n}{n^2} = \lim_{n\to\infty} \frac{1/n}{2n} = \lim_{n\to\infty} \frac{1}{2n^2} = 0.$$

Therefore, $\{a_n\}$ converges to 0.

(ii) Both numerator and denominator tend to $\infty$ as $x \to \infty$. Applying L'Hospital's rule twice, we have

$$\lim_{n\to\infty} a_n = \lim_{n\to\infty} \frac{n^2}{e^n} = \lim_{n\to\infty} \frac{2n}{e^n} = \lim_{n\to\infty} \frac{2}{e^n} = 0.$$

Therefore, $\{b_n\}$ converges to 0.

(iii) Both numerator and denominator tend to 0 as $x \to \infty$. Using L'Hospital's rule, we have

$$\lim_{n\to\infty} a_n = \lim_{n\to\infty} \frac{e^{\pi/n} - 1}{1/n} = \lim_{n\to\infty} \frac{e^{\pi/n}(-\pi/n^2)}{(-1/n^2)} = \pi \lim_{n\to\infty} e^{\pi/n} = \pi.$$

Therefore, $\{c_n\}$ converges to $\pi$.

## 3.2. Summations and series

Summations and series are used to describe algebraic patterns. Series are of great interest because computers and smart phones use them internally to calculate many functions.

If we add the terms of an infinite sequence $\{a_k\}_{k=0}^{\infty}$ we get

$$a_0 + a_1 + a_2 + \cdots + a_n + \cdots . \tag{3.1}$$

We write

$$\sum_{k=0}^{\infty} a_k \tag{3.2}$$

(read "the sum of $a$ sub $k$ from $k$ equals 0 to $k$ equals infinity") to indicate the sum (3.1). The infinite sum (3.2) is called an infinite series. The corresponding sequence $\{s_n\}$ defined by the finite sum

$$s_n = a_0 + a_1 + a_2 + \cdots + a_n = \sum_{k=0}^{n} a_k$$

is called the sequence of partial sums of the series (3.2).

---

**Definition 3.2** *Given the infinite series $\{a_k\}_{k=0}^{\infty}$. If the sequence of partial sums $\{s_n\}$, with $s_n = \sum_{k=0}^{n} a_k$, converges to a finite limit L, then the series $\sum_{k=0}^{\infty} a_k$ is said to converge to L, written as*

$$\sum_{k=0}^{\infty} a_k = L.$$

*The number L is called the sum of the series. If the sequence of partial sums diverges, then the series $\sum_{k=0}^{\infty} a_k$ diverges.*

---

Using Definition 3.2, we can prove the following theorem.

---

**Theorem 3.6** *If $\sum_{k=0}^{\infty} a_k$ and $\sum_{k=0}^{\infty} b_k$ are convergent series, then their sum, difference and scalar multiplication by any constant are convergent series. Moreover,*

$$(i) \ \sum_{k=0}^{\infty} a_k \pm \sum_{k=0}^{\infty} b_k = \sum_{k=0}^{\infty} (a_k \pm b_k). \qquad (ii) \ \sum_{k=0}^{\infty} \alpha a_k = \alpha \sum_{k=0}^{\infty} a_k, \ for \ any \ \alpha \in \mathbb{R}.$$

---

It is important to point out that the convergence or divergence of an infinite series is not affected by where you start the summation because the limit of the sequence of partial sums does not depend on where you begin the summation. Therefore $\sum_{k=0}^{\infty} a_k$ converges if and only if $\sum_{k=m}^{\infty} a_k$ converges, for any positive integer $m$.

**Example 3.6**  Test the following series for convergence. If it converges, find its sum.

$$\sum_{k=1}^{\infty} \frac{1}{k(k+1)}.$$

**Solution**  Note that

$$\frac{1}{k(k+1)} = \frac{(k+1)-k}{k(k+1)} = \frac{1}{k} - \frac{1}{k+1}.$$

It follows that the sequence of partial sums is

$$
\begin{aligned}
\sum_{k=1}^{\infty} \frac{1}{k(k+1)} &= \sum_{k=1}^{n} \left( \frac{1}{k} - \frac{1}{k+1} \right) \\
&= \left( \frac{1}{1} - \frac{1}{2} \right) + \left( \frac{1}{2} - \frac{1}{3} \right) + \cdots + \left( \frac{1}{n-1} - \frac{1}{n} \right) + \left( \frac{1}{n} - \frac{1}{n+1} \right) \\
&= 1 - \frac{1}{2} + \frac{1}{2} - \frac{1}{3} + \cdots + \frac{1}{n-1} - \frac{1}{n} + \frac{1}{n} - \frac{1}{n+1} \\
&= 1 - \frac{1}{n+1}.
\end{aligned}
$$

Now, as $n \to \infty$, $s_n \to 1$. This means that the series converges to 1. Therefore

$$\sum_{k=0}^{\infty} \frac{1}{k(k+1)} = 1.$$

Infinite series with the property that its terms can be arranged in pairs with opposite signs, except for the first and last term, are called telescoping series. For instance, the infinite series given in Example 3.6 is a telescoping series.

The following theorem states that $k$th term of a convergent series tends to 0.

**Theorem 3.7**

$$\text{If } \sum_{k=m}^{\infty} a_k \text{ converges, then } a_k \to 0 \text{ as } k \to \infty.$$

**Proof**  Let

$$s_n = \sum_{k=m}^{n} a_k.$$

Since the series $\sum_{k=m}^{\infty} a_k$ converges, its sequence of partial sums, $\{s_n\}$, converges to some number $L$. That is, $s_n \to L$. Hence, $s_{n-1} \to L$ as well. Since $a_n = s_n - s_{n-1}$, we have $a_n \to L - L = 0$. A change in notation gives $a_k \to 0$. The proof is complete. ∎

The truth of the following corollary follows from the contrapositive law (see Table 1.8) and Theorem 3.7.

---

**Corollary 3.2 (The divergence test)**

$$\text{If } a_k \nrightarrow 0 \text{ as } k \to \infty, \text{ then } \sum_{k=m}^{\infty} a_k \text{ diverges.}$$

---

For example, since $\lim_{k\to\infty}(k/(k+1)) = 1 \neq 0$, using the divergence test we conclude that the series $\sum_{k=0}^{\infty} \frac{k}{k+1}$ diverges.

We now introduce the geometric series, which is the most famous convergent series. The series is first presented in its finite version.

---

**Theorem 3.8 (The finite geometric series)** *Let $x \in \mathbb{R} - \{0\}$, and $n$ and $m$ be integers such that $0 \leq m \leq n$. Then*

$$\sum_{k=m}^{n} x^k = \begin{cases} \dfrac{x^m - x^{n+1}}{1-x} & \text{if } x \neq 1, \\ n - m + 1 & \text{if } x = 1. \end{cases} \tag{3.3}$$

---

**Proof**   Let

$$s_n = \sum_{k=m}^{n} x^k.$$

To compute $s_n$, first multiply both sides of the equality by $r$ and then manipulate the resulting sum as follows.

$$\begin{aligned} xs_n &= x\sum_{k=m}^{n} x^k \\ &= \sum_{k=m}^{n} x^{k+1} \\ &= \sum_{j=m+1}^{n+1} x^j = \left(\sum_{j=m}^{n} x^j\right) + (x^{n+1} - x^m). \end{aligned}$$

It follows that

$$xs_n = s_n + (x^{n+1} - x^m).$$

Solving for $s_n$ shows that if $x \neq 1$, then

$$s_n = \frac{x^m - x^{n+1}}{1-x}.$$

If $x = 1$, then $s_n = \sum_{k=m}^{n} 1 = n - m + 1$ as desired. This proves the theorem.   ∎

**Corollary 3.3** *Let $x \in \mathbb{R}$, and $n$ and $m$ be integers such that $0 \leq m \leq n$. Then*

$$\sum_{k=m}^{n} f(x) = (n - m + 1)\, f(x), \tag{3.4}$$

*where $f(\cdot)$ is a real-valued function that does not depend on k.*

The following example is a direct application of (3.3) and (3.4).

**Example 3.7** Find

(i) $\displaystyle\sum_{k=1}^{9} \left(\frac{3}{4}\right)^k$.

(ii) $\displaystyle\sum_{k=1}^{9} \sum_{j=0}^{9} \left(\frac{3}{4}\right)^k$.

**Solution**    (i) Using (3.3), we have

$$\sum_{k=1}^{9} \frac{3^k}{4} = \frac{\left(\frac{3}{4}\right)^{10} - \frac{3}{4}}{\frac{3}{4} - 1} = 3\left(1 - \left(\frac{3}{4}\right)^9\right) \approx 2.775.$$

(ii) This is an example of a double summation. To evaluate this, first compute the inner summation and then continue by computing the outer summation as follows.

$$\sum_{k=1}^{9} \sum_{j=0}^{9} \left(\frac{3}{4}\right)^k = \sum_{k=1}^{9} \left((9 - 0 + 1)\left(\frac{3}{4}\right)^k\right) = 10 \sum_{k=1}^{9} \left(\frac{3}{4}\right)^k \approx (10)(2.775) = 27.75,$$

where we used (3.4) to obtain the first equality, and used item (i) to obtain the approximate equality.

Now, we state and prove the infinite version of the geometric series.

**Theorem 3.9 (The infinite geometric series)** *If $x \in \mathbb{R}$, then*

$$\sum_{k=m}^{\infty} x^k = \begin{cases} \dfrac{x^m}{1-x} & \text{if } |x| < 1, \\ \text{diverges} & \text{if } |x| \geq 1. \end{cases} \tag{3.5}$$

**Proof** Let

$$s_n = \sum_{k=m}^{n} x^k.$$

We break the proof into two cases:

Case 1: When $|x| = 1$. If $x = 1$, then $s_n = n - m + 1$, and hence $\{s_n\}$ diverges as $n \to \infty$. Therefore $\sum_{k=m}^{\infty} x^k$ diverges. If $x = -1$, then $x^k = (-1)^k \not\to 0$ as $k \to \infty$, and hence by the divergence test, the series $\sum_{k=m}^{\infty} x^k$ diverges.

Case 2: When $|x| \neq 1$, the same steps followed in the proof of Theorem 3.8 can be used to prove that

$$s_n = \frac{x^{n+1} - x^m}{x - 1}.$$

If $|x| > 1$, then $\{x^{n+1}\}$ is unbounded, and hence $\{s_n\}$ diverges as $n \to \infty$. Therefore, $\sum_{k=m}^{\infty} x^k$ diverges. If $|x| < 1$, then $\{x^{n+1}\} \to 0$ as $n \to \infty$, and hence

$$s_n \to \frac{x^m}{1 - x}.$$

Therefore

$$\sum_{k=m}^{\infty} x^k = \lim_{n \to \infty} s_n = \lim_{n \to \infty} \frac{x^m}{1 - x} = \frac{x^m}{1 - x}.$$

This proves the theorem. ∎

The following example is a direct application of (3.5).

**Example 3.8** Test the series for convergence. If it converges, find its sum.

(i) $\displaystyle\sum_{k=1}^{\infty} \left(\frac{5}{4}\right)^k.$

(ii) $\displaystyle\sum_{k=1}^{\infty} \left(\frac{5}{4}\right)^{3-k}$

**Solution**    (i) The series diverges as $5/4 > 1$.

(ii) We have

$$
\begin{aligned}
\sum_{k=1}^{\infty} \left(\frac{5}{4}\right)^{3-k} &= \left(\frac{5}{4}\right)^3 \sum_{k=1}^{\infty} \left(\frac{4}{5}\right)^k \\
&= \left(\frac{5}{4}\right)^3 \frac{(4/5)^3}{1 - 4/5} \\
&= \frac{1}{1/5} = 5.
\end{aligned}
$$

Thus the series converges to 5. ∎

We can use the geometric series to derive an explicit formula, also-called a closed form, for some series.

**Example 3.9** Use the geometric series to derive a closed form for the following series.

$$\sum_{k=0}^{\infty} (-1)^k x^k, \quad |x| < 1.$$

**Solution**   Since $|-x| = |x| < 1$, using (3.5), we have

$$\sum_{k=0}^{\infty}(-1)^k x^k = \sum_{k=0}^{\infty}(-x)^k = \frac{1}{1-(-x)} = \frac{1}{1+x}.$$

The proof of the following theorem is left as an exercise for the reader.

---

**Theorem 3.10** *Let $f(x) = \sum_{i=0}^{n} a_i x^i$ and $g(x) = \sum_{j=0}^{m} b_j x^j$. Then*

$$f(x)g(x) = \sum_{k=0}^{m+n}\left(\sum_{r=0}^{k} a_r b_{k-r}\right)x^k, \tag{3.6}$$

*where we use the convention that $a_i = 0$ for all integers $i > m$ and $b_j = 0$ for all integers $j > n$.*

---

The following is the infinite version of Theorem 3.10.

---

**Theorem 3.11** *Let $f(x) = \sum_{k=0}^{\infty} a_k x^k$ and $g(x) = \sum_{k=0}^{\infty} b_k x^k$. Then*

$$f(x)g(x) = \sum_{k=0}^{\infty}\left(\sum_{r=0}^{k} a_r b_{k-r}\right)x^k. \tag{3.7}$$

---

The following example shows how Theorem 3.11 is applied.

**Example 3.10**   Find a series expansion for the expression $1/(1-x)^2$, $|x| < 1$.

**Solution**   Since $|x| < 1$, we have

$$\frac{1}{1-x} = \sum_{k=0}^{n} x^k.$$

Using Theorem 3.11, we have

$$\frac{1}{(1-x)^2} = \sum_{k=0}^{\infty}\left(\sum_{r=0}^{k} 1\right)x^k = \sum_{k=0}^{\infty}(k+1)x^k.$$

Another way to approach the series expansion for the expressions, such as that in Example 3.10, is to use Theorem 3.12. The proof of the following theorem can be found in any standard calculus textbook.

| Sum | Closed form | Sum | Closed form |
|---|---|---|---|
| $\displaystyle\sum_{k=1}^{n} 1$ | $n$ | $\displaystyle\sum_{k=0}^{n} x^k, \; x \neq 1$ | $\dfrac{1 - x^{n+1}}{1 - x}$ |
| $\displaystyle\sum_{k=1}^{n} k$ | $\dfrac{n(n + 1)}{2}$ | $\displaystyle\sum_{k=0}^{\infty} x^k, \; |x| < 1$ | $\dfrac{1}{1 - x}$ |
| $\displaystyle\sum_{k=1}^{n} k^2$ | $\dfrac{n(n + 1)(2n + 1)}{6}$ | $\displaystyle\sum_{k=0}^{\infty} (-1)^k x^k, \; |x| < 1$ | $\dfrac{1}{1 + x}$ |
| $\displaystyle\sum_{k=1}^{n} k^3$ | $\dfrac{n^2(n + 1)^2}{4}$ | $\displaystyle\sum_{k=0}^{\infty} (k + 1)x^k, \; |x| < 1$ | $\dfrac{1}{(1 - x)^2}$ |

Table 3.1: Some useful summation formulas.

---

**Theorem 3.12** *If $\sum_{k=0}^{\infty} a_k x^k$ converges on $(-c, c)$, then $\sum_{k=0}^{\infty} \frac{d}{dx}(a_k x^k)$ also converges on $(-c, c)$. Moreover, if*

$$f(x) = \sum_{k=0}^{\infty} a_k x^k \quad \text{for all } x \in (-c, c),$$

*then $f$ is differentiable on $(-c, c)$ and*

$$f'(x) = \sum_{k=0}^{\infty} \frac{d}{dx}(a_k x^k) \quad \text{for all } x \in (-c, c).$$

---

The result in Example 3.10 can be derived from Theorem 3.12 by differentiation as follows.

$$
\begin{aligned}
\frac{1}{(1 - x)^2} &= \frac{d}{dx}\left(\frac{1}{1 - x}\right) \\
&= \frac{d}{dx}\left(\sum_{k=0}^{n} x^k\right) \\
&= \sum_{k=0}^{n} \frac{d}{dx}(x^k) \\
&= \sum_{k=1}^{n} kx^{k-1} = \sum_{k=0}^{n} (k + 1)x^k.
\end{aligned}
$$

It is important to point out that the above differentiation is valid provided that $|x| < 1$.

Table 3.1 provides closed forms for commonly occurring summations. Some summations given in the left-hand column of Table 3.1 were proven in Section 2.1 by induction (see Corollary 2.1 and Exercises 2.1 and 2.3).

## 3.3. Matrices, subspaces, and bases

In this section, we review some notions and concepts from elementary linear algebra.

### Matrices

Throughout this and other subsequent chapters, we will use vectors and matrices. A matrix of dimension $n \times m$ is an array of numbers $a_{ij}$:

$$
A = \begin{bmatrix}
a_{11} & a_{12} & \cdots & a_{1m} \\
a_{21} & a_{22} & \cdots & a_{2m} \\
\vdots & \vdots & \ddots & \vdots \\
a_{n1} & a_{n2} & \cdots & a_{nm}
\end{bmatrix}.
$$

We assume that the entries of $A$ are real. A row vector is a matrix with $n = 1$, and a column vector is a matrix with $m = 1$. The word *vector* will always mean *column vector* unless the contrary is explicitly stated. So, a vector of dimension $n$ is an array of numbers $x_i$:

$$
x = \begin{bmatrix}
x_1 \\
x_2 \\
\vdots \\
x_m
\end{bmatrix}.
$$

Scalars will be always denoted by lower case characters, vectors will be always denoted by lower case boldface characters, and matrices will be always denoted by upper case characters.

We use $\mathbb{R}^n$ to denote the set of all $n$th-dimensional vectors. The standard inner product of $\mathbb{R}^n$ is defined as

$$
\langle x, y \rangle \triangleq x^\mathsf{T} y = \sum_{i=1}^{n} x_i y_i,
$$

for $x, y \in \mathbb{R}^n$. Here $x^\mathsf{T}$ denotes the transpose of the vector $x$. The Euclidean norm (also called the 2-norm) of $x \in \mathbb{R}^n$ is denoted as $\|\cdot\|$, and is defined to be the square root of the inner product of a vector with itself. That is

$$
\|x\| \triangleq \sqrt{x^\mathsf{T} x} = \sqrt{\sum_{i=1}^{n} x_i^2},
$$

for $x \in \mathbb{R}^n$.

The transpose $A^\mathsf{T}$ of an $n \times m$ matrix $A$ is the $m \times n$ matrix

$$A^\mathsf{T} = \begin{bmatrix} a_{11} & a_{12} & \cdots & a_{1n} \\ a_{21} & a_{22} & \cdots & a_{2n} \\ \vdots & \vdots & \ddots & \vdots \\ a_{m1} & a_{m2} & \cdots & a_{mn} \end{bmatrix}.$$

A square matrix is a matrix with the same number of rows and columns. A square matrix $A$ is called symmetric if it is equal to its own transpose matrix, i.e., $A = A^\mathsf{T}$. We use $I$ to denote the identity matrix, which is a square matrix whose diagonal entries are ones and its off-diagonal entries are zeros.

A square matrix is called a diagonal matrix all of the entries off of the main diagonal are zero. That is, a square matrix $D$ is diagonal if $d_{ij} = 0$ whenever $i \neq j$. A square matrix is called upper triangular if all the entries above the main diagonal are zero. Therefore, an upper triangular matrix has the form:

$$\begin{bmatrix} \times & \times & \times & \times & \times \\ & \times & \times & \times & \times \\ & & \times & \times & \times \\ & \mathbf{0} & & \times & \times \\ & & & & \times \end{bmatrix}$$

Likewise, a square matrix is called lower triangular if all the entries above the main diagonal are zero.

Let $A$ be a square matrix. If there exists a square matrix $B$ of the same dimensions satisfying $AB = BA = I$, we say that $A$ is invertible or nonsingular. Such a matrix $B$, called the inverse of $A$, is unique and is denoted by $A^{-1}$. The matrix $A$ is called an orthogonal matrix if $A^\mathsf{T} A = I$. In other words, a square matrix $A$ is orthogonal if its transpose is equal to its inverse matrix, i.e., $A^\mathsf{T} = A^{-1}$.

The determinant is a function whose input is a square matrix $A$ and whose output is a number. We use the notation $\det(A)$ to denote the determinant of a square matrix $A$. The reader can consult any linear algebra textbook, such as Anton and Rorres [AR14], to see how to compute $\det(\cdot)$.

If a system of linear equations $Ax = b$ has at least one solution, it is said to be consistent. A consistent system has either exactly one solution or an infinite number of solutions. Therefore, for a system of linear equations, we have three possibilities: The system has a unique solution, it has infinitely many solutions, or it is inconsistent. We give, without proof, the following standard result in this context. For a proof, see, for example, Anton and Rorres [AR14].

**Theorem 3.13** *Let A be a square matrix. Then, the following statements are equivalent:*

(a) *The matrix A is invertible.*

(b) *The matrix $A^T$ is invertible.*

(c) *The determinant of A is nonzero.*

(d) *The rows of A are linearly independent.*

(e) *The columns of A are linearly independent.*

(f) *For every vector $b$, the linear system $Ax = b$ has a unique solution.*

(g) *There exists some vector $b$ such that the linear system $Ax = b$ has a unique solution.*

(h) $\det(A) \neq 0$.

In this section (and throughout the book), we use $\mathbb{R}^{n \times m}$ to denote the set of all real matrices of dimension $n \times m$.

**Definition 3.3** *A matrix $A \in \mathbb{R}^{n \times n}$ is called positive semidefinite (respectively, positive definite) if it is symmetric and $x^T A x \geq 0$ for all $x \in \mathbb{R}^n$ (respectively, $x^T A x > 0$ for all $x \in \mathbb{R}^n - \{0\}$).*

The eigenvalues of a matrix are the roots of its characteristic polynomial (for more detail, consult any linear algebra text; for example refer to [Ren96]). As an alternative to Definition 3.3, we say that a square matrix is positive semidefinite (respectively, positive definite) if it is symmetric and all its eigenvalues are nonnegative (respectively, positive). Note that every positive definite matrix is invertible.

## Subspaces and bases

Subspaces and bases are introduced in this part. We have the following definition.

**Definition 3.4** *A set $S \subset \mathbb{R}^n$ is a subspace if $ax + by \in S$ for every $x, y \in S$ and every $a, b \in \mathbb{R}$. If, in addition, $S \neq \mathbb{R}^n$, we say that $S$ is a proper subspace.*

Note that the zero vector, $0$, must belong to every subspace (take $a = b = 0$). As an example, every line passing the origin is a subspace of $\mathbb{R}^2$. As another example, every plane passing the origin is a subspace of $\mathbb{R}^3$.

**Definition 3.5** *We say the vectors $x^{(1)}, x^{(2)}, \ldots, x^{(m)} \in \mathbb{R}^n$ are linearly dependent if there exists $a_1, a_2, \ldots, a_m \in \mathbb{R}$, not all of them zero, such that $\sum_{k=1}^{m} a_k x^{(k)} = 0$; otherwise, they are called linearly independent.*

For example, clearly the two vectors $(1, 0)$ and $(0, 1)$ are linearly independent in $\mathbb{R}^2$. For another example, it is also clear that the three vectors $(1, 0, 0), (0, 1, 0)$ and

$(0, 0, 1)$ are linearly independent in $\mathbb{R}^3$. Note that the maximum number of linearly independent points in $\mathbb{R}^n$ is $n$.

As a basic fact in linear algebra, the maximum number of linearly independent rows of a matrix $A$ is equal to the maximum number of linearly independent columns of $A$.

---

**Definition 3.6** *The rank of a matrix $A$, denoted by rank($A$), is the maximum number of linearly independent rows (columns) of $A$.*

---

**Definition 3.7** *The span of a finite number of vectors $x^{(1)}, x^{(2)}, \ldots, x^{(m)}$ in $\mathbb{R}^n$ is the subspace of $\mathbb{R}^n$ defined as the set of all vectors $y$ of the form $y = \sum_{k=1}^{m} a_k x^{(k)}$, where $a_1, a_2, \ldots, a_m \in \mathbb{R}$. Any such vector $y$ is called a linear combination of $x^{(1)}, x^{(2)}, \ldots, x^{(m)}$.*

---

**Definition 3.8** *A basis of a nonzero subspace $S \subset \mathbb{R}^n$ is a collection of vectors that are linearly independent and whose span is $S$. The dimension of a subspace is the number of vectors in any basis for the subspace.[a]*

---
[a] Every basis of a given subspace has the same number of vectors and this number is its dimension.

---

For example, the rank of the $3 \times 3$ identity matrix is 3. The space $\mathbb{R}^3$ has $\{(1, 0, 0), (0, 1, 0), (0, 0, 1)\}$ as a span. This spanning set is also a basis. Therefore, the dimension of $\mathbb{R}^3$ is 3. In general, the dimension of $\mathbb{R}^n$ is $n$.

---

**Definition 3.9** *Let $S$ be a subspace of $\mathbb{R}^n$ and $x_0 \notin S$ be a vector. The set*

$$S_0 = x_0 + S = \{x_0 + x : x \in S\}$$

*is called an affine subspace parallel to $S$.*

---

It is not hard to see that the dimension of $S_0$ is equal to the dimension of the subspace $S$. If $S$ is an $m$-dimensional subspace of $\mathbb{R}^n$ with $m < n$, there will be $n - m$ linearly independent vectors orthogonal to $S$. The set of such orthogonal vectors is indeed a subspace of $\mathbb{R}^n$.

---

**Definition 3.10** *If $S \subseteq \mathbb{R}^n$ is a subspace, then the subspace*

$$S^\perp \triangleq \left\{ x \in \mathbb{R}^n : x^\mathsf{T} y = 0 \text{ for } y \in S \right\}$$

*is called the orthogonal subspace of $S$.*

---

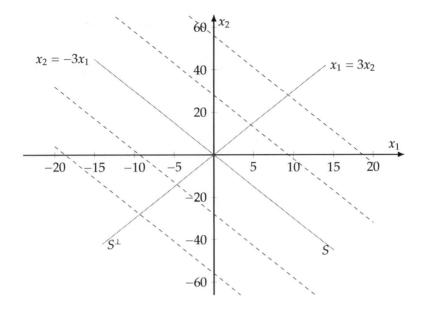

Figure 3.1: A graphical illustration for the subspace in Example 3.11 and affine subspaces parallel to it.

We give, without proof, the following standard result in this context.

---

**Proposition 3.1** *Let $A$ be an $m \times n$ matrix. The orthogonal of the subspace $S = \{x \in \mathbb{R}^n : Ax = 0\}$ is $S^\perp = \{x \in \mathbb{R}^n : x = A^\mathsf{T} u, u \in \mathbb{R}^m\}$.*

---

■ **Example 3.11**   The line $x_2 = -3x_1$, which passes through the origin, is a subspace of $\mathbb{R}^2$. The equation of the line can be written as

$$Ax = 0, \text{ where } A = \begin{bmatrix} 3 & 1 \end{bmatrix} \text{ and } x = \begin{bmatrix} x_1 \\ x_2 \end{bmatrix}.$$

All points on this line can be described by

$$\begin{bmatrix} -1 \\ 3 \end{bmatrix} u, \text{ for } u \in \mathbb{R}.$$

So, the set $\{(-1, 3)^\mathsf{T}\}$ is a basis for the given subspace, and hence its dimension equals 1. Note that the vector $(-1, 3)^\mathsf{T}$ is orthogonal to this subspace and every affine subspace parallel to it. To see this, note that by Proposition 3.1 we have

$$
\left\{ \begin{bmatrix} x_1 \\ x_2 \end{bmatrix} \in \mathbb{R}^2 : 3x_1 + x_2 = 0 \right\}^\perp = \left\{ \begin{bmatrix} x_1 \\ x_2 \end{bmatrix} \in \mathbb{R}^2 : \begin{bmatrix} 3 & 1 \end{bmatrix} \begin{bmatrix} x_1 \\ x_2 \end{bmatrix} = 0 \right\}^\perp
$$

$$
= \left\{ \begin{bmatrix} x_1 \\ x_2 \end{bmatrix} \in \mathbb{R}^2 : x = \begin{bmatrix} 3 \\ 1 \end{bmatrix} u, u \in \mathbb{R} \right\}
$$

$$
= \left\{ \begin{bmatrix} x_1 \\ x_2 \end{bmatrix} \in \mathbb{R}^2 : x_1 = 3x_2 \right\}
$$

$$
= \left\{ \begin{bmatrix} x_1 \\ x_2 \end{bmatrix} \in \mathbb{R}^2 : \begin{bmatrix} -1 \\ 3 \end{bmatrix}^T \begin{bmatrix} x_1 \\ x_2 \end{bmatrix} = 0 \right\}.
$$

## 3.4. Convexity, polyhedra and cones

This section introduces the definitions of convex functions, convex sets, convex hulls, polyhedra, cones, and related notions.

**Definition 3.11** *A function $f : \mathbb{R}^n \to \mathbb{R}$ is called convex if for every $x, y \in \mathbb{R}^n$ and every $\lambda \in [0, 1]$, we have*

$$
f(\lambda x + (1 - \lambda)y) \le \lambda f(x) + (1 - \lambda)f(y).
$$

To visually illustrate the above definition, see Figure 3.2. Examples of convex functions on their domains are $x^2, e^x$, and $-\log x$.

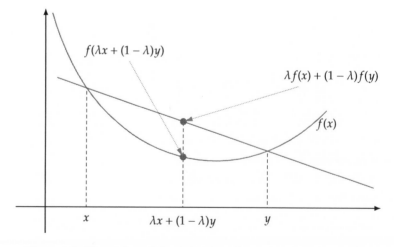

Figure 3.2: Illustration of the definition of a convex function on $\mathbb{R}$.

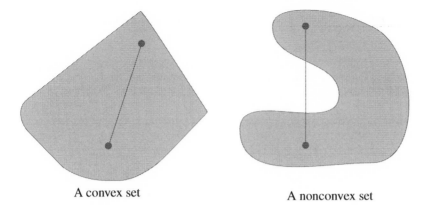

A convex set                      A nonconvex set

Figure 3.3: A convex set versus a nonconvex set in $\mathbb{R}^2$.

---

**Definition 3.12** *A set $T \subseteq \mathbb{R}^n$ is said to be convex if $\lambda x + (1 - \lambda)y \in T$ for any $x, y \in T$ and $\lambda \in [0, 1]$.*

---

See Figure 3.3, which shows convex and nonconvex sets in $\mathbb{R}^2$. We now introduce the convex hull of points in $\mathbb{R}^n$.

---

**Definition 3.13** *A point $x \in \mathbb{R}^n$ is a convex combination of points of $S \subseteq \mathbb{R}^n$ if there exist a finite set of points $\{x^{(i)}\}_{i=1}^t$ in $S$ and $\lambda \in \mathbb{R}_+^t$ with $\sum_{i=1}^t \lambda_i = 1$ such that $x = \sum_{i=1}^t \lambda_i x^{(i)}$. The convex hull of $S$, denoted by $conv(S)$, is the set all points that are convex combinations of points in $S$.*

---

Figure 3.4 shows a one-dimensional convex hull and three two-dimensional convex hulls.

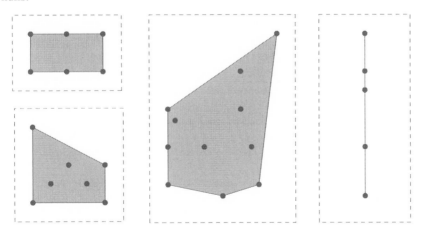

Figure 3.4: Four convex hulls; three in $\mathbb{R}^2$ and one in $\mathbb{R}$.

> **Lemma 3.1** *The convex hull of a finite number of vectors is a convex set.*

**Proof** Let $H$ be the convex hull of a finite number of vectors, say $x^{(1)}, x^{(2)}, \ldots, x^{(k)}$. Let also $y = \sum_{i=1}^{t} \alpha_i x^{(i)} \in H$, $z = \sum_{i=1}^{t} \beta_i x^{(i)} \in H$, where $\alpha_i, \beta_i \geq 0$, and $\sum_{i=1}^{t} \alpha_i = \sum_{i=1}^{t} \beta_i = 1$. Then, for any $\lambda \in [0, 1]$, we have

$$\lambda y + (1 - \lambda)z = \lambda \left( \sum_{i=1}^{t} \alpha_i x^{(i)} \right) + (1 - \lambda) \left( \sum_{i=1}^{t} \beta_i x^{(i)} \right) = \sum_{i=1}^{t} (\lambda \alpha_i + (1 - \lambda)\beta_i) x^{(i)}.$$

Note that $\sum_{i=1}^{t} (\lambda \alpha_i + (1 - \lambda)\beta_i) = \lambda \sum_{i=1}^{t} \alpha_i + (1 - \lambda) \sum_{i=1}^{t} \beta_i = \lambda + (1 - \lambda) = 1$. Hence, $\alpha y + \beta z \in H$. Thus, $H$ is a convex set. The proof is complete. ∎

A hyperplane is a subspace whose dimension is one less than that of its surrounding space. More formally, we have the following definition.

> **Definition 3.14** *A hyperplane is the set of points that satisfy a linear equation. That is, if $a \in \mathbb{R}^n$ and $b \in \mathbb{R}$, then the subspace $\{x \in \mathbb{R}^n : a^T x = b\}$ is a hyperplane in $\mathbb{R}^n$.*

A half-space is either of the two parts into which a hyperplane divides an affine space. More formally, we have the following definition.

> **Definition 3.15** *A half-space is the set of points that satisfy a linear inequality. That is, if $a \in \mathbb{R}^n$ and $b \in \mathbb{R}$, then the subspace*
>
> $$H = \{x \in \mathbb{R}^n : a^T x \geq b\}$$
>
> *is a half-space in $\mathbb{R}^n$.*

A polyhedron is the intersection of a finite number of half-spaces. More formally, we have the following definition.

> **Definition 3.16** *A polyhedron is the set of points that satisfy a finite number of linear inequalities. That is, if $A$ is an $m \times n$ matrix and $b \in \mathbb{R}^m$, then the subspace*
>
> $$P = \{x \in \mathbb{R}^n : Ax \geq b\}$$
>
> *is a polyhedron in $\mathbb{R}^n$.*

> **Definition 3.17** *A polyhedron $P \subseteq \mathbb{R}^n$ is said to be bounded if there exists a positive constant $K$ such that*
>
> $$P \subseteq \{x \in \mathbb{R}^n : -K \leq x_i \leq K \text{ for } i = 1, 2, \ldots, n\}.$$

> **Definition 3.18** *A polytope is a bounded polyhedron. That is, if A is an $m \times n$ matrix, $\mathbf{b} \in \mathbb{R}^m$ and K is a positive constant, then the subspace $\{\mathbf{x} \in [-K, K]^n : A\mathbf{x} \geq \mathbf{b}\}$ is a polytope in $\mathbb{R}^n$.*

> **Lemma 3.2** *A polyhedron is a convex set.*

**Proof** Let $P = \{\mathbf{x} \in \mathbb{R}^n : A\mathbf{x} \geq \mathbf{0}\}$ be a given polyhedron. Let also $\mathbf{x}, \mathbf{y} \in P$ and $\lambda \in [0, 1]$. Then $A\mathbf{x} \geq \mathbf{b}$ and $A\mathbf{y} \geq \mathbf{b}$. It follows that

$$A(\lambda \mathbf{x} + (1 - \lambda)\mathbf{y}) = \lambda A\mathbf{x} + (1 - \lambda)A\mathbf{y} \geq \lambda \mathbf{b} + (1 - \lambda)\mathbf{b} = \mathbf{b}.$$

Hence, $\alpha \mathbf{x} + \beta \mathbf{y} \in P$. Thus, $P$ is a convex set. The proof is complete. ∎

> **Definition 3.19** *A set $C \subseteq \mathbb{R}^n$ is called a cone if $\lambda \mathbf{x} \in C$ for any $\mathbf{x} \in C$ and $\lambda > 0$.*

> **Lemma 3.3** *The polyhedron $P = \{\mathbf{x} \in \mathbb{R}^n : A\mathbf{x} \geq \mathbf{0}\}$ is a cone.*

**Proof** Let $\mathbf{x}, \mathbf{y} \in P$ and $\alpha > 0$. Then $A\mathbf{x} \geq \mathbf{0}$. It follows that $A(\alpha \mathbf{x}) = \alpha A\mathbf{x} \geq \alpha \mathbf{0} = \mathbf{0}$. Hence, $\alpha \mathbf{x} \in P$. Thus, $P$ is a cone. The proof is complete. ∎

A convex cone is a cone that is also convex, i.e., a cone that is closed under addition. So, a cone $C$ is convex if $C + C \subseteq C$. The following definition combines Definitions 3.12 and 3.19.

> **Definition 3.20** *A set $C \subseteq \mathbb{R}^n$ is a convex cone if $\alpha \mathbf{x} + \beta \mathbf{y} \in C$ for any $\mathbf{x}, \mathbf{y} \in C$ and $\alpha, \beta > 0$.*

The following is a corollary to Lemmas 3.2 and 3.3. It can be also proven directly using Definition 3.20.

> **Corollary 3.4** *The polyhedron $P = \{\mathbf{x} \in \mathbb{R}^n : A\mathbf{x} \geq \mathbf{0}\}$ is a convex cone.*

We now introduce the conic hull of points in $\mathbb{R}^n$.

> **Definition 3.21** *A point $\mathbf{x} \in \mathbb{R}^n$ is a conic combination of points of $S \subseteq \mathbb{R}^n$ if there exist a finite set of points $\{\mathbf{x}^{(i)}\}_{i=1}^t$ in S and $\boldsymbol{\lambda} \in \mathbb{R}_+^t$ such that $\mathbf{x} = \sum_{i=1}^t \lambda_i \mathbf{x}^{(i)}$. The conic hull of S, denoted by cone(S), is the set all points that are conic combinations of points in S.*

Let $v_1, v_2, \ldots, v_t \in \mathbb{R}^n$. According to Definitions 3.13 and 3.21, we have

$$\text{conv}(v_1, \ldots, v_t) \triangleq \left\{ \mathbf{x} \in \mathbb{R}^n : \mathbf{x} = \sum_{i=1}^t \lambda_i v_i, \sum_{i=1}^t \lambda_i = 1, \lambda_i \geq 0, i = 1, \ldots, t \right\},$$

$$\text{cone}(v_1, \ldots, v_t) \triangleq \left\{ \mathbf{x} \in \mathbb{R}^n : \mathbf{x} = \sum_{i=1}^t \lambda_i v_i, \lambda_i \geq 0, i = 1, \ldots, t \right\}.$$

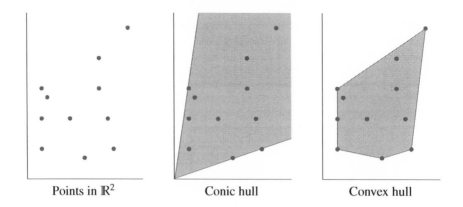

| Points in $\mathbb{R}^2$ | Conic hull | Convex hull |

Figure 3.5: The conic and convex hulls of some points in $\mathbb{R}^2$.

Note that a convex combination differs from a conic combination in the sense that we do not require that $\sum_{i=1}^{t} \lambda_i = 1$ in the latter. Figure 3.5 shows conic and convex hulls of points in $\mathbb{R}^2$.

A cone is said to be regular if it is a closed, convex, pointed, solid cone.

---

**Definition 3.22** *Let* $\mathcal{V}$ *be a finite-dimensional Euclidean space over* $\mathbb{R}$ *with an inner product* "$\langle \cdot, \cdot \rangle$". *The dual cone of a regular cone* $\mathcal{K} \subset \mathcal{V}$ *is defined as*

$$\mathcal{K}^\star \triangleq \{s \in \mathcal{V} : \langle x, s \rangle \geq 0, \ \forall x \in \mathcal{K}\}.$$

*A regular cone* $\mathcal{K}$ *is said to be self-dual if* $\mathcal{K} = \mathcal{K}^\star$.

---

Examples of self-dual cones include:

- The cone of nonnegative orthant of $\mathbb{R}^n$ (i.e., the polyhedron $P = \{x \in \mathbb{R}^n : Ax \geq 0\}$ with $A$ equals the identity matrix);

- The second-order cone (see Lemma 11.3);

- The cone of real symmetric positive semidefinite matrices;

- The cone of complex Hermitian positive semidefinite matrices;

- The cone of quaternion Hermitian positive semidefinite matrices.

Examples of non-self-dual cones include the $p$th-order cones, the hyperbolic cones (when $p \neq 2$), the cone of copositive matrices, the doubaly nonnegative cone, the power cone, and the exponential cone.

## 3.5.  Farkas' lemma and its variants

Farkas' lemma plays a central role in the development of the field of optimization, and more specifically, linear programming duality (see Chapter 10). In this section, we state and prove two versions of Farkas' lemma. First, we give, without proof, the following result which will be used in the proof of Farkas' lemma.

---

**Theorem 3.14 (Separating hyperplane)** *Let $Q \subseteq \mathbb{R}^n$ be a closed, nonempty and convex set. Let $b \in \mathbb{R}^n, b \notin Q$. Then there exist $0 \neq v \in \mathbb{R}^n$ and $\beta \in \mathbb{R}$ such that $v^\mathsf{T} b > \beta$ and $v^\mathsf{T} q < \beta$ for all $q \in Q$.*

---

For a proof of Theorem 3.14, see, for example, Bertsimas and Tsitsiklis [BT97].

---

**Theorem 3.15 (Farkas' lemma (Version I))** *Let $A \in \mathbb{R}^{m \times n}$ and $b \in \mathbb{R}^m$. Then exactly one of the following two conditions holds for given $A$ and $b$:*

*(1)  $\exists x \in \mathbb{R}^n$ such that $Ax = b$ and $x \geq 0$;*

*(2)  $\exists y \in \mathbb{R}^m$ such that $A^\mathsf{T} y \geq 0$ and $b^\mathsf{T} y < 0$.*

---

**Proof** The proof consists of two steps. At first, we prove that we cannot have both (1) and (2) simultaneously, then we prove that if (1) does not hold then (2) does.

Suppose, in the contrary, that we can have both (1) and (2) simultaneously. That is, there are $x \in \mathbb{R}^n$ and $y \in \mathbb{R}^m$ satisfying $Ax = b$ and $y^\mathsf{T} b < 0$, with $0 \leq x$ and $0 \leq A^\mathsf{T} y$. Then

$$0 \leq \left( y^\mathsf{T} A \right) x = y^\mathsf{T} (Ax) = y^\mathsf{T} b < 0,$$

giving the desired contradiction. Thus, we cannot have both (1) and (2) together.

Now we prove that (2) holds if (1) does not hold. Let $a_1, a_2, \ldots, a_n$ be the columns of $A$. Then $Ax = \sum_{j=1}^n x_j a_j$. Note that the vector $b$ satisfies $Ax = b$ with $x \geq 0$ if and only if $b \notin Q$, where $Q$ is the nonempty (as it contains $0$), closed, convex set:

$$Q \triangleq \mathrm{cone}(a_1, \ldots, a_n) = \left\{ \sum_{j=1}^n x_j a_j : x_j \geq 0, j = 1, \ldots, n \right\}.$$

Therefore, if Condition (1) does not hold then $b \notin Q$. The separating hyperplane theorem (Theorem 3.14) implies that there exists $0 \neq v \in \mathbb{R}^n$ and $\beta \in \mathbb{R}$ such that $v^\mathsf{T} b > \beta$ and $v^\mathsf{T} q < \beta$ for all $q \in Q$. In particular, because $0, x_1 a_1, \ldots, x_n a_n \in Q$, we have $0 = v^\mathsf{T} 0 < \beta$ and $v^\mathsf{T} (x_j a_j) < \beta$ for all $j = 1, 2, \ldots, n$. Letting $x_j \longrightarrow \infty$, we have $v^\mathsf{T} a_j < \beta / x_j \longrightarrow 0$, which implies that $v^\mathsf{T} a_j \leq 0$ for all $j = 1, 2, \ldots, n$. Picking $y = -v$, we get

$$A^\mathsf{T} y = -A^\mathsf{T} v = - \begin{bmatrix} v^\mathsf{T} a_1 \\ v^\mathsf{T} a_2 \\ \vdots \\ v^\mathsf{T} a_n \end{bmatrix} \geq 0, \text{ and } b^\mathsf{T} y = -b^\mathsf{T} v < -\beta < 0.$$

Thus Condition (2) holds. The proof is complete. ∎

> **Theorem 3.16 (Farkas' lemma (Version II))** *Let $A \in \mathbb{R}^{m \times n}$ and $c \in \mathbb{R}^n$. Then exactly one of the following two conditions holds for given $A$ and $c$:*
>
> *(1) $\exists y \in \mathbb{R}^m$ such that $A^\mathsf{T} y \leq c$;*
>
> *(2) $\exists x \in \mathbb{R}^n$ such that $Ax = 0, c^\mathsf{T} x < 0$ and $x \geq 0$.*
>
> *The following condition is equivalent to (2):*
>
> *(2') $\exists x \in \mathbb{R}^n$ such that $Ax = 0, c^\mathsf{T} x = -1$ and $x \geq 0$.*

**Proof** The proof consists of three steps or parts. At first, we prove that (2) and (2') are equivalent. Then, we prove that we cannot have both (1) and (2) simultaneously. And finally, we prove that if (2) does not hold then (1) does.

The proof of the first part is left as an exercise for the reader (see Exercise 3.11). Now, we prove that we cannot have both (1) and (2) simultaneously. Suppose, in the contrary, that we can have both (1) and (2) simultaneously. That is, there are $y \in \mathbb{R}^m$ and $x \in \mathbb{R}^n$ satisfying $A^\mathsf{T} y \leq c$ and $Ax = 0$, with $c^\mathsf{T} x < 0$ and $x \geq 0$. Then

$$0 = (Ax)^\mathsf{T} y = x^\mathsf{T} (A^\mathsf{T} y) \leq x^\mathsf{T} c < 0,$$

giving the desired contradiction. Thus, we cannot have both (1) and (2) together.

Finally, we prove if (2) is not true (and hence (2') is not true as well), then (1) is true. Assume that (2') does not hold, i.e., there is no $x$ satisfying $Ax = 0, c^\mathsf{T} x = -1$, and $x \geq 0$, or equivalently, satisfying $\hat{A}x = b$ and $x \geq 0$, where

$$\hat{A} = \begin{bmatrix} A \\ c^\mathsf{T} \end{bmatrix}, \quad \text{and} \quad b = \begin{bmatrix} 0 \\ -1 \end{bmatrix}.$$

This means that Condition (1) of Farkas' lemma (Version I) does not hold. It follows that Condition (2) of Farkas' lemma (Version I) holds, i.e., there is $\hat{y}$ satisfying $\hat{A}^\mathsf{T} \hat{y} \geq 0$ and $b^\mathsf{T} \hat{y} < 0$. Let $\alpha$ be the last component of $\hat{y}$, and $p$ be the subvector of its remaining components. That is, $\hat{y} = (p^\mathsf{T}, \alpha)^\mathsf{T}$. Then $b^\mathsf{T} \hat{y} < 0$ implies that $\alpha > 0$. Also $\hat{A}^\mathsf{T} \hat{y} \geq 0$ implies that

$$\begin{bmatrix} A \\ c^\mathsf{T} \end{bmatrix}^\mathsf{T} \begin{bmatrix} p \\ \alpha \end{bmatrix} = \begin{bmatrix} A^\mathsf{T} & \vdots & c \end{bmatrix} \begin{bmatrix} p \\ \alpha \end{bmatrix} \geq 0. \tag{3.8}$$

Note that (3.8) can be written as $A^\mathsf{T} p + \alpha c \geq 0$ or $A^\mathsf{T}(\frac{-p}{\alpha}) \leq c$. Thus, Condition (1) holds with $y = \frac{-1}{\alpha} p$. The proof is complete. ∎

# EXERCISES

**3.1**   Choose the correct answer for each of the following multiple-choice questions/items.

(*a*)  Which of the following sequences is bounded above?

(*i*)  $\left\{\dfrac{n^2}{n+2}\right\}.$

(*ii*)  $\left\{\dfrac{3^n}{2^n+32}\right\}.$

(*iii*)  $\left\{(-1)^{2n+1}\sqrt{n}\right\}.$

(*iv*)  $\left\{\dfrac{2^n}{(n+1)^2}\right\}.$

(*b*)  Which one of the following sequences diverges?

(*i*)  $\left\{\dfrac{n^3}{n^3+1}\right\}.$

(*ii*)  $\left\{\dfrac{1}{n}\ln\dfrac{1}{n}\right\}.$

(*iii*)  $\left\{\dfrac{n^2\ln n}{e^n}\right\}.$

(*iv*)  $\left\{\dfrac{n^2+1}{n+1000}\right\}.$

(*c*)  Which one of the following sequences converges to a number that is not 4/9?

(*i*)  $\left\{\dfrac{4n-1}{9n}\right\}.$

(*ii*)  $\left\{\dfrac{4n-1}{9n^2}\right\}.$

(*iii*)  $\left\{\dfrac{(2n+1)^2}{(3n-1)^2}\right\}.$

(*iv*)  $\left\{\dfrac{4n^2-1}{9n^2}\right\}.$

(*d*)  $\sum_{k=0}^{\log_{3/2} n} n$ equals

(*i*)  $(1+\log_{3/2}n)n.$

(*ii*)  $n\log_{3/2}n.$

(*iii*)  $(1+\log n)n.$

(*iv*)  $n\log n.$

(*e*)  Which of the following is a subspace of $\mathbb{R}^3$?

(*i*)  All vectors of the form $(a,0,0)$.

(*ii*)  All vectors of the form $(a,1,1)$.

(*iii*)  All vectors of the form $(a,b,c)$ where $b=a+c+1$.

(*iv*)  None of the above.

(*f*) The value of $\alpha$ such that the vector $(\alpha, 7, -4)$ is a linear combination of vectors $(-2, 2, 1)$ and $(2, 1, -2)$ is

  (*i*) 2.                  (*ii*) -2.                  (*iii*) 0.                  (*iv*) -1.

(*g*) The composition of two convex functions is

  (*i*) convex.                              (*ii*) nonconvex.

(*h*) The intersection of two convex sets is convex.

  (*i*) convex.                              (*ii*) nonconvex.

**3.2**   Use induction and L'Hospital's rule to show that, for each positive $k$,

$$\lim_{n \to \infty} \frac{(\ln n)^k}{n} = 0.$$

**3.3**   Find $\sum_{k=30}^{60} k^2$.

**3.4**   Test the following series for convergence. If it converges, find its sum.

$$\sum_{k=1}^{\infty} \frac{1}{(2k + 1)(2k - 1)}.$$

**3.5**   Test the following series for convergence.

(*i*) $\displaystyle\sum_{k=1}^{\infty} \frac{1}{2 + 3^{-k}}.$          (*ii*) $\displaystyle\sum_{k=1}^{\infty} \frac{1}{2 + (0.3)^k}.$

**3.6**   Test the following series for convergence. If it converges, find its sum.

(*i*) $\displaystyle\sum_{k=1}^{\infty} \left(\frac{2}{3}\right)^k.$          (*iii*) $\displaystyle\sum_{k=2}^{\infty} (-1)^k \left(\frac{2}{5}\right)^{k-2}.$

(*ii*) $\displaystyle\sum_{k=1}^{\infty} \left(\frac{5}{4}\right)^{3-k}.$          (*iv*) $\displaystyle\sum_{k=1}^{\infty} k \left(\frac{2}{3}\right)^k.$

**3.7**   Prove that the inverse of a positive definite matrix is also positive definite.

**3.8**   Find the orthogonal of the subspace $H = \{x \in \mathbb{R}^2 : x_1 = 2x_2\}$.

**3.9**    Let $f_1, f_2, \ldots, f_m : \mathbb{R}^n \to \mathbb{R}$ be convex functions. Show that the function $f$ defined by $f(x) = \max_{i=1,\ldots,m} f_i(x)$ is also convex.

**3.10**    Prove that the intersection of a finite number of convex sets is convex. Argue if the intersection of an infinite number of convex sets is convex.

**3.11**    Prove that Conditions (2) and (2') in Farkas' lemma (Version II) are equivalent.

# COMBINATORICS

# CHAPTER 4

# GRAPHS

## Contents

© 2022 by Baha Alzalg | Kindle Direct Publishing, Washington, United States 2022/9
B. Alzalg, *Combinatorial and Algorithmic Mathematics: From Foundation to Optimization*,
DOI 10.5281/zenodo.7110909

Suppose that we want to visualize flight routes in North America. A natural representation is with a graph as follows.

- Create a set of nodes (also-called vertices), with each node representing
  a city. Each node is labeled by the
  city's three-letter airport code.

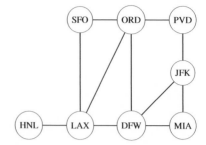

- Connect any 2 cities (nodes) which
  have a flight route between them
  with a line (called an edge). We can
  also label this edge with the mileage
  of the route.

Some flight routes in North America.

A way to think about graphs:

- Vertices (nodes) specify some entities we are interested in.

- Edges specify the relationships between the entities (vertices).

Note that this graph is a completely different concept from the graphs used for plotting functions.

After we represent our data (flight) with a graph, we can answer a lot of interesting questions such as:

- Can we reach one city from another city?

- What is the route with the minimum number of connections between two cities?

- What is the minimum mileage route between two cities?

Many interesting questions can be answered efficiently, but for some, there is no hope of an efficient answer. This chapter is devoted to introducing graph concepts and terminology.

## 4.1.  Basic graph definitions

In this section, we introduce some basic graph definitions. This includes directed and undirected graphs, simple and multi-graphs, subgraphs, paths, cycles, connected components, trees, spanning trees, complete graphs and bipartite graphs.

> **Definition 4.1** *A graph, G = (V, E), is a finite set of vertices, V, and a finite set of edges, E, where each edge (u, v) connects two vertices, u and v.*

### Example 4.1

In the graph shown to the right, we have $G = (V, E)$ where $V = \{u, v, w\}$ is the set of the vertices and $E = \{(u, w), (u, v), (w, v)\}$ is the set of edges.

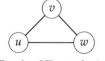

Graph of Example 4.1.  ■

### Example 4.2

In the graph shown to the right, we have $G = (V, E)$ where $V = \{a, b, c, d, e\}$ is the set of the vertices and $E = \{(a, c), (b, c), (c, e), (b, d), (d, e), (c, d)\}$ is the set of edges.

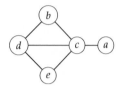

Graph of Example 4.2.  ■

***Directed and undirected graphs***    We classify the edges into two distinct categories: Directed edges and undirected edges. The directed edge has an origin vertex and a destination vertex, whereas for the undirected edge, there is no designated beginning or end to the edges. An easy way to think about a directed edge is a flight going from one vertex to another. If there is a flight going from San Francisco international airport (SFO) to Chicago O'Hare international airport (ORD), it does not necessarily mean that there is a flight going from ORD to SFO. On the other hand, an undirected edge can be thought of as a network of "friends". For example, if Noah is a friend of Zaid, then this necessarily means that Zaid is also a friend of Noah. See Figure 4.1.

This leads us to classify the graphs into two distinct categories: Directed graphs and undirected graphs. In directed graphs, all edges are directed. In undirected graphs, all edges are undirected (see Figure 4.1). Note that $(u, v)$ and $(v, u)$ are two different edges if the graph is directed, but they are the same edge if the graph is undirected.

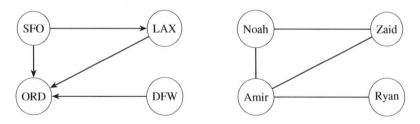

Figure 4.1: Directed edges forming a directed graph (left) and undirected edges forming a undirected graph (right).

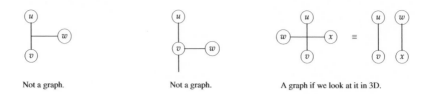

Figure 4.2: Examples of graphs and non-graphs.

In this chapter, we first study the undirected graphs, then we study the directed graphs. Applications of directed and undirected graphs include, but not limited to:

- Transportation networks: City map, highway network, flight network, etc.

- Computer networks: Local area network, internet, web, etc.

- Linguistics, physics, chemistry, biology, etc.

Two vertices $u$ and $v$ of a graph $G$ are adjacent if there exists an edge $(u, v)$ in $G$. In Figure 4.2, left-hand and middle side pictures do not represent graphs, while the right-hand picture represents a graph because looking at it in three-dimensional space one finds that the vertices $u$ and $v$ are adjacent and the vertices $w$ and $x$ are adjacent.

***Simple and multi-graphs***   A self-loop is an edge that connects a vertex to itself. Multi-edges (or parallel edges) are edges that have the same endpoints in undirected graphs, or the same origin and destination in directed graphs.

In the graph shown to the right, there is a self-loop at vertex $u$ and there are two parallel edges between the vertices $u$ and $v$. If a graph does not have parallel edges and self-loops, then it is said to be simple. A multi-graph can have multiple edges between the same two vertices and self-loops.

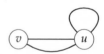

A non-simple graph.

In this chapter, we will deal almost exclusively with simple graphs. In other words, unless otherwise specified, when we say "graph" we mean "simple graph".

***The vertex degree***   Let $G = (V, E)$ be a graph. If $u, v \in V$ and $(u, v) \in E$, we say that the edge $(u, v)$ is incident to vertices $u$ and $v$. The degree of a vertex $v \in V$, denoted as $\deg(v)$, is the number of edges incident on $v$. A vertex whose degree is zero is called isolated.

For example, in the graph shown to the right, we have $\deg(i) = 1, \deg(b) = \deg(e) = 2, \deg(c) = 3, \deg(d) = 4$ and $\deg(a) = 0$. So, $a$ is an isolated vertex. The degree list of an undirected graph is the non-decreasing sequence of its vertex degrees. For instance, the degree list of the graph shown to the right is $0, 1, 2, 2, 3, 4$.

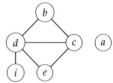

A graph with a degree sequence $0, 1, 2, 2, 3, 4$.

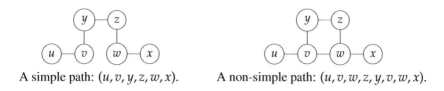

A simple path: $(u, v, y, z, w, x)$.    A non-simple path: $(u, v, w, z, y, v, w, x)$.

Figure 4.3: Simple and non-simple paths.

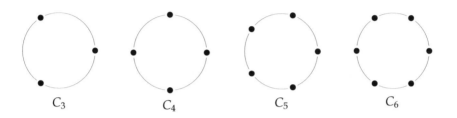

$C_3$        $C_4$        $C_5$        $C_6$

Figure 4.4: Cycle graphs.

**Paths and cycles**    A path $P_n$ from a vertex $u$ to a vertex $u'$ is a graph written as a sequence $(v_0, v_1, v_2, \ldots, v_n)$ of $n + 1$ vertices, such that $u = v_0$ and $u' = v_n$, and $n$ edges $(v_{i-1}, v_i)$ for $i = 1, 2, \ldots, n$. The length of a path is the number of its edges. So, the path $P_n$ has length $n$. A subpath of a path is a contiguous subsequence of its vertices. A path is simple if all vertices in the path are distinct. Revisiting vertices and/or edges are allowed in non-simple paths. Figure 4.3 shows simple and non-simple paths. The simple path shown to the left has length 5 and has, for instance, the path $(v, y, z, w)$ as a subpath.

A cycle is a path on which the first vertex is equal to the last vertex and all edges are distinct. So, a cycle $C_n$ is a path $P_n$ with $v_0 = v_n$. In Figure 4.4, we show some cycles. A cycle is simple if all its vertices, except the first and the last one, are distinct. A graph with no simple cycles is called acyclic. In Figure 4.5, we show an acyclic graph, a simple cycle graph, and a non-simple cycle graph.

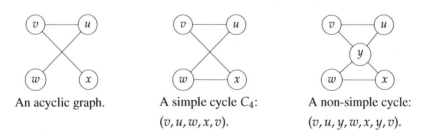

An acyclic graph.    A simple cycle $C_4$:    A non-simple cycle:
                     $(v, u, w, x, v)$.        $(v, u, y, w, x, y, v)$.

Figure 4.5: Acyclic and (simple and non-simple) cycle graphs.

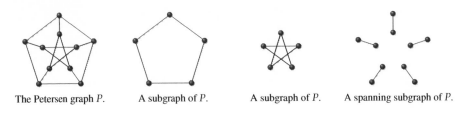

| The Petersen graph $P$. | A subgraph of $P$. | A subgraph of $P$. | A spanning subgraph of $P$. |

Figure 4.6: The Petersen graph $P$ and some subgraphs of $P$.

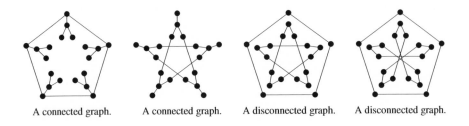

| A connected graph. | A connected graph. | A disconnected graph. | A disconnected graph. |

Figure 4.7: Connected and disconnected graphs.

**_Subgraphs and connected components_**    A graph $H = (V', E')$ is called a sub-graph of a graph $G = (V, E)$ if $V' \subseteq V$ and $E' \subseteq E$. A spanning subgraph of a graph $G$ is a subgraph of $G$ that contains all the vertices of $G$. That is, $H = (V', E')$ is a spanning subgraph of a graph $G = (V, E)$ if $V' = V$ and $E' \subseteq E$. The Petersen graph is well-known graph in graph theory.[1] Figure 4.6 shows three subgraphs of the Petersen graph, $P$, and only one of them is a spanning subgraph.

A graph is called connected if every vertex is reachable from all other vertices, and disconnected otherwise. That is, there is a path between every pair of distinct vertices of a connected graph. In Figure 4.7, we show four graphs; each one has 20 vertices; we see that the two left-hand side ones are connected, whereas the two right-hand side ones are disconnected because the five external vertices are not reachable from any of other internal vertices.

A connected component $G'$ of a graph $G$ is a maximal connected subgraph of $G$. By maximality we mean that there is no way to add into $G'$ any vertices and/or edges of $G$ which are not currently in $G'$ in such a way that the resulting subgraph is connected.

**Example 4.3**  The graph, say $G$, shown to the right in Figure 4.8 has 9 vertices and 7 edges. Note that the subgraph surrounded by the blue dashed lines is not a connected component of $G$. The reason for that is if we add a new vertex $v_4$ and a new edge $(v_3, v_4)$, the resulting subgraph is still connected.

---

[1]The Petersen graph serves as a useful example and counterexample for many problems in graph theory.

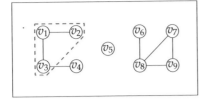 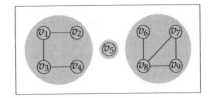

Figure 4.8: A graph with three connected components.

The graph $G$ has 3 connected components, where each corrected component is framed by a blue disk background shown to the left in Figure 4.8. Note that, for instance, if we add the vertex $v_5$ to the right- or left-hand side connected component, the resulting subgraph is not connected anymore.     ∎

**Example 4.4** The graph shown in Figure 4.9 has 18 vertices, 12 edges, and 7 connected components. These components are framed by blue ellipsoid backgrounds.

∎

The following remark can be directly proven on the basis of Definitions 2.7 and 2.8.

**Remark 4.1** *The connected components of a graph define equivalence classes over its vertices under the "is reachable from" relation.*

From Theorem 2.1, the "is reachable from" relation defines a partition on the vertices, with two vertices being in the same partition (or the same equivalence class) if and only if they are in the same connected component.

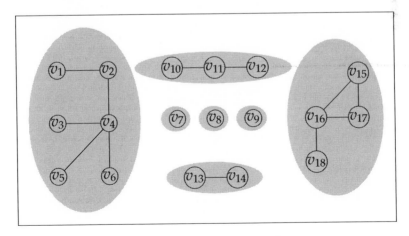

Figure 4.9: A graph with seven connected components.

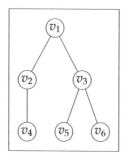 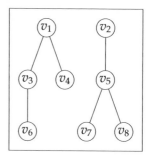

Figure 4.10: A tree graph (left) and a forest graph (right).

***Trees and spanning trees***   One of the important classes of graphs is the trees. The importance of trees is evident from their applications in various areas, especially theoretical computer science and molecular evolution. A tree is a connected undirected graph with no cycles (i.e., connected, acyclic, undirected graph). A forest is any undirected graph without cycles. Figure 4.10 shows a tree and a forest. Note that the connected components of a forest are trees.

A spanning tree of a connected graph is a spanning subgraph that is a tree. Note that a spanning tree is not unique unless the graph is a tree. Figure 4.11 shows a graph and two of its spanning trees. Spanning trees have applications to the design of communication networks.

A spanning forest of a graph is a spanning subgraph that is a forest. Note that disconnected graphs have spanning forests and do not have spanning trees, and that only connected graphs have spanning trees.

***Complete and bipartite graphs***   A complete graph of $n$ vertices, denoted as $K_n$, is a graph in which every pair of vertices is adjacent. In Figure 4.12, we show some complete graphs. An empty graph is a graph with no edges, i.e., $E = \emptyset$.

A bipartite graph is a graph in which its vertices can be partitioned into two sets such that, for every edge $(u, v)$ in the graph, $u$ is in one of the sets and $v$ is in the other. A complete bipartite graph, denoted as $K_{m,n}$, is a bipartite graph in which each vertex in one of the sets is joined to each vertex in the other. Figure 4.13 shows bipartite and complete bipartite graphs.

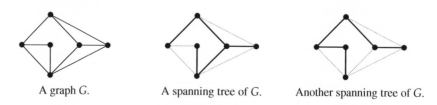

A graph $G$.                A spanning tree of $G$.         Another spanning tree of $G$.

Figure 4.11: A graph with spanning trees.

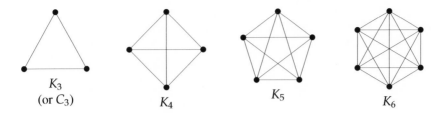

Figure 4.12: Complete graphs.

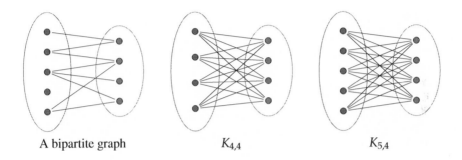

Figure 4.13: Bipartite and complete bipartite graphs.

Figure 4.14: A planar graph (left) and a nonplanar graph (right).

A planar graph is a graph that can be embedded in the plane, i.e., it can be drawn on the plane in such a way that its edges intersect only at their endpoints. In other words, it can be drawn in such a way that no edges cross each other. See Figure 4.14.

The complement of a graph $G = (V, E)$ is the graph $\bar{G}(V, \bar{E})$ on the same set of vertices as of $G$ such that there will be an edge between two vertices $u$ and $v$ in $V$, if and only if there is no edge in between $u$ and $v$ in $V$. See Figure 4.15.

Figure 4.15: A graph $G$ (left) and its complement $\bar{G}$ (right).

A graph with three edges and three bridge edges.

A graph with three edges and no bridge edges.

Figure 4.16: Graphs with/without bridges.

A dense graph is a graph in which the number of edges is close to the maximal number of edges. The opposite, a graph with only a few edges, is a sparse graph. The distinction between sparse and dense graphs is rather vague, and depends on the context. Formally speaking, a graph $G = (V, E)$ is dense if $|E|$ is close to $|V^2|$ (i.e., $|E| \approx |V|^2$), and $G$ is sparse if $|E|$ is much less than $|V^2|$ (i.e., $|E| \approx |V|$).

The matching (or assignment) problem is the problem of choosing an optimal assignment of a number of applicants to a number of jobs. We can model this with a bipartite graph. Here we assume that there are $n_1$ job applicants and $n_2$ job openings. Each applicant has a subset of the job openings s/he is interested in. Conversely, each job opening can only accept one applicant out of some subset of the applicants. In other words, there are certain allowed or "compatible" pairings of applicants to jobs. We can find an assignment of jobs to applicants in such a way all the pairings are allowed and as many applicants as possible get jobs. This can be modeled with a bipartite graph as follows: On the left side we have the applicants and on the right side we have the jobs. We draw an edge between applicant $u$ and job $v$ if they are compatible. The problem of finding a maximum matching in a given graph is an NP-hard problem in general (the class of NP-hard problems will be introduced briefly in Section 7.7), but if the graph is bipartite, the problem can be solved efficiently.

A bridge or cut-edge is an edge of a graph whose deletion increases its number of connected components. In Figure 4.16, we show two graphs, one has three bridges while the other has no bridges.

The first characterization of trees is that a graph is a tree if and only if it is connected and every edge is a bridge. Further characterizations of trees will be given in the next section.

## 4.2.  Isomorphism and properties of graphs

In this section, we discuss graph isomorphism and present some graph properties.

### Graph isomorphism

What does it mean for two graphs to be "the same"? We say that two graphs are isomorphic when the vertices of one can be relabeled to match the vertices of the other in a way that preserves adjacency. Formally, we have the following definition.

> **Definition 4.2** *We say that two graphs $G$ and $H$ are isomorphic if there is a bijection $f : V(G) \to V(H)$ such that any two vertices $u$ and $v$ of $G$ are adjacent in $G$ if and only if $f(u)$ and $f(v)$ are adjacent in $H$.*

See Figure 4.17 which shows an example of two isomorphic graphs. It is also not hard to see that the two graphs with solid vertices in Figure 4.18 are another example of an isomorphic pair of graphs.

In fact, given two graphs, it is often really hard to tell if they are isomorphic, but it is usually easier to see if they are not isomorphic. We have the following remark which shows that isomorphic graphs have the same number of vertices and the same number of edges.

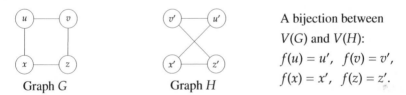

A bijection between
$V(G)$ and $V(H)$:
$f(u) = u', \quad f(v) = v',$
$f(x) = x', \quad f(z) = z'.$

Figure 4.17: An isomorphism between two graphs.

Figure 4.18: A pair of isomorphic graphs.

> **Remark 4.2** *If two graphs have different numbers of vertices or different numbers of edges, then they are not isomorphic.*

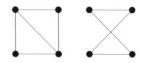

For example, the two connected graphs shown to the right have 4 vertices, but they are not isomorphic because they have different numbers of edges.

Having the same number of vertices and the same number of edges is not a guarantee that two graphs will be isomorphic. We have the following remark which shows that isomorphic graphs have the same degree lists.

> **Remark 4.3** *If two graphs have different degree lists, then they are not isomorphic.*

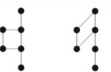

As an example, the two connected graphs shown to the right have 6 vertices and 6 edges, but they are not isomorphic because they have different degree lists.

Having the same number of vertices, the same number of edges, and the same degree lists is not a guarantee that two graphs will be isomorphic. We have the following remark which shows that isomorphic graphs have the same number of connected components.

> **Remark 4.4** *If two graphs have different number of connected components (for instance, one graph is connected, and the other is not), then they are not isomorphic.*

As an example, the two graphs shown to the right have 6 vertices, 6 edges and the degree list is the same, but they are not isomorphic because one of them is connected while the other is not.

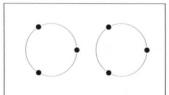 

$C_3$ duplicated. (Disconnected graph).    The cycle $C_6$. (Connected graph).

Finally, we also point out that having the same number of vertices, the same number of edges, the same degree lists, and the same number of connected components is not a guarantee that two graphs will be isomorphic.

As an example, both the connected graphs shown to the right have 6 vertices, 8 edges, 1 connected component, and the degree list is the same, but they are not isomorphic (why?).

## Graph properties

In this part, we present some graph properties which are stated in the following theorems.

---

**Theorem 4.1** *In an undirected graph $G = (V, E)$, we have*

$$\sum_{v \in V} \deg(v) = 2|E|.$$

---

**Proof** Every edge $(w, u)$ has 2 endpoints ($w$ and $u$). The first endpoint $w$ contributes exactly 1 to $\deg(u)$. The second endpoint $u$ contributes exactly 1 to $\deg(w)$. Thus, each edge contributes exactly 2 to the sum on the left. This proves the theorem. ■

The following example is a direct application of Theorem 4.1.

■ **Example 4.5**

In the graph, $G = (V, E)$, shown to the right, we have

$$\sum_{v \in V} \deg(v) = \sum_{i=1}^{4} \deg(v_i) = 3 + 2 + 3 + 2 = 10 = 2|E|.$$

This matches the result in Theorem 4.1 above.

Graph of Ex. 4.5.

■

Throughout the rest of this section, for a given undirected graph $G = (V, E)$, we let $n = |V|$ and $m = |E|$. The following is the bipartite version of Theorem 4.1.

---

**Theorem 4.2** *In a bipartite graph $G = (A \cup B, E)$, we have*

$$\sum_{v \in A} \deg(v) = \sum_{v \in B} \deg(v) = m.$$

---

**Proof** Since $A$ and $B$ are partition sets of the vertex set, $V$, of the bipartite graph $G$ (i.e., $A \cup B = V$ and $A \cap B = \emptyset$), each vertex in $V$ is either in $A$ or $B$, but not both. Then, from Theorem 4.1, we have

$$\sum_{v \in V} \deg(v) = \sum_{v \in A} \deg(v) + \sum_{v \in B} \deg(v) = 2m.$$

Note that each edge has one endpoint in $A$ and one endpoint in $B$. So, we have $\sum_{v \in A} \deg(v) = \sum_{v \in B} \deg(v)$. Thus,

$$2 \sum_{v \in A} \deg(v) = 2 \sum_{v \in B} \deg(v) = 2m.$$

The result follows by dividing each term by 2. The proof is complete. ■

For example, in the bipartite graph shown on the left-hand side of Figure 4.13, we have

$$\sum_{v \text{ is a blue vertex}} \deg(v) = \sum_{v \text{ is a red vertex}} \deg(v) = 8 = |E|.$$

**Theorem 4.3** *In an undirected graph, we have $m \leq n(n-1)/2$.*

**Proof** Form Theorem 4.1, we have $2m = \sum_{v \in V} \deg(v)$. Note that $\deg(v) \leq n-1$ because each vertex has degree at most $n-1$. It follows that

$$m = \frac{1}{2} \sum_{v \in V} \deg(v) \leq \frac{1}{2} \sum_{v \in V} (n-1) = \frac{1}{2} \sum_{i=1}^{n} (n-1) = \frac{1}{2} n(n-1).$$

The proof is complete. ∎

From Theorem 4.3, we can conclude the following corollary.

**Corollary 4.1** *In an undirected graph $G = (V, E)$, we have $|E| = O(V^2)$[a].*

---
[a]To be consistent with the notations in the literature, throughout this book, we use $O(V)$, $O(V^2)$ and $O(V + E)$ to mean that $O(|V|)$, $O(|V|^2)$ and $O(|V| + |E|)$, respectively.

The following theorem tells us when the inequality in Theorem 4.3 is actually an equality.

**Theorem 4.4** *An $n$-vertex complete graph has $n(n-1)/2$ edges.*

**Proof** We prove that $|E(K_n)| = n(n-1)/2$ by induction on $n$. The base case is trivial: $K_1$ has one vertex and hence no edges and $n(n-1)/2 = (1)(0)/2 = 0$. Assume that the statement is true for $n = \ell$, i.e., $|E(K_\ell)| = \ell(\ell-1)/2$, for some $\ell \in \mathbb{N}$. Now, we prove that $|E(K_{\ell+1})| = (\ell+1)\ell/2$. Let $H$ be a subgraph of $K_{\ell+1}$ created by removing any vertex $v$ and all edges incident to $v$. Then $H = K_\ell$ and

$$|E(K_{\ell+1})| = |E(K_\ell)| + \deg(v) = \ell(\ell-1)/2 + \ell = (\ell+1)\ell/2.$$

Thus, any $n$-vertex complete graph has $n(n-1)/2$ edges. The proof is complete. ∎

For example, in the complete graph $K_6$ (see $K_6$ at the right-hand side of Figure 4.12), we have $6(6-1)/2 = 15 = |E|$.

**Theorem 4.5** *The number of odd degree vertices is always even.*

**Proof** From Theorem 4.1, the sum of all the degrees is equal to twice the number of edges. Since the sum of the degrees is even and the sum of the degrees of vertices with even degrees is even, the sum of the degrees of vertices with odd degrees must be even. If the sum of the degrees of vertices with odd degrees is even, there must be an even number of those vertices. The proof is complete. ∎

Figure 4.19: Graph of Example 4.6 ($b$).

For instance, in Example 4.5, the number of odd degree vertices is two, which is even. We have also the following example.

▨ **Example 4.6** Determine whether each of the following sequences is a degree sequence of a graph? Answer by 'yes' or 'no' and justify your answer.

($a$) $(1,1,2,2,2,3,3,3)$.     ($b$) $(0,0,1,1,2,2,3,3)$.     ($c$) $(0,0,1,1,1,1,6)$.

**Solution**  ($a$) The answer is no, because we have an odd number of odd degree vertices.

($b$) Yes. To justify this answer, a graph that represents this degree sequence is given in Figure 4.19.

($c$) No. The justification for this answer is that there is a vertex of degree 6. So, the minimum number of vertices that this vertex must be adjacent to is 6. Therefore, all 7 vertices must be connected. However, the degree sequence has two vertices of degree 0, contradicting this.

▨

The most important formula for studying planar graphs is Euler's formula.[2]

---

**Theorem 4.6 (Euler's formula)** *Let $G = (V, E)$ be a finite, connected, planar graph that is drawn in the plane without any edge intersections. Let also $F$ be the set of faces (regions bounded by edges, including the outer, infinitely large region) of $G$, then*

$$|V| - |E| + |F| = 2.$$

---

**Proof**  We prove the theorem by induction on $m$ where $m = |E|$. If $m = 0$, then it consists of a single vertex with a single region surrounding it, and hence $|V| - |E| + |F| = 1 - 0 + 1 = 2$. This proves the base case.

For the inductive step, we assume that for some graph, the formula is true for any $m = k$ for some $k \in \mathbb{N}$. To prove that the formula is true for $m = k + 1$, choose any edge $e \in E$. If $e$ connects two vertices, then contracting it would reduce $|V|$ and $|E|$ by one. Otherwise, the edge $e$ separates two faces, then removing it would reduce $|F|$ and $|E|$ by one. In either case, the result follows the inductive hypothesis. This proves the theorem. ▨

---

[2]Euler's formula was first proved by Leonhard Euler (1707 - 1783), a Swiss mathematician who made important and influential discoveries in many branches of mathematics.

As an illustration, in the graph shown to the right, we have

$$|V| - |E| + |F| = 6 - 8 + 4 = 2.$$

We end this section with the following theorem which characterizes the tree graph.

---

**Theorem 4.7** *Let G be an undirected graph on n vertices and m edges. The following statements are equivalent.*

(*a*) *G is a tree.*

(*b*) *Any two vertices in G are connected by a unique simple path.*

(*c*) *G is connected but each edge is a bridge.*

(*d*) *G is connected and m = n − 1.*

(*e*) *G is acyclic and m = n − 1.*

(*f*) *G is acyclic but if any edge is added to G it creates a cycle.*

---

**Proof**  We need to show that $(a) \rightarrow (b) \rightarrow (c) \rightarrow (d) \rightarrow (e) \rightarrow (f) \rightarrow (a)$.

$(a) \rightarrow (b)$: Since every tree is connected, there is at least one path between any two vertices in $G$. To show that there is a unique path between any two vertices in $G$, we use a contradiction. Suppose that there are at least two paths between some pair of vertices of $G$. The union of these two paths contains a cycle. So, $G$ contains a cycle. This contradicts the fact that $G$ is a tree. Thus, there is a unique path between any two vertices in $G$.

$(b) \rightarrow (c)$: Because any two vertices in $G$ are connected by a unique path, $G$ is connected. Let $(x, y)$ be any edge in $G$. Then, $P = (x, y)$ is a path from $x$ to $y$. So, it must be a unique path from $x$ to $y$. If we remove $(x, y)$ from $G$, then there is no path from $x$ to $y$. Hence, $G - (x, y)$ is disconnected. Because $(x, y)$ is arbitrary edge of $G$, $G - e$ is disconnected for every edge $e$ of $G$. Thus, $G$ is connected but $G - e$ is disconnected for every edge $e$ of $G$.

$(c) \rightarrow (d)$: By assumption, $G$ is connected. So we need only to show that $m = n - 1$. We prove this by induction. It is clear that a connected graph with $n = 1$ or $n = 2$ vertices has $n - 1$ edges. Assume that every graph with fewer than $n$ vertices satisfying (*c*) also satisfies (*d*). Let $G$ be an $n$-vertex connected graph but $G - e$ is disconnected for every edge $e$ of $G$. Let $e'$ be any edge of $G$. Now, $G - e'$ is disconnected. Hence $G - e'$ has two connected components. Let $G_1$ and $G_2$ be the connected components of $G$. Let $n_i$ and $m_i$ be the number of vertices and edges in $G_i$, for $i = 1, 2$. Now, each component satisfies (*c*) because $n_i < n$ for $i = 1, 2$. By induction hypothesis, we have

$m_i = n_i - 1$, for $i = 1, 2$. So, $m = m_1 + m_2 + 1 = (n_1 - 1) + (n2 - 1) + 1 = n - 1$. Thus, by induction principle, $G$ has exactly $n - 1$ edges.

$(d) \rightarrow (e)$: We have to show that every connected graph $G$ with $n$ vertices and $n - 1$ edges is acyclic. We prove this by induction. For $n = 1, 2$, it is clear that all connected graphs with $n$ vertices and $n - 1$ edges are acyclic. Assume that every connected graph with fewer than $n$ vertices satisfying $(d)$ is acyclic. Let $G$ be an $n$-vertex connected graph with $n - 1$ edges. Because $G$ is connected and has $n - 1$ edges, $G$ has a vertex, say $x$, of degree 1. Let $G' = G - \{x\}$. Then, $G'$ is connected and has $n - 1$ vertices and $n - 2$ edges. By induction hypothesis, $G'$ is acyclic. Because $x$ is a 1-degree vertex in $G$, $x$ can not be in any cycle of $G$. Since $G' = G - \{x\}$ is acyclic, $G$ must be acyclic. So, by induction, every $n$-vertex connected graph with $n - 1$ edges is acyclic.

$(e) \rightarrow (f)$: Assume that $G$ is acyclic and that $m = n - 1$. Let $G_i$, $1 \le i \le k$, be the connected components of $G$. Because $G$ is acyclic, $G_i$ is acyclic for $i = 1, 2, \ldots, k$. Hence, each $G_i$, $1 \le i \le k$, is a tree. Let $n_i$ and $m_i$ be the number of vertices and edges in $G_i$ for each $i = 1, 2, \ldots, k$. Since $(a)$ implies $(e)$, we have

$$ m = \sum_{i=1}^{k} m_i = \sum_{i=1}^{k} (n_i - 1) = n - k. $$

So, $k = 1$. This means that $G$ is connected. Hence, $G$ must be a tree. Since $(a)$ implies $(b)$, any two vertices in $G$ are connected by a unique path. Thus, adding any edge to $G$ creates a cycle.

$(f) \rightarrow (a)$: The proof of this direction is left to the reader as an exercise (see Exercise 4.5). ∎

## 4.3. Eulerian and Hamiltonian graphs

In this section, we introduce Eulerian and Hamiltonian paths and cycles. Eulerian graphs are motivated by the following 18th century problem about the seven bridges of Königsberg.

**Königsberg bridge problem**   Graph theory began in 1736 when Leonhard Euler solved the well-known Königsberg bridge problem.

The city of Königsberg in Prussia (now Kaliningrad in Russia) is set on both sides of the Pregel river, which has seven bridges across it (see Figure 4.20). In 1735, citizens of the city were wondering if it is possible to:

- Walk over all the bridges exactly once returning to where you started.

- Walk over all the bridges exactly once ending at a different place than where you started.

The negative resolution by Euler in 1736 laid the foundations of graph theory.

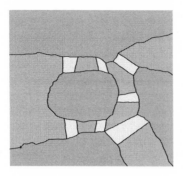 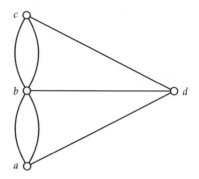

Figure 4.20: The graph of Königsberg bridge problem has four vertices $a, b, c$ and $d$ (representing land masses) and seven edges (representing the bridges).

 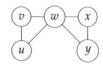

An Eulerian path: $(u, v, w, x, y, w)$.     An Eulerian cycle: $(u, v, w, x, y, w, u)$.

Figure 4.21: An Eulerian path and cycle.

**Eulerian paths and cycles**   A graph is called Eulerian when it contains an Eulerian cycle. We have the following definition.

> **Definition 4.3** *An Eulerian path in a graph is a path that visits every edge exactly once (allowing for revisiting vertices). An Eulerian cycle in a graph is an Eulerian path that starts and ends on the same vertex.*

Note that, from Definition 4.3, every Eulerian cycle is an Eulerian path. Figure 4.21 shows an Eulerian path and cycle. Note that, in the Eulerian path shown to the left in Figure 4.21, all vertices have even degrees. Note also that, in the Eulerian cycle shown to the right in Figure 4.21, there are exactly two vertices of odd degrees. The following two theorems, which are due to Euler, are regarded as an excellent characterization for Eulerian paths and cycles.

> **Theorem 4.8** *A connected undirected graph has an Eulerian cycle if and only if the degrees of all the vertices are even.*

**Proof**   An Eulerian cycle is a traversal of all the edges of a simple graph once and only once, staring at one vertex and ending at the same vertex. We can repeat vertices as many times as we want, but you can never repeat an edge once it is traversed. We prove that if a connected undirected graph has an Eulerian cycle, then the degrees of all the vertices are even. Let $G$ be a graph that has an Eulerian cycle. Every time we arrive at a vertex during our traversal of $G$, we enter via one edge and exit via another. Thus, there must be an even number of edges incident to every vertex. Therefore, every vertex of $G$ has an even degree. This proves one direction of the theorem. The proof of the other direction is left to the reader as an exercise. ∎

> **Theorem 4.9** *A connected undirected graph has an Eulerian path that is not a cycle if and only if there are exactly two vertices of odd degrees.*

**Proof**  We prove that if a connected undirected graph has exactly two vertices of odd degrees, then it has an Eulerian path. Let $G$ be a graph with exactly two vertices, say $u$ and $v$, of odd degrees. Adding the edge $(u, v)$ to $G$ forms a graph having all even degrees. By Theorem 4.8, the resulting graph contains an Eulerian cycle. This cycle uses the edge $(u, v)$. Thus, deleting the edge $(u, v)$ from the resulting graph yields an Eulerian path in the graph $G$. This proves one direction of the theorem. The proof of the other direction is left to the reader as an exercise. ∎

■ **Example 4.7**  Determine if the following graphs have an Eulerian cycle, Eulerian path, or neither? If there is one, give it. Otherwise, justify why there is none.

(a)

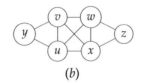

(b)

(c)

**Solution**  (a) Note that all vertices in the graph are of even degrees. From Theorem 4.8, the graph has an Eulerian cycle. Also, from Theorem 4.9, the graph does not have an Eulerian path. The Eulerian cycle is $(u, v, w, x, u)$.

(b) Note that all vertices in the graph are of even degrees. From Theorem 4.8. the graph has an Eulerian cycle, Also, from Theorem 4.9, the graph does not have an Eulerian path. The Eulerian cycle is $(y, v, w, z, x, u, v, x, w, u, y)$.

(c) Note that all vertices in the graph are of odd degrees. From Theorem 4.8, the graph does not have an Eulerian cycle. Also, from Theorem 4.9, the graph does not have an Eulerian path. ■

To illustrate the usefulness Theorems 4.8 and 4.9, we present some examples.

**Example 4.8** Determine if $K_{5,7}$ has an Eulerian cycle, Eulerian path, or neither? Justify your answer.

**Soltion.** The complete bipartite graph $K_{5,7}$ has 12 vertices: 7 vertices of degree 5, and 5 vertices of degree 7. By Theorem 4.8, because not all vertices of $K_{5,7}$ are of even degrees, the graph $K_{5,7}$ does not have an Eulerian cycle. In addition, by Theorem 4.8, because $K_{5,7}$ does not have exactly two vertices of odd degrees, the graph $K_{5,7}$ does not have an Eulerian path.

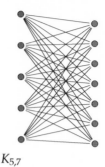

$K_{5,7}$

Finally, revisiting the Königsberg bridge problem, note that the degrees of all 4 vertices of the graph shown in Figure 4.20 are odd. So, applying Theorems 4.8 and 4.9 to the Königsberg bridge graph, we understand now why Euler was led to a negative resolution of the problem.

**Hamiltonian paths and cycles**   A graph is called Hamiltonian when it contains a Hamiltonian cycle. We have the following definition.

**Definition 4.4** *A Hamiltonian path in a graph is a path that visits every vertex in the graph exactly once. A Hamiltonian cycle in a graph is a Hamiltonian path that starts and ends on the same vertex.*

See Figure 4.22, which shows a Hamiltonian path and Hamiltonian cycles.

Note that if a graph has a Hamiltonian cycle, then it automatically has a Hamiltonian path by deleting any edge of the cycle.

The following theorem is due to Dirac. The proof of this theorem is beyond the scope of this book and is omitted. The proof can be found in any standard textbook on graph theory.

**Theorem 4.10** *If each vertex of a connected graph with $n \geq 3$ vertices is adjacent to at least $n/2$ vertices, then the graph has a Hamiltonian cycle (hence, also has a Hamiltonian path).*

A Hamiltonian path.

A Hamiltonian cycle.

A Hamiltonian cycle.

Figure 4.22: A graph with a Hamiltonian path and graphs with Hamiltonian cycles.

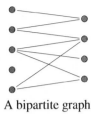
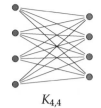
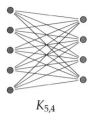

A bipartite graph          $K_{4,4}$          $K_{5,4}$

Figure 4.23: Some bipartite graphs. Among them, only $K_{4,4}$ satisfies the hypothesis of Dirac's theorem, hence it is Hamiltonian.

Note that the converse of Theorem 4.10 is not necessarily true, i.e., if a connected graph with $n \geq 3$ vertices has a Hamiltonian cycle, then it is not necessarily that each vertex of the graph is adjacent to at least $n/2$ vertices. The graph shown on the right-hand side of Figure 4.22 is an example, and, indeed, many other graphs can also be taken to verify that the converse of Theorem 4.10 is not true in general. For example, any circle $C_n$, where $n \geq 5$, can be considered.

Note also that, if $n$ is odd in Theorem 4.10, we take the ceiling of $n/2$. In the graph shown to the right, we give a 5-vertex graph, in which the degree of each vertex is greater than or equal $\lfloor 5/2 \rfloor = 2$, but the graph is non-Hamiltonian. (Why?).

A non-Hamiltonian graph.

Figure 4.23 shows some bipartite graphs. Among them, only the complete bipartite graph $K_{4,4}$ satisfies the condition that each vertex has a degree at least half the number of vertices. Based on Theorem 4.10, $K_{4,4}$ has a Hamiltonian cycle. It is also not hard to see that the complete bipartite graph $K_{5,4}$ has a Hamiltonian path but does not have a Hamiltonian cycle.

Generally, from Theorem 4.10, any complete bipartite graph $K_{n,n}$, where $n \geq 2$, has a Hamiltonian cycle (hence, has also a Hamiltonian path). Additionally, the complete bipartite graphs $K_{n,n+1}$ and $K_{n+1,n}$, where $n \geq 1$, have Hamiltonian paths but have no Hamiltonian cycles. Finally, from Theorem 4.10, any complete graph $K_n$, where $n \geq 2$, has a Hamiltonian cycle (hence, also has a Hamiltonian path).

## 4.4. Graph coloring

We start with the following example which introduces graph coloring and motivates the reader to some of its applications.

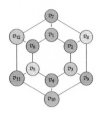   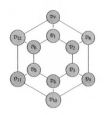

Figure 4.24: Scheduling final exams for twelve classes by coloring a graph of 12 vertices using four colors (left) and coloring the same graph using only two colors (right).

**Example 4.9** Every semester, the registrar schedule final exams in a way so that no student must take two exams at the same time.

We begin modeling this situation by constructing a graph $G = (V, E)$. The vertex set $V$ corresponds to the set of all classes (one vertex per class). The set $E$ is defined as follows: There is an edge between $u$ and $v$ if there exists one student taking both class $u$ and class $v$. The goal is to assign each vertex to a time slot that the two endpoints of every edge are assigned to different slots.

The simplest solution is to assign each final exam to a unique tie slot. If there are $n$ classes, this results in $n$ time slots, but this may lead to an extremely long final exam period. So, we want a solution that is not only feasible, but also minimizes the number of total time slots.

One way to visualize the notion of "assigning a time slot" is to consider each time slot as a color. In other words, given $G$, we must color every vertex of $G$ so that the two endpoints of every edge receive different colors.

In Figure 4.24, we color a graph of 12 vertices to show how to schedule final exams for 12 classes. Each class is represented by a vertex $c_i$ for some $i = 1, 2, \ldots, 12$. It is given that each class $c_i$ contains exactly three students taking three classes different from the class $c_i$ for each $i = 1, 2, \ldots, 12$. The coloring shown to the left of Figure 4.24 uses four colors, while the coloring shown to the right uses only two colors.   ∎

Now we are ready to introduce the properly colored graphs. Formally, we have the following definition.

> **Definition 4.5** *A proper coloring is an assignment of colors to the vertices of a graph so that no two adjacent vertices have the same color.*

Figure 4.25 shows a properly colored graph and an improperly colored graph. Note that the two colorings used for the 12-vertex graph shown in Figure 4.24 are both proper.

> **Definition 4.6** *A (proper) k-coloring of a graph is a proper coloring using at most k colors. A graph that has a k-coloring is said to be k-colorable.*

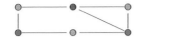 

Figure 4.25: Proper graph coloring (left) versus improper graph coloring (right).

Clearly, every graph $G = (V, E)$ is $|V|$-colorable because we can assign a different color to each vertex. We are usually interested in the minimum number of colors we can get away with and still properly color a graph. We have the following theorems.

**Theorem 4.11** *A graph is 2-colorable if and only if it is bipartite.*

**Proof** Let $G$ be a 2-colorable graph. This means that we can color every vertex either red or blue, and no edge will have both endpoints colored with the same color. Let $A$ denote the subset of vertices colored red, and $B$ denote the subset of vertices colored blue. Since all vertices of $A$ are red, there is no edge within $A$, and similarly for $B$. This implies that every edge has one endpoint in $A$, and the other in $B$, which means $G$ is bipartite.

Conversely, assume that $G$ is bipartite, that is, we can partition the vertices into two subsets $A$ and $B$, and every edge has one endpoint in $A$ and the other in $B$. Then coloring every vertex of $A$ red and every vertex of $B$ blue yields a proper coloring. So, $G$ is 2-colorable. The proof is complete. ∎

**Theorem 4.12** *The complete graph $K_n$ is not $(n-1)$-colorable.*

**Proof** Consider any color assignment on the vertices of $K_n$ that uses at most $n - 1$ colors. Since there are $n$ vertices, there exist two vertices $u$ and $v$ that share a color. However, since $K_n$ is complete, $(u, v)$ is an edge of the graph. This edge has two endpoints with the same color, so this coloring is improper. Thus, $K_n$ is not $n - 1$-colorable. The proof is complete. ∎

We now introduce the notion of the chromatic number of a graph.

**Definition 4.7** *The chromatic number of a graph $G$ is denoted by $\chi(G)$ and is defined to be the minimum positive integer $k$ such that $G$ is $k$-colorable. A graph $G$ is called $k$-chromatic if $\chi(G) = k$.*

From Definition 4.7, a graph is called $k$-chromatic if it is $k$-colorable, but not $(k-1)$-colorable. For example, by Theorem 4.12, the complete graph $K_n$ is $n$-chromatic (hence $\chi(K_n) = n$). Note that, from Theorem 4.11, a graph $G$ is bipartite iff $\chi(G) \leq 2$, and the equality holds iff $G$ has at least one edge. In particular, the complete bipartite graph is 2-chromatic (hence $\chi(K_{m,n}) = 2$). See Figure 4.23. We also have the following theorem.

> **Theorem 4.13** *If H is a subgraph of a graph G, then $\chi(H) \le \chi(G)$.*

**Proof** Let $G$ be a graph that has $H$ as a subgraph. Let also $\chi(G) = k$, then G is $k$-colorable, hence so is $H$. This implies that $\chi(H) \le k = \chi(G)$. The proof is complete. ∎

Finding the chromatic number of a graph is an optimization problem. In general, when $\chi(G) \ge 3$, computing the chromatic number of $G$ is an NP-hard problem, which belongs to a class of problems that will be introduced briefly in Section 7.7.

## 4.5. Directed graphs

All graphs that we have looked at so far are undirected graphs. Their edges are also said to be undirected. Sometimes it is necessary to associate directions with edges and get a directed graph instead.

Before formally introducing the definition of a directed graph, it is worth mentioning that we can use relations in this definition. The representation of a relation $\mathcal{R}$ on a set $A$ can be introduced with a directed graph as follows: A collection of vertices, one for each element in $A$, and directed edges, where an edge exists from vertex $u$ to vertex $v$ if and only if $u \, \mathcal{R} \, v$. It is also standard to treat a directed graph as we define it below.

> **Definition 4.8** *A directed graph, or digraph, is a graph whose edges have a defined direction, usually edges are represented as ordered pairs where the ordered pair $(u, v)$ indicates that there is a directed edge from vertex $u$ to vertex $v$.*

A slightly different set of definitions are used for directed graphs. A simple digraph has no loops and no multiple edges. See Figure 4.26. All directed graphs that we consider in the book are simple.

***Vertex in-degree and out-degree*** If $e = (u, v)$ is a directed edge in a digraph, then $e$ is called incident from $u$ and incident to $v$. The in-degree of a vertex $v$, denoted as $\deg_{in}(v)$, is the number of edges incident to $v$ (i.e., those entering the vertex $v$). The out-degree of a vertex $v$, denoted as $\deg_{out}(v)$, is the number of edges incident from $v$ (i.e., those leaving the vertex $v$). For example, in the graph shown on the left-hand side of Figure4.26, we have $\deg_{in}(v) = \deg_{out}(u) = 2$ and $\deg_{out}(v) = \deg_{in}(u) = \deg_{out}(w) = \deg_{in}(w) = 1$.

Figure 4.26: A simple digraph (left) versus a non-simple digraph (right).

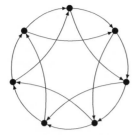 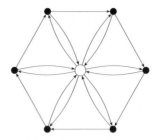

Figure 4.27: Balanced digraphs.

A vertex $v$ is called balanced if its in-degree is equal to its out-degree. A directed graph is balanced if every vertex is balanced. In Figure 4.27, we show two balanced directed graphs in which each vertex has 2 incoming and 2 outgoing edges, except the non-solid vertex which has 6 incoming and 6 outgoing edges.

The following is the directed version of Theorem 4.1.

---

**Theorem 4.14** *In a directed graph $G = (V, E)$, we have*

$$\sum_{v \in V} \deg_{in}(v) = \sum_{v \in V} \deg_{out}(v) = |E|.$$

---

**Proof** Every directed edge $(w, u)$ has 2 endpoints ($w$ and $u$). The first endpoint $w$ contributes exactly 1 to $\deg_{out}(w)$. The second endpoint $u$ contributes exactly 1 to $\deg_{in}(u)$. Thus, each directed edge contributes exactly 1 to exactly one of the sums on the left. The proof is complete. ∎

The simple calculations in following example match the result in Theorem 4.14.

**Example 4.10**
In the digraph, $G = (V, E)$, shown to the right, we have

$$\sum_{v \in V} \deg_{in}(v) = \sum_{i=1}^{4} \deg_{in}(v_i) = 3 + 0 + 1 + 1 = 5 = |E|, \text{ and}$$

$$\sum_{v \in V} \deg_{out}(v) = \sum_{i=1}^{4} \deg_{out}(v_i) = 0 + 2 + 2 + 1 = 5 = |E|.$$

Digraph of Ex. 4.10.

**Directed paths, cycles and trees** A directed path, or dipath, in a directed graph is a sequence of edges joining a sequence of distinct vertices where the edges are all directed in the same direction. See Figure 4.28.

Figure 4.28: A valid directed path from $u$ to $z$ (left) versus a "non-valid path" from $u$ to $z$ (right).

A directed cycle, or dicycle, is a directed path (with at least one edge) whose first and last vertices are the same. A directed tree, or ditree, is a directed graph whose underlying graph is a tree. A directed forest is a family of disjoint directed trees. See Figure 4.29.

A semicycle is a directed cycle in which some of its edges have been reversed. In other words, a semicycle constitutes an (undirected) cycle when neglecting its directionality. A cycle in a digraph is either a directed cycle or a semicycle. For example, the circle given by the sequence $(v, w, x, y, v)$ in the digraph shown to the right in Figure 4.30 is a directed cycle, whereas that given by the same sequence in the digraph shown in the middle of Figure 4.30 is a semicircle. After introducing the weakly connectedness in this section, we can see that a directed tree can now be redefined to be a weakly connected digraph that has no cycles. We also have the following definition.

**Definition 4.9** *A directed acyclic graph (DAG) is a digraph that has no directed cycles.*

Figure 4.30 shows a directed tree, a DAG, and a non-DAG. In the next chapter, we will see that a DAG can be redefined to be a digraph that has the so-called topological ordering.

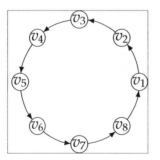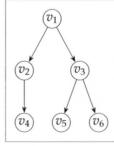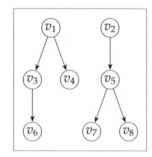

Figure 4.29: A directed cycle (left), a directed tree (middle), and a directed forest (right).

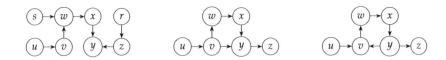

Figure 4.30: A directed tree (left), a DAG (middle), and a non-DAG (right).

***Connectedness*** A directed graph is called weakly connected if replacing all of its directed edges with undirected edges produces a connected (undirected) graph. A directed graph is called strongly connected if every two vertices are reachable from each other. In other words, a digraph strongly connected if there is a path in each direction between each pair of vertices of the digraph. See Figure 4.31.

The vertex $y$ is not reachable from the vertex $w$, but the underlying undirected graph is connected.

In the graph shown on the right, every two vertices are reachable from each other.

Figure 4.31: A weakly connected digraph (left) versus a strongly connected digraph (right).

A strongly connected component $G'$ of a digraph $G$ is a maximal strongly connected directed subgraph of $G$. By maximality we mean that there is no way to add into $G'$ any vertices and/or directed edges of $G$ which are not currently in $G'$ in such a way that the resulting directed subgraph is strongly connected.

For example, the digraph shown in Figure 4.32 has three strongly connected components which are surrounded by blue dashed polygons. For another example, the digraph shown in Figure 4.33 has five strongly connected components which are surrounded by blue dashed polygons.

Note that a strongly connected digraph has only one strongly connected component.

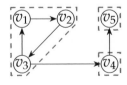

Figure 4.32: A digraph with three strongly connected components.

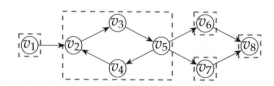

Figure 4.33: A digraph with five strongly connected components.

> **Remark 4.5** *The strongly connected components of a digraph define equivalence classes over its vertices under the "are mutually reachable from" relation.*

Remark 4.5 can be directly proven on the basis of Definitions 2.7 and 2.8.

# EXERCISES

**4.1**  Choose the correct answer for each of the following multiple-choice questions/items.

(*a*) If a simple graph has 3 vertices, then the number of vertices with odd degrees can only be:

   (*i*) 0.  (*ii*) 1.  (*iii*) 2.  (*iv*) 0 or 2.

(*b*) What is the number of edges present in a graph having 244 degrees total?

   (*i*) 61.  (*ii*) 122.  (*iii*) 244.  (*iv*) 488.

(*c*) Let $G = (V, E)$ be an undirected graph. The following statements are equivalent, except for one that is not. Which one?

   (*i*) Any two vertices in $G$ are connected by a simple path.
   (*ii*) $G$ is connected but each edge is a bridge.
   (*iii*) $G$ is connected and $|V|^2 = |E|^2 + 2|E| + 1$.
   (*iv*) $G$ is acyclic but if any edge is added to $G$ it creates a cycle.

(*d*) Let $G$ be a simple connected graph with $n$ vertices, where $n$ is an odd number greater than 2. Which one of the following statements is the contrapositive of Dirac's theorem for graphs?

   (*i*) If every vertex of $G$ is adjacent to at least $(n + 1)/2$ vertices, then $G$ has a Hamiltonian cycle.

   (*ii*) If $G$ has no Hamiltonian cycle, then every vertex of $G$ is adjacent to at most $(n + 1)/2$ vertices.

   (*iii*) If $G$ has no Eulerian cycle, then each vertex of $G$ is adjacent to less than $(n + 1)/2$ vertices.

   (*iv*) Every vertex of $G$ has degree at most $n/2$ if $G$ has no Hamiltonian cycle.

(*e*) Let $p$ and $q$ be even numbers greater than two. Which one of the following statements is false?

   (*i*) The complete bipartite graph $K_{2,q}$ has an Eulerian cycle.

   (*ii*) The complete bipartite graph $K_{2,q}$ has an Eulerian path.

   (*iii*) The complete graph $K_p$ has no Eulerian cycle.

   (*iv*) The complete bipartite graph $K_{p,q}$ has an Eulerian cycle.

(*f*) Let $p$ be an integer number greater than one. Which one of the following graphs has no Hamiltonian cycle?

   (*i*) The complete graph $K_p$.  (*iii*) The complete bipartite $K_{p,p+1}$.

   (*ii*) The complete bipartite $K_{p,p}$.  (*iv*) The cycle $C_p$.

(*g*) The chromatic number of an even cycle (i.e., a cycle with even length) equals:

   (*i*) 2.  (*ii*) 3.  (*iii*) 2 or 3.  (*iv*) its length.

(*h*) Let $G = (V, E)$ be a digraph, where $V = \{a, b, c, d, e, f, g, h, i\}$ and $E = \{(a, d), (d, a), (b, c), (c, f), (f, b), (g, h), h, g), (b, e), (e, d), (e, i), (h, e)\}$. What is the number of strongly connected components in G?

   (*i*) 3.  (*ii*) 4.  (*iii*) 5.  (*iv*) 6.

(*i*) Which one of the following statements is true?

   (*i*) There are two isomorphic graphs with different degree sequences.

   (*ii*) There are no two non-isomorphic graphs with the same degree sequence.

   (*iii*) There are no two isomorphic graphs with the same degree sequences.

   (*iv*) There are two non-isomorphic graphs with the same degree sequence.

**4.2**    Find a spanning tree for the following graph so that no vertex has a degree of 4.

**4.3**    Are the graphs $G_1$ and $G_2$ shown below isomorphic? Why or why not?

$G_1$

$G_2$

**4.4**    Determine whether each of the following sequences is a degree sequence of a graph? Answer by 'yes' or 'no' and justify your answer.

(*a*) $(1, 2, 3, 3, 4, 5, 5)$.        (*b*) $(0, 0, 1, 1, 1, 1, 4, 4)$.        (*c*) $(0, 2, 2, 2, 2, 4, 4, 4)$.

**4.5**    Prove that the direction "$(f) \rightarrow (a)$" in Theorem 4.7.

**4.6**    Give an undirected graph with 5 vertices, all of which must be at least degree 3. It must have an Eulerian cycle and a Hamiltonian cycle. Can you give another undirected graph?

**4.7**    Does the bipartite graph $K_{5,7}$ have a Hamiltonian cycle or path? Justify your answer.

**4.8**    Let $G$ be a $k$-colorable simple graph. Provide a tight upper bound[3] on the maximum number of colors that one could need to properly color the graph if

(*a*) one adds an edge between two vertices in $G$.

(*b*) one removes an edge from $G$.

**4.9**    Is the statement that "Any graph with a vertex of degree $d$ is $d + 1$-colorable" true or false? If it is false, explain why.

---

[3]By a tight upper bound, we look for an "interesting answer". For example, the upper bound $|V|$ is not an interesting answer in terms of credit.

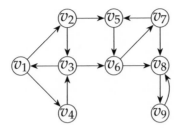

Figure 4.34: Digraph of Exercise 4.10.

**4.10**    Find the strongly connected components of the digraph shown in Figure 4.34.

# CHAPTER 5

# RECURRENCES

## Contents

© 2022 by Baha Alzalg | Kindle Direct Publishing, Washington, United States 2022/9
B. Alzalg, *Combinatorial and Algorithmic Mathematics: From Foundation to Optimization*,
DOI 10.5281/zenodo.7110942

In this chapter, we present some techniques for solving recurrences. There are five efficient methods for solving recurrences: the guess-and-confirm method, the iteration method, the recursion-tree method, the generating functions method, and the master method. In this chapter, we present the first four methods. The fifth method (master method) is beyond the scope of this book. However, there are a number of good references for solving recurrences using the master method, see for example [CLRS01, Section 4.5].

## 5.1. Guess-and-confirm

In this section, we present the guess-and-confirm method for solving recurrences. As a matter of example, it can be shown (see Exercise 5.2 (*b*)) that the following recurrence

$$T(1) = 1, \quad T(n) = 3T(n-1) + 4, \quad n = 1, 2, 3, \ldots, \tag{5.1}$$

describes the function

$$T(n) = 3^n - 2, \quad n = 0, 1, 2, \ldots. \tag{5.2}$$

A formal definition of a recurrence looks like:

> **Definition 5.1** *A recurrence is an equation or inequality that describes a function in terms of its value on smaller inputs.*

An explicit formula (also-called a closed formula) is called a solution of a recurrence relation if its terms satisfy the recurrence relation. For example, the formula (7.10) is the solution of the recurrence relation (5.1). The initial conditions for a recurrence relation specify the terms that precede the first term where the recurrence relation takes effect. For example, $T(1) = 1$ is the initial condition for the recurrence relation (5.1).

The following are some instances for recurrence formulas.

- A recurrence formula that loops through the input to eliminate one item. For example:
$$T(n) = T(n-1) + 1, \quad T(n) = T(n-1) + n, \text{ etc.}$$

- A recurrence formula that halves the input. For example:
$$T(n) = T(n/2) + 1, \quad T(n) = T(n/2) + n, \text{ etc.}$$

- A recurrence formula that splits the input into two halves. For example:
$$T(n) = 2T(n/2) + 1, \text{ etc.}$$

Each remaining section of this chapter describes a method for solving recurrences.

As we mentioned earlier, this section describes the guess-and-confirm method. In this method, we first use repeated substitutions to guess an explicit expression for the recurrence formula, and then we use mathematical induction to prove and confirm our guess. So, this method is also known as the substitution/induction method. We have the following examples.

**Example 5.1** Use the guess-and-confirm method to solve the following recurrences.

(a) $T(n)^1 = \begin{cases} c_1, & \text{if } n = 1; \\ T(n-1) + c_2, & \text{if } n > 1. \text{ Here, } c_1 \text{ and } c_2 \text{ are constants.} \end{cases}$

(b) $T(n)^2 = \begin{cases} 0, & \text{if } n = 1; \\ T(n/2) + 1, & \text{if } n > 1, \text{ and } n \text{ is a power of 2.} \end{cases}$

**Solution** (a) Using repeated substitutions, we have

$$
\begin{aligned}
T(1) &= c_1, \\
T(2) &= T(1) + c_2 = c_1 + c_2, \\
T(3) &= T(2) + c_2 = c_1 + 2c_2, \\
T(4) &= T(3) + c_2 = c_1 + 3c_2, \\
&\vdots \\
T(n) &= T(n-1) + c_2 = c_1 + (n-1)c_2, \text{ for all } n \geq 1.
\end{aligned}
$$

We prove the $n$th-term guess by mathematical induction. The base case is trivial: $T(1) = c_1$. For the inductive hypothesis: we assume that $T(k) = c_1 + (k-1)c_2$, for $k < n$. We now use the inductive hypothesis to prove the statement for $k+1$. Note that

$$T(k+1) = c_2 + T(k) = c_2 + c_1 + (k-1)c_2 = c_1 + kc_2,$$

which completes the inductive step. Thus $T(n) = c_1 + (n-1)c_2$ for $n \geq 1$.

[1]This recurrence formula is the one that arises from the factorial function algorithm (Algorithm 7.28).
[2]This recurrence formula is the one that arises from the binary search algorithm (Algorithm 8.6).

(b) Using repeated substitutions, we have

$$
\begin{aligned}
T(1) &= 0 = \log 1, \\
T(2) &= T(2/2) + 1 = T(1) + 1 = 0 + 1 = 1 = \log 2, \\
T(4) &= T(4/2) + 1 = T(2) + 1 = 1 + 1 = 2 = \log 4, \\
T(8) &= T(8/2) + 1 = T(4) + 1 = 2 + 1 = 3 = \log 8, \\
&\vdots \\
T(n) &= \log n, \text{ where } n > 1 \text{ and } n \text{ is a power of 2.}
\end{aligned}
$$

We prove the $n$th-term guess by mathematical induction. The base case is trivial: $T(1) = 0 = \log 1$. For the inductive hypothesis: we assume that $T(k) = \log k$, for $k < n$. We now use the inductive hypothesis to prove the statement for $n$. Note that

$$
T(n) = T(n/2) + 1 = \log(n/2) + 1 = \log n - \log 2 + 1 = \log n.
$$

This completes the inductive step and hence completes the proof. ∎

**Example 5.2** Use the guess-and-confirm method to determine a good upper bound on the following recurrence.

$$
T(n) = \begin{cases} 1, & \text{if } n = 0, 1; \\ T(n-1) + T(n-2) + 1, & \text{if } n > 1. \end{cases}
$$

**Solution**   Using repeated substitutions, we have

$$
\begin{aligned}
T(0) &= 1 \leq 2^0, \\
T(1) &= 1 \leq 2^1, \\
T(2) &= T(1) + T(0) + 1 = 3 \leq 2^2, \\
T(3) &= T(2) + T(1) + 1 = 5 \leq 2^3, \\
T(4) &= T(3) + T(2) + 1 = 9 \leq 2^4, \\
T(5) &= T(4) + T(3) + 1 = 15 \leq 2^5, \\
&\vdots \\
T(n) &= T(n-1) + T(n-2) + 1 \leq 2^n, \text{ for all } n \geq 1.
\end{aligned}
$$

We prove the $n$th-term guess, which is $T(n) \leq 2^n$, by induction on $n$. The base case is trivial: we have $T(0) \leq 2^0 = 1$ (same for $T(1)$). For the inductive hypothesis: we assume that $T(k) \leq 2^k$. We now use the inductive hypothesis to prove the statement for $k + 1$. Note that

$$T(k+1) = T(k) + T(k-1) + 1 \le 2^k + 2^{k-1} + 1 \le 2^k + 2^k = 2(2^k) = 2^{k+1},$$

where the last inequality follows from the fact that $2^{k-1} + 1 \le 2^k$ for all $k > 0$. This completes the inductive step and hence completes the proof. In conclusion, we have $T(n) \le 2^n$ for $n = 0, 1, 2, \dots$. ∎

## 5.2. Recursion-iteration

In this section, we present the iteration method for solving recurrences. In the iteration method, we decompose the recurrence into a series of terms, and derive the $n$th expression from the previous ones. Other methods for solving recurrences will be the substance of the upcoming sections.

In the following examples, we solve recurrence formulas using the iteration method.

**Example 5.3**  Use the iteration method to solve the following recurrence formula.

$$
\begin{aligned}
T(1) &= c_1, \\
T(n) &= c_2 + T(n-1), \quad n > 1,
\end{aligned}
\tag{5.3}
$$

where $c_1$ and $c_2$ are constants.

**Solution**  From (5.3), we have

$$
\begin{aligned}
T(n) &= c_2 + T(n-1) \\
&= c_2 + (c_2 + T(n-2)) \\
&= c_2 + (c_2 + (c_2 + T(n-3))) \\
&= 3c_2 + T(n-3) \\
&\ \ \vdots \\
&= kc_2 + T(n-k).
\end{aligned}
$$

Let $k = n - 1$, then $n - k = 1$ and we have

$$T(n) = (n-1)c_2 + T(1) = (n-1)c_2 + c_1.$$

∎

**Example 5.4**   Use the iteration method to solve the following recurrence formula.

$$
\begin{aligned}
T(1) &= c, \\
T(n) &= c + T(n/2),
\end{aligned}
\tag{5.4}
$$

where $c$ is a constant.

**Solution**   From (5.4), we have

$$
\begin{aligned}
T(n) &= c + T(n/2) \\
&= c + (c + T(n/4)) \\
&= c + (c + (c + T(n/8))) \\
&= 4c + T(n/2^4) \\
&\ \ \vdots \\
&= ck + T(n/2^k).
\end{aligned}
$$

Let $n = 2^k$, then $k = \log n$. It follows that

$$
T(n) = ck + T(n/2^k) = c \log n + T(1) = c \log n + c.
$$

**Example 5.5**   Use the iteration method to solve the following recurrence formula.

$$
T(n) = n + 2T(n/2).
\tag{5.5}
$$

**Solution**   From (5.5), we have

$$
\begin{aligned}
T(n) &= n + 2T\left(\frac{n}{2}\right) \\
&= n + 2\left(\frac{n}{2} + 2T\left(\frac{n}{4}\right)\right) \\
&= n + 2\left(\frac{n}{2} + 2\left(\frac{n}{4} + 2T\left(\frac{n}{8}\right)\right)\right) \\
&= 3n + 2^3 T\left(\frac{n}{2^3}\right) \\
&\ \ \vdots \\
&= kn + 2^k T\left(\frac{n}{2^k}\right).
\end{aligned}
$$

Let $n = 2^k$, then $k = \log n$. It follows that

$$
T(n) = nk + 2^k T(n/2^k) = n \log n + nT(1).
$$

***Change of variables***    Sometimes we need to transform the recurrence to one that we solved before. We have the following example.

**Example 5.6** Solve the following recurrence.

$$T(n) = 2T(\sqrt{n}) + \log n.$$

**Solution**    Set $m = \log n$, then $n = 2^m$ and

$$T(2^m) = 2T(2^{m/2}) + m. \tag{5.6}$$

After renaming $m$ to $k$ and letting $S(k) = T(2^k)$, the formula (5.6) becomes

$$S(k) = 2S(k/2) + k,$$

which is the same recurrence formula (5.6). In Algorithm 8.9, we used the iteration method to solve this recurrence and found that $S(k) = k \log k + kS(1)$. It immediately follows that

$$T(n) = T(2^m) = S(m) = m \log m + mS(1) = \log n \log(\log n) + c \log n,$$

where $c = T(2)$.                                                                                    ■

## 5.3. Generating functions

Recurrence relations can be solved by finding a closed form for the associated generating function.

> **Definition 5.2** *Let $\{a_n\}$ be a sequence of real numbers. The generating function for $\{a_k\}$ is the infinite series*
>
> $$g(x) = a_0 + a_1 x + a_2 x^2 + \cdots + a_n x^n + \cdots = \sum_{k=1}^{\infty} a_k x^k.$$

Below are some examples of generating functions.

**Example 5.7** The generating function for the sequence $1, 1, 1, \ldots$ is

$$1 + x + x^2 + \cdots = \sum_{k=1}^{\infty} x^k = \frac{1}{1-x} \quad \text{for } |x| < 1,$$

where the last equality was obtained by the geometric series (3.5).                                ■

▮ **Example 5.8** The generating function for the sequence $1, a, a^2, \ldots$ is

$$1 + ax + a^2 x^2 + \cdots = \sum_{k=1}^{\infty} a^k x^k = \sum_{k=1}^{\infty} (ax)^k = \frac{1}{1 - ax} \quad \text{for } |ax| < 1^3, \qquad (5.7)$$

where the last equality was obtained by the geometric series (3.5). ∎

We can define generating functions for finite sequences of real numbers by extending a finite sequence $a_1, a_2, \ldots, a_n$ into an infinite sequence by setting $a_j = 0$ for $j \geq n + 1$. We have the following example.

▮ **Example 5.9** The generating function for the sequence $\binom{n}{0}, \binom{n}{1}, \binom{n}{2}, \ldots, \binom{n}{n}$ is the finite series

$$\binom{n}{0} + \binom{n}{1} x + \binom{n}{2} x^2 + \cdots + \binom{n}{n} x^n = \sum_{k=1}^{n} \binom{n}{k} x^k = (1 + x)^n,$$

where the last equality was obtained by the binomial theorem (Theorem 6.1). ∎

In the generating function method for solving recurrences, we transform the recurrence relation for the terms of a sequence into an equation involving a generating function. This equation can then be solved to find a closed form for the generating function. We have the following examples.

▮ **Example 5.10** Use the generating function method to solve the following recurrences.

(a) $T(n) = \begin{cases} 4, & \text{if } n = 0; \\ 4T(n - 1), & \text{if } n \geq 1. \end{cases}$

(b) $T(n) = \begin{cases} 1, & \text{if } n = 0; \\ 8T(n - 1) + 10^{n-1}, & \text{if } n \geq 1. \end{cases}$

**Solution** (a) Let $g(x)$ be the generating function for the sequence $\{a_n\} = \{T(n)\}$. Then

$$g(x) = \sum_{k=0}^{\infty} T(k) x^k.$$

---

[3]Equivalently, $|x| < 1/|a|$ for $a \neq 0$.

Using the recurrence relation, we have

$$
\begin{aligned}
g(x) &= T(0) + \sum_{k=1}^{\infty} T(k)x^k \\
&= 4 + \sum_{k=1}^{\infty} 4T(k-1)x^k \\
&= 4 + 4x \sum_{k=1}^{\infty} T(k-1)x^{k-1} \\
&= 4 + 4x \sum_{k=0}^{\infty} T(k)x^k = 4 + 4xg(x).
\end{aligned}
$$

It follows that

$$
g(x) = 4 + 4xg(x).
$$

Solving for $g(x)$, we get

$$
g(x) = \frac{4}{1 - 4x} = 4 \sum_{k=0}^{\infty} (4x)^k = \sum_{k=0}^{\infty} \underbrace{4^{k+1}}_{T(k)} x^k,
$$

where the second equality was obtained by the geometric series (5.7). Therefore, $T(n) = 4^{n+1}$ for any $n \geq 0$.

(b) Let $g(x)$ be the generating function for the sequence $\{a_n\} = \{T(n)\}$. Then

$$
g(x) = \sum_{k=0}^{\infty} T(k)\, x^k.
$$

Using the recurrence relation, we have

$$
\begin{aligned}
g(x) &= T(0) + \sum_{k=1}^{\infty} T(k)x^k \\
&= 1 + \sum_{k=1}^{\infty} (8T(k-1) + 10^{k-1})x^k \\
&= 1 + 8x \sum_{k=1}^{\infty} T(k-1)x^{k-1} + x \sum_{k=0}^{\infty} 10^{k-1}x^{k-1} \\
&= 1 + 8x \sum_{k=0}^{\infty} T(k)x^k + x \sum_{k=0}^{\infty} 10^k x^k = 1 + 8xg(x) + \frac{x}{1 - 10x},
\end{aligned}
$$

where the last equality was obtained by the geometric series (5.7). It follows that

$$
g(x) = 1 + 8xg(x) + \frac{x}{1 - 10x}.
$$

Solving for $g(x)$, we get

$$g(x) = \frac{1}{(1 - 8x)}\left(1 + \frac{x}{1 - 10x}\right) = \frac{1 - 9x}{(1 - 10x)(1 - 8x)}.$$

Note that

$$1 - 9x = \frac{1}{2}(1 + 1) - \frac{1}{2}(8x + 10x) = \frac{1}{2}(1 - 8x) + \frac{1}{2}(1 - 10x).$$

Hence, the function $g(x)$ can be written as

$$g(x) = \frac{\frac{1}{2}(1 - 8x) + \frac{1}{2}(1 - 10x)}{(1 - 10x)(1 - 8x)} = \frac{1/2}{1 - 10x} + \frac{1/2}{1 - 8x} = \frac{1}{2}\left(\frac{1}{1 - 10x} + \frac{1}{1 - 8x}\right).$$

Using the geometric series (5.7), we have

$$g(x) = \frac{1}{2}\left(\sum_{k=0}^{\infty}(10x)^k + \sum_{k=0}^{\infty}(8x)^k\right) = \sum_{k=0}^{\infty}\underbrace{\frac{1}{2}\left(10^k + 8^k\right)}_{T(k)}x^k.$$

Therefore, $T(n) = \frac{1}{2}(10^k + 8^k)$ for any $n \geq 0$. ▪

## 5.4. Recursion-tree

In the recursion-tree method for solving recurrences, we convert the recurrence into a tree so that each node represents the cost incurred at various levels of recursion, and then we sum up the costs of all levels. Perhaps the best way to understand how to use this method is to give the following illustrative example.

**Example 5.11**   Use the recursion tree method to determine a good upper bound on each of the following recurrences. (Here $c$ is a constant).

(a) $T(n) = 2T(n/2) + n^2$.               (c) $T(n) = T(n/3) + T(2n/3) + n$.

(b) $T(n) = 3T(n/4) + cn^2$.

**Solution**   (a) For convenience, we assume that $n$ is an exact power of 2 so that all subproblem sizes are integers. In Figure 5.1, we show in detail the construction of a recursion tree for the recurrence $T(n) = 2T(n/2) + n^2$. Note that we continue expanding each node in the tree by breaking it into its constituent parts as determined by the recurrence.

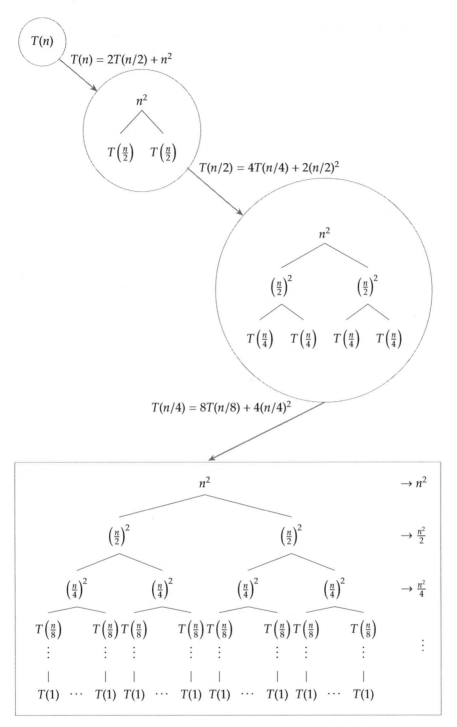

Figure 5.1: Constructing a recursion tree for $T(n) = 2T(n/2) + n^2$.

Let $h$ be the height of the recursion tree. At each level $i = 0, 1, 2, \ldots, h$, we have

- subproblem size = $n/2^i$,
- cost of each node = $(n/2^i)^2$,

- number of nodes = $2^i$.

The total cost at all levels is $T(n) = \sum_{i=0}^{h}$ total cost at level $i$, where

$$
\begin{aligned}
\text{total cost at level } i &= (\text{\# of node at level } i) \cdot (\text{\# cost of each node at level } i) \\
&= 2^i \left(\frac{n}{2^i}\right)^2 = \frac{n^2}{2^i}.
\end{aligned}
$$

Now, we want to know how to find the height of the tree $h$. Because subproblem sizes decrease by a factor of 2 each time we go down one level, we eventually must reach a boundary condition. Essentially, we want to know how far from the root do we reach one. Note that the subproblem size for a node at level $h$ is $n/2^h$. Therefore, starting from the root, we reach one when $n/2^h = 1$ (as $T(n/2^h) = T(1)$). Hence, the height of the tree is $h = \log n$.

As a result, the total cost at all levels is

$$
T(n) = \sum_{i=0}^{\log n} \frac{n^2}{2^i} = n^2 \sum_{i=0}^{\log n} \left(\frac{1}{2}\right)^i \le n^2 \sum_{i=0}^{\infty} \left(\frac{1}{2}\right)^i \le n^2 \frac{1}{1 - \frac{1}{2}} = 2n^2,
$$

where the last inequality follows from the geometric series (3.5). Thus $T(n) \le 2n^2$.

(b) For convenience, we assume that $n$ is an exact power of 4 so that all subproblem sizes are integers. In Figure 5.2, we show the recursion tree for the recurrence $T(n) = 3T(n/4) + cn^2$.

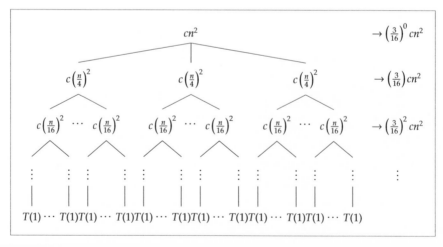

Figure 5.2: The recursion tree for $T(n) = 3T(n/4) + cn^2$.

Let $h$ be the height of the recursion tree. At each level $i = 0, 1, 2, \ldots, h$, we have

- subproblem size $= n/4^i$,
- cost of each node $= c(n/4^i)^2$,
- number of nodes $= 3^i$.

The total cost at all levels is $T(n) = \sum_{i=0}^{h}$ total cost at level $i$, where

$$
\begin{aligned}
\text{total cost at level } i \; &= \; (\text{\# of node at level } i) \cdot (\text{\# cost of each node at level } i) \\
&= \; c\left(\frac{n}{4^i}\right)^2 3^i \\
&= \; cn^2 \left(\frac{3}{16}\right)^i .
\end{aligned}
$$

To know how far from the root do we reach one, we note that the subproblem size for a node at level $h$ is $n/4^h$. Therefore, starting from the root, we reach one when $n/4^h = 1$. Hence, the height of the tree is $h = \log_4 n$.

As a result, the total cost at all levels is

$$
T(n) = \sum_{i=0}^{\log_4 n} cn^2 \left(\frac{3}{16}\right)^i \le cn^2 \sum_{i=0}^{\infty} \left(\frac{3}{16}\right)^i = cn^2 \frac{1}{1 - \left(\frac{3}{16}\right)} = \frac{16}{13} cn^2,
$$

where the last inequality follows from the geometric series. Thus, we have $T(n) \le (16/3)cn^2$.

(c) Figure 5.3 shows the recursion tree for $T(n) = T(n/3) + T(2n/3) + n$.

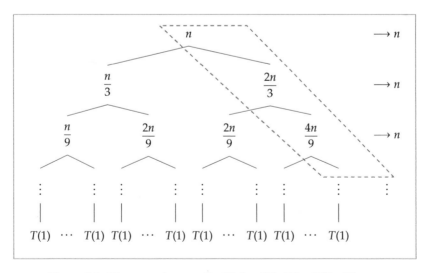

Figure 5.3: The recursion tree for $T(n) = T(n/3) + T(2n/3) + n$.

Note that, unlike in the trees of ($a$) and ($b$), in this tree the paths from the root to the leaves have different lengths. Let $h$ be the height of the recursion tree. Note also that the total cost at level each level $i$ ($i = 0, 1, 2, \ldots, h$) equals $n$, and that the longest path from the root to a leaf is

$$n \longrightarrow \left(\frac{2}{3}\right)n \longrightarrow \left(\frac{2}{3}\right)^2 n \longrightarrow \left(\frac{2}{3}\right)^3 n \longrightarrow \cdots \longrightarrow 1.$$

Starting from the root, we reach one when $(2/3)^h n = 1$. Hence, the height of the tree is $h = \log_{3/2} n$. It follows that the total cost at all levels is

$$T(n) \leq \sum_{i=0}^{\log_{3/2} n} n = \left(1 + \log_{3/2} n\right) n = \left(1 + \frac{\log n}{\log 3/2}\right) n \leq k n \log n,$$

for some positive constant $k$. Thus, $T(n) \leq k n \log n$. ∎

Note that in Example 5.11 ($c$), we were looking for the longest path from the root to a leaf because we need an asymptotic upper bound. However, if an asymptotic lower bound is required, we should be looking the shortest path from the root to a leaf as in Exercise 5.9 ($b$).

## EXERCISES

**5.1**    Choose the correct answer for each of the following multiple-choice questions/items.

($a$) If the guess-and-confirm method is used to solve the recurrence relation $T(n) = \sqrt{n}\, T(\sqrt{n}) + n$, we find that $T(n)$ is (here $c$ is some constant)

    ($i$) $cn$.
                      ($iii$) $cn \log n$.

    ($ii$) $cn \sqrt{\log n}$.
             ($iv$) $cn \log(\log n)$.

($b$) If we use the iteration method to solve the following recurrence.

$$T(n) = \begin{cases} T(n-1) + n, & \text{if } n > 1; \\ 1, & \text{if } n = 1, \end{cases}$$

we find that $T(n)$ is

    ($i$) $1 + 2 \cdots + (n-2) + (n-1) = n(n-1)/2.$

    ($ii$) $1 + 2 + \cdots + (n-1) + n = n(n+1)/2.$

(*iii*)  $1 + 2 + \cdots + n + (n + 1) = (n + 1)(n + 2)/2$.

(*iv*)  $1 + 2 + \cdots + (n - 1) + n + n^2 = n(3n + 1)/2$.

(*c*)  The generating function for the sequence $1, -1, 1, -1, 1, -1, \ldots$ is

(*i*)  $\sum_{k=0}^{\infty} x^k = \frac{1}{1-x}$  for $|x| < 1$.
    (*iii*)  $\sum_{k=0}^{\infty} (-1)^k x^k = \frac{1}{1+x}$  for $|x| < 1$.

(*ii*)  $\sum_{k=0}^{\infty} x^k = \frac{1}{1+x}$  for $|x| < 1$.
    (*iv*)  $\sum_{k=0}^{\infty} (-1)^{k+1} x^k = \frac{1}{1+x}$  for $|x| < 1$.

(*d*)  The height of the recursion tree drawn to give a good asymptotic upper bound on the recurrence relation $T(n) = T(n/5) + T(4n/5) + n$ is

(*i*)  $\log_4 n$.
    (*ii*)  $\log_5 n$.
    (*iii*)  $\log_{5/4} n$.
    (*iv*)  $\log_{4/5} n$.

(*e*)  The height of the recursion tree drawn to give a good lower bound on the recurrence relation $T(n) = T(n/3) + T(2n/3) + n$ is

(*i*)  $\log_{2/3} n$.
    (*ii*)  $\log_2 n$.
    (*iii*)  $\log_{3/2} n$.
    (*iv*)  $\log_3 n$.

**5.2**    Use the guess-and-confirm method to solve the following recurrences.

(*a*)  $T(n) = 2T(n/2) + n$, and $T(1) = 1$.      (*b*)  $T(n) = 3T(n - 1) + 4$, and $T(1) = 1$.

**5.3**    Use the guess-and-confirm method to solve the following recurrence.

$$T(n) = \begin{cases} 1, & \text{if } n = 0; \\ 0, & \text{if } n = 1, 2; \\ 5T(n - 1) - 8T(n - 2) + 4T(n - 3), & \text{if } n \geq 3. \end{cases}$$

**5.4**    Use the iteration method to solve the following recurrences.

(*a*)  $T(n) = 5T(n - 1)$, where $T(0) = 3$.

(*b*)  $T(n) = 2T(n/2) + n \log n$, where $T(1) = c$ for some constant $c$.

**5.5**    Use the generating function method to solve the following recurrence.

$$T(n) = \begin{cases} 1, & \text{if } n = 0; \\ 3, & \text{if } n = 1; \\ -T(n - 1) + 6T(n - 2), & \text{if } n \geq 2. \end{cases}$$

**5.6**    Use the generating function method to solve the Fibonacci recurrence:

$$T(n) = \begin{cases} 0, & \text{if } n = 0; \\ 1, & \text{if } n = 1; \\ T(n-1) + T(n-2), & \text{if } n \geq 2. \end{cases}$$

**5.7**    Use the recursion tree method to solve the recurrence $T(n) = 2T(n-1) + 1$.

**5.8**    Use the recursion tree method to determine a good upper bound on each of the following recurrences.

(a)  $T(n) = T(n/2) + n^2$.                   (b)  $T(n) = T(n-1) + T(n/2) + n$.

**5.9**    Use the recursion tree method to determine a good lower bound on each of the following recurrences.

(a)  $T(n) = 4T((n/2) + 2) + n$.

(b)  $T(n) = T(n/3) + T(2n/3) + cn$, where $c$ is a constant.

# CHAPTER 6

# COUNTING

## Contents

© 2022 by Baha Alzalg | Kindle Direct Publishing, Washington, United States 2022/9
B. Alzalg, *Combinatorial and Algorithmic Mathematics: From Foundation to Optimization*,
DOI 10.5281/zenodo.7110960

In this chapter, we present some important counting principles, permutations and combinations. The first section in this chapter introduces binomial coefficients and identities which will be used throughout the chapter.

# 6.1. Binomial coefficients and identities

In this section, we introduce the binomial series, which is needed before we start with the counting methods. This also includes the essence of a combinatorial proof and some binomial coefficients and identities.

A polynomial is an expression consisting of variables and coefficients, that involves only the operations of addition, subtraction, multiplication, and nonnegative integer exponentiation of variables. For example, $x^3 + 4xyz^2 + 3y^2z$ is a polynomial of three variables. A polynomial of one variable has the form

$$P(x) = a_n x^n + a_{n-1} x^{n-1} + \cdots + a_2 x^2 + a_1 x + a_0$$

where $n$ is a nonnegative integer and the constant numbers $a_0, a_1, a_2, \ldots, a_n$ are the coefficients. A monomial is a polynomial which has only one term. For example, $2x$, $4x^3$ and $5x^2y^4$ are monomials. A binomial is a polynomial that is the sum of two terms, each of which is a monomial. For example, $2x + y$ is a binomial. The function $3x^2 + 4y^3z$ is also a binomial where the constants 3 and 4 are the binomial coefficients. A term can be a product of constants and variables, such as $5x^2y^4z^3$, but that does not concern us in our context here. In this section, we establish some identities that express relationships among the so-called binomial coefficients.

Most proofs that are given in this section are combinatorial proofs. A combinatorial proof of an identity is a proof that uses either two types of mathematical proof:

- A double counting proof: A proof that uses counting arguments to prove that both sides of the identity count the same objects but in different ways.

- A bijective proof: A proof that is based on showing that there is a bijection between the sets of objects counted by the two sides of the identity.

An example of a combinatorial proof is the proof of Theorem 6.1 presented in the following subsection, which is variously known as the binomial theorem.

### The binomial theorem and coefficients

The binomial theorem gives a formula for expanding the binomial power $(x + y)^n$ for any positive integer $n$. The symbol $\binom{n}{k}$ that appears in the following theorem is read as "$n$ choose $k$", where $n$ and $k$ are integers that satisfy $n \geq k \geq 0$. The symbol $\binom{n}{k}$ represents the number of ways to choose $k$ objects from a set of $n$ objects, and is given by the formula[1]

$$\binom{n}{k} = \frac{n!}{k!(n-k)!} = \frac{n(n-1)\cdots(n-k+1)}{k(k-1)\cdots 1}.$$

It is easy to see that

$$\binom{n}{k} = \binom{n}{n-k} \tag{6.1}$$

for any integers $n \geq k \geq 0$. Note also that $\binom{k}{1} = k$ and $\binom{k}{k} = \binom{k}{0} = 1$ for any integer $k \geq 0$. This includes $\binom{1}{1} = \binom{0}{0} = 1$.

The symbol $\binom{n}{k}$ occurs as coefficients in the binomial theorem stated below. These coefficients are commonly called the binomial coefficients.

---

**Theorem 6.1 (The binomial theorem)** *Let $x$ and $y$ be variables and $n$ be a nonnegative integer, then*

$$(x+y)^n = \sum_{k=0}^{n} \binom{n}{k} x^{n-k} y^k = \binom{n}{0} x^n + \binom{n}{1} x^{n-1} y + \cdots + \binom{n}{n-1} xy^{n-1} + \binom{n}{n} y^n.$$

---

**Proof** The terms in the product when it is expanded are of the form $x^{n-k}y^k$ for $k = 0, 1, 2, \ldots, n$. To count the number of terms of the form $x^{n-k}y^k$, note that to obtain such a term it is necessary to choose $n - k$ $x$'s from the $n$ sums (so that the other $k$ terms in the product are $y$'s). Therefore, the coefficient of $x^{n-k}y^k$ is $\binom{n}{n-k}$, which is equal to $\binom{n}{k}$. This proves the theorem. ∎

The series presented in the binomial theorem is known as the binomial series. Some computational uses of the binomial theorem are illustrated in the following examples.

**Example 6.1** Find the expansion of each of the following expressions.

(a) $(x+y)^3$.   (b) $(x+y)^4$.   (c) $(x+y)^5$.

**Solution** (a) From the binomial theorem it follows that

$$\begin{aligned} (x+y)^3 &= \sum_{k=0}^{3} \binom{3}{k} x^{3-k} y^k \\ &= \binom{3}{0} x^3 + \binom{3}{1} x^2 y + \binom{3}{2} xy^2 + \binom{3}{3} y^3 = x^3 + 3x^2 y + 3xy^2 + y^3. \end{aligned}$$

---

[1]The factorial of a positive integer $n$, denoted by $n!$, is defined as $n! = n(n-1)\cdots 2 \cdot 1$. For example, $4! = 4 \cdot 3 \cdot 2 \cdot 1 = 24$. Note that $0!=1$ and $1!=1$.

The same result can be obtained by simple computations as follows.

$$
\begin{aligned}
(x+y)^3 &= (x+y)(x+y)^2 \\
&= (x+y)(x^2+2xy+y^2) = x^3+3x^2y+3xy^2+y^3.
\end{aligned}
$$

(b) From the binomial theorem it follows that

$$
\begin{aligned}
(x+y)^4 &= \sum_{k=0}^{4} \binom{4}{k} x^{4-k} y^k \\
&= \binom{4}{0}x^4 + \binom{4}{1}x^3 y + \binom{4}{2}x^2 y^2 + \binom{4}{3}xy^3 + \binom{4}{4}y^4 \\
&= x^4 + 4x^3 y + 6x^2 y^2 + 4xy^3 + y^4.
\end{aligned}
$$

(c) The proof of this item is left as an exercise for the reader (see Exercise 6.2).

■

■ **Example 6.2** Find the coefficient of $x^7 y^{12}$ in each of the following expressions.

(a) $(x+y)^{19}$.                              (b) $(3x-2y)^{19}$.

Solution   (a) From the binomial theorem it follows that

$$
\binom{19}{12} = \frac{19!}{12!\,7!} = 503,88.
$$

(b) By the binomial theorem, we have

$$
(3x-2y)^{19} = (3x+(-2y))^{19} = \sum_{k=0}^{19} \binom{19}{k}(3x)^{19-k}(-2y)^k.
$$

Consequently, the coefficient of $x^7 y^{12}$ in the expression is obtained when $k=12$, namely,

$$
\binom{19}{12} 3^7 (-2)^{12} = \frac{19!}{12!\,7!} 3^7 2^{12} = 451,373,285,376.
$$

■

## Binomial identities

The binomial theorem can be used to prove some binomial identities. We have the following corollary of the binomial theorem.

---

**Corollary 6.1** *Let n be a nonnegative integer. We have*

(a) $\displaystyle\sum_{k=0}^{n}\binom{n}{k} = 2^{n}.$

(c) $\displaystyle\sum_{k=0}^{n}\binom{n}{2k} = \sum_{k=0}^{n}\binom{n}{2k+1}.$

(b) $\displaystyle\sum_{k=0}^{n}(-1)^{k}\binom{n}{k} = 0.$

(d) $\displaystyle\sum_{k=0}^{n}2^{k}\binom{n}{k} = 3^{n}.$

---

**Proof**  (a) Using the binomial theorem with $x = 1$ and $y = 1$, we have

$$2^{n} = (1+1)^{n} = \sum_{k=0}^{n}\binom{n}{k}1^{k}1^{n-k} = \sum_{k=0}^{n}\binom{n}{k}.$$

(b) Using the binomial theorem with $x = -1$ and $y = 1$, we have

$$0 = 0^{n} = ((-1)+1)^{n} = \sum_{k=0}^{n}\binom{n}{k}(-1)^{k}1^{n-k} = \sum_{k=0}^{n}\binom{n}{k}(-1)^{k}.$$

(c) From item (b), we have

$$\binom{n}{0} - \binom{n}{1} + \binom{n}{2} - \binom{n}{3} + \binom{n}{4} - \binom{n}{5} + \cdots = 0,$$

and hence

$$\binom{n}{0} + \binom{n}{2} + \binom{n}{4} + \cdots = \binom{n}{1} + \binom{n}{3} + \binom{n}{5} + \cdots.$$

This proves item (c).

(d) The proof of this item is left as an exercise for the reader (see Exercise 6.4). ∎

The binomial coefficients satisfy some important recurrences and identities. We first introduce the so-called Pascal's identity or Pascal's recurrence formula, which is defined as $T(n,k) = T(n-1,k-1) + T(n-1,k)$, where $T(n,0) = T(n,n) = 1$. It can be shown that the solution of this recurrence is the binomial coefficient $T(n,k) = \binom{n}{k}$. We have the following theorem.

---

**Theorem 6.2 (Pascal's identity)** *Let n and k be integers with $n \geq k \geq 0$. Then*

$$\binom{n}{k} = \binom{n-1}{k-1} + \binom{n-1}{k}.$$

---

**Proof** Note that

$$\begin{aligned}
\binom{n-1}{k-1}+\binom{n-1}{k} &= \frac{(n-1)!}{(n-k)!\,(k-1)!}+\frac{(n-1)!}{(n-1-k)!\,k!} \\
&= \frac{(n-1)!}{(n-k)!\,(k-1)!}\,\frac{k}{k}+\frac{(n-1)!}{(n-1-k)!\,k!}\,\frac{n-k}{n-k} \\
&= \frac{(n-1)!\,k}{(n-k)!\,k!}+\frac{(n-1)!\,(n-k)}{(n-k)!\,k!} \\
&= \frac{(n-1)!\,(n-k)}{(k+n-k)!\,k!}=\binom{n}{k}.
\end{aligned}$$

This proves the theorem. ■

Pascal's identity is the basis for a geometric arrangement of the binomial coefficients in a triangle, as shown in Figure 6.1.

The $i$th row in the triangle consists of the binomial coefficients

$$\binom{i}{0},\binom{i}{1},\binom{i}{2},\ldots,\binom{i}{i}.$$

This triangle is known as Pascal's triangle which is due to Blaise Pascal (1623-1662). Pascal's identity shows that when two adjacent binomial coefficients in this triangle are added, the binomial coefficient in the next row between these two coefficients is produced.

$$\begin{array}{cccccccccccc}
 & & & & & \binom{0}{0} & & & & & \\
 & & & & \binom{1}{0} & & \binom{1}{1} & & & & \\
 & & & \binom{2}{0} & & \binom{2}{1} & & \binom{2}{2} & & & \\
 & & \binom{3}{0} & & \binom{3}{1} & & \binom{3}{2} & & \binom{3}{3} & & \\
 & \binom{4}{0} & & \binom{4}{1} & & \binom{4}{2} & & \binom{4}{3} & & \binom{4}{4} & \\
\binom{5}{0} & & \binom{5}{1} & & \binom{5}{2} & & \binom{5}{3} & & \binom{5}{4} & & \binom{5}{5} \\
 & \vdots & & \vdots & & \vdots & & \vdots & & \vdots &
\end{array}
\qquad
\begin{array}{ccccccccccc}
 & & & & & 1 & & & & & \\
 & & & & 1 & & 1 & & & & \\
 & & & 1 & & 2 & & 1 & & & \\
 & & 1 & & 3 & & 3 & & 1 & & \\
 & 1 & & 4 & & 6 & & 4 & & 1 & \\
1 & & 5 & & 10 & & 10 & & 5 & & 1 \\
 & \vdots & & \vdots & & \vdots & & \vdots & & \vdots &
\end{array}$$

Figure 6.1: Pascal's triangle.

The identity in the following theorem is due to Alexandre-Theophile Vandermonde (1735-1796).

**Theorem 6.3 (Vandermonde's identity)** *Let $m, n$ and $r$ be nonnegative integers with $r$ not exceeding either $m$ or $n$. Then*

$$\binom{m+n}{r} = \sum_{k=0}^{r} \binom{m}{r-k}\binom{n}{k}.$$

**Proof**  Using the binomial theorem, we have

$$
\begin{aligned}
\sum_{r=0}^{m+n} \binom{m+n}{r} x^r &= (1+x)^{m+n} \\
&= (1+x)^m (1+x)^n \\
&= \left( \sum_{r=0}^{m} \binom{m}{i} x^i \right) \left( \sum_{j=0}^{n} \binom{n}{j} x^j \right) = \sum_{r=0}^{m+n} \left( \sum_{k=0}^{r} \binom{m}{k} \binom{n}{r-k} \right) x^r,
\end{aligned}
$$

where we used (3.6) to obtain the last equality.[2]

By comparing coefficients of $x^r$, the identity follows for all integers $r$ with $0 \le r \le m + n$. For larger integers $r$, both sides of the identity are zero due to the definition of binomial coefficients. The proof is complete. ∎

The following is a corollary of Vandermonde's identity.

**Corollary 6.2**  *For any nonnegative integer $n$, we have*

$$\binom{2n}{n} = \sum_{k=0}^{n} \binom{n}{k}^2.$$

**Proof**  The result follows immediately from Vandermonde's identity with $m = r = n$ and the identity $\binom{n}{k} = \binom{n}{n-k}$. ∎

## 6.2.  Fundamental principles of counting

Counting has applications in a variety of areas, including computer science, probability and statistics. In this section, we present some fundamental counting principles.

---

[2]Note that (3.6) can be used because the binomial coefficients $\binom{m}{i}$ and $\binom{n}{j}$ give zero for all $i > m$ and $j > n$, respectively.

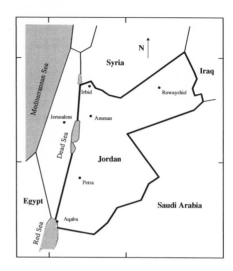

Figure 6.2: The possible routes that one can take to get from Irbid city to Amman city, and those to get from Amman city to Petra city in the country of Jordan. The source LaTeX code of the Jordan map is publicly available at *http://sites.ju.edu.jo/sites/alzalg/pages/jordanmap.aspx*.

### The product principle of counting

We start with the product principle which is stated below.

**Principle 6.1** *Assume there is a procedure that can be broken down into a sequence of two tasks. If there are $n_1$ ways to do the first task and for each of these ways of doing the first task, there are $n_2$ ways to do the second task, then the procedure can be done in $n_1 \times n_2$ ways.*

The following example shows how the product principle is used.

**Example 6.3** In the country of Jordan, there are some amazing restaurants worth to visit in some cities, such as Amman city, Petra city, and Irbid city. Suppose that one can take 3 routes from Irbid to Amman, and can take 5 routes from Amman to Petra. See Figure 6.2. How many possible routes can one take to get from Irbid to Petra if every Irbid-Petra route consists of two routes that are linked in Amman?

**Solution** Every Irbid-Petra route consists of an Irbid-Amman route, which can be done using 3 possible routes, and an Amman-Petra route, which can be done using 5 possible routes. Therefore, by the product principle, there are $3 \times 5 = 15$ possible routes can one take to get from Irbid city to Petra city. ∎

The following theorem is the extended version of Principle 6.1.

---

**Theorem 6.4** *Assume that a procedure can be broken down into a sequence of $m$ tasks, say $T_1, T_2, \ldots, T_m$. If each task $T_i, i = 1, 2, \ldots, m$, can be done in $n_i$ ways, regardless of how the previous task were done, then the procedure can be done in $n_1 \times n_2 \times \cdots \times n_m$ ways.*

---

Theorem 6.4 can be proved using induction on $m$ (see Exercise 6.7). The following examples show how Theorem 6.4 is applied.

**Example 6.4** Each Jordanian national ID card contains an ID number in the back, which consists of three uppercase English letters followed by five digits. What is the number of possible Jordanian national ID cards?

**Solution** There are 26 choices for each of the three uppercase English letters and ten choices for each of the five digits (see the graph shown below). Hence, by Theorem 6.4, the number of possible Jordanian national ID cards is

$$26 \times 26 \times 26 \times 10 \times 10 \times 10 \times 10 \times 10 = 1,757,600,000.$$

26 choices for each letter.       10 choices for each digit.

■

**Example 6.5** Use the product principle of counting to prove that the cardinality of the powerset of a finite set $A$ is equal to $2^n$ if the cardinality of $A$ is $n$.[3]

**Solution** If the set $A$ has $n$ elements, then forming a subset of $A$ amounts to carrying out $n$ independent tasks, where each task is to decide whether to place the element in the subset. Since each task has two outcomes, there are $2^n$ ways this process can be carried out, so $\mathcal{P}(A)$ has $2^n$ elements. This completes the desired proof. ■

**Example 6.6** What is the value of "sum" after the fragment[4] given in Algorithm 6.1, where $n, m, p$ and $q$ are positive integers, has been executed?

---

[3]The reader is asked to prove this fact in Exercise 2.4 by mathematical induction.
[4]"For-statement" written in the fragment is formally introduced in Section 7.1.

---

**Algorithm 6.1:** The algorithm of Example 6.6

---

1: sum = 0
2: **for** $(i = 1; i \le n; i + +)$ **do**
3:     **for** $(j = 0; j < m; j + +)$ **do**
4:         **for** $(k = p; k \ge 1; k - -)$ **do**
5:             **for** $(r = q; r > 0; r - -)$ **do**
6:                 | sum = sum + 1
7:             **end**
8:         **end**
9:     **end**
10: **end**

---

**Solution**   Let $T_s$ be the number of times of traversing the statement (and only the statement) given in line $(s)$ for each $s = 1, 2, \ldots, 6$. The number of times of traversing the statement given in line $(s)$ is the number of ways to do the task $T_s$ for $s = 1, 2, \ldots, 6$. These numbers are listed in Table 6.1.

Also let $P_v$ be the procedure of executing the body of the "for" loop in line $(r)$ for each $r = 2, \ldots, 5$. Let us add a column for the number of ways to carry out the procedure $P_v, v = 2, \ldots, 5$ (see the most-right column in Algorithm 6.2). The initial value of "sum" is zero. Each time the "sum" statement in line (6) is executed, 1 is added to "sum". Therefore, the value of "sum" after the fragment given in Algorithm 6.1 has been executed is equal to the number of times to execute the body of the "for" loop in line (5), which is the number of ways to carry out the procedure $P_5$.

The following points are noted, which lead us to conclude the number of ways to carry out the procedure $P_5$.

- How many times do we execute the body of the "for" statement in line (2)? The answer is from 1 to $n$, which is $n$ times. Note that when $i = n + 1$ is checked, the inequality is false. Hence, the number of ways to carry out the procedure $P_2$ is $n$.

| Line | Statement | Task | # ways to do task $T_s$ |
|------|-----------|------|-------------------------|
| 1 | sum=0 | $T_1$ | 1 |
| 2 | for (i=1; i<=n; i++) | $T_2$ | n |
| 3 | for (j=0; j<m; j++) | $T_3$ | m |
| 4 | for (k=p; k>=1; k-) | $T_4$ | p |
| 5 | for (r=q; r>0; r-) | $T_5$ | q |
| 6 | sum=sum+1 | $T_6$ | 1 |

Table 6.1: Numbers of ways to do the tasks of going over simple and for statements.

---

**Algorithm 6.2:** Algorithm 6.1 revisited

---

```
1:  sum = 0
2:  for (i = 1; i ≤ n; i + +) do          // Procedure P₂ is done in n ways
3:      for (j = 0; j < m; j + +) do       // Procedure P₃ is done in n × m ways
4:          for (k = p; k ≥ 1; k − −) do    // P₄ is done in n × m × p ways
5:              for (r = q; r > 0; r − −) do  // P₅ is done in n × m × p × q ways
6:                  sum = sum + 1
7:              end
8:          end
9:      end
10: end
```

---

- How many times do we execute the body of the "for" statement in line (3)? If the "for" statement in line (3) was not nested, it would execute $m$ times just like the "for" statement in line (2). Since the "for" statement in line (3) is nested, its body is executed $m$ times for *each* time we execute the body of the "for" statement in line (2) (which is $n$ times). In other words, by the product principle of counting, the number of ways to carry out the procedure $P_3$ is equal to the number of ways to do the task $T_2$ multiplied by the number of ways to do the task $T_3$, which is $n \times m$.

- Similarly, since the "for" statement in line (4) is nested, its body is executed $p$ times for *each* time we execute the body of the "for" statement in line (3) (which is $n \times m$ times). In other words, by the product principle of counting, the number of ways to carry out the procedure $P_4$ equals the number of ways to do the task $T_2$ times the number of ways to do the task $T_3$ times the number of ways to do the task $T_4$ which is $n \times m \times p$.

- Again and similarly, since the "for" statement in line (5) is nested, its body is executed $q$ times for *each* time we execute the body of the "for" statement in line (4) (which is $n \times m \times p$ times). In other words, by the product principle of counting, the number of ways to carry out the procedure $P_5$ equals the number of ways to do the task $T_2$ times the number of ways to do the task $T_3$ times the number of ways to do the task $T_4$ times the number of ways to do the task $T_5$, which is $n \times m \times p \times q$.

Thus, the value of "sum" after the fragment given in Algorithm 6.1 has been executed is equal to $n \times m \times p \times q$. ∎

The product principle is also given in terms of sets. We have the following remark.

> **Remark 6.1** *The number of elements in the Cartesian product of finite sets is the product of the number of elements in each set.*

**Proof** Let $A_1, A_2, \ldots, A_m$ be finite sets. Note that the task of choosing an element in the Cartesian product $A_1 \times A_2 \times \cdots \times A_m$ is done by choosing an element in $A_1$, an element in $A_2, \ldots$, and an element in $A_m$. By the product principle, it follows that

$$|A_1 \times A_2 \times \cdots \times A_m| = |A_1| \times |A_2| \times \cdots \times |A_m|.$$

The proof is complete. ∎

### The sum principle of counting

In this part, we present the sum principle of counting.

**Principle 6.2** *If a task can be done either in one of $n_1$ ways or in one of $n_2$ ways, where none of the set of $n_1$ ways is the same as any of the set of $n_2$ ways, then there are $n_1 + n_2$ ways to do the task.*

The following example shows how the sum principle is used.

**Example 6.7** Sara has decided to study a Bachelor's degree in Mathematics at one of the Jordanian universities, either in Irbid city or in Amman city. If Sara decides to go to Irbid, she will study at either Yarmouk University, Jordan University of Science and Technology, Irbid National University, or Jadara University. If Sara decides to go to Amman, she will study at either University of Jordan, Philadelphia University, Petra University, Al-Zaytoonah University, or Al-Ahliyya Amman University. What is the possible number of Jordanian universities Sara could choose?

**Solution** There are 4 possible universities (4 ways) Sara could go to Irbid city, and 5 possible universities (5 ways) Sara could go to in Amman city. Thus, by the sum principle of counting, there are 4+5=9 possible universities Sara could choose to study in Irbid or Amman. ∎

The following theorem is the extended version of Principle 6.2.

**Theorem 6.5** *Assume that a procedure can be done in one of $n_1$ ways, in one of $n_2$ ways, ..., or in one of $n_m$ ways, where none of the set of $n_i$ ways of doing the procedure is the same as any of the set of $n_j$ ways, for all pairs $i$ and $j$ with $1 \leq i < j \leq m$, then the total number of ways to do the procedure is $n_1 + n_2 + \cdots + n_m$ ways.*

Theorem 6.5 can be proved using induction on $m$ (see Exercise 6.8). The following examples show how Theorem 6.5 is applied.

**Example 6.8** Zaid won a scholarship to study an undergraduate degree at one of the following universities in Ohio: Case Western Reserve University (CWRU), Kent State University (KSU), Ohio State University (OSU), and University of Cincinnati (UC). At CWRU, the scholarship requires that Zaid chooses either veterinary medicine, biomedical engineering, or biology. At KSU, it requires that Zaid studies clinical nutrition. At OSU, it requires that Zaid chooses either medicine, dentistry, pharmacy, or public health. At UC, it requires that Zaid chooses either nursing or chemistry. How many possible undergraduate degree programs that Zaid can choose from?

**Solution** Zaid can choose an undergraduate degree program by selecting a program at CWRU, KSU, OSU, or UC. There are 3 ways to choose a program at CWRU, one way to choose a program at KSU, 4 ways to choose a program at OSU, and 2 ways to choose a program at UC. Therefore, by the sum principle of counting, there are $3 + 1 + 4 + 2 = 10$ ways to choose a program.   ∎

**Example 6.9** What is the value of "sum" after the fragment given in Algorithm 6.3, where $n, m, p$ and $q$ are positive integers, has been executed?

**Solution** Let $T_s$ be the task of executing the statement in line $(s)$ for each $s = 1, 2, 3, 5, 6, 8, 9, 11, 12$. Let us add comments showing the number of ways to carry out the task $T_s, s = 1, 2, 3, 5, 6, 8, 9, 11, 12$ (see the comments in gray in Algorithm 6.1). The initial value of "sum" is zero. For each time each "sum" statement in lines (3), (6), (9) and (12) is executed, 1 is added to "sum". Therefore, by the sum principle of counting, the value of "sum" after the fragment given in Algorithm 6.1 has been executed is equal to the number of ways to carry out the task $T_3$, plus the number of ways to carry out the task $T_6$, plus the number of ways to carry out the task $T_9$, plus the number of ways to carry out the task $T_{12}$.

Note that the number of ways to carry out the task $T_3$ equals the number of times to execute the body of the "for" statement in line (2). How many times the "for" statement in line (2) executes? 1 to $n$, which is $n$ times, plus 1 for the last time that $i$ is checked and the inequality is false, that is, $n + 1$ times. Hence, the number of times the statement in line (3) (i.e., body of the "for" statement in line (2)) is $n$. Thus, the number of ways to carry out the task $T_3$ is $n$. Similarly, the tasks $T_6, T_9$ and $T_{12}$ are carried out in $m, p$ and $q$ ways, respectively.

Therefore, by the sum principle of counting, the value of "sum" after the fragment given in Algorithm 6.1 has been executed is equal to $n + m + p + q$.   ∎

---

**Algorithm 6.3:** The algorithm of Example 6.9

---

```
1:  sum = 0                              // Task T₁ is carried out in 1 way
2:  for (i = 1; i ≤ n; i + +) do         // Task T₂ is carried out in n + 1 ways
3:  |   sum = sum + 1                     // Task T₃ is carried out in n ways
4:  end
5:  for (j = 0; j < m; j + +) do         // Task T₅ is carried out in m + 1 ways
6:  |   sum=sum+1                         // Task T₆ is carried out in m ways
7:  end
8:  for (k = p; k ≥ 1; k − −) do         // Task T₈ is carried out in p + 1 ways
9:  |   sum = sum + 1                     // Task T₉ is carried out in p ways
10: end
11: for (r = q; r > 0; r − −) do         // Task T₁₁ is carried out in q + 1 ways
12: |   sum = sum + 1                     // Task T₁₂ is carried out in q ways
13: end
```

---

The sum principle is also given in terms of sets. We have the following remark.

**Remark 6.2**  *The number of elements in the union of pairwise disjoint finite sets is the sum of the number of elements in each set.*

**Proof**  Let $A_1, A_2, \ldots, A_m$ be pairwise disjoint sets (i.e., $A_i \cap A_j = \emptyset$ for all $i, j$). Note that there are $|A_i|$ ways to choose an element from $A_i$ for $i = 1, 2, \ldots, m$. Because the sets are pairwise disjoint, when we select an element from one of the sets $A_i$, we do not also select an element from a different set $A_j$. Consequently, by the sum principle, because we cannot select an element from two of these sets at the same time, the number of ways to choose an element from one of the sets, which is the number of elements in the union, is

$$|A_1 \cup A_2 \cup \cdots \cup A_m| = |A_1| + |A_2| + \cdots + |A_m| \text{ when } A_i \cap A_j = \emptyset \text{ for all } i, j.$$

The proof is complete. ∎

## The subtraction principle of counting

In this part, we present the subtraction principle which is also known as the principle of inclusion-exclusion.

**Principle 6.3**  *If a task can be done either in one of $n_1$ ways or in one of $n_2$ ways, then the total number of ways to do the task is $n_1 + n_2$ minus the number of ways to do the task that are common to the two different ways.*

The subtraction principle is also given in terms of sets. The following remark follows immediately from Principle 6.3.

**Remark 6.3 (Inclusion-exclusion for two sets)**  *The number of elements in the union of finite sets $A_1$ and $A_2$ is the sum of the number of elements in each set, minus the number of ways to select an element that is in both $A_1$ and $A_2$. That is,*

$$|A_1 \cup A_2| = |A_1| + |A_2| - |A_1 \cap A_2|. \tag{6.2}$$

We have the following example.

**Example 6.10**  Let $S$ be a space or universe of two sets $A$ and $B$, where $|S| = 60, |A| = 34, |B| = 22$ and $|A \cap B| = 8$. Find $|(A \cup B)'|$.

**Solution**  By using (6.2), we have

$$|(A \cup B)'| = |S| - |A \cup B| = |S| - (|A| + |B| - |A \cap B|) = 60 - (34 + 22 - 8) = 12.$$

Note that Equation (6.2) can follow directly from Venn diagrams. Also note that the following basic formula is also followed directly from Venn diagrams.

$$|A_1 - A_2| = |A_1| - |A_1 \cap A_2|.$$

The subtraction principle can be extended to find the number of ways to do one of $n$ different tasks or, equivalently, to find the number of elements in the union of $n$ finite sets (see [Ros02] for more detailed presentations).

We have introduced the product, sum, and subtraction principles of counting. There is also a division principle of counting which is out the scope of this book. See [Ros02] for a good presentation of the division principle of counting.

# 6.3. The pigeonhole principle

Suppose that there are 10 pigeons but only 9 pigeonholes, at least one of these 9 pigeonholes must have at least two pigeons in it. To see why this is true, note that if each pigeonhole had at most one pigeon in it, then at most 9 pigeons, one per hole, could be accommodated. See Figure 6.2.

The pigeonhole principle can be stated as follows.

**Principle 6.4** *If there are more pigeons than pigeonholes, then there must be at least one pigeonhole with at least two pigeons in it.*

Table 6.2: Illustrating the pigeonhole principle: There are 10 pigeons but only 9 pigeonholes, at least one of these 9 pigeonholes must have at least two pigeons in it.

More generally, we have the following theorem.

> **Theorem 6.6** *Let k be a positive integer. If k + 1 or more objects are placed into k boxes, then there is at least one box containing two or more of the objects.*

**Proof** Suppose, in the contrary, that none of the $k$ boxes contains more than one object. Then the total number of objects would be at most $k$. This contradicts the fact that there are at least $k + 1$ objects. ∎

The following examples show how Theorem 6.6 is applied.

**Example 6.11** If 20 different algorithms exist to solve 19 different problems, then there is at least one problem that can be solved by two different algorithms. ∎

**Example 6.12** Among any group of 367 people infected by COVID-19 in 2020, there must be at least two cases who were diagnosed on the same day, because there are only 366 possible days in the year. In addition, by the same reasoning, there must be at least two of them with the same birthday. ∎

The following theorem is the extended version of Theorem 6.6.

> **Theorem 6.7** *Let n and k be positive integers. If $k(n - 1) + 1$ or more objects are placed into k boxes, then there is at least one box containing n or more objects.*

**Proof** Suppose, in the contrary, that none of the $k$ boxes contains more than $n - 1$ objects. Then the total number of objects would be at most $k(n - 1)$. This contradicts the fact that there are a total of $k(n - 1) + 1$ objects. ∎

Another version of Theorem 6.7 is the following theorem, which can also be proved using contradiction (see Exercise 6.9).

> **Theorem 6.8** *Let m and k be positive integers. If m objects are placed into k boxes, then there is at least one box containing at least $\lceil m/k \rceil$ objects.[a]*
>
> ---
> [a]For a real number $x$, the ceiling of $x$ is denoted as $\lceil x \rceil$ and is defined as the smallest integer that is not smaller than $x$. For example, $\lceil 5 \rceil = 5$ and $\lceil \sqrt{5} \rceil = 3$.

The following examples show how Theorems 6.8 and 6.7 are applied.

**Example 6.13** According to the Pew Research Center, a millennial is defined as a person born between 1981 and 1996. This means millennials aged 24 to 40 in 2021. Among 145 millennials there are at least 10 who were born in the same year. This can be seen using Theorem 6.7 and noting that $145 = 16(10 - 1) + 1$. ∎

🖥 **Example 6.14** A Nobel Prize-winning scientist was invited to give plenary talks at 25 international conferences which were held in 2019. From Theorem 6.8, among these 25 conferences, there are at least $\lceil 25/12 \rceil = 3$ which were held in the same month.  ∎

The pigeonhole principle is also given in terms of sets. We have the following remark.

> **Remark 6.4** *Let $k$ be a positive integer. If $A$ is a finite set with $k + 1$ or more elements and $A_1, A_2, \ldots, A_k$ are subsets of $A$ that form a partition of $A$, then there exists at least one subset $A_i$, where $i \in \{1, 2, \ldots, k\}$, such that $|A_i| \geq 2$.*

Rather than proving Remark 6.4, we state and prove the following theorem which generalizes Remark 6.4.

> **Theorem 6.9** *Let $r_1, r_2, \ldots, r_k$ be positive integers. If $A$ is a finite set with $(\sum_{i=1}^{k} r_i) - k + 1$ or more elements and $A_1, A_2, \ldots, A_k$ are subsets of $A$ that form a partition of $A$, then we have $|A_i| \geq r_i$ for some $i \in \{1, 2, \ldots, k\}$.*

**Proof** We prove the theorem by contradiction. Suppose that for each $i \in \{1, 2, \ldots, k\}$ we have that $|A_i| \leq r_i - 1$. By the sum principle, we have

$$
\begin{aligned}
|A| &= |A_1| + |A_2| + \cdots + |A_k| \\
&\leq (r_1 - 1) + (r_2 - 1) + \ldots + (r_k - 1) \\
&= (\textstyle\sum_{i=1}^{k} r_i) - k,
\end{aligned}
$$

which contradicts the fact that $A$ has more than $(\sum_{i=1}^{k} r_i) - k$ elements. Thus, $|A_i| \geq 2$ for some $i \in \{1, 2, \ldots, k\}$. The proof is complete.  ∎

Note that Remark 6.4 is a special case of Theorem 6.9 which occurs when $r_1 = r_2 = \cdots = r_k = 2$.

Ramsey Theory refers to the study of partitions of large structures, and generalizes the pigeonhole principle. The Ramsey number $R(m, n)$ gives the solution to the party problem, which asks the minimum number of guests $R(m, n)$ that must be invited so that at least $m$ will know each other or at least $n$ will not know each other. For example, one can show that $R(3, 3) = 6$. By symmetry, it is true that $R(m, n) = R(n, m)$. It is possible to prove some useful properties about Ramsey numbers, but for most part it is difficult to find their exact values. See, for example, [Rad11, LP21, JM07, JM08] and the references contained therein.

## 6.4. Permutations

In order to find the number of possible arrangements of a set of objects, we use a concept called permutations. There are methods for calculating permutations, and it is important to understand the difference between a set with and without repetition.

### Permutations without repetition

Many counting problems can be solved by finding the number of ways to arrange a specified number of distinct elements of a set of a particular size, where the order of these elements matters. For example, in how many ways can we choose three singers from a group of five singers to perform three different songs in a concert, where each singer will perform exactly one song individually? In this section, we develop present rules to solve counting problems such as this.

To answer the question posed in the previous paragraph, note that the order in which we choose the singers matters. There are five ways to choose the first singer for the first song that will be performed at the start of the concert. Once this singer has been chosen, there are four ways to choose the second singer for the second song that will be performed in the middle of the concert. Once the first and second singers have been chosen, there are three ways to choose the third singer who will perform at the end of the concert. By the product rule, there are $5 \cdot 4 \cdot 3 = 60$ ways to choose three singers from a group of five singers to perform three different songs in a concert, where each singer will perform exactly one song individually.

A permutation of a set of distinct objects is an ordered arrangement of some or all of these objects. Formally, we have the following definition.

> **Definition 6.1** Let $n$ and $r$ be integers with $0 \le r \le n$. An ordered arrangement of $r$ elements of a set with $n$ distinct elements is called an r-permutation and is denoted by $P(n, r)$.

The product principle of counting can be used to find a formula for $P(n, r)$.

> **Theorem 6.10** Let $n$ and $r$ be integers with $0 \le r \le n$. Then
> $$P(n, r) = \frac{n!}{(n-r)!} = n(n-1)(n-2)\cdots(n-r+1).$$

**Proof**  We will use the product rule to prove that this formula is correct. The first element of the permutation can be chosen in $n$ ways because there are $n$ elements in the set. There are $n - 1$ ways to choose the second element of the permutation, because there are $n - 1$ elements left in the set after using the element picked for the first position. Similarly, there are $n - 2$ ways to choose the third element, and so on, until there are exactly $n - (r - 1) = n - r + 1$ ways to choose the $r$th element. Consequently, by the product rule, there are

$$n(n-1)(n-2)\cdots(n-r+1) = \frac{n!}{(n-r)!}$$

$r$-permutations of the set. This proves the theorem.  ∎

Let $n$ be a nonnegative integer. Note that $P(n, n) = n!$. Note also that $P(n, 0) = 1$ because there is exactly one way to order zero elements. That is, there is exactly one list with no elements in it, namely the empty set.

we give some examples as direct applications of Theorem 6.10.

**Example 6.15** In how many ways can we arrange all singers of a group of five singers to perform different songs in a concert, where each singer will perform exactly one song individually?

**Solution** By Theorem 6.10, there are

$$P(5,5) = 5! = 5 \cdot 4 \cdot 3 \cdot 2 \cdot 1 = 120$$

ways to select five singers from a group of five singers to perform different songs in a concert, where each singer will perform exactly one song individually. ∎

**Example 6.16** In how many ways can we form four distinct letter passwords from the letters:

(*a*) A, B, C, D?  (*b*) A, B, C, D, E, F?

**Solution** (*a*) By Theorem 6.10, there are $P(4,4) = 4 \cdot 3 \cdot 2 \cdot 1 = 24$ ways to form four distinct letter passwords from the letters $A, B, C, D$.

(*b*) By Theorem 6.10, there are $P(6,4) = 6 \cdot 5 \cdot 4 \cdot 3 = 360$ ways to form four distinct letter passwords from the letters $A, B, C, D, E, F$.

∎

## Permutations with repetition

In many counting problems, elements may be used repeatedly. For instance, a letter may be used more than once on a password. Permutations when repetition of elements is allowed can be easily counted by using the product principle of counting. We have the following theorem.

> **Theorem 6.11** *The number of r-permutation of a set of n objects with repetition allowed is $n^r$.*

**Proof** There are $n$ ways to select an element of the set for each of the $r$ positions in the $r$-permutation when repetition is allowed, because for each choice all $n$ objects are available. Hence, by the product rule there are $n^r$ $r$-permutation when repetition is allowed. This proves the theorem. ∎

The following examples are direct applications of Theorems 6.10 and 6.11.

**Example 6.17**

(*a*) In how many ways can ten boys have ten different birthdays?

(*b*) In how many ways can ten boys have ten birthdays?

Ignore the existence of the leap year, so a year has 365 days.

**Solution** (*a*) By Theorem 6.10, there are

$$P(365, 10) = 365 \cdot 364 \cdot 363 \cdot 362 \cdot 361 \cdot 360 \cdot 359 \cdot 358 \cdot 357 \cdot 356$$

ways ten boys have ten different birthdays.

(*b*) By Theorem 6.11, there are $365^{10}$ ways ten boys have ten birthdays.

∎

**Example 6.18** Suppose that there are three daily round-trip bus routes between Columbus and Cleveland, five daily round-trip bus routes between Cleveland and Buffalo, and four daily round-trip bus routes between Buffalo and Toronto. See Figure 6.3. A passenger would like to take a round trip that departs from (and returns to) Columbus, and runs as follows: Columbus - Cleveland - Buffalo - Toronto - Buffalo - Cleveland - Columbus. In how many ways can the passenger take such a round trip in each of the following two cases:

(*a*) The passenger cannot use the same bus more than once.

(*b*) The passenger may use the same bus more than once.

**Solution** (*a*) If the passenger cannot use the same bus more than once, then there are

$$3 \cdot 5 \cdot 4 \cdot 3 \cdot 4 \cdot 2 = 1440$$

ways to take the planed round trip. This is a direct consequence of Theorem 6.10 since $1440 = 3 \cdot (3 - 1) \cdot 5 \cdot (5 - 1) \cdot 4 \cdot (4 - 1)$.

(*b*) If the passenger may use the same bus more than once, then there are

$$3 \cdot 5 \cdot 4 \cdot 4 \cdot 5 \cdot 3 = 3600$$

ways to take the planed round trip. This is a direct consequence of Theorem 6.11 since $3600 = 3^2 \cdot 5^2 \cdot 4^2$.

∎

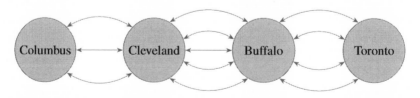

Figure 6.3: Daily round-trips between Columbus and Toronto, with connections in Cleveland and Buffalo.

# 6.5. Combinations

In order to find the number of ways to select a particular number of elements from a set of a particular size, we use a concept called combinations. There are methods for calculating combinations, and it is important to understand the difference between a set with and without repetition.

## Combinations without repetition

Many counting problems can be solved by finding the number of ways to select a particular number of elements from a set of a particular size, where the order of the elements selected does not matter. For example, how many different bands of three singers can be formed from a group of four singers? In this section, we develop present rules to solve counting problems such as this.

To answer the question posed in the previous paragraph, note that the order in which we choose the singers does not matter. We need to find the number of subsets with three elements from the set containing the four singers. We see that there are four such subsets, one for each of the four singers, because choosing three singers is the same as choosing one of the four singers to leave out the group. This means that there are four ways to choose three singers for the band, where the order in which these singers are chosen does not matter.

A combination of a set of distinct objects is an unordered arrangement of some or all of these objects. Formally, we have the following definition.

> **Definition 6.2** *Let $n$ and $r$ be integers with $0 \leq r \leq n$. An unordered selection of $r$ elements from a set with $n$ distinct elements is called an $r$-combination and is denoted by $C(n, r)$.*

The product principle of counting can be used to find a formula for $C(n, r)$.

> **Theorem 6.12** *Let $n$ and $r$ be integers with $0 \leq r \leq n$. Then*
>
> $$C(n, r) = \binom{n}{r} = \frac{n!}{(n-r)!\, r!}.$$

**Proof** The $P(n, r)$ $r$-permutations of the set can be obtained by forming the $C(n, r)$ $r$-combinations of the set, and then ordering the elements in each $r$-combination, which can be done in $P(r, r)$ ways. Consequently, by the product rule, $P(n, r) = C(n, r) \cdot P(r, r)$. This implies that

$$C(n, r) = \frac{P(n, r)}{P(r, r)} = \frac{n!/(n-r)!}{r!/(r-r)!} = \frac{n!}{(n-r)!\, r!}.$$

This proves the theorem.    ■

Returning to the question posed at the beginning of this section, we find that we cam form $C(4,3) = \binom{4}{3} = 4$ different bands of three singers from a group of four singers. As we mentioned earlier, choosing three singers is the same as choosing one of the four singers to leave out the group. That is, $C(4,3) = C(4,1) = \binom{4}{1} = 4$. More generally, we have the following corollary.

---

**Corollary 6.3** *Let n and r be integers with $0 \le r \le n$. Then $C(n,r) = C(n,n-r)$.*

---

**Proof** The result follows immediately from Theorem 6.12 and the identity given in (6.1).                                                                            ∎

The following examples are direct applications of Theorem 6.12.

**Example 6.19** How many pairs can be chosen from a group of six people?

**Solution** By Theorem 6.12, we can choose $C(6,2) = \binom{6}{2} = 15$ pairs from a group of six people.                                                            ∎

**Example 6.20** Suppose that a class consists of six boys and four girls.

(*a*) How many ways are there to choose a group of five students?

(*b*) How many ways are there to choose a group of five students if the group consists three boys and two girls?

**Solution**   (*a*) By Theorem 6.12, the number of ways to select the group is

$$C(10,5) = \binom{10}{5} = 252.$$

(*b*) By the product principle of counting, the answer is the product of the number of 3-combinations of boys and the 2-combinations of girls. By Theorem 6.12, the number of ways to select the group is

$$C(6,3) \cdot C(4,2) = \binom{6}{3}\binom{4}{2} = (20)(6) = 120.$$

                                                                            ∎

## Combinations with repetition

Let us say there are five flavors of ice cream: banana (B), chocolate (C), lemon (L), strawberry (S), and vanilla (V). We can have three scoops. Example selections include:

- {C, C, C}, which means that three scoops of chocolate were selected.

Figure 6.4: Five different kinds of ice cream in five boxes (containers).

- {B, L, V}, which means that one each of banana, lemon and vanilla was selected.

- {B, V, V}, which means that one of banana, two of vanilla were selected.

We are interested in how many variations there will be? This is an example of a combination with repetition of elements allowed. Note that there are five things to choose from, and we choose three of them. Also note that the order in which the scoops are selected does not matter, and we can repeat! Therefore, this example involves counting 3-combinations with repetition allowed from a 5-element set. Below we show a technique for solving this counting problem, which leads us to a general method for counting the $r$-combinations with repetition allowed from an $n$-element set.

Let us think about the ice cream problem posed above as ice cream being in boxes (see Figure 6.4) and a robot being ordered remotely to get the desired selections. For instance:

- To select {C, C, C}, we send the orders "move past the first box, then take three scoops, then move along three more boxes to the end". In other words, we send the orders "move once, then take three scoops, then move thrice to the right".

- To select {B, L, V}, we send the orders "take a scoop, then move twice to the right, then take a scoop, then move twice to the right, then take a scoop".

- To select {B, V, V}, we send the orders "take a scoop, then move quarce to the right, then take two scoops".

Let us write "→" to mean that "move once to the right", and "○" to mean that "take one scoop". The above three selections can be written as follows:

- The selection {C, C, C} is represented as "→ ○○○ → → →".

- The selection {B, L, V} is represented as "○ → → ○ → → ○".

- The selection {B, V, V} is represented as "○ → → → → ○ ○".

From the above discussion, the above counting question can be simplified as follows: How many different ways can we arrange arrows and circles?

Note that there are always 3 circles we always have 3 scoops of ice cream, and there are always 4 arrows because we need to move 4 times to go from the first container to the fifth container. Therefore, there are 7 positions, and we want to choose 3 of them to have circles. That is, we have $3 + (5-1)$ positions and want to choose 3 of them. So, the number of ways of having 3 scoops from 5 flavors of ice cream is

$$\binom{3 + 5 - 1}{3} = \binom{7}{3} = \frac{7 \cdot 6 \cdot 5}{1 \cdot 2 \cdot 3} = 35.$$

We can also look at the arrows instead of the circles, and say that we have $3 + (5-1)$ positions and want to choose $(5-1)$ of them to have arrows. The following theorem generalizes this discussion.

---

**Theorem 6.13** *There are*

$$C(n + r - 1, n - 1) = C(n + r - 1, r) = \binom{r + n - 1}{r}$$

*$r$-combinations from a set with $n$-elements when repetition of elements is allowed.*

---

**Proof** Each $r$-combination of a set with $n$ elements when repetition is allowed can be represented by $n - 1$ right arrows and $r$ circles. The $n - 1$ right arrows are used to mark off $n$ different cells, with the $i$th cell containing a circle for each time the $i$th element of the set occurs in the combination. For instance, a 5-combination of a set with four elements is represented with three arrows and five circles. Here

$$\bigcirc\, \bigcirc \rightarrow\, \rightarrow \bigcirc\, \rightarrow\, \bigcirc\, \bigcirc$$

represents the combination containing exactly two of the first element, none of the second element, one of the third element, and two of the fourth element of the set.

As we have seen, each different list containing $n - 1$ arrows and $r$ circles corresponds to an $r$-combination of the set with $n$ elements, when repetition is allowed. The number of such lists is $C(n - 1 + r, r)$, because each list corresponds to a choice of the $r$ positions to place the $r$ circles from the $n - 1 + r$ positions that contain $r$ circles and $n - 1$ arrows. The number of such lists also equals to $C(n - 1 + r, n - 1)$, because each list corresponds to a choice of the $n - 1$ positions to place the $n - 1$ arrows. The proof is complete. ∎

The following examples show how Theorem 6.13 is applied.

**Example 6.21** Suppose there are four varieties of donuts: Chocolate (C), Glazed (G), Pumpkin (P), and Raspberry (R).

(a) Find the number of ways one can select 30 donuts.

(b) Find the number of ways one can select 30 donuts such that the selection includes at least 2 C donuts, 2 G donuts, 3 P donuts, and 4 R donuts.

**Solution** Note that the order in which the donuts can be selected does not matter, and the donuts can be repeated. Total number of varieties of donuts is $n = 4$. The number of donuts to be selected is $r = 30$.

(a) This count would be the number of 30-combinations with repetition allowed from a set with six elements. From Theorem 6.13, the donuts can be selected in the following different ways:

$$C(4 + 30 - 1, 30) = C(33, 30) = \frac{33 \cdot 32 \cdot 31}{1 \cdot 2 \cdot 3} = 5456.$$

(*b*) To satisfy the requirements, we preselect the minimum number of each type: 2 C, 2 G, 3 P, 4 R. We then have at most $30 - 2 - 2 - 3 - 4 = 19$ donuts left to choose, now without any restrictions. This count would be

$$C(4 + 19 - 1, 19) = C(22, 19) = \frac{22 \cdot 21 \cdot 20}{1 \cdot 2 \cdot 3} = 1540.$$

**Example 6.22** How many solutions does the equation $x_1 + x_2 + x_3 + x_4 = 13$ have, where $x_1, x_2, x_3$, and $x_4$ are nonnegative integers?

**Solution** To count the number of solutions, we note that a solution corresponds to a way of selecting 13 items from a set with four elements so that $x_1$ items of type one, $x_2$ items of type two, $x_3$ items of type three, and $x_4$ items of type four are chosen. Here, the number of solutions is equal to the number of 13-combinations with repetition allowed from a set with four elements. From Theorem 6.13, it follows that there are

$$C(4 + 13 - 1, 13) = C(16, 13) = C(16, 3) = \frac{16 \cdot 15 \cdot 14}{1 \cdot 2 \cdot 3} = 560$$

solutions.

**Example 6.23** What is the value of "sum" after the fragment given in Algorithm 6.4, where $n, m, p$ and $q$ are positive integers, has been executed?

**Solution** Note that the initial value of "sum" is 0 and that 1 is added to "sum" each time the nested loop is traversed with a sequence of four letters $i, j, k$, and $r$ such that

$$1 \le m \le k \le j \le i \le n.$$

The number of such sequences of integers is the number of ways to choose 4 integers from $\{1, 2, \ldots, n\}$, with repetition allowed.

---

**Algorithm 6.4:** The algorithm of Example 6.23

---

```
1:  sum = 0
2:  for (i = 1; i ≤ n; i + +) do
3:      for (j = 1; j ≤ i; j + +) do
4:          for (k = 1; k ≤ j; k + +) do
5:              for (m = 1; m ≤ k; m + +) do
6:                  sum = sum + 1
7:              end
8:          end
9:      end
10: end
```

Hence, from Theorem 6.13, it follows that

$$\text{sum} = C(n+4-1,4) = \binom{n+3}{4} = \frac{1}{24}(n^4 + 6n^3 + 11n^2 + 6n)$$

after this code has been executed.  ∎

The formulas for the numbers of ordered and unordered selections of $r$ elements, chosen with and within repetition allowed from a set with $n$ elements, are shown in Table 6.3.

***Using combinations to find permutations with indistinguishable objects*** Some elements may be indistinguishable in counting problems. When this is the case, care must be taken to avoid counting things more than once. In particular, we can use combinations to find the number of different permutations with indistinguishable objects. We have the following theorem.

> **Theorem 6.14** *The number of different permutations of n objects, where there are $n_1$ indistinguishable objects of type 1, $n_2$ indistinguishable objects of type 2, ..., $n_k$ indistinguishable objects of type k, is*
>
> $$\frac{n!}{n_1! \, n_2! \, \cdots \, n_k!}.$$

**Proof** To determine the number of permutations, first note that the $n_1$ objects of type 1 can be placed among the $n$ positions in $C(n,n_1)$ ways, leaving $n - n_1$ positions free. Then the objects of type 2 can be placed in $C(n - n_1, n_2)$ ways, leaving $n - n_1 - n_2$ positions free. Continue placing the objects of type 3, ..., type $k - 1$, until at the last stage, $n_k$ objects of type $k$ can be placed in $C(n - n_1 - n_2 - \cdots - n_{k-1}, n_k)$ ways. Hence, by the product principle of counting, the total number of different permutations is

$$C(n,n_1)C(n-n_1,n_2)\cdots C(n-n_1-n_2-\cdots-n_{k-1},n_k)$$
$$= \frac{n!}{n_1!\,(n-n_1)!} \frac{(n-n_1)!}{n_2!\,(n-n_1-n_2)!} \cdots \frac{(n-n_1-\cdots n_{k-1})!}{n_k!\,0!} = \frac{n!}{n_1!\,n_2!\cdots n_k!}.$$

The proof is complete.  ∎

|  | Without repetition | With repetition |
|---|---|---|
| $r$-permutations | $\dfrac{n!}{(n-r)!\,r!}$ | $n^r$ |
| $r$-combinations | $\dfrac{n!}{(n-r)!\,r!}$ | $\dfrac{(n+r-1)!}{(n-1)!\,r!}$ |

Table 6.3: Permutations and combinations with and without repetition.

The following example shows how Theorem 6.14 is applied.

**Example 6.24** Determine the number of ways to arrange 6 letters of the word BANANA.

**Solution**  We count permutations of 3 A's, 2 N's, and 1 B, a total of 6 symbols. By Theorem 6.14, the number of these is $\frac{6!}{3!\,2!\,1!} = 60$. ∎

## Distributing objects into distinguishable boxes

Many counting problems can be solved by enumerating the ways objects, distinguishable (i.e., different) or indistinguishable (i.e., identical), can be placed into boxes that are distinguishable (often called labeled) or indistinguishable (often called unlabeled).

There are closed formulas for counting the ways to distribute objects, distinguishable or indistinguishable, into distinguishable boxes, but there are no closed formulas for counting the ways to distribute objects, distinguishable or indistinguishable, into indistinguishable boxes.

***Distributing distinguishable objects into distinguishable boxes***  Counting problems that involve distributing distinguishable objects into distinguishable boxes can be solved using the following theorem.

---

**Theorem 6.15** *The number of ways to distribute n distinguishable objects into k distinguishable boxes so that $n_i$ objects are placed into box i, $i = 1, 2, \ldots, k$, equals*

$$\frac{n!}{n_1!\,n_2! \cdots n_k!}.$$

---

The proof of Theorem 6.15 uses the product principle, similar to the proof of Theorem 6.14, and it is therefore omitted. It can also be proved by setting up a one-to-one correspondence between the permutations counted in Theorem 6.14 and the ways to distribute distinguishable objects counted by Theorem 6.15.

**Example 6.25** How many ways are there to distribute hands of 5 cards to each of four players from a card game of 52 cards[5]?

**Solution**  This counts the number of ways to distribute 52 distinguishable objects into 5 distinguishable boxes so that 5 objects are placed into the first box, 5 objects are placed into the second box, 5 objects are placed into the third box, 5 objects are placed into the fourth box, and 32 objects (which represent $52 - 4 \times 5$ remaining cards) are placed into the fifth box. By Theorem 6.15, these are $52!/(5!\,5!\,5!\,5!\,32!)$ ways. ∎

---

[5]The standard deck consists of 52 cards.

*Distributing indistinguishable objects into distinguishable boxes* Counting problems that involve distributing indistinguishable objects into distinguishable boxes can be solved using the following theorem.

---

**Theorem 6.16** *There are*

$$C(n + r - 1, n - 1) = C(n + r - 1, r) = \binom{n + r - 1}{r}$$

*ways to place r indistinguishable objects into n distinguishable boxes.*

---

Theorem 6.16 is similar to the proof Theorem 6.13 and it is therefore omitted. It can also be proved by setting up a one-to-one correspondence between the combinations counted in Theorem 6.13 and the ways to place distinguishable objects counted by Theorem 6.16.

**Example 6.26** How many ways to place 12 indistinguishable homemade doughnuts into 9 distinguishable plates?

**Solution** This counts the number of ways to place 12 indistinguishable objects into 9 distinguishable boxes. By Theorem 6.16, these are

$$C(9 + 12 - 1, 9 - 1) = C(20, 8) = \binom{20}{8} = 125970$$

ways.

# EXERCISES

**6.1** Choose the correct answer for each of the following multiple-choice questions/items.

(*a*) The coefficient of $x^{13}y^{12}$ in the expression $(2x - 3y)^{25}$ is

(*i*) $-\dfrac{25!}{12!\,13!}2^{12}3^{13}$.

(*iii*) $\dfrac{25!}{12!\,13!}2^{12}3^{13}$.

(*ii*) $-\dfrac{25!}{13!\,12!}2^{13}3^{12}$.

(*iv*) $\dfrac{25!}{13!\,12!}2^{13}3^{12}$.

(*b*) There are cats in three rooms. The first room has 4 cats, the second room has 5 cats, and the third room has 3 cats. In how many ways to choose a cat from these three rooms.

    (*i*) 12.           (*ii*) 24.           (*iii*) 48.           (*iv*) 60.

(*c*) Let $A$ and $B$ be two disjoint sets. If $|A| = 3|B| = 6$, then the cardinality of the powerset of $A \times B$ is

    (*i*) 8.           (*ii*) 12.           (*iii*) 256.           (*iv*) 4096.

(*d*) In an exam, there are 7 true/false questions and 8 multiple-choice questions for which the answers can be (*i*), (*ii*), (*iii*), (*iv*). The number of different ways of answering the exam are:

    (*i*) $2^7 \times 4^7$.     (*ii*) $2^7 \times 4^8$.     (*iii*) $2^8 \times 4^7$.     (*iv*) $2^8 \times 4^8$.

(*e*) How many even 4 digit whole numbers are there?

    (*i*) 256.           (*ii*) 625.           (*iii*) 4500.           (*iv*) 5000.

(*f*) A professor gives a multiple-choice quiz that has eight questions, each with four possible answers (*i*), (*ii*), (*iii*), (*iv*). What is the minimum number of students that must be in the professor's class in order to guarantee that at least four answer sheets must be identical? (Assume that no answers are left blank.)

    (*i*) $4^8$.           (*ii*) $4^8 + 1$.     (*iii*) $2 \times 4^8 + 1$.     (*iv*) $3 \times 4^8 + 1$.

(*g*) How many ways can we assign four problems to four students to solve them so that each student solves one problem?

    (*i*) 12.           (*ii*) 16.           (*iii*) 24.           (*iv*) 48.

(*h*) How many numbers of two digits can be formed with digits 1, 3, 5, 7 and 9?

    (*i*) 60.           (*ii*) 120.           (*iii*) 180.           (*iv*) 240.

(*i*) What is the number of diagonals can be drawn in an octagon?

    (*i*) 20.           (*ii*) 28.           (*iii*) 34.           (*iv*) 42.

(*j*) Let $A, B$ and $C$ be three disjoint sets. If $|A| = |B| + 1 = |C| + 2 = 4$, then the cordiality of $A \cup B \cup C$ is

    (*i*) 9.               (*ii*) 24.               (*iii*) 512.               (*iv*) 4096.

(*k*) The number of subsets of a 99-element set is:

    (*i*) $2^{99}$.         (*ii*) $\binom{99}{2}$.         (*iii*) $9^{99}$.         (*iv*) $\binom{99}{9}$.

(*l*) The number of 9-element subsets of a 99-element set is:

    (*i*) $2^{99}$.         (*ii*) $\binom{99}{2}$.         (*iii*) $9^{99}$.         (*iv*) $\binom{99}{9}$.

(*m*) The number of 7-letter upper-case words is:

    (*i*) $26^{7}$.         (*ii*) $25^{7}$.         (*iii*) $26^{7} - 25^{7}$.     (*iv*) $\binom{26}{7}$.

(*n*) The number of 7-letter upper-case words that do not contain the letter B is:

    (*i*) $26^{7}$.         (*ii*) $25^{7}$.         (*iii*) $26^{7} - 25^{7}$.     (*iv*) $\binom{26}{7}$.

(*o*) The number of 7-letter upper-case words that contain the letter B is:

    (*i*) $26^{7}$.         (*ii*) $25^{7}$.         (*iii*) $26^{7} - 25^{7}$.     (*iv*) $\binom{26}{7}$.

(*p*) The number of 7-letter upper-case words whose letters are distinct and occur in alphabetically increasing order is[6]:

    (*i*) $26^{7}$.         (*ii*) $25^{7}$.         (*iii*) $26^{7} - 25^{7}$.     (*iv*) $\binom{26}{7}$.

**6.2**    Use the binomial theorem to find the expansion of $(x + y)^{5}$.

**6.3**    Give a combinatorial proof for item (*a*) of Corollary 6.1.

**6.4**    Prove item (*d*) of Corollary 6.1.

**6.5**    Give a combinatorial proof for Pascal's identity (Theorem 6.2).

**6.6**    Give a combinatorial proof for Vandermonde's identity (Theorem 6.3).

**6.7**    Use mathematical induction to prove Theorem 6.4.

---

[6]E.g., ABEL is counted, but not ABLE (since L and E are not in alphabetical order), nor APPEL (since P is repeated).

---
**Algorithm 6.5:** The algorithm of Exercise 6.12
---
1: sum = 1
2: **for** $(i = 1; i \leq n; i + +)$ **do**
3:     **for** $(j = 0; j < m; j + +)$ **do**
4:         sum = sum + 1
5:         **for** $(k = p; k \geq 1; k - -)$ **do**
6:             sum = sum + 1
7:         **end**
8:         **for** $(r = q; r > 0; r - -)$ **do**
9:             sum = sum + 1
10:         **end**
11:     **end**
12: **end**
---

**6.8** Use mathematical induction to prove Theorem 6.5.

**6.9** Use contradiction to prove Theorem 6.8

**6.10** Let $A, B$ and $C$ be three sets such that $|B \cap C| = 19$ and $|A \cap B \cap C| = 11$. Find $|A' \cap B \cap C|$.

**6.11** A survey of 240 people showed that 91 like tea, 70 like coffee, 31 like tea and coffee, 91 like neither coffee nor tea, and in addition do not like milk, and 7 like coffee, tea, and milk. How many like milk only? Justify your answer. (Hint: Use a Venn diagram).

**6.12** What is the value of "sum" after the fragment given in Algorithm 6.5, where n, m, p and q are positive integers, has been executed?

**6.13** Prove that a function $f$ from a set with $k + 1$ or more elements to a set with $k$ elements is not an injection.

**6.14** Zaid has six different colored shirts. In how many ways can he hang the four shirts in the cupboard?

**6.15** Determine the number of ways to arrange:

(*a*) The 7 letters of the word ONEWORD. (Note that this word has 2 O's, 1 N, 1 E, 1 W, 1 R, and 1 D).

(*b*) The word ONEWORD if the 2 O's must be adjacent (as in NEWDOOR).

**6.16** Determine the number of ways to arrange:

(*a*) The 9 letters of the word PINEAPPLE. (Note that this word has 3 P's, 2 E's, 1 I, 1 N, 1 A, and 1 L).

(*b*) The 9 letters of the word PINEAPPLE if the 3 P's must be adjacent (as in APPPLENIE).

**Part III**

# ALGORITHMS

# CHAPTER 7

---

# ANALYSIS OF ALGORITHMS

## Contents

© 2022 by Baha Alzalg | Kindle Direct Publishing, Washington, United States 2022/9
B. Alzalg, *Combinatorial and Algorithmic Mathematics: From Foundation to Optimization*,
DOI 10.5281/zenodo.7110969

An algorithm is a finite set of precise instructions that are performed to solve a problem. So, every algorithm is constructed using a finite sequence of statements. The origin of the word algorithm goes back to Muhammad Ibn Musa Al-Khwarizmi (ca. 780–840), whose book on the Indian number system and reckoning methods (addition, subtraction, multiplication, division, working with fractions, computing roots) revolutionized occidental mathematics (see [KVL63]).

In the first half of this chapter, we present the fundamentals of runtime analysis, and study the asymptotic analysis of algorithms. In the second half of this chapter, we analyze sequential programs, which are programs without function calls. We also analyze recursive and nonrecursive programs, which are programs with function calls. Finally, we briefly introduce the complexity classes NP and NP-complete.

In mathematics, asymptotic analysis is a method of describing limiting behavior. For instance, the study of the properties of a function $f(x)$ as $x$ becomes very large. As an example, the graph of the function $f(x) = \frac{x}{x-2}$ approaches 1 as $x$ goes to infinity. See Figure 7.1.

In fact,

$$\lim_{x \to \infty} \frac{x}{x - 2} = \lim_{x \to \infty} \frac{x}{x(1 - \frac{2}{x})} = \lim_{x \to \infty} \frac{1}{1 - \frac{2}{x}} = 1.$$

Therefore, mathematically speaking, we call the line $y = 1$ a horizontal asymptotic for $f(x)$.

In computer science, asymptotic analysis is the evaluation of the performance of an algorithm in terms of the input size $(n)$, where $n$ is very large. What is an algorithm? How can we construct an algorithm for performing certain tasks? And how can we compare between various algorithms for solving a certain problem? This is the substance of the first section.

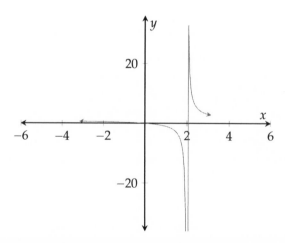

Figure 7.1: The graph of $f(x) = \frac{x}{x-2}$.

# 7.1. Constructing and comparing algorithms

In the first part of this section, we present some decision making statements that are used to build blocks of algorithms, and introduce the running time of an algorithm.

## Basic tools for constructing algorithms

Decision making statements can be placed into the following categories:

***Simple statements:***   The following are simple statements in the programming language C.

- Expressions: This includes

  - `printf;` (write statement – print formatted data).
  - `scanf;` (read statement – read formatted data).
  - assignment statements (such as "set a=1;").

- Jump statements: This includes `goto; break; continue; return;` etc.

- The null statement: This contains only a semicolon ";". It is known that nothing happens when a null statement is executed.

We note that, in the programming languages such as C, simple statements end in a semicolon.

***If-statement:***   The "if-statement" and "if-else-statement" are also-called selection statements. In Algorithm 7.1, we present a simple code for "if-statement". Here statement(s) is (are) executed only if condition is true. See Figure 7.8 (a).

In Algorithm 7.2, we present a simple code for "if-else-statement", where $S_1$ denotes a set of statement(s), and $S_2$ denotes another set of statement(s). Here, if condition is true, then $S_1$ is executed. Otherwise, $S_2$ is executed. See Figure 7.8 (b).

---

**Algorithm 7.1:** Writing an if-statement

1: **if** some condition is true **then**
2: | do some statement(s) $S$
3: **end**

---

**Algorithm 7.2:** Writing an if-else-statement

1: **if** some condition is true **then**
2: | do some statement(s) $S_1$
3: **end**
4: **else**
5: | do some different statements $S_2$
6: **end**

---

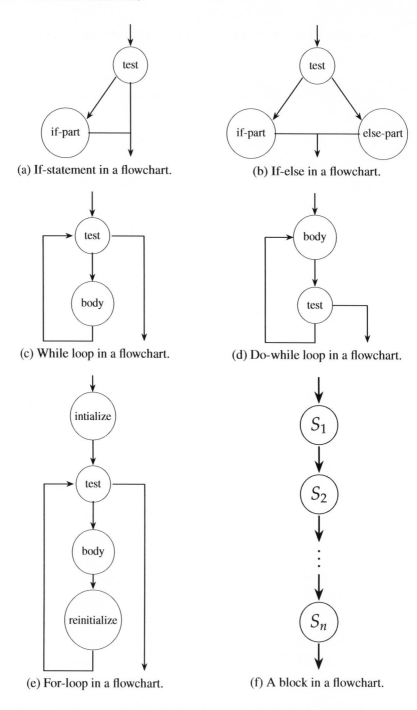

(a) If-statement in a flowchart.

(b) If-else in a flowchart.

(c) While loop in a flowchart.

(d) Do-while loop in a flowchart.

(e) For-loop in a flowchart.

(f) A block in a flowchart.

Figure 7.8: Flowcharts showing basic statements.

📱 **Example 7.1  H** In Algorithm 7.3, we present a code for finding the largest of two integers. ▪

---

**Algorithm 7.3:** Finding the largest of two integers

---

    **Input:** Integers $a, b$
    **Output:** Bigger value of $a, b$
1: **if** $(a > b)$ **then**
2:   |  printf("bigger value = "$a$)
3: **end**
4: **else**
5:   |  printf("bigger value = "$b$)
6: **end**

---

**For-statement:**  Algorithm 7.4 shows a simple code that represents "for-statement". See also Figure 7.4.

---

**Algorithm 7.4:** Writing a for-statement

---

1: **for** (initialize; test; in/decrement) **do**
2:   |  do some statement(s)
3: **end**

---

📱 **Example 7.2** In Algorithm 7.5, we present a code for finding the maximum of integers. ▪

---

**Algorithm 7.5:** Finding the maximum of integers

---

    **Input:** Integers $a_1, a_2, \ldots, a_n$
    **Output:** Largest value of $a_1, a_2, \ldots, a_n$
1: max $= a_1$
2: **for** $(i = 0; i < n; i++)$ **do**
3:   |  **if** $(\text{max} < a_i)$ **then**
4:   |   |  max $= a_i$
5:   |  **end**
6: **end**
7: printf("largest value = "max)

---

**While-statement:**  In Algorithm 7.6, we present a simple code for "while statement". Here the body (statement(s)) is executed as long as condition is true. See Figure 7.8 (c).

---

**Algorithm 7.6:** Writing a while-statement

---

1: **while** some condition is true **do**
2:   |   do some statement(s)
3: **end**

---

**Example 7.3** In Algorithm 7.7, we present a while code that prints the numbers $1, 2, \ldots, 5$.    ■

---

**Algorithm 7.7:** Using a while-statement to print numbers 1 to 5

---

**Input:** Integer $i$
**Output:** Print numbers from 1 to 5
1: int $i = 1$
2: **while** $(i \leq 5)$ **do**
3:   |   printf$(i)$                 // Output is 1 2 3 4 5
4:   |   $++i$
5: **end**
6: **return** 0

---

***Do-while-statement:*** In Algorithm 7.8, we present a simple code for "do-while statement". Note that the do-while is similar to the while-statement except that the body (statement(s)) is executed at least once.

---

**Algorithm 7.8:** Writing a do-while-statement

---

1: **do**
2:   |   do some statement(s)
3: **while** some condition is true

---

**Example 7.4** In Algorithm 7.9, we present a do-while code that writes the numbers $1, 2, \ldots, 5$.    ■

---

**Algorithm 7.9:** Using a do-while-statement to print numbers from 1 to 5

---

**Input:** Integer $i$
**Output:** Print numbers from 1 to 5
1: int $i = 1$
2: **do**
3:   |   printf$(i)$                 // Output is 1 2 3 4 5
4:   |   $++i$
5: **while** $(i \leq 5)$
6: **return** 0

---

***Block:*** If $S_1, S_2, \ldots, S_n$ are statements, then the code in Algorithm 7.10 is called a block. See Figure 7.8 (e).

---

**Algorithm 7.10:** Writing a block

1: do a statement $S_1$
2: do a statement $S_2$

3: $\vdots$

4: do a statement $S_n$

---

The following example shows a block with three statements.

**Example 7.5** In Algorithm 7.11, we present a code that writes the letters $a, b$ and $c$ in three lines. Here, the command "$\backslash n$" means jumping to a newline. ∎

---

**Algorithm 7.11:** Printing the letters $a, b$ and $c$

1: printf("$a\backslash n$")                                          // Output is $a$
2: printf("$b\backslash n$")                                          // Output is $b$
3: printf("$c$")                                                // Output is $c$

---

**Example 7.6** A very simple algorithm called find-max is designed for the following problem:

PROBLEM: Given an array of positive numbers, return the largest number of the array.
INPUT: An array A of positive numbers. This list must contain at least one number. (Asking for the largest number in a list of no numbers is not a meaningful question.)
OUTPUT: A number that will be the largest number of the list.

(*a*) Write an algorithmic code of the following algorithm which solves the above problem.

    1. Set max to 0.
    2. For each x in the list L, compare it to max. If x is larger, set max to x.
    3. max is now set to the largest number in the list.

(*b*) Does the algorithm derived in item (*a*) have defined inputs and outputs? Is it guaranteed to terminate? Does it produce the correct result?

**Solution** (*a*) An algorithmic code that solves the given problem is stated in Algorithm 7.12.

(*b*) The answer to all questions is yes. Note that the input for Algorithm 7.12 is a finite array of positive numbers whose length is n, and the output is the maximum in the array. Note also that Algorithm 7.12 terminates after visiting all n elements,

---

**Algorithm 7.12:** Finding the largest number of an array: `find-max`$(A, n)$

---

**Input:** An array $A[0 : n - 1]$ of positive integers and length $n$
**Output:** The largest number of the array

1: max $= 0$
2: **for** $(i = 0; i < n; i + +)$ **do**
3:     **if** $(A[i] > $ max$)$ **then**
4:        max $= A[i]$
5:     **end**
6: **end**
7: **return** max
8: }

---

and that the "for" loop from line (2) to line (6) searches through the array and assigns any value larger than `max` to `max`.

It is not in the scope of this book to provide the reader a comprehensive list of the control statements because this is an asymptotic analytically-oriented book chapter. Such a list can be found in any book about C programming. The switch statement, for instance, is known as one of the selection statements available in computer programming languages.

## Choosing and comparing algorithms

Some problems can be solved by more than one algorithm. How, then, should we choose an algorithm to solve a given problem? If we need to write a program that will be used once on small amounts of data (and then discarded), then we should select the easiest-to-implement algorithm we know. However, when we need to write a program that is to be used and maintained by many people over a long period of time, other issues arise:

- Simplicity: A simple (more understandable) algorithm is easier to implement correctly than a complex one.

- Clarity: An algorithm that is written clearly and documented carefully can be maintained by others.

- Efficiency: An efficient algorithm becomes very important when the size of the problem increases.

Generally, the efficiency is associated with the time it takes a program to run, although there are other resources that a program sometimes must conserve, such as the storage space. For large problems, however, the running time determines whether a given algorithm should be chosen.

We shall, in fact, take the efficiency of an algorithm to be its running time.

> **Definition 7.1** *The running time of an algorithm is the amount of time it takes, measured as a function of the size of its input.*

Examples of input size (number of elements in the input) are:

- Size of an array.
- Vertices and edges of a graph.

- Degree of a polynomial.
- Number of elements in a matrix.

Given $n$, the size of the input, we can express the running time as a function of the input, say $f(n)$.

**Example 7.7**   For each of the two fragments in Algorithms 7.13 and 7.14, we associate a "cost" with each statement as follows: $c_0$ is the cost of checking the for-statement statement's condition at every iteration in Fragment B shown in Algorithm 7.14, and $c_1$ is the cost of executing the assignment statement array$[i] = 0$ for each $i = 0, 1, \ldots, n - 1$.

We find the "total cost" by finding the total number of times each statement is executed. The total cost of performing Fragment A shown in Algorithm 7.13 is $f(n) = c_1 n$.

The body of the "for" statement is executed n times (from 0 to n$-1$). The "for" statement is executed n times (from 0 to n$-1$), plus 1 for the last time that i is checked and the inequality is false, that is, n+1 times. Thus, the total cost of performing Fragment B shown in Algorithm 7.14 is $g(n) = (n + 1)c_0 + nc_1 = (c_0 + c_1)n + c_0$.

Now, when $c_0 > 0$, can we say that Algorithm 7.13 is more efficient than Algorithm 7.14? Note that both fragments have linear cost-time. We may find an answer to this question at the end of Example 7.8.  ∎

---

**Algorithm 7.13:** Fragment A of Example 7.7

---

1: array$[0] = 0$                                      // Cost is $c_1$
2: array$[1] = 0$                                      // Cost is $c_1$
3: array$[2] = 0$                                      // Cost is $c_1$

4: $\vdots$
5: array$[n - 1] = 0$                                  // Cost is $c_1$
   // Total cost is $c_1 + c_1 + \cdots + c_1 = nc_1$

---

**Algorithm 7.14:** Fragment B of Example 7.7

---

1: **for** $(i = 0; i < n; i + +)$ **do**                 // Cost is $c_0$
2: $\quad |$   array$[i] = 0$                            // Cost is $c_1$
3: **end**
   // Total cost is $(n + 1)c_0 + nc_1 = (c_0 + c_1)n + c_0$

---

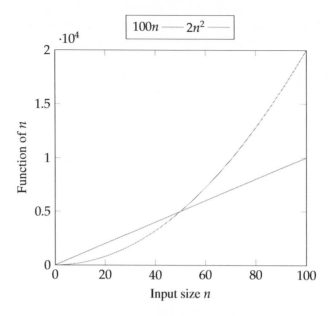

Figure 7.9: $100n$ versus $2n^2$.

In general, to compare two algorithms, we compare the running time for each function. We will see a rough measure that characterizes how fast function grows, that is, the rate of growth. We compare functions for large values of $n$, that is, we compare functions asymptotically (in the limit).

**Example 7.8**   Suppose that for some problem we have the choice of using a linear-time program whose running time is $f(n) = 100n$ and a quadratic-time program whose running time is $g(n) = 2n^2$. The graphs of the running times are shown in Figure 7.9.

The natural question that arises now is which algorithm is faster? Indeed, we are always interested in which is fastest asymptotically as $n$ gets very large (we can think of $n$ as going to infinity). So, clearly the linear-time program is faster than the quadratic-time program.

Another question that also arises is what impact do the constants have on the running time? Asymptotically speaking, it is clear that the constants have little or no impact on the running time. This might answer the question that we had at the end of Example 7.7. ∎

## 7.2. Running time of algorithms

In Section 7.1, we defined the running time of an algorithm to the amount of time it takes, measured as a function of the size of its input, such as the number of elements in a matrix, number vertices and edges of a graph, etc. We also took the efficiency of an algorithm to be its running time. Given $n$, the size of the input, we can express the running time as a function of the input, $f(n)$.

In this section, we study line-by-line runtime analysis of algorithmic fragments, present types of runtime analysis of algorithms, determine upper/lower bounds for running time, and provide more examples on running time.

### Line-by-line runtime analysis

The runtime analysis that we did in Example 7.7 is called a line-by-line analysis. In this kind of runtime analysis, we multiply the cost of execution of each line by the number of times we execute that line and sum the results of each line. In this part, we present more examples on performing line-by-line runtime analysis.

**Example 7.9**  Find the "total cost" needed to perform the fragment given in Algorithm 7.15 by finding the total number of times each statement is executed.

**Solution**  Let us add comments showing the number of times each statement is executed (see the comments in gray in Algorithm 7.15). The following points are noted in order to find the total running time:

- It is clear that the assignment statement in line (1) is executed once.

- How many times the "for" statement in line (2) executes? 0 to $n - 1$, which is $n$ times, plus 1 for the last time that $i$ is checked and the inequality is false, that is, $n + 1$ times.

- How many times the "for" statement in line (3) executes? If the "for" statement in line (3) was not nested it would execute $n+1$ times just like the "for" statement in line (2). Since the "for" statement in line (3) is nested it executes $n + 1$ times for *each* time we execute the body of the "for" statement in line (2) which is $n$ times.

- What about the sum statement in line (4)? If the second loop was not nested, the sum statement in line (4) would execute $n$ times since that is the number of times the body of the second loop would execute. Since the second loop is nested, the sum statement in line (4) executes $n$ times for each time we execute the body of the first loop which is $n$ times. Therefore, $n \times n = n^2$.

---

**Algorithm 7.15:** The algorithm of Example 7.9

| | |
|---|---|
| 1: sum $= 0$ | // Cost $= c_1$, # times $= 1$ |
| 2: **for** $(i = 0; i < n; i + +)$ **do** | // Cost $= c_2$, # times $= n + 1$ |
| 3:     **for** $(j = 0; j < n; j + +)$ **do** | // Cost $= c_2$, # times $= n(n + 1)$ |
| 4:         sum += array$[i][j]$ | // Cost $= c_3$, # times $= n^2$ |
| 5:     **end** | |
| 6: **end** | |

---

Now, the total cost or running time of the algorithm is the sum of the cost multiplied by the number of times for each line.

Running time is $c_1(1) + c_2(n + 1) + c_2 n(n + 1) + c_3 n^2$. Simplifying to polynomial form, we conclude that the running time is

$$(c_2 + c_3)n^2 + 2c_2 n + (c_1 + c_2),$$

or $\bar{c}n^2 + \hat{c}n + \tilde{c}$, where $\bar{c} = c_2 + c_3, \hat{c} = 2c_2, \tilde{c} = c_1 + c_2$.

We use the predominate term or in other words, the degree of the polynomial, to express the running time. So, we conclude the running time is $\bar{c}n$, where $\bar{c} = c_2 + c_3$.
∎

In Example 7.9, we found that the condition in the for-statement "for $(i = 0; i < n; i++)$" was checked $n+1$ times. Similarly, the condition in each of the for-statements "for $(i = n; i > 0; i--)$", "for $(i = 1; i <= n; i++)$", and "for $(i = n; i >= 1; i--)$" is checked $n + 1$ times. We have the following example.

**Example 7.10** Find the running time of the fragment given in Algorithm 7.16 by finding the total number of times each statement is executed.

**Solution** Note that this algorithm has two independent loops, each of which does slightly different things. We have added comments in gray showing the number of times each statement is executed.

The running time of the fragment is the sum of the cost multiplied by the number of times for each line. Therefore,

$$\begin{aligned} \text{Running time} &= c_1 + c_2(n + 1) + c_3 n + c_2(n + 1) + c_4 n \\ &= (2c_2 + c_3 + c_4)n + (c_1 + 2c_2) = \bar{c}n + \hat{c}, \end{aligned}$$

where $\bar{c} = 2c_2 + c_3 + c_4$ and $\hat{c} = c_1 + 2c_2$. Using the predominate term, the running time is $\bar{c}n$.
∎

In Example 7.10, we found that the condition in the for-statement "for $(i = 1; i <= n; i++)$" is checked $n + 1$ times. One can also find that the condition in each of the for-statements "for $(i = 1; i <= \log n; i++)$", "for $(i = 1; i < n; i* = 2)$", and "for $(i = n; i > 1; i/ = 2)$" is checked $(\log n) + 1$ times.

---

**Algorithm 7.16:** The algorithm of Example 7.10

---

| | |
|---|---|
| 1: int $q = 0$ | // Cost $= c_1$, # times $= 1$ |
| 2: **for** $(i = 1; i \leq n; i++)$ **do** | // Cost $= c_2$, # times $= n + 1$ |
| 3: $\quad \mid \quad q = q + i^2$ | // Cost $= c_3$, # times $= n$ |
| 4: **end** | |
| 5: **for** $(j = n; j \geq 1; j--)$ **do** | // Cost $= c_2$, # times $= n + 1$ |
| 6: $\quad \mid \quad q = q + j$ | // Cost $= c_4$, # times $= n$ |
| 7: **end** | |

---

---

**Algorithm 7.17:** The algorithm of Example 7.11

| | |
|---|---|
| 1: int $q = 0$ | // Cost $= c_1$, # times $= 1$ |
| 2: **for** $(i = 1; i \leq \log n; i + +)$ **do** | // Cost $= c_2$, # times $= (\log n) + 1$ |
| 3:   $\mid$   $q = q + i^2$ | // Cost $= c_3$, # times $= \log n$ |
| 4: **end** | |
| 5: **for** $(j = 1; j < n; j* = 2)$ **do** | // Cost $= c_4$, # times $= (\log n) + 1$ |
| 6:   $\mid$   $q = q + j^2$ | // Cost $= c_3$, # times $= \log n$ |
| 7: **end** | |
| 8: **for** $(k = n; k > 1; k/ = 2)$ **do** | // Cost $= c_4$, # times $= (\log n) + 1$ |
| 9:   $\mid$   $q = q + k^2$ | // Cost $= c_3$, # times $= \log n$ |
| 10: **end** | |

---

🖥 **Example 7.11**   Find the running time of the fragment given in Algorithm 7.17 by finding the total number of times each statement is executed.

**Solution**   Note that this algorithm has two independent loops, each of which do slightly different things. We have added comments in gray showing the number of time each statement is executed.

The running time of the fragment is the sum of the cost multiplied by the number of times for each line. Therefore,

$$
\begin{aligned}
\text{Running time} &= c_1 + c_2(1 + \log n) + c_3 \log n + 2c_4(1 + \log n) + 2c_3 \log n \\
&= (c_1 + c_2 + 2c_4) + (c_2 + 3c_3 + 2c_4) \log n = \bar{c} \log n + \hat{c},
\end{aligned}
$$

where $\bar{c} = c_2 + 3c_3 + 2c_4$ and $\hat{c} = c_1 + c_2 + 2c_4$. Using the predominate term, the running time is $\bar{c} \log n$.   ∎

## Types of runtime analysis

A question that suggests itself is how do we evaluate the running time given there is conditional execution (i.e., selection statement)? Here, we will look at the worst-case running time. Why not the best-case? What would be the worst-case scenario be? There are three types of runtime analysis for a given algorithm:

- Worst-case analysis: This type provides an upper bound on running time guaranteeing that the algorithm would not run longer.

- Average-case analysis: This type provides a prediction about running time, assuming input is random.

- Best-case analysis: This type provides a lower bound on running time on input for which algorithm runs the fastest.

Note that the average running time is sometimes a more realistic measure of what performance one will see in practice, but it is often much harder to compute than the worse-case running time.

---

**Algorithm 7.18:** The algorithm of Example 7.12

---

1:   sum $= i$                              // Cost $= c_1$, # times $= 1$
2:   **for** $(j = i + 1; j < n; j + +)$ **do**          // Cost $= c_2$, # times $= n - i$
3:      **if** $(A[i] < A[\text{small}])$ **then**        // Cost $= c_3$, # times $= n - i - 1$
4:         sum $= j$           // Cost $= c_1$, # times $= n - i - 1$ [worst-case]
5:      **end**
6: **end**

---

🖼 **Example 7.12** The code in Algorithm 7.18 sets small to the index of the smallest element found in the portion of the array $A$ from $A[i]$ through $A[n-1]$. Find the running time equipped with worst-case performance.

**Solution** We are looking at the worst-case running time. We have added comments in gray showing the number of times each statement is executed.

We consider the body of the "for" statement, the "if" statement (lines (3) and (5)). The test of line (3) is always executed, but the assignment at line (4) is executed only if the test succeeds. Thus, the body (lines (3) and (4)) takes either $c_3 + c_1$ or $c_3$ time costs, depending on the data in array $A$. If we want to take the worst-case, then we can assume that the body takes $c_3 + c_1$ time cost.

The (worst-case) total cost or running time of the code is the sum of the cost multiplied by the number of times for each line. Therefore

$$
\begin{aligned}
\text{Running time} &= c_1 + c_2(n - i) + c_3(n - i - 1) + c_1(n - i - 1) \\
&= (c_1 + c_2 + c_3)(n - i) - c_3.
\end{aligned}
$$

It is natural to regard the size "$\bar{n}$" of the data on which the code operates as $\bar{n} = n - i$, since that is the length of the array $A[i : n-1]$ on which it operates. Then the running time, which is $(c_1 + c_2 + c_3)(n - i) - c_3$, equals $\bar{c}\bar{n} - c_3$ where $\bar{c} = c_1 + c_2 + c_3$. To sum up, using the predominate term, the running time is $\bar{c}\bar{n}$ where $\bar{n} = n - i$ and $\bar{c} = c_1 + c_2 + c_3$. ∎

🖼 **Example 7.13** Compute the worst-case and best-case runtime complexities of the program snippet shown in Algorithm 7.19.

**Solution** First, we find the worst-case running time. The most-right column adds the number of times each statement is executed.

The worst-case running time of the fragment is the sum of the cost multiplied by the number of times for each line. Therefore,

$$
\begin{aligned}
\text{Running time} &= c_1 m + c_2(m - 1) + c_3(m - 1)n + (c_4 + c_5)(m - 1)(n - 1) \\
&\quad + c_6(m - 1)(1 + \log_{10} n) + c_7(m - 1)\log_{10} n + c_8(m - 1)\log_{10} n \\
&= c_1 m + (c_2 + c_6)(m - 1) + c_3(m - 1)n + (c_4 + c_5)(m - 1)(n - 1) \\
&\quad + (c_6 + c_7 + c_8)(m - 1)\log_{10} n.
\end{aligned}
$$

---

**Algorithm 7.19:** The algorithm of Example 7.13

---

1: **for** $(i = 1; i < m; i + +)$ **do**       // Cost $= c_1$, # times $= m$

2:    **if** $(i < n/2)$ **then**       // Cost $= c_2$, # times $= m - 1$

3:      **while** $(n > 1)$ **do**      // Cost $= c_3$, # times $= (m - 1)n$ [worst case]

4:        $\mathtt{printf}(n)$    // Cost $= c_4$, # times $= (m - 1)(n - 1)$ [worst case]

5:        $n - -$       // Cost $= c_5$, # times $= (m - 1)(n - 1)$ [worst case]

6:      **end**

7:    **end**

8:    **while** $(n > 1)$ **do**      // Cost $= c_6$, # times $= (m - 1)((\log_{10} n) + 1)$

9:      $\mathtt{printf}(n)$      // Cost $= c_7$, # times $= (m - 1)\log_{10} n$

10:      $n/ = 10$      // Cost $= c_8$, # times $= (m - 1)\log_{10} n$

11:    **end**

12: **end**

---

Using the predominate term, the running time is $\bar{c}nm$ where $\bar{c} = c_3 + c_4 + c_5$.

The best-case running time of the fragment is the sum of the cost multiplied by the number of times for each line. Therefore,

$$
\begin{aligned}
\text{Running time} &= c_1 m + c_2(m - 1) + c_6(m - 1)(1 + \log_{10} n) + c_7(m - 1)\log_{10} n \\
&\quad + c_8(m - 1)\log_{10} n \\
&= c_1 m + (c_2 + c_6)(m - 1) + (c_6 + c_7 + c_8)(m - 1)\log_{10} n.
\end{aligned}
$$

Using the predominate term, the running time is $\hat{c}m \log n$ where $\hat{c} = (c_6 + c_7 + c_8)/\log 10$. ∎

Another example for computing different case complexities is shown in Insertion Sort which will be introduced and analyzed in Chapter 8.

## Summation representations for looping

In all examples that we have seen so far on running time, we performed a line-by-line runtime analysis. In this part, we present more examples on running time for loop programs without doing a line-by-line analysis, but instead representing their running time as summations.

Note that the inner loop in Algorithm 7.15 of Example 7.9 is the part that contributed most to the running time. In fact, there is an informal rule called the 90/10 complexity rule which states that 90% of the running time is spent in 10% of the code. However, the exact percentage varies from a program to another. Instead of analyzing every statement, we can save time and effort by focusing on the statement that contributes most to the running time.

**Example 7.14** Consider the code in Algorithm 7.20. It does not really matter what the code does, but what really matters is finding the running time of its execution.

---

**Algorithm 7.20:** The algorithm of Example 7.14

---

1: $x = 0$
2: **for** $(i = 1; i \leq n; i + +)$ **do**
3:    |   **for** $(j = 1; j \leq n; j + +)$ **do**
4:    |    |   $x = x + (i - j)$                      // Execution cost is $c$
5:    |   **end**
6: **end**
7: **return** $x$

---

Instead of analyzing very statement, we can save time and effort by focusing on the statement in line (4).

Assuming $n$ is a constant, say 3, how many times would the statement in line (4) be executed? It is clear that the answer here is $3 \times 3 = 3^2 = 9$ times. Generalizing this for any $n$, we find that the statement in line (4) would be executed $n^2$ times.

Let $c$ be the execution cost of the statement in line (4). Then we can represent the running time using summations and obtain $\sum_{i=1}^{n} \sum_{j=1}^{n} c = c \sum_{i=1}^{n} n = cn^2$.

Thus, the running time of the code in Algorithm 7.20 is $cn^2$.

The following points are noted when finding the running time of an algorithm:

- One can analyze algorithms by focusing on the part(s) where most of the execution occurs.

- The specific input makes a difference in the running time of an algorithm.

**Example 7.15** Find the running time of the code in Algorithm 7.21 without doing a line-by-line analysis.

---

**Algorithm 7.21:** The algorithm of Example 7.15

---

1: $x = 0$
2: **for** $(i = 1; i \leq n; i + +)$ **do**
3:    |   **for** $(j = 1; j \leq i; j + +)$ **do**
4:    |    |   $x = x + (i - j)$                      // Execution cost is $c$
5:    |   **end**
6: **end**
7: **return** $x$

---

**Solution**  To save time and effort, we focus on the statement in line (4).

Assume that $n$ is a constant, say 3. We determine the number of times the statement in line (4) is executed by tracing the code and find that

| i | j = 1 to i |
|---|---|
| 1 | 1 to 1 = 1 |
| 2 | 1 to 2 = 2 |
| 3 | 1 to 3 = 3 |

So, the number of times the statement in line (4) is executed is $1 + 2 + 3 = 6$ times.

Now, generally speaking, for any $n$, we determine the number of times the statement in line (4) is executed by tracing the code and find that

| i | j = 1 to i |
|---|---|
| 1 | 1 to 1 = 1 |
| 2 | 1 to 2 = 2 |
| ⋮ | ⋮ |
| n | 1 to n = n |

Therefore, the number of times the statement in line (4) is executed is

$$1 + 2 + \cdots + n = \sum_{i=1}^{n} i = \frac{n(n+1)}{2} \text{ times,}$$

where we used the arithmetic series to obtain the last equality.

Let $c$ be the execution cost of the statement in line (4). Then, the running time of the code in Algorithm 7.21 is $cn(n+1)/2 \approx cn^2$.  ■

**Example 7.16**  Find the running time of the code in Algorithm 7.22 without doing a line-by-line analysis.

**Algorithm 7.22:** The algorithm of Example 7.16

```
1: x = 0
2: for (i = 1; i ≤ n; i + +) do
3:     for (j = i; j ≤ n; j + +) do
4:         x = x + (i − j)                    // Execution cost is c
5:     end
6: end
7: return x
```

**Solution**   To save time and effort, we focus on the statement in line (4). Assume that $n$ is a constant, say 3. We determine the number of times the statement in line (4) is executed by tracing the code and find that

| i | j = i to n |
|---|---|
| 1 | 1 to 3 = 3 |
| 2 | 2 to 3 = 2 |
| 3 | 3 to 3 = 1 |

So, the number of times the statement in line (4) is executed is $3 + 2 + 1 = 6$ times.
In general, for any $n$, we have

| i | j = i to n |
|---|---|
| 1 | 1 to n = n |
| 2 | 2 to n = n-1 |
| 3 | 3 to n = n-2 |
| ⋮ | ⋮ |
| n-1 | n-1 to n = 2 |
| n | n to n = 1 |

Therefore, the number of times the statement in line (4) is executed is

$$n + (n - 1) + (n - 2) + \cdots + 2 + 1 = \sum_{i=1}^{n} (n - i + 1) = \frac{n(n + 1)}{2} \text{ times.}$$

Let $c$ be the execution cost of the statement in line (4). Then, the running time of the code in Algorithm 7.21 is $cn(n + 1)/2 \approx cn^2$. ∎

Other examples of (double and triple) summation representations for looping are the matrix-vector multiplication and the matrix-matrix multiplication.

## Upper and lower bounds for running time

Some summations are difficult to work with. In this part, we find upper and lower bounds for running time. This allows us to simplify the summations.

In adopting this approach, we require that the upper and lower bounds must be the same function, only differing by a constant. The constant for the upper bound must be larger than or equal to the constant for the lower bound.

To obtain an upper bound in this approach, we remove terms of expression being subtracted, if helpful, and substitute terms (usually, but not necessarily, the upper/top bound of the summation) into expression. To obtain a lower bound, we split summations to reduce size, if helpful, and substitute terms (usually, but not necessarily, the lower/bottom bound of the summation) into expression.

**Example 7.17 (Example 7.15 revisited)** We find the upper and lower bounds for the running time of the code in Algorithm 7.21. Let $c$ be the execution cost of the statement in line (4). Then we can represent the running time using summations and obtain

$$\sum_{i=1}^{n} \sum_{j=i}^{n} c = \sum_{i=1}^{n} c(n - i + 1).$$

Finding an upper bound is straightforward:

$$\sum_{i=1}^{n} c(n - i + 1) \leq \sum_{i=1}^{n} cn = cn^2.$$

We now find a lower bound:

$$\sum_{i=1}^{n} \sum_{j=i}^{n} c \geq \sum_{i=1}^{n} c(n - i) \geq \sum_{i=\frac{n}{2}}^{n} c(n - i). \tag{7.1}$$

Our goal is to make the right-hand summation in (7.1) at least as large as a quadratic function. Let us try to substitute $\frac{n}{2}$ or $n$ for $i$ and see what happens.

Substituting $\frac{n}{2}$ for $i$ in (7.1), we get

$$\sum_{i=\frac{n}{2}}^{n} c(n - i) \leq \sum_{i=\frac{n}{2}}^{n} c\left(n - \frac{n}{2}\right),$$

but this does not work because it makes the lower bound larger instead of smaller (for lower bounds, we should not make the lower bound larger).

Substituting $n$ for $i$ in (7.1), we get

$$\sum_{i=\frac{n}{2}}^{n} c(n - i) \geq \sum_{i=\frac{n}{2}}^{n} c(n - n) = 0,$$

but this does not work either because it makes the lower bound larger very small, it is zero.

Splitting the summation differently and using the lower split, we get

$$\sum_{i=1}^{n} c(n - i) \geq \sum_{i=1}^{\frac{n}{2}} c(n - i) \geq \sum_{i=1}^{\frac{n}{2}} c\left(n - \frac{n}{2}\right) = c\left(\frac{n}{2}\right)\left(\frac{n}{2}\right) = \left(\frac{c}{4}\right)n^2.$$

Thus, the running time is $\bar{c}n^2$ where $\frac{c}{4} \leq \bar{c} \leq c$. ∎

📝 **Example 7.18** The running time of an algorithm is represented by the summation

$$\sum_{i=1}^{n} \sum_{j=i}^{n^2} c, \qquad (7.2)$$

where $n$ is the number of times the outer loop is iterated and $c$ is the execution cost of the statement that contributed most to the running time of the algorithm. Find upper and lower bounds for the running time of the algorithm. Can your upper/lower bound(s) be sharpened to obtain a tight bound for the running time?

**Solution** Finding an upper bound is straightforward:

$$\sum_{i=1}^{n} \sum_{j=i}^{n^2} c = c \sum_{i=1}^{n} (n^2 - i + 1) \leq c \sum_{i=1}^{n} n^2 = cn^3.$$

For a lower bound, we need to make the summation representation in (7.2) at least as large as a cubic function. This is immediately obtained by noting that

$$\sum_{i=1}^{n} \sum_{j=i}^{n^2} c \geq c \sum_{i=1}^{n} (n^2 - i) \geq c \sum_{i=\frac{n}{2}}^{n} (n^2 - i) \geq c \sum_{i=\frac{n}{2}}^{n} (n^2 - n) = \frac{c}{2}(n^3 - n^2) = \left(\frac{c}{2}\right) n^3. \quad (7.3)$$

Although a lower bound was successfully obtained, the splitting was not helpful in (7.3). In fact, the above lower bound can be sharpened as follows.

$$\sum_{i=1}^{n} \sum_{j=i}^{n^2} c \geq c \sum_{i=1}^{n} (n^2 - i) \geq c \sum_{i=1}^{n} (n^2 - n) = c(n^3 - n^2) \geq cn^3. \qquad (7.4)$$

Therefore, a tight bound for the running time of the algorithm represented by the summation $\sum_{i=1}^{n} \sum_{j=i}^{n^2} c$ is $cn^3$. ∎

Note that, in obtaining lower bounds in Example 7.18 (where second inequality in (7.3) and the first inequality in (7.4) follow), we substituted $n$ for $i$ and did not get 0. This is the difference between Examples 7.17 and 7.18.

📝 **Example 7.19** Find upper and lower bounds for the running time of the algorithm represented by the summation

$$\sum_{i=1}^{n} \sum_{j=1}^{i^2} \sum_{k=j}^{i^2} c, \qquad (7.5)$$

where $n$ is the number of times the outer loop iterated and $c$ is the execution cost of the statement that contributed most to the running time of the algorithm.

**Solution**  For an upper bound, we have

$$\sum_{i=1}^{n}\sum_{j=1}^{i^2}\sum_{k=j}^{i^2} c = c\sum_{i=1}^{n}\sum_{j=1}^{i^2}(i^2 - j + 1) \leq c\sum_{i=1}^{n}\sum_{j=1}^{i^2} i^2 = c\sum_{i=1}^{n} i^4 \leq c\sum_{i=1}^{n} n^4 = cn^5.$$

For a lower bound, we need to make the summation representation in (7.5) at least as large as a quintic function. This is immediately obtained by noting that

$$
\begin{aligned}
\sum_{i=1}^{n}\sum_{j=1}^{i^2}\sum_{k=j}^{i^2} c \;&\geq\; c\sum_{i=1}^{n}\sum_{j=1}^{i^2}(i^2 - j) \\
&\geq\; c\sum_{i=1}^{n}\sum_{j=1}^{i^2/2}(i^2 - j) \\
&\geq\; c\sum_{i=1}^{n}\sum_{j=1}^{i^2/2}\left(i^2 - \frac{i^2}{2}\right) \\
&=\; c\sum_{i=1}^{n}\left(\frac{i^2}{2}\right)\left(i^2 - \frac{i^2}{2}\right) \\
&=\; c\sum_{i=1}^{n}\frac{i^4}{4} \\
&\geq\; c\sum_{i=\frac{n}{2}}^{n}\frac{i^4}{4} \\
&\geq\; c\sum_{i=\frac{n}{2}}^{n}\frac{(n/2)^4}{4} \\
&\geq\; c\left(\frac{n}{2}\right)\left(\frac{n^4/16}{4}\right) = \left(\frac{c}{128}\right)n^5.
\end{aligned}
$$

Therefore, the running time of the algorithm represented by the summation in (7.5) is $\bar{c}n^5$, where $\frac{c}{128} \leq \bar{c} \leq c$. ∎

# 7.3. Asymptotic notation

The asymptotic notation allows us to express the behavior of a function as the input approach infinity. In other words, it is connected about what happens to a function $f(n)$ as $n$ gets larger, and is not concerned about the value of $f(n)$ for small values of $n$. In this section, we introduce the notations, present their properties and criteria of selection, characterize the definitions of the notations using limits, and give a complexity classification of algorithms based on these notations.

## The notations

We present the definitions of Big-Oh, Big-Omega, and Big-Theta which are all asymptotic notations to describe the running time of algorithms. Throughout this section, unless it is stated explicitly otherwise, we assume that $f$ and $g$ are two asymptotically nonnegative functions on $\mathbb{R}$, that is, $f(n)$ and $g(n)$ are nonnegative whenever $n$ is sufficiently large.

**Definition 7.2 (Big-O)** *We say that $f(n)$ is Big-O of $g(n)$, written as $f(n) = O(g(n))$, if there are positive constants $c$ and $n_0$ such that $f(n) \leq cg(n)$ for all $n \geq n_0$.*

According to Definition 7.2, $f(n) = O(g(n))$ means that $f(n)$ grows no faster than $g(n)$. See Figure 7.10.

**Remark 7.1** *The "=" in the statement "$f(n) = O(g(n))$" should be read and thought of as "is", not "equals". The reader can think of it as a one-way equals. An alternative notation is to write $f(n) \in O(g(n))$ instead of $f(n) = O(g(n))$.*

Bearing in mind Remark 7.1, Definition 7.2 can be rewritten as follows.

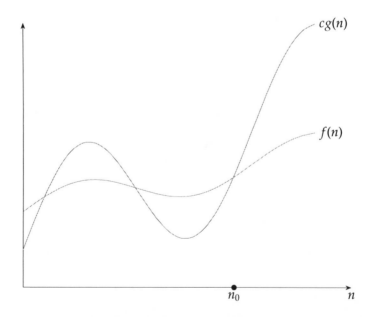

Figure 7.10: $f(n) = O(g(n))$.

> **Definition 7.3 (Big-O revisited)** *Big-Oh of g(n), written as $O(g(n))$, is the set of functions with smaller or same order of growth as $g(n)$. More specifically,*
>
> $$O(g(n)) \triangleq \{f(n) : \exists \text{ positive constants } c \text{ and } n_0 \text{ s.t. } 0 \leq f(n) \leq cg(n) \, \forall n \geq n_0\}.$$
>
> *Here $g(n)$ is called an asymptotic upper bound for $f(n)$.*

The following example illustrates Definitions 7.2 and 7.3.

**Example 7.20** Prove the following asymptotic statements.

(a) $n^2 + n = O(n^2)$.

(c) $n^{\frac{7}{2}} + n^3 \log n = O(n^4)$.[1]

(b) $3n^3 - 2n^2 + 13n - 15 = O(n^3)$.

**Solution** (a) If $n \geq 1$, we have $n \leq n^2$. It follows that $n^2 + n \leq n^2 + n^2 = 2n^2$ for all $n \geq 1$. Therefore, according to Definition 7.2, we have $n^2 + n = O(n^2)$ with $n_0 = 1$ and $c = 2$.

(b) If $n \geq 1$, we have that $3n^3 - 2n^2 + 13n - 15 \leq 3n^3 + 13n \leq 3n^3 + 13n^3 = 16n^3$ for all $n \geq 1$. Thus $3n^3 - 2n^2 + 13n - 15 = O(n^3)$ with $n_0 = 1$ and $c = 16$.

(c) If $n \geq 1$, we have $n^{\frac{7}{2}} \leq n^4$ and $\log n \leq n$, which implies that

$$n^{\frac{7}{2}} + n^3 \log n \leq n^4 + (n^3)(n) = 2n^4.$$

Thus, with $n_0 = 1$ and $c = 2$, we have $n^{\frac{7}{2}} + n^3 \log n = O(n^4)$.

∎

Below are some relationships between the growth rates of common functions:

$$c \ll \log n \ll \log^2 n \ll \sqrt{n} \ll n \ll n \log n \ll n^{1.1} \ll n^2 \ll 2^n \ll 3^n \ll n! \ll n^n.$$

For instance, $n^3 \ll 2^n$ means that $n^3$ is asymptotically much less than $2^n$. In fact, when $n = 10$, we have $2^{10} = 1024$ and $10^3 = 1000$. Now, each time we add 1 to $n$, $2^n$ doubles, while $n^3$ is multiplied by the quantity $(\frac{n+1}{n})^3$ which is less than 2 when $n \geq 10$.

**Example 7.21** Prove the following asymptotic statements.

(a) $2^n + n^3 = O(2^n)$.    (b) $\log(n!) = O(n \log n)$.    (c) $(\sqrt{2})^{\log n} = O(\sqrt{n})$.

**Solution** (a) Note that for $n \geq 10$, we have $n^3 \leq 2^n$. It follows that, for $n \geq 10$, we have $2^n + n^3 \leq 2^n + 2^n = 2(2^n)$. Thus, $2^n + n^3 = O(2^n)$ with $n_0 = 10$ and $c = 2$.

---

[1]Computer scientists generally think of "$\log n$" as meaning $\log_2 n$ rather than $\log_{10} n$ and $\log_e n$.

(b) If $n \geq 1$, we have $n! = n(n-1) \cdots 2 \cdot 1 \leq \overbrace{n \cdot n \cdots n \cdot n}^{n \text{ times}} = n^n$, and hence $\log n! \leq \log n^n = n \log n$. Thus, $\log(n!) = O(n \log n)$ with $n_0 = 1$ and $c = 1$.

(c) Note that

$$\left(\sqrt{2}\right)^{\log n} = \left(2^{\frac{1}{2}}\right)^{\log n} = 2^{\left(\frac{1}{2}\right)(\log n)} = 2^{\log n^{\frac{1}{2}}} = n^{\frac{1}{2}} = \sqrt{n}.$$

Then it is clear that $\left(\sqrt{2}\right)^{\log n} = O(\sqrt{n})$.

---

**Definition 7.4 (Big-$\Omega$)** *Big-Omega of $g(n)$, written as $\Omega(g(n))$, is the set of functions with larger or same order of growth as $g(n)$. More specifically,*

$$\Omega(g(n)) \triangleq \{f(n) : \exists \text{ positive constants } c \text{ and } n_0 \text{ such that}$$
$$0 \leq cg(n) \leq f(n) \, \forall n \geq n_0\}.$$

*Here $g(n)$ is called an asymptotic lower bound for $f(n)$.*

---

Figure 7.11 illustrates Definition 7.4. We also present Example 7.22.

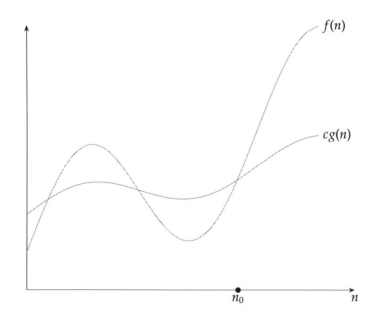

Figure 7.11: $f(n) = \Omega(g(n))$.

**Example 7.22**  Prove the following asymptotic statements.

(a) $n^3 + 4n^2 = \Omega(n^2)$. 

(b) $n \log n - 2n + 13 = \Omega(n \log n)$.

**Solution**  (a) If $n \geq 1$, we have $n^2 \leq n^3 \leq n^3 + 4n^2$. Thus, according to Definition 7.4, we conclude that $n^3 + 4n^2 = \Omega(n^2)$ with $n_0 = 1$ and $c = 1$.

(b) We need to show that there exist positive constants $c$ and $n_0$ such that $cn \log n \leq n \log n - 2n + 13$ for all $n \geq n_0$. Since $n \log n - 2n \leq n \log n - 2n + 13$, we will instead show that

$$cn \log n \leq n \log n - 2n, \quad \text{or equivalently} \quad c \leq 1 - \frac{2}{\log n}.$$

If $n \geq 8$, then $\frac{2}{\log n} \leq \frac{2}{3}$ (as $\log 8 = 3$), or equivalently $\frac{1}{3} \leq 1 - \frac{2}{\log 8}$. So, picking $c = \frac{1}{3}$ suffices. In other words, we have just shown that if $n \geq 8$,

$$\frac{1}{3} n \log n \leq n \log n - 2n.$$

Thus, if $c = \frac{1}{3}$ and $n_0 = 8$, then for all $n \geq n_0$, we have

$$cn \log n \leq n \log n - 2n \leq n \log n - 2n + 13.$$

Therefore, $(n \log n - 2n + 13) = \Omega(n \log n)$.

---

**Definition 7.5 (Big-$\Theta$)**  *Big-Theta of $g(n)$, written as $\Theta(g(n))$, is the set of functions with the same order of growth as $g(n)$. More specifically,*

$$\Theta(g(n)) \quad \triangleq \quad \{f(n) : \exists \text{ positive constants } c_1, c_2 \text{ and } n_0 \text{ such that} \\ 0 \leq c_1 g(n) \leq f(n) \leq c_2 g(n) \ \forall n \geq n_0\}.$$

*Here $g(n)$ is called an asymptotically tight bound for $f(n)$.*

---

Figure 7.12 illustrates Definition 7.5. We also present the following example.

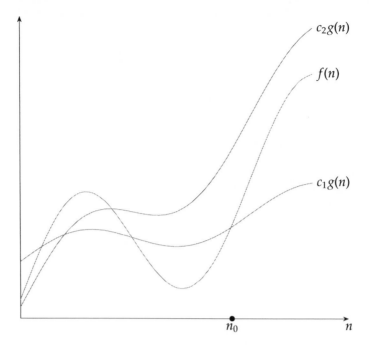

Figure 7.12: $f(n) = \Theta(g(n))$.

**Example 7.23** Prove that $n^2 + 5n + 7 = \Theta(n^2)$.

**Solution**   When $n \geq 1$, we have $n^2 \leq n^2 + 5n + 7 \leq n^2 + 5n^2 + 7n^2 \leq 13n^2$. Thus, according to Definition 7.4, we conclude that $n^2 + 5n + 7 = \Theta(n^2)$ with $n_0 = 1$ and $c_1 = 1$ and $c_2 = 13$.   ■

What if some function is not Big-Oh, not Big-Omega, or not Big-Theta of some other function? The method of proving this is to assume that witnesses $n_0$ and $c$ exist, and derive a contradiction. We have the following example.

**Example 7.24** Show that $n^2$ is not $O(n)$.

**Solution**   Suppose that $n^2 = O(n)$. Then there exists $n_0$ and $c$ such that $n^2 \leq cn$ for all $n \geq n_0$. Picking $n_1 = \max\{n_0, 2c\}$, we have $n_1^2 \leq cn_1$, and dividing both sides by $n_1$, we get $n_1 \leq c$. But we chose $n_1$ to be at least $2c$, a contradiction. This concludes that $n^2$ is not $O(n)$.   ■

## Properties of the notations

There are a lot of properties for Big-Oh, Big-Omega, and Big-Theta notations. In this part, we present a few of the most important ones.

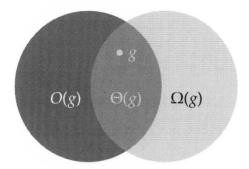

Figure 7.13: Relationships among Big-Oh, Big-Omega, and Big-Theta notations.

The following property follows almost immediately from Definitions 7.3, 7.4 and 7.5, and therefore its proof is left as an exercise for the reader.

**Property 7.1** $f(n) = \Theta(g(n))$ *iff* $f(n) = O(g(n))$ *and* $f(n) = \Omega(g(n))$.

Figure 7.13 shows a conceptual relationship among the notations.

**Example 7.25** Prove the following asymptotic statements. Do not prove by a direct use of Big-Theta definition (Definition 7.5).

(a) $\frac{1}{2}n^2 + 3n = \Theta(n^2)$.

(b) $n^8 + 7n^7 - 10n^5 - 2n^4 + 3n^2 - 17 = \Theta(n^8)$.

**Solution**   (a) If $n \geq 1$, we have $\frac{1}{2}n^2 + 3n \leq \frac{1}{2}n^2 + 3n^2 = \frac{7}{2}n^2$. So, $\frac{1}{2}n^2 + 3n^2 = O(n^2)$. On the other hand, when $n \geq 1$, we have $\frac{1}{2}n^2 \leq \frac{1}{2}n^2 + 3n$. So, $\frac{1}{2}n^2 + 3n^2 = \Omega(n^2)$. Therefore, by Theorem 7.1, we conclude that $\frac{1}{2}n^2 + 3n^2 = \Theta(n^2)$.

(b) Let $f(n) = n^8 + 7n^7 - 10n^5 - 2n^4 + 3n^2 - 17$. To prove that $f(n) = \Theta(n^8)$, we show that $f(n) = O(n^8)$ and $f(n) = \Omega(n^8)$.

It is clear that when $n \geq 1$, we have

$$f(n) \leq n^8 + 7n^7 + 3n^2 \leq n^8 + 7n^8 + 3n^8 = 11n^8.$$

Thus, $f(n) = O(n^8)$ as desired.

Next, we prove that $f(n) = \Omega(n^8)$. That is, we need to show that there exist positive constants $n_0$ and $c$ such that $cn^8 \leq f(n)$ for all $n \geq n_0$.

Now, because

$$
\begin{aligned}
f(n) &= n^8 + 7n^7 - 10n^5 - 2n^4 + 3n^2 - 17 \\
&\geq n^8 - 10n^5 - 2n^4 - 17 \\
&\geq n^8 - 10n^7 - 2n^7 - 17n^7 = n^8 - 29n^7,
\end{aligned}
$$

we will instead show that $cn^8 \leq n^8 - 29n^7$ for some $c > 0$ and for all $n \geq n_0$.

Note that

$$cn^8 \leq n^8 - 29n^7 \iff (1-c)n^8 \geq 29n^7 \iff c \leq 1 - \frac{29}{n}.$$

So, if $n \geq 58$, then $c = \frac{1}{2}$ suffices.

Thus, we have shown that if $n \geq 58$, then $f(n) \geq \frac{1}{2}n^8$. Thus, $f(n) = \Omega(n^8)$. This completes the proof.

∎

The following property follows almost immediately from Definition 7.3, and therefore its proof is also left as an exercise for the reader.

---

**Property 7.2 (Transitivity)** *If we have $f(n) = O(g(n))$ and $g(n) = O(h(n))$, then $f(n) = O(h(n))$. Same for Big-Omega and Big-Theta.*

---

**Example 7.26** It is easy to see that $4n^2 + 3n + 17 = O(n^3)$ and that $n^3 = O(n^4)$. Based on Property 7.2, we conclude that $4n^2 + 3n + 17 = O(n^4)$. ∎

---

**Property 7.3 (Scaling by a constant)** *If $f(n) = O(g(n))$, then $kf(n) = O(g(n))$ for any $k > 0$. Same for Big-Omega and Big-Theta.*

---

**Proof** Assume that $f(n) = O(g(n))$, then from Definition 7.3, there are positive constants $c$ and $n_0$ such that $f(n) \leq cg(n)$ for $n \geq n_0$. Multiplying with $k > 0$, we get $kf(n) \leq \bar{c}g(n)$, for $n \geq n_0$, where $\bar{c} = kc$. The result is established for Big-O. Similar arguments can be made for Big-Omega and Big-Theta. ∎

**Example 7.27** Let $a$ and $b$ be positive numbers that are different from 1. Bearing in mind that $\log_a n = \log_b n / \log_b a$. Then, according to Property 7.3, it follows that $\log_a n = O(\log_b n)$. ∎

---

**Property 7.4 (Sums)** *If we have $f_1(n) = O(g_1(n))$ and $f_2(n) = O(g_2(n))$, then $f_1(n) + f_2(n) = O(g_1(n) + g_2(n)) = O(\max\{g_1(n), g_2(n)\})$. Same for Big-Omega and Big-Theta.*

---

**Proof** We prove the result for Big-O, and similar arguments can be made for Big-Omega and Big-Theta.

Let $f_i(n) = O(g_i(n))$ for $i = 1, 2$, then from Definition 7.3, there are positive constants $c_i$ and $n_i$ such that $f_i(n) \leq c_i g_i(n)$ for $n \geq n_i$.

Now, let $n_0 = \max\{n_1, n_2\}$, then for all $n \geq n_0$, we have

$$
\begin{aligned}
f_1(n) + f_2(n) &\leq c_1 g_1(n) + c_2 g_2(n) \\
&\leq c_1 \max\{g_1(n), g_2(n)\} + c_2 \max\{g_1(n), g_2(n)\} \\
&= (c_1 + c_2) \max\{g_1(n), g_2(n)\} \\
&\leq \bar{c} \max\{g_1(n), g_2(n)\} \leq \bar{c}(g_1(n) + g_2(n)),
\end{aligned}
$$

where $\bar{c} = 2\max\{c_1, c_2\}$. The proof is complete. ∎

We leave the proof of the following property as an exercise for the reader.

> **Property 7.5 (Products)** *If we have $f_1(n) = O(g_1(n))$ and $f_2(n) = O(g_2(n))$, then $f_1(n)f_2(n) = O(g_1(n)g_2(n))$. Same for Big-Omega and Big-Theta.*

**Example 7.28** In Example 7.20, we showed that

$$
n^2 + n = O(n^2) \quad \text{and} \quad n^{\frac{7}{2}} + n^3 \log n = O(n^4).
$$

Applying Property 7.4, we conclude that

$$
n^{\frac{7}{2}} + n^3 \log n + n^2 + n = O\left(n^4 + n^2\right) = O(n^4).
$$

Applying Property 7.5, we conclude that

$$
\left(n^{\frac{7}{2}} + n^3 \log n\right)(n^2 + n) = O\left(\left(n^4\right)\left(n^2\right)\right) = O(n^6).
$$

∎

We also leave the proofs of Properties 7.6 and 7.7 as exercises for the reader.

> **Property 7.6 (Symmetry)** *We have $f(n) = \Theta(g(n))$ iff $g(n) = \Theta(f(n))$. We also have $f(n) = O(g(n))$ iff $g(n) = \Omega(f(n))$.*

> **Property 7.7 (Reflexivity)** $f(n) = O(f(n))$. *Same for Big-Omega and Big-Theta.*

> **Property 7.8** *If $f(n) = O(g(n))$ and $g(n) = O(f(n))$, then $f(n) = \Theta(g(n))$.*

**Proof** Let $f(n) = O(g(n))$, then there are positive constants $c_1$ and $n_1$ such that

$$
f(n) \leq c_1 g(n), \forall n \geq n_1. \tag{7.6}
$$

Let also $f(n) = O(g(n))$, then there are positive constants $c_2$ and $n_2$ such that

$$
g(n) \leq c_2 f(n), \forall n \geq n_2. \tag{7.7}
$$

Combining (7.6) and (7.7), we obtain

$$\frac{1}{c_2}g(n) \le f(n) \le c_1 g(n), \forall n \ge n_0,$$

where $n_0 = \max\{n_1, n_2\}$. The proof is complete. ∎

The following remark follows immediately from Properties 7.2, 7.6 and 7.7.

> **Remark 7.2** *The asymptotic notation* $\Theta$ *defines an equivalence relation over the set of nonnegative functions on* $\mathbb{N}$.

From Theorem 2.1, $\Theta$ defines a partition on these functions, with two functions being in the same partition (or the same equivalence class) if and only if they have the same growth rate. For instance, the functions $n^2$, $n^2 + \log n$ and $3n^2 + n + 1$ are all $\Theta(n^2)$. So, they all belong to the same equivalence class.

## The notations in terms of limits

In this part, we characterize Big-Oh, Big-Omega and Big-Theta notations in terms of limits. The limits are used as another technique to prove asymptotic statements, which is often much easier than that of definitions. It is not hard to prove the following theorem.

> **Theorem 7.1 (Limit characterization of notations)** *Let L be a finite real number, and* $f(n)$ *and* $g(n)$ *be two asymptotically nonnegative functions on* $\mathbb{R}$ *such that* $\lim_{n \to \infty} \frac{f(n)}{g(n)} = L$. *Then*
>
> *(i) If* $L = 0$, *then* $f(n) = O(g(n))$.     *(iii) If* $0 < L < \infty$, *then* $f(n) = \Theta(g(n))$.
>
> *(ii) If* $L = \infty$, *then* $f(n) = \Omega(g(n))$.

If the limit in Theorem 7.1 does not exist, we need to resort using definitions or some other technique.

**Example 7.29** Use limits to prove the following asymptotic statements.

(a) $n^2 = O(n^3)$.

(b) $n^2 = \Omega(n)$.

(c) $7n^2 = \Theta(n^2)$.

(d) $n^4 - 23n^3 + 12n^2 + 15n - 21 = \Theta(n^4)$.

(e) $\frac{n(n+1)}{2} = O(n^3)$.

(f) $\log n = O(n)$.

(g) $n^3 = O(2^n)$.

(h) $\sqrt{5n^2 - 4n + 12} = \Theta(n)$.

**Solution** We simply apply Theorem 7.1.

(a) $\lim_{n \to \infty} \frac{n^2}{n^3} = \lim_{n \to \infty} \frac{1}{n} = 0$. By Theorem 7.1(*i*), we have $n^2 = O(n^3)$.

(b) $\lim_{n \to \infty} \dfrac{n^2}{n} = \lim_{n \to \infty} n = \infty$. By Theorem 7.1(*ii*), we have $n^2 = \Omega(n)$.

(c) $\lim_{n \to \infty} \dfrac{7n^2}{n^2} = \lim_{n \to \infty} 7 = 7 > 0$. By Theorem 7.1(*iii*), we have $7n^2 = \Theta(n^2)$.

(d) We have

$$\lim_{n \to \infty} \frac{n^4 - 23n^3 + 12n^2 + 15n - 21}{n^4} = \lim_{n \to \infty} \left(1 - \frac{23}{n} + \frac{12}{n^2} + \frac{15}{n^3} - \frac{21}{n^4}\right)$$
$$= 1 - 0 + 0 + 0 - 0$$
$$= 1.$$

Thus, by Theorem 7.1(*iii*), we have $n^4 - 23n^3 + 12n^2 + 15n - 21 = \Theta(n^4)$.

(e) We have

$$\lim_{n \to \infty} \frac{n(n+1)/2}{n^3} = \lim_{n \to \infty} \frac{n^2 + n}{2n^3} = \lim_{n \to \infty} \frac{2n + 1}{6n^2} = \lim_{n \to \infty} \frac{2}{12n} = 0,$$

where we used the L'Hospital's rule twice to find the limit. Therefore, by Theorem 7.1(*i*), we have $\frac{n(n+1)}{2} = O(n^3)$.

(f) We have

$$\lim_{n \to \infty} \frac{\log n}{n} = \lim_{n \to \infty} \frac{1/n}{1} = \lim_{n \to \infty} \frac{1}{n} = 0,$$

where we used the L'Hospital's rule to find the limit. Therefore, by Theorem 7.1(*i*), we have $\log n = O(n)$.

(g) We have

$$\lim_{n \to \infty} \frac{n^3}{2^n} = \lim_{n \to \infty} \frac{3n^2}{2^n \ln 2} = \lim_{n \to \infty} \frac{6n}{2^n (\ln 2)^2} = \lim_{n \to \infty} \frac{6}{2^n (\ln 2)^3} = 0,$$

where we used the L'Hospital's rule three times to find the limit. Therefore, by Theorem 7.1(*i*), we have $n^3 = O(2^n)$.

(h) We have

$$\lim_{n \to \infty} \frac{\sqrt{5n^2 - 4n + 12}}{n} = \lim_{n \to \infty} \sqrt{\frac{5n^2 - 4n + 12}{n^2}} = \sqrt{\lim_{n \to \infty} \left(5 - \frac{4}{n} + \frac{12}{n^2}\right)} = \sqrt{5}.$$

Thus, by Theorem 7.1(*iii*), we have $\sqrt{5n^2 - 4n + 12} = \Theta(n)$.

### Complexity classification of algorithms

In this part, we give the computational complexity classification of algorithms based on these notations. First, we discuss the properties that we should consider to choose the Big-Oh of algorithms.

***Choosing Big-Oh for algorithms*** For any algorithm, our Big-Oh choice will have two major properties: simplicity and tightness.

- Simplicity: Our choice of a Big-Oh bound is simplicity in the expression of the function. We have the following definition.

> **Definition 7.6** *Let $g(n)$ be a Big-Oh bound on $f(n)$, i.e., $f(n) = O(g(n))$. The function $g(n)$ is said to be simple if the following two conditions hold:*
>
> *(i) It is a single term.*      *(ii) The coefficient of that term is one.*

The following example illustrates Definition 7.6.

**Example 7.30** In Example 7.9, we showed that the running time of the code in Algorithm 7.15 is

$$f(n) = \bar{c}n^2 + \hat{c}n + \tilde{c},$$

where $\bar{c} = c_2 + c_3, \hat{c} = 2c_2, \tilde{c} = c_1 + c_2$ and $c_i$ is the execution cost of line (i) in Algorithm 7.15 for $i = 1, \ldots, 4$.

Define

$$g_1(n) = n^2, \ g_2(n) = \bar{c}n^2, \text{ and } g_3(n) = \bar{c}n^2 + \hat{c}n.$$

Then, we can show that $f(n) = O(g_i(n))$ for each $i = 1, 2, 3$. Note that $g_1$ is simple, but $g_2$ and $g_3$ are not simple functions. So, choosing $g_1(n)$ as a Big-Oh bound on $f(n)$ is better than choosing $g_2(n)$ and $g_3(n)$. ∎

- Tightness: We generally want the "tightest" Big-Oh upper bound we can prove. We have the following definition.

> **Definition 7.7** *Let $g(n)$ be a Big-Oh bound on $f(n)$, i.e., $f(n) = O(g(n))$. The function $g(n)$ is said to be a tight bound on $f(n)$ if the following condition holds:*
>
> $$\left(\exists h(n), f(n) = O(h(n))\right) \longrightarrow \left(g(n) = O(h(n))\right). \qquad (7.8)$$

The implication in (7.8) means that if we can find a function $h(n)$ that satisfies the statement that $f(n) = O(h(n))$, then the statement $g(n) = O(h(n))$ is also satisfied. In other words, according to Definition 7.7, $g(n)$ be a tight Big-Oh bound on $f(n)$ if $g(n)$ is a Big-Oh bound on $f(n)$ and we cannot find a function that grows at least as fast as $f(n)$ but grows slower than $g(n)$. The following example illustrates Definition 7.7.

**Example 7.31 (Example 7.30 revisited)** In Example 7.30, we found that the function $n^2$ is a simple Big-Oh bound on $f(n) = \bar{c}n^2 + \hat{c}n + \tilde{c}$. In this example, we use Definition 7.7 to show that $n^2$ is a tight bound on $f(n)$, while $n^3$ is not a tight bound on $f(n)$.

To show that $n^2$ is a tight bound on $f(n)$, suppose that there exists a function $h(n)$ such that $f(n) = O(h(n))$. Then there are positive constants $c$ and $n_0$ such that $f(n) = \bar{c}n^2 + \hat{c}n + \tilde{c} \le ch(n)$ for all $n \ge n_0$. Then $h(n) \ge (\frac{\bar{c}}{c})n^2$ for all $n \ge n_0$, which in turn implies that $n^2 \ge (\frac{c}{\bar{c}})h(n)$ for all $n \ge n_0$. This means that $n^2 = O(h(n))$. According to Definition 7.7, we conclude that $n^2$ is a tight Big-Oh bound on $f(n)$.

To see why the function $n^3$ is not bound on $f(n)$, we pick $h(n) = n^2$. We have see that $f(n) = O(h(n))$, but it is clear that $n^3$ is not $O(h(n))$.     ∎

One can prove the following theorem.

---

**Theorem 7.2** *If $f(n) = \Theta(g(n))$, then $g(n)$ is a tight Big-Oh bound on $f(n)$.*

---

**Example 7.32** In Example 7.25, we proved that $\frac{1}{2}n^2 + 3n = \Theta(n^2)$. By Theorem 7.2, we conclude that $n^2$ is a tight Big-Oh bound on $\frac{1}{2}n^2 + 3n$.     ∎

Based on the asymptotic notations introduced in this section, we can now express the worst-, average- and best-case time complexities (i.e., running times) of algorithms using Big-Oh, Big-Theta and Big-Omega notations. In the following chapters, we will analyze different programs by expressing their running times using the asymptotic notations.

***Classification of algorithms based on the notations***    The time complexity of an algorithm signifies the total time required by the program to run till its completion. Determining a precise formula for the total running time $f(n)$ of a program is a difficult, if not impossible, task. Frequently, we can simplify matters considerably by using the Big-Oh notation. In other words, the time complexity of algorithms is most commonly expressed using the Big-Oh expression $O(g(n))$ as an upper bound on $f(n)$.

Algorithmic complexities are classified according to the type of function appearing in the asymptotic notation. In this part, we classify the algorithms according to this. We have the following definition.

| Running time | Name |
| --- | --- |
| $O(1)$ | Constant time |
| $O(\log n)$ | Logarithmic time |
| $O(\log^k n)$ | Polylogarithmic time |
| $O(n)$ | Linear time |
| $O(n \log n)$ | Linearithmic time |
| $O(n^2)$ | Quadratic time |
| $O(n^3)$ | Cubic time |
| $O(n^k)$ | Polynomial time |
| $O(2^n)$ | Exponential time |
| $O(n!)$ | Factorial time |
| $O(n^n)$ | Super-exponential time |

Table 7.1: Some common time complexities.

**Definition 7.8** *Algorithms with running time:*

(i) $\Theta(n)$ *are said to have linear complexity and they are called linear-time algorithms.*

(ii) $\Theta(n^2)$ *are said to have quadratic complexity and they are called quadratic-time algorithms.*

(iii) $\Theta(n^k)$, *for some constant k, are said to have polynomial complexity and they are called polynomial-time algorithms.*

Note that, as $n$ doubles, the run time doubles in the linear-time algorithms and quadruples in the quadratics-time algorithms. In Example 7.13, we found that, when the best-case analysis is considered, the algorithm has linear-time, while when the worst-case analysis is considered, the algorithm has linearithmic-time. Note also that linear and quadratic-time algorithms are special cases of polynomial-time algorithms. The following definition identifies precisely the efficiency of an algorithm.

**Definition 7.9** *An algorithm is called efficient if it runs in polynomial time.*

Hence, when we say that an efficient algorithm exists to solve a problem, we typically mean a polynomial-time algorithm. In Table 7.1, we list some of the more common running times for programs and their names.

The set of all decision problems that can be solved with worst-case polynomial time-complexity is said to be the complexity class P. There are other complexity classes to be introduced. One of the most important among them is so-called NP-complete class, which will be introduced in Section 7.7.

## 7.4. Analyzing decision making statements

In this section, we analyze and identify the time complexity of the decision making statements given in Section 7.1. We consider only basic statements such as if-statement, for-statement, while-statement and others.

**Simple statements:**    The bound for a simple statement – that is, for an assignment, a read, a write, or a jump statement – is $O(1)$.

**If-statement:**    A first look at Figure 7.14 indicates that an upper bound on the running time of an if-statement is $O(1 + \max\{f_1(n), f_2(n)\})$. The "1+" represents the test. Because at least one of $f_1(n)$ and $f_2(n)$ is positive for all $n$, the "1+" can be omitted to conclude that the running time of the if-statement is $O(\max\{f_1(n), f_2(n)\})$ as stated in Figure 7.14.

**For-statement:**    A first look at Figure 7.15 indicates that an upper bound on the running time of a for loop is $O(1 + (f(n) + 1)g(n))$. The factor $f(n) + 1$ represents the cost of going around once, including the body, the test, and the reinitialization. The "1+" at the beginning represents the first initialization and the possibility that the first test is negative, resulting in zero iterations of the loop. In the common case that $f(n)$ and $g(n)$ are both at least 1, then the running time of the for-statement is $O(g(n)f(n))$ as stated in Figure 7.15.

**While-statement:**    A first look at Figure 7.16 indicates that an upper bound on the running time of a while loop is $O(1 + (f(n) + 1)g(n))$. The $O(f(n) + 1)$ is an upper bound on the running time of the body plus the test after the body. The additional 1 at the beginning of the formula accounts for the test, before entering the loop. In the common case $f(n)$ and $g(n)$ are at least 1, so the running time of the while statement is $O(g(n)f(n))$ as stated in Figure 7.16.

**Do-while-statement:**    A first look at Figure 7.17 indicates that an upper bound on the running time of a do-while loop is $O((f(n) + 1)g(n))$. The "+1" represents the time to compute and test the condition at the end of each iteration of the loop. Note that, here, $g(n)$ is always at least 1. In the common case $f(n) \geq 1 \; \forall n$. So the running time of the do-while statement is $O(g(n)f(n))$ as stated in Figure 7.17.

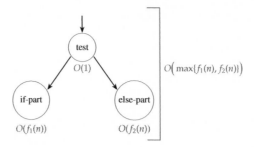

Figure 7.14: The running time complexity of an if-statement.

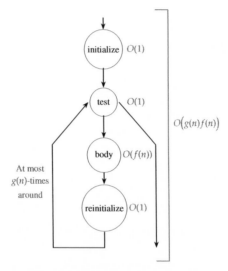

Figure 7.15: The running time complexity of a for-statement.

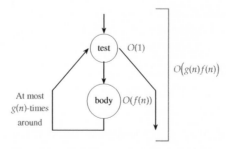

Figure 7.16: The running time complexity of a while statement.

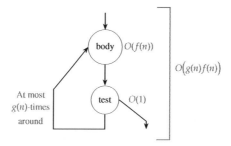

Figure 7.17: The running time complexity of a do-while statement.

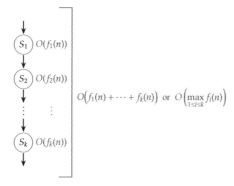

Figure 7.18: The running time complexity of a block.

**Block:** A first look at Figure 7.18 indicates that an upper bound on the running time of a lock is $O\big(f(n) + \cdots + f_k(n)\big)$. Using the Big-Oh rules, this can be simplified as the one stated in Figure 7.18.

In summary, below are the major points we should take from above:

- The bound for a simple statement is $O(1)$.

- The running time of a selection statement is the time to decide which branch to take plus the larger of the running times of the branches.

- The running time of a loop is computed by taking the running time of the body, plus any control steps (e.g., reinitializing the index and comparing it to the limit). Multiply this running time by an upper bound on the number of times the loop can be iterated. Then, add anything that is done just once, such as the initialization or the first test for termination, if the loop might be iterated zero times.

- The running time of a sequence of statements is the sum of the running times of the statements. Often, one of the statements dominates the others. By the summation rule the running time of the sequence is just the running time of the dominant statement.

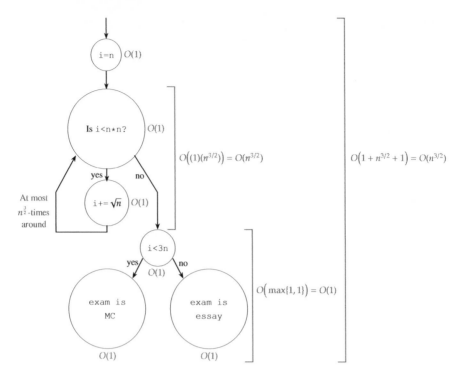

Figure 7.19: The graph structure with running time complexity for the code in Algorithm 7.23.

## 7.5. Analyzing programs without function calls

In this section, we analyze the running time of programs that do not contain function calls (other than library functions such as "printf"), leaving the matter of analyzing programs with function calls to the next section.

We use a graph- or tree-based running time analysis to derive Big-Oh upper bounds on the running time of programs without function calls (also known as sequential programs).

A graph structure, or a flowchart, is a graphical representation of the separate steps of a program in sequential order. A tree structure is a graph structure where loops are clustered as nodes instead of cycles. In this section, we create and use graph and tree structures to analyze and identify the running time of sequential programs (i.e., programs without function calls). We have the following examples.

**Example 7.33**  Draw the graph structure for the code in Algorithm 7.23. Give a Big-Oh upper bound on the running time of each compound statement in Algorithm 7.23, as a function of $n$. What is the running time of the function as a whole?

---

**Algorithm 7.23:** The algorithm of Example 7.33

---

1: $i = n$
2: **while** $(i < n^2)$ **do**
3:    $|$   $i = i + \sqrt{n}$
4: **end**
5: **if** $(i < 3n)$ **then**
6:    $|$   printf("The exam is multiple choice.")
7: **end**
8: **else**
9:    $|$   printf("The exam is essay.")
10: **end**

---

**Solution**  The graph structure is shown in Figure 7.19. The while loop of lines (2) and (3) in Algorithm 7.23 may be executed as many as $n^{3/2}$-times, but no more. To see this, note that the initial value of $i$ is $n$, and that in each iteration we increment $i$ by $\sqrt{n}$. We keep executing the while loop until $i$ is no longer smaller than $n^2$. Therefore, the number of times we go around the while loop is $k$ where $k$ is the largest positive integer so that

$$n + \underbrace{\sqrt{n} + \sqrt{n} + \cdots + \sqrt{n}}_{k-\text{time}} = n + k\sqrt{n} < n^2.$$

Hence $k < (n^2 - n)/\sqrt{n} = O(n^{3/2})$.

From the summation rule, the running time of Algorithm 7.23 is $O(n^{3/2})$, because that is the maximum of the time for the assignment of line (1), the time for the while loop in lines (2) and (3), and the time for the selection statement in lines (5)–(10). The running time found in this example is the worst-case performance of the code in Algorithm 7.23. ∎

**Example 7.34**  A positive integer n is said to be prime if it is divisible by only 1 and itself. For example, 7 is a prime number. So, if n is not a prime, then it is divisible evenly by some integer i between 2 and $\sqrt{n}$. Algorithm 7.24 determines whether a given integer n is prime or not prime.

Draw the tree structure for this algorithm. Give a Big-Oh upper bound on the running time of each compound statement in Algorithm 7.24, as a function of n. What is the running time of the function as a whole?

**Solution**  The tree structure is shown in Figure 7.20. As detailed in Figure 7.20, the running time of the if-else-statement is $O(1)$, thus making the running time of the while-statement $O(\sqrt{n})$. Since all the other lines are single statements and thus take $O(1)$, the runtime of the function as a whole is $O(\sqrt{n})$. ∎

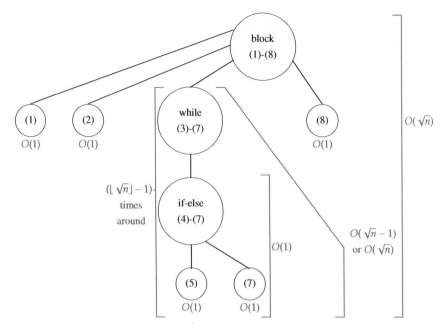

Figure 7.20: The tree structure for Algorithm 7.24 with running time complexity.

---

**Algorithm 7.24:** Testing for primeness: `prime`(int $n$)

---

**Input:** A positive integer $n$

**Output:** A TRUE/FALSE answer to Question "Is $n$ a prime number?"

1: int $i$

2: $i = 2$

3: **while** ($i^2 \leq n$) **do**

4:    **if** ($n\%i == 0$) **then**

5:       **return** FALSE

6:    **end**

7:    **else**

8:       $i + +$

9:    **end**

10: **end**

11: **return** TRUE

---

---

**Algorithm 7.25:** Interior-point linear optimization algorithm

---

**Input:** Linear program with an $n \times m$ matrix $A$ and a barrier function $f$

**Output:** An $\epsilon$-optimal solution to the linear program

1: initialize $\epsilon > 0, \gamma \in (0,1), \theta > 0, \beta > 0, x^{(0)}, \mu^{(0)} > 0, \lambda^{(0)}$

2: set $x \triangleq x^{(0)}, \mu \triangleq \mu^{(0)}, \lambda = \lambda^{(0)}$

3: **while** ($\mu \geq \epsilon$) **do**

4:    compute $g = \nabla f(\mu, x) - A^{\mathsf{T}}\lambda$

5:    compute$\Delta x = -\left((\nabla^2 f)^{-1} - (\nabla^2 f)^{-1} A^{\mathsf{T}} \left(A (\nabla^2 f)^{-1} A^{\mathsf{T}}\right)^{-1} A (\nabla^2 f)^{-1}\right)(g)$

6:    compute $\Delta\lambda = \left(A \left(\nabla^2 f(\mu, x)\right)^{-1} A^{\mathsf{T}}\right)^{-1} A \left(\nabla^2 f(\mu, x)\right)^{-1} (g)$

7:    compute $\delta(\mu, x) = \sqrt{\dfrac{1}{\mu}(\Delta x)^{\mathsf{T}} \nabla^2 f(\mu, x)(\Delta x)}$

8:    **while** ($\delta > \beta$) **do**

9:       set $x \triangleq x + \theta\Delta x$

10:       set $\lambda \triangleq \lambda + \theta\Delta\lambda$

11:       compute $g = \nabla f(\mu, x) - A^{\mathsf{T}}\lambda$

12:       compute
$$\Delta x = -\left(\left(\nabla^2 f\right)^{-1} - \left(\nabla^2 f\right)^{-1} A^{\mathsf{T}} \left(A \left(\nabla^2 f\right)^{-1} A^{\mathsf{T}}\right)^{-1} A \left(\nabla^2 f\right)^{-1}\right)(g)$$

13:       compute $\Delta\lambda = \left(A \left(\nabla^2 f(\mu, x)\right)^{-1} A^{\mathsf{T}}\right)^{-1} A \left(\nabla^2 f(\mu, x)\right)^{-1} (g)$

14:       compute $\delta(\mu, x) = \sqrt{\dfrac{1}{\mu}(\Delta x)^{\mathsf{T}} \nabla^2 f(\mu, x)(\Delta x)}$

15:    **end**

16:    set $\mu \triangleq \gamma\mu$

17: **end**

---

🔲 **Example 7.35**   Algorithm 7.25 is an example of the so-called interior-point algorithm (also referred to as a barrier algorithm) and is used for solving linear (and nonlinear) optimization problems (see [NN94]).[2]

(*a*) No matter how the algorithm works and how the mathematical notations therein look, the graph structure (flowchart) of the algorithm can be visualized. Draw the graph structure for Algorithm 7.25.

(*b*) Let $N_{\mathrm{in}}$ be the number of times we go around the inner while loop of line (8). In view of the graph structure of the algorithm drawn in item (*a*), determine a good

---

[2]Linear optimization (also-called linear programming, or LP for short) is a method to achieve the best outcome (such as maximum profit or lowest cost) in a mathematical model whose requirements are represented by linear relationships (see Chapter 10 for more detail).

asymptotic upper bound on the number of iterations needed to obtain a desired solution using Algorithm 7.25 in each of the following cases.

(*i*) $N_{in} = O(nm)$ and $\gamma \in (0, 1)$ is an arbitrarily chosen constant.

(*ii*) $N_{in} = O(1)$ and $\gamma = 1 - \sigma/\sqrt{nm}$, where $\sigma > 0$.

In each case, give your answer as a function of $n, m, \mu^{(0)}, \epsilon$ using Big-oh, and justify your determination.

**Solution** (*a*) The graph structure of Algorithm 7.25 is shown in Figure 7.21.

(*b*) Let $N_{out}$ (respectively, $N_{in}$) be the number of times we go around the outer (respectively, inner) while loop, and $N$ be the number of iterations needed to obtain a desired solution using Algorithm 7.25. Then, in view of the graph structure of the algorithm drawn in item (*a*), we have

$$N = N_{in} N_{out}.$$

We now estimate $N_{out}$. Let $\mu^{(i)}$ be the parameter at the $i^{th}$ iteration. Then we have

$$\mu^{(i)} = \gamma \mu^{(i-1)} = \gamma^2 \mu^{(i-2)} = \cdots = \gamma^i \mu^{(0)}.$$

Thus $\mu^{(i)}$ will be less than the given $\epsilon > 0$ if $\gamma^i \mu^{(0)} < \epsilon$. Note that

$$\gamma^i \mu^{(0)} < \epsilon \iff \gamma^i < \epsilon/\mu^{(0)} \iff i \log \gamma = \log(\gamma^i) < \log(\epsilon/\mu^{(0)}).$$

Since $\gamma \in (0, 1)$, we have

$$i > \frac{\log(\epsilon/\mu^{(0)})}{\log \gamma} = \frac{\log(\mu^{(0)}/\epsilon)}{-\log \gamma}.$$

Hence,

$$N_{out} \le \frac{\log(\mu^{(0)}/\epsilon)}{-\log \gamma}.$$

(*i*) It is given that $N_{in} = O(nm)$. Now, if $\gamma \in (0, 1)$ is an arbitrarily chosen constant, then $\gamma = O(1)$, and hence

$$N_{out} \le \log\left(\frac{\mu^{(0)}}{\epsilon}\right) O(1).$$

Thus, the number of iterations needed to obtain a desired solution is

$$N = N_{in} N_{out} = O\left(nm \log\left(\frac{\mu^{(0)}}{\epsilon}\right)\right).$$

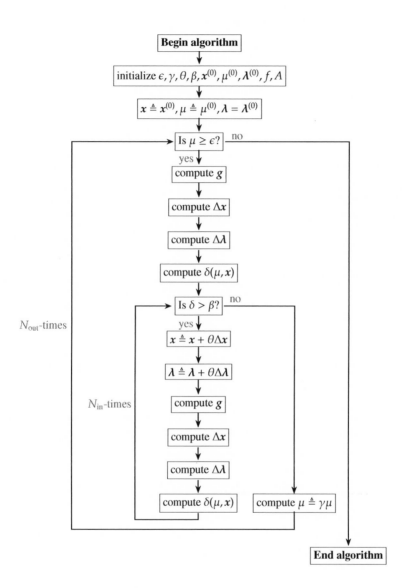

Figure 7.21: The graph structure of Algorithm 7.25.

(*ii*) It is given that $N_{in} = O(1)$. Now, if $\gamma = 1 - \sigma/\sqrt{nm}$ ($\sigma > 0$), then[3]

$$\log \gamma = \log\left(1 - \sigma/\sqrt{nm}\right) \approx -\sigma/\sqrt{nm},$$

and hence

$$N_{out} \leq \frac{\log(\mu^{(0)}/\epsilon)}{-\log\gamma} \approx \frac{\log(\mu^{(0)}/\epsilon)}{\sigma/\sqrt{nm}} = \sqrt{nm}\log\left(\frac{\mu^{(0)}}{\epsilon}\right) O(1).$$

Thus, the number of iterations needed to obtain a desired solution is

$$N = N_{in}N_{out} = O\left(\sqrt{nm}\,\log\left(\frac{\mu^{(0)}}{\epsilon}\right)\right).$$

■

In Example 7.35, having a good asymptotic bound on the number of iterations required to obtain a desired solution using Algorithm 7.21, we can find a good asymptotic time complexity of the algorithm if the running times of lines (1)-(17) in Algorithm 7.21 are provided.

Other examples of sequential programs are Linear Search and Selection Sort which will be introduced and analyzed in Chapter 8. More specifically, a graph structure will be drawn to analyze Linear Search, and a tree structure will be drawn to analyze Selection Sort.

# 7.6. Analyzing programs with function calls

In this section, we analyze the running time of programs or program fragments that contain function calls. The function call can be recursive or nonrecursive. We have the following definition.

> **Definition 7.10** *A function that calls itself recursively is called a recursive function. A function that does not call itself but calls other functions is called a nonrecursive function.*

We start with analyzing nonrecursive programs.

## Analyzing nonrecursive programs

A nonrecursive program is the one that contains nonrecursive functions. The following example analyzes a nonrecursive program that contains a nonrecursive function.

---

[3]For small positive values of $x$, we have $\log(1 + x) \approx (1 + x) - 1 = x$.

**Algorithm 7.26:** The algorithm of Example 7.36

1: sum $= 0$
2: **for** $(i = 1; i \le f(n); i + +)$ **do**
3: $\quad |$ sum $+ = i$
4: **end**

■ **Example 7.36** Consider the fragment in Algorithm 7.26 shown below. Here f(n) is a function call. Give a simple and tight Big-Oh upper bound on the running time of Algorithm 7.26, as a function of n, on the assumption that

(*a*) the running time of f(n) is $O(1)$, and the value of f(n) is 0.

(*b*) the running time of f(n) is $O(n^2)$, and the value of f(n) is n.

Solution   The first statement is an assignment statement with constant time. The inside of the for loop is also a simple statement with constant time. So, the part that will affect overall runtime is the runtime and value of $f(n)$.

(*a*) If the value of $f(n)$ is 0, then the loop will never execute the body. It will only evaulate the condition once. Since the runtime of $f(n)$ (and thus the runtime of checking the condition) is $O(1)$, the overall runtime is $O(1)$.

(*b*) In Figure 7.15, we see how the runtime of for loops is calculated. We add the initialization runtime (usually $O(1)$) plus the cost of going around the loop once multiplied by the number of times we go around the loop, represented $O(1 + (f(n) + 1)g(n))$. We keep in mind "the cost of going around the loop once" is represented in Figure 7.15 as the runtimes of "test" plus "body" plus "reinitialize", with "test" being the condition of the for loop and "body" obviously being the body. Now, we can apply this to our problem. The"test" runtime in the for loop is $O(n^2)$, since that is how long it takes $f(n)$ to run. The "body" of the for loop takes $O(1)$, as it is just a simple statement. The "reinitialization" runtime still takes $O(1)$. We perform the loop $O(n)$ times, since the value of $f(n)$ is $n$. Thus, putting it all together, we have $O(1 + (n^2 + 1 + 1)n) = O(n^3)$.

■

To compute the running time of a program that contains nonrecursive functions, we perform the following steps:

- Firstly, we evaluate the running times of those functions that do not call any other functions.

- Secondly, we evaluate the running times of the functions that call only functions whose running time we have already determined.

- Finally, we proceed in this manner until we have evaluated the running time for all functions.

---

**Algorithm 7.27:** The algorithm of Example 7.37

---

1: # include <stdio. $h$ >

2: int cat(*int x, int n*)

3: int cow(*int x, int n*)

   // *This line is left intentionally blank*

4: main( )

5: int $a, n$

6: scanf(*"%d", & n*)

7: $a$=cow(*0, n*);

8: printf(*"%d n",* cat(*a, n*))

   // *This line is left intentionally blank*

9: int cat(*int x, int n*)

10: int $i$

11: **for** $(i = 1; i \leq n; i + +)$ **do**

12: | $x+ = i;$

13: **end**

14: **return** $x$

   // *This line is left intentionally blank*

15: int cow(*int x, int n*)

16: int $i$

17: **for** $(i = 1; i \leq n; i + +)$ **do**

18: | $x+ =$ cat(*i, n*);

19: **end**

20: **return** $x$

---

For more illustration of the given steps, we present the following example.

**Example 7.37**   Given the (meaningless) program in Algorithm 7.27. What is the time complexity of this program?

**Solution**   Note that this program is nonrecursive. Figure 7.22 shows a basic scheme of the program shown in Algorithm 7.27. In view of Figure 7.22, we first analyze the function cat, which does not call any function, then analyze the function cow, which calls the function cat only, and finally analyze the function main which calls both the functions cow and cat.

It is clear that the function cat is $O(n)$. Now, if the function call is in the body of a for loop, we add its cost to the bound on the time for each iteration. It follows that the running time of a call to cow is $O(n^2)$. Next, when the function call is within a simple statement, we add its cost to the cost of that statement. Thus, the function main takes $O(n^2)$ times. Therefore, the running time of this program is $O(n^2)$. ∎

The function `main` calls both the functions `cow` and `cat`, the function `cow` calls the function `cat`, and the function `cat` does not call any function. We first analyze the function `cat`, then to analyze the function `cow`, and finally to analyze the function `main`.

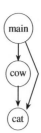

Figure 7.22: A basic scheme of the program shown in Algorithm 7.27.

## Analyzing recursive programs

A recursive program is the one that contains recursive functions. Determining the running time of a function that calls itself recursively requires more work than analyzing nonrecursive functions. Running time of a recursive program is represented by a recurrence.

Recall that a recurrence is an equation or inequality that describes a function in terms of its value on smaller inputs.

For example, the following recurrence

$$T(1) = 1, \quad T(n) = 3T(n-1) + 4, \quad n = 1, 2, 3, \ldots, \qquad (7.9)$$

describes the function (see Exercise 5.2 (b))

$$T(n) = 3^n - 2, \quad n = 0, 1, 2, \ldots. \qquad (7.10)$$

The following are some instances for recurrence formulas of recursive algorithms.

- A recursive algorithm that loops through the input to eliminate one item. For example:
$$T(n) = T(n-1) + 1, \; T(n) = T(n-1) + n, \text{ etc.}$$

- A recursive algorithm that halves the input. For example:
$$T(n) = T(n/2) + 1, \; T(n) = T(n/2) + n, \text{ etc.}$$

- A recursive algorithm that splits the input into two halves. For example:
$$T(n) = 2T(n/2) + 1, \text{ etc.}$$

The question that naturally arises now is: how to determine the actual running time of a recursive program? The answer to this question is by following three steps:

- Firstly, derive a recurrence formula from the given recursive program.

- Secondly, solve the recurrence formula to find an explicit expression of the recurrence. In this section, we present and use the so-called **iteration method** to solve recurrences. In the iteration method, we decompose the recurrence into a series of terms, and derive the $n$th expression from the previous ones. Other methods for solving recurrences will be the substance of the next section.

- Finally, bound the recurrence explicit formula by an asymptotic expression that involves $n$. For instance, from (7.10), it is clear that the running time for the recursive program of the recurrence formula (7.9) is $O(3^n)$.

In the following examples, we introduce three recursive programs, derive their recurrence formulas, and analyze their running times.

**Example 7.38 (Factorial)** The recursive program in Algorithm 7.28 computes the factorial function $n!$. Assuming the cost times of the simple statements `return 1` in line (3) and `return n` in line (6) are $c_1$ and $c_2$, respectively, and observing that the function `fact(n-1)` that the program calls in line (6) is the same problem but with size $(n-1)$, we conclude that the cost time of executing line (6) is the constant time $c_2$ plus the time taken by the function `fact(n-1)`. Let $T(n)$ be the running time of Algorithm 7.28 on an integer $n$. It follows that the recurrence formula for this recursive program is

$$
\begin{aligned}
T(1) &= c_1, \\
T(n) &= c_2 + T(n-1), \quad n > 1.
\end{aligned}
\tag{7.11}
$$

From Example 5.3, we have

$$T(n) = (n-1)c_2 + T(1) = (n-1)c_2 + c_1.$$

Thus, the recurrence formula (7.11) of Algorithm 7.28 describes the linear function $T(n) = c_2 n + (c_1 - c_2)$ for $n \geq 1$. As a result, the factorial function algorithm is linear, running in $O(n)$ time. ∎

---

**Algorithm 7.28:** Factorial function algorithm: `fact(int n)`

---

**Input:** A positive integer $n$
**Output:** The factorial $n!$

1: `fact`(*int n*)
2: **if** $(n \leq 1)$ **then**
3: | **return** 1
4: **end**
5: **else**
6: | **return** $n *$`fact`$(n-1)$
7: **end**

---

---

**Algorithm 7.29:** Integer power algorithm: pow(int $x$,int $n$)

---

    **Input:** Positive integers $x$ and $n$
    **Output:** The $n$th power of $x$
1: pow(*int x, int n*)
2: **if** $(n == 0)$ **then**
3:    |   **return** 1
4: **end**
5: **else**
6:   |   **if** $(n\%2 == 0)$ **then**                            // Checking if $n$ is even
7:   |   |   int $p = $ pow$(x, n/2)$
8:   |   |   **return** $p * p$
9:   |   **else**
10:   |   |   int $p = $ pow$(x, (n-1)/2)$
11:   |   |   **return** $x * p * p$
12:   |   **end**
13:   |   **end**
14: **end**

---

🔲    **Example 7.39 (Integer Power)**  The recursive program in Algorithm 7.29 computes $x^n$. We assume that the cost time of the simple statement return 1 in line (3) is $c_1$, and that the cost time of each of the simple statements return p*p in line (8) and return x*p*p in line (11) is $c_2$. Note that the function pow (x,n/2) that the program calls in line (7) is the same problem but with size n/2, and that the function pow (x, (n-1) /2) that the program calls in line (10) is the same problem but with size (n-1)/2. Let $T(n)$ be the running time of Algorithm 7.29 on an integer $n$, then the recurrence formula for this recursive program is

$$T(n) = \begin{cases} c_1, & \text{if } n = 1; \\ T(n/2) + c_2, & \text{if } n > 1 \text{ and is even;} \\ T((n-1)/2) + c_2, & \text{if } n > 1 \text{ and is odd.} \end{cases} \quad (7.12)$$

Since the function $T$ is monotonically increasing, the recurrence formula (7.12) can be simplified to

$$T(n) = T(\lfloor n/2 \rfloor) + c_2 \text{ if } n > 1,$$

and hence

$$T(n) \leq T(n/2) + c_2.$$

The same method that was used in Example 5.4 can be used to show that

$$T(n) \leq kc_2 + T(n/2^k) \leq c_2 \log n + T(1) = c_2 \log n + c_1.$$

Thus, the recurrence formula (7.12) of Algorithm 7.29 describes the logarithmic function $T(n) = c_2 \log n + c_1$. As a result, the integer power algorithm is logarithmic, running in $O(\log n)$ time. ∎

Other examples of recursive programs are Binary Search and Merge Sort which will be introduced and analyzed in Chapter 8.

At the end of this section, we note that sometimes it is convenient to distinguish common recursive algorithms as in Table 7.2.

## 7.7. The complexity class NP-complete

In 1971, Leonid Levin, working independently with Stephen Cook, discovered the existence of so-called NP-complete problems by showing that satisfiability problem (see Section 1.5) is NP-complete. NP-complete problems occupy an important place in computational complexity theory, and they are confined to the realm of decision problems. So, in this section, we shift our focus from "algorithms" to "problems". We have the following definitions.

> **Definition 7.11** *An abstract problem is a binary relation on a set of problem instances and a set of problem solutions.*

For example, in a problem that we call the shortest path problem (this problem will be formally introduced in Section 9.3), we are given an undirected graph $G$ and two vertices $u$ and $v$ of $G$, and we wish to find a path between $u$ and $v$ that uses the fewest edges. This problem itself is a binary relation that associates each instance of a graph and two vertices with a shortest path in the graph that connects the two vertices. Shortest paths are not necessarily unique, so a given problem instance may have more than one solution.

> **Definition 7.12** *A decision problem is any problem to which the answer is simply yes or no (or, more formally, 1 or 0).*

Examples of decision problems are: Is this a satisfiable proposition? Is this a Hamiltonian graph? Is this an Eulerian graph? (See Section 4.3 for the definitions of Eulerian and Hamiltonian graphs) etc.

Many problems are not decision problems but rather optimization problems (these problems will be formally introduced and studied in Part IV), for which we find the optimal (maximal or minimal) solution to a problem from all the feasible solutions (a set of values for the decision variables that satisfies all of the constraints in an optimization problem). For example, in the shortest path problem, we wish to find a feasible path with the fewest number of edges. NP-completeness applies directly not to optimization problems, but to decision problems. However, we can usually recast an optimization problem to a decision problem that is no harder by imposing a bound on the value to be optimized. For example, a decision problem related to the shortest path problem is the following: Given an undirected graph $G$, two vertices $u$ and $v$ of

| Divide and conquer algorithms $T(n) = aT(n/b) + f(n), \ (a \geq 1 \text{ and } b > 1)$ | Running complexity time |
|---|---|
| $T(n) = T(n/2) + c$ | $\Theta(\log n)$ |
| $T(n) = T(n/3) + c$ | $\Theta(\log n)$ |
| $T(n) = T(n/2) + cn$ | $\Theta(n)$ |
| $T(n) = T(n/3) + cn$ | $\Theta(n)$ |
| $T(n) = 2T(n/2) + cn$ | $\Theta(n \log n)$ |
| $T(n) = 3T(n/3) + cn$ | $\Theta(n \log n)$ |
| $T(n) = 3T(n/2) + cn$ | $\Theta(n^{\log 2(3)})$ |
| $T(n) = 4T(n/2) + cn$ | $\Theta(n^2)$ |
| $T(n) = 2T(n/2) + cn^2$ | $\Theta(n^2)$ |
| $T(n) = 4T(n/2) + cn^2$ | $\Theta(n^2 \log n)$ |

| Chip and conquer algorithms $T(n) = T(n - a) + f(n), \ (a \geq 1)$ | Running complexity time |
|---|---|
| $T(n) = T(n - 1) + c$ | $\Theta(n)$ |
| $T(n) = T(n - 1) + cn$ | $\Theta(n^2)$ |
| $T(n) = T(n - 1) + cn^2$ | $\Theta(n^3)$ |

| Exponential algorithms | Running complexity time |
|---|---|
| $T(n) = 2T(n - 1) + f(n)$ | $\Omega(2^n)$ |
| $T(n) = 3T(n - 1) + f(n)$ | $\Omega(3^n)$ |
| $T(n) = 4T(n - 1) + f(n)$ | $\Omega(4^n)$ |
| $T(n) = 2T(n - 2) + f(n)$ | $\Omega(2^{n/2})$ |
| $T(n) = 2T(n - 3) + f(n)$ | $\Omega(2^{n/3})$ |
| $T(n) = T(n - 1) + T(n - 2) + f(n)$ | $\Omega(2^{n/2})$ |
| $T(n) = T(n - 1) + T(n - 2) + T(n - 3) + f(n)$ | $\Omega(2^{n/2})$ |
| $T(n) = \sum_{i=1}^{n-1} T(i) + f(n)$ | $\Omega(2^{n/2})$ |

Table 7.2: Some common recurrence formulas and their complexities.

$G$ and a positive integer $k$, does a path exist between $u$ and $v$ consisting of at most $k$ edges?

In order to introduce the NP-complete problems, we must first define the class NP. Before that, we define the following notions and terminology in the computational complexity theory. Recall that an efficient algorithm the one that can run in polynomial time. We also have the following definitions.

**Definition 7.13** *A problem is said to be tractable if there exists an efficient algorithm that solves all instances of it, and is intractable otherwise.*

**Definition 7.14** *A certificate of a solution to a decision problem is a scheme that is used to check and certify whether a decision problem gives the answer yes or no.*

**Definition 7.15** *A decision problem is called verifiable if a certificate of a solution to the problem exists.*

In particular, a decision problem is verifiable in polynomial time if for every instance with the answer yes, there is a solution $S$ which we can use to check in polynomial time that the answer is yes.

The notion of reduction from one problem to another explains to us how problems relate.

**Definition 7.16** *A reduction of a decision problem $Q_1$ to a decision problem $Q_2$ is a mapping of every instance $q_1$ of problem $Q_1$ to an instance $q_2$ of problem $Q_2$ such that $q_1$ is yes if and only $q_2$ is yes.*

A reduction allows us to say "If I can solve problem $Q_1$, then I can also solve problem $Q_2$". In this case, if for example $Q_1$ is solvable in polynomial time, then $Q_2$ can be solved in polynomial time. This is helpful to know how difficult problem $Q_1$ is. A useful property is that reductions are transitive. In particular, we can show that if $Q_1$ reduces to $Q_2$ in polynomial time and $Q_2$ reduces to $Q_3$ in polynomial time, then $Q_1$ reduces to $Q_3$ in polynomial time.

**Definition 7.17** *A problem is classified as P if it can be solved in polynomial time.*

All of the actual problems we have thus far solved in this class have been in P. This

It is common for many problems, including all of the problems we will study in Chapters 8 and 9, are in P.

Because P stands for polynomial, many people assume that NP must stand for "non-polynomial". This is incorrect. The set NP refers to "non-deterministic polynomial". We need some more definitions.

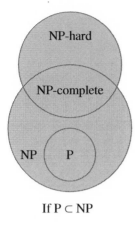 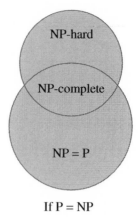

Figure 7.23: Graphical relationships among the complexity classes P, NP, NP-hard, and NP-complete.

> **Definition 7.18** *A problem is classified as NP if it can be verified in polynomial time.*

Note that $P \subseteq NP$. In fact, being able to solve a problem in polynomial ensures that we can also verify the problem by simply solving it.

> **Definition 7.19** *A problem is classified as NP-hard if any NP problem can be reduced to this problem in polynomial time.*

In other words, a problem is classified as NP-hard if an algorithm for solving it can be translated to solve any NP problem. In light of Definition 7.19, an NP-hard problem is at least as hard as any NP problem (it could be even much harder).

NP-Complete problems are problems that reside in both the NP and NP-hard classes.

> **Definition 7.20** *A problem is classified as NP-complete if it can be verified in polynomial time and that any NP problem can be reduced to this problem in polynomial time.*

Examples of NP-Complete problems include the Hamiltonian path problem, the Boolean satisfiability problem, and the integer programming problem.

It is natural question to ask how do these sets relate to each other. A million dollar question is: Does P=NP? This is a well-known unsolved problem in mathematics and theoretical computer science. Figure 7.23 shows a Venn diagram of the different classes of problems.

# EXERCISES

**7.1**    Choose the correct answer for each of the following multiple-choice questions/items.

(*a*)  Which loop is guaranteed to execute at least one time?

    (*i*)  For loop.                        (*iii*)  Do-while loop.

    (*ii*)  While loop.                    (*iv*)  None of the above.

(*b*)  What is the use of for loop?

    (*i*)  To repeat the statement finite number of times.

    (*ii*)  To repeat the statement until any condition holds true.

    (*iii*)  To repeat the statements for infinite time.

    (*iv*)  To repeat statements inside until any condition is false.

(*c*)  We have three algorithms to solve a problem, Algorithms 1, 2, and 3. The running time function of Algorithm 2 is the function $g(n) = n^2$, that of Algorithm 1 is the function $f(n) = \frac{999999}{n} g(n)$, and that of Algorithm 3 is the function $h(n) = \frac{n}{999999} g(n)$. Here, n is the input size. Which algorithm is asymptotically the fastest?

    (*i*)  Algorithm 1.        (*ii*)  Algorithm 2.        (*iii*)  Algorithm 3.

    (*iv*)  All algorithms have the same performance.

(*d*)  We have three algorithms to solve a problem. The running time function of Algorithm 1 is $f(n) = 1000n$, that of Algorithm 2 is $g(n) = n^2$, and that of Algorithm 3 is $h(n) = \frac{1}{1000}n^3$, where *n* is the input size. Which algorithm is asymptotically the fastest?

    (*i*)                              (*ii*)  Algorithm 2.        (*iii*)  Algorithm 3.

    (*iv*)  All algorithms have the same performance because when $n = 1000$, which is a large input size, $f(1000) = g(1000) = h(1000) = 10^6$.

(*e*)  Suppose that we have an algorithm of seven lines and the cost of execution each line is the constant *c*. By doing a line-by-line analysis, we found that the running time of the algorithm is $c[(n - n^8)/(1 - n)]$.

How many times line $i$ executes (for $i = 1, 2, \ldots, 7$)?

(*i*) $n^i$-times.

(*iii*) $n^{(i+1)}$-times.

(*ii*) $n^{(i-1)}$-times.

(*iv*) None of the above.

(*f*) In the fragment given in Algorithm 7.15, if the for-statement in line (2) is replaced with `for (i=1; i<=n; i++)`, and that in line (3) is replaced with with `for (j=1; j<=n; j++)`, then what is the new running time?

(*i*) The running time will not change.

(*ii*) It is $c_1 + c_2 n + c_2(n-1)n + c_3(n-1)^2$.

(*iii*) It is $c_1 + c_2(n+2) + c_2(n+1)(n+2) + c_3(n+1)^2$.

(*iv*) It is $c_1 + c_2(n+1) + c_2(n-1)(n+1) + c_3(n-1)^2$.

(*g*) Suppose that we have an algorithm of five lines, the cost of execution each line is the constant $c$, and the line $i$ executes $n^i$ times for each $i = 1, 2, \ldots, 5$. If a line-by-line runtime analysis is used, what is the running time of the algorithm?

(*i*) $\dfrac{c(n - n^5)}{1 - n}$.

(*ii*) $\dfrac{c(n - n^6)}{1 - n}$.

(*iii*) $\dfrac{c(1 - n^5)}{1 - n}$.

(*iv*) $\dfrac{c(1 - n^6)}{1 - n}$.

(*h*) Let $f(n)$ be the running time function of the code given in Algorithm 7.30. Which one of the following summations represents $f(n)$?

(*i*) $f(n) = \displaystyle\sum_{i=1}^{n} \sum_{j=i+1}^{n} \sum_{k=i}^{j} c.$

(*iii*) $f(n) = \displaystyle\sum_{i=1}^{n} \sum_{j=i}^{n} \sum_{k=i}^{j} c.$

(*ii*) $f(n) = \displaystyle\sum_{i=1}^{n-1} \sum_{j=i}^{n} \sum_{k=i}^{j} c.$

(*iv*) $f(n) = \displaystyle\sum_{i=1}^{n-1} \sum_{j=i+1}^{n} \sum_{k=i}^{j} c.$

---

**Algorithm 7.30:** The algorithm of Exercise 7.1 (h)

---

```
1: x = 0
2: for (i = 1; i < n; i + +) do
3:      for (j = i + 1; j ≤ n; j + +) do
4:          for (k = i; k ≤ n; k + +) do
5:              x = x + (i − j − k)          // Execution cost is c
6:          end
7:      end
8: end
9: return x
```

---

**Algorithm 7.31:** The algorithm of Exercise 7.1 (i)

---

1: $x = 0$
2: **for** $(i = 1; i \le n; i + +)$ **do**
3:     **for** $(j = \log n; j \ge 1; j - -)$ **do**
4:     |    $x = x + (i - j)$              // Execution cost is $c(i, j)$
5:     **end**
6: **end**
7: **return** $x$

---

(*i*) Given the program snippet in Algorithm 7.31. Which one of the following summations represents the running time of this program?

(*i*) $\displaystyle\sum_{i=1}^{n}\sum_{j=1}^{\log n} c(i, j).$

(*ii*) $\displaystyle\sum_{i=1}^{n}\sum_{j=1}^{\log n} c(i, \log n - j).$

(*iii*) $\displaystyle\sum_{i=1}^{n}\sum_{j=1}^{\log n} c(i, \log n - j + 1).$

(*iv*) None of the above.

(*j*) If the running time of an algorithm is a polynomial in $n$ and is represented by the sum $\sum_{i=1}^{n}\sum_{j=1}^{i^2}(j + (j + 1) + (j + 2) + \cdots + (i^2 - 1) + i^2)$, which one of the following correctly describes the algorithm's running time?

(*i*) quadratic.      (*ii*) cubic.      (*iii*) quartic.      (*iv*) quintic.

(*k*) Which of the following asymptotic statements is/are true?

(*i*) $n^2 + n = O(n^2)$.

(*ii*) $n^2 + n = O(n^3)$.

(*iii*) $n^2 + n = O(n^4)$.

(*iv*) All of the above.

(*l*) Let $f(n) = \Theta(g(n))$ and $k < 0$. Which of the following asymptotic statements is/are true?

(*i*) $|kf(n)| = \Theta(g(n))$.

(*ii*) $|kf(n)| = O(g(n))$.

(*iii*) $|kf(n)| = \Omega(g(n))$.

(*iv*) All of the above.

(*m*) Which one of the following statements about Theorem 7.1 is incorrect?

(*i*) The converse of item (*i*) is false.

(*ii*) The converse of item (*i*) is true.

(*iii*) The converse of item (*ii*) is false.

(*iv*) The contrapositive of (*ii*) is true.

---

**Algorithm 7.32:** The algorithm of Exercise 7.1 (o)

---

1: cnt = 0
2: **for** $(i = 0; i < 64; i + +)$ **do**
3:     **for** $(j = 0; j^2 < 64; j + +)$ **do**
4:         **for** $(k = 1; k^2 < 64/k; k + +)$ **do**
5:             **if** $(2i + j \geq 3k)$ **then**
6:                 cnt ++                     // Execution cost is $c$
7:             **end**
8:         **end**
9:     **end**
10: **end**
11: **return** $x$

---

(*n*) Which one of the following is true?

    (*i*) $a^n$ is $O(b^n)$ if $1 < b < a$.         (*iii*) $n^a$ is $O(b^n)$ for any $a$ and $b > 1$.

    (*ii*) $n^a$ is $O(n^b)$ if $1 < b < a$.         (*iv*) $a^n$ is $O(n^b)$ for any $b$ and $a > 1$.

(*o*) Let $f(n)$ be the worst-case running time function of the code given in Algorithm 7.32. Which one of the following summations represents $f(n)$?

    (*i*) $\displaystyle\sum_{i=0}^{64}\sum_{j=0}^{64}\sum_{k=1}^{64} c.$                (*iii*) $\displaystyle\sum_{i=0}^{64}\sum_{j=0}^{8}\sum_{k=0}^{4} c.$

    (*ii*) $\displaystyle\sum_{i=0}^{64}\sum_{j=0}^{8}\sum_{k=1}^{4} c.$                (*iv*) $\displaystyle\sum_{i=0}^{63}\sum_{j=0}^{7}\sum_{k=1}^{3} c.$

(*p*) If the running time of an algorithm is a polynomial of degree 5 in $n$ and is represented by the sum

$$\sum_{m=1}^{n}\sum_{k=1}^{m^2}[(k + (k + 1) + (k + 2) + \cdots + (m^2 - 1) + m^2)(64c)].$$

Which of the following is the greatest lower bound[4] for the algorithm's running time?

    (*i*) $\frac{7c}{8}n^5$.         (*ii*) $\frac{3c}{4}n^5$.         (*iii*) $\frac{c}{2}n^5$.         (*iv*) $\frac{c}{4}n^5$.

---

[4]By the greatest lower bound, we mean the largest polynomial bounds the running time from below.

($q$) Which one of the following is false?

    ($i$) $n^a$ is $O(n^b)$ if $a \le b$.            ($iii$) $a^n$ is $O(b^n)$ if $1 < b < a$.

    ($ii$) $n^a$ is not $O(n^b)$ if $b < a$.      ($iv$) $a^n$ is not $O(b^n)$ if $1 < b < a$.

($r$) Let $f(n) = (n^2 - 1)^5$. Which of the following is true about $f(n)$?

    ($i$) $f(n) = \Theta(n^2)$.               ($iii$) $f(n) = \Theta(n^7)$.

    ($ii$) $f(n) = \Theta(n^5)$.             ($iv$) $f(n) = \Theta(n^{10})$.

($s$) The tight Big-Oh bound is best expressed by

    ($i$) $O$.                     ($iii$) $\Theta$.

    ($ii$) $\Omega$.                   ($iv$) All the above.

($t$) Which one of the following functions is a simple and tight bound on the function $f(n) = (5.01)^{n-1} + 5^{n+1}$?

    ($i$) $(5.01)^{n-1}$.      ($ii$) $(5.01)^n$.      ($iii$) $5^n$.             ($iv$) $5^{n+1}$.

($u$) Which one of the following functions is a simple and tight bound on the function $f(n) = (9.99)^{n+1} + (10)^n + (10.01)^{n-1}$?

    ($i$) $(9.99)^{n+1}$.      ($ii$) $(10)^n$.      ($iii$) $(10.01)^{n-1}$.      ($iv$) $(10.01)^n$.

($v$) Which one of the following is/are true?

    ($i$) If a logarithmic-time algorithm exists to solve a problem, then we can say that the problem is solved efficiently.

    ($ii$) If a polynomial-time algorithm exists to solve a problem, then we can say that the problem is solved efficiently.

    ($iii$) If a polylogarithmic-time algorithm exists to solve a problem, then we can say that the problem is solved efficiently.

    ($iv$) All the above.

**7.2** Write an algorithmic code of the following algorithm which solves the same problem in Example 7.6.

(i) If L is of length 1, return the first item of L.

(ii) Set v1 to the first item of L.

(iii) Set v2 to the output of performing find-max() on the rest of L.

(iv) If v1 is larger than v2, return v1. Otherwise, return v2.

**7.3**    For the algorithm in Exercise 7.2, answer the following questions by "Yes" or "No" with any necessary comments.

(*a*)  Does it have defined inputs?                (*c*)  Is it guaranteed to terminate?

(*b*)  Does it have defined outputs?             (*d*)  Does it produce the correct result?

**7.4**    Find the running time equipped with worst-case performance for Algorithm 7.12 by carrying out a line-by-line analysis. Assume that the cost of the statement in line (i) is $c_i$ for i=1,2,...,5.

**7.5**    Do a line-by-line analysis to find the running time of the fragment given in Algorithm 7.33.

**7.6**    Let $p$ and $q$ be two positive integers such that $p \le q$. How many times do we go around each of the following loops, as a function of $p, q$, and possibly $k$? (Here $k$ is a positive integer. Also, in item (f), we assume that $q$ is divisible by $k$ and its powers).

(*a*)  for $(i = p; i \le q; i + +)$.                (*d*)  for $(i = q; i \ge p; i- = k)$.

(*b*)  for $(i = q; i \ge p; i - -)$.                (*e*)  for $(i = p; i \le q; i* = k)$.

(*c*)  for $(i = p; i \le q; i+ = k)$.             (*f*)  for $(i = q; i \ge p; i/ = k)$.

**7.7**    Given the code in Algorithm 7.34, write the summations that represent the running time of the code and solve the summations to find an approximate running time. You must show all your work without doing a line-by-line analysis.

---
**Algorithm 7.33:** The algorithm of Exercise 7.5
---
```
1: sum = 0                              // Cost = c₁
2: for (i = log n; i ≥ 1; i − −) do      // Cost = c₂
3:     for (j = n; j > 1; j/ = 2) do     // Cost = c₃
4:         sum + = array[i][j]            // Cost = c₄
5:     end
6: end
```

$$\text{1: } sum = 0 \qquad // \text{ Cost} = c_1$$
$$\text{2: } \textbf{for } (i = \log n; i \ge 1; i - -) \textbf{ do} \qquad // \text{ Cost} = c_2$$
$$\text{3: } \quad \textbf{for } (j = n; j > 1; j/ = 2) \textbf{ do} \qquad // \text{ Cost} = c_3$$
$$\text{4: } \quad\quad sum + = array[i][j] \qquad // \text{ Cost} = c_4$$
$$\text{5: } \quad \textbf{end}$$
$$\text{6: } \textbf{end}$$

---
**Algorithm 7.34:** The algorithm of Exercise 7.7
---
$$\text{1: } x = 0$$
$$\text{2: } \textbf{for } (i = n^2; i \le n^2 + 5; i + +) \textbf{ do}$$
$$\text{3: } \quad \textbf{for } (j = 4; j \le n; j + +) \textbf{ do}$$
$$\text{4: } \quad\quad x = x + (i - j) \qquad // \text{ Execution cost is } c$$
$$\text{5: } \quad \textbf{end}$$
$$\text{6: } \textbf{end}$$
$$\text{7: } \textbf{return } x$$

---

**Algorithm 7.35:** The algorithm of Exercise 7.8

---

1: $x = 0$
2: **for** $(i = 1; i \leq n; i + +)$ **do**
3:    | **for** $(j = 1; j \leq 3i^3; j + +)$ **do**
4:    |   | $x = x + (i - j)$                      // Execution cost is $c$
5:    | **end**
6: **end**
7: **return** $x$

---

**7.8**   Given Algorithm 7.35, write the summations that represent the running time of the algorithm and solve them to determine an upper bound. You must show all your work without doing a line-by-line analysis.

**7.9**   Find upper and lower bounds for the running time of the algorithm represented by the following summation $f(n) = \sum_{i=n}^{4n^3} \sum_{j=i}^{8n^3} c$, where $c$ is the execution cost of the statement that contributed most to the running time of the algorithm.

**7.10**   Given Algorithm 7.30 which has a running time function $f(n)$. Use bounding to solve the summations that represent $f(n)$ and find a tight asymptotic bound for $f(n)$. You must show all your work.

**7.11**   Use the Big-Oh definition to prove the following asymptotic statements.

(a) $5n^2 - 3n + 20 = O(n^2)$.

(b) $4n^2 - 12n + 10 = O(n^2)$.

(c) $5n^5 - 4n^4 - 2n^2 + n = O(n^5)$.

(d) $n^{3/2} + \sqrt{n} \sin n + n \log n = O(n^2)$.

**7.12**   Use the Big-Omega definition to prove the following asymptotic statements.

(a) $4n^2 + n + 1 = \Omega(n^2)$.

(b) $n \log n - 2n + 13 = \Omega(n \log n)$.

**7.13**   Use the Big-Theta definition to prove the following asymptotic statements.

(a) $n^5 + n^3 + 7n + 1 = \Theta(n^5)$.

(b) $\frac{1}{2}n^2 - 3n = \Theta(n^2)$.

**7.14**   Let $f_1, f_2, \ldots, f_k, g_1, g_2, \ldots, g_k$ be asymptotically nonnegative functions. Assume that $f_i(n) = O(g_i(n))$ for each $i = 1, 2, \ldots, k$, prove that

(a) $\displaystyle\sum_{i=1}^{k} f_i(n) = O\left(\max_{1 \leq i \leq k}\{g_i(n)\}\right)$.[5]

(b) $\displaystyle\prod_{i=1}^{k} f_i(n) = O\left(\prod_{i=1}^{k} g_i(n)\right)$.[6]

---

[5] It is also true that $\sum_{i=1}^{k} f_i(n) = O\left(\sum_{i=1}^{k} g_i(n)\right)$.
[6] Recall that $\sum_{i=1}^{k} f_i(n) = f_1(n) + f_2(n) + \cdots + f_k(n)$. Likewise, $\prod_{i=1}^{k} f_i(n) = f_1(n)f_2(n)\cdots f_k(n)$.

**7.15**   Prove Properties 7.1 and 7.6. Prove also Properties 7.2 and 7.7 for Big-Oh.

**7.16**   Use limits to prove the following asymptotic statements.

(a)  $\sqrt{4n^2 + 1} = \Theta(n)$.      (b)  $n^n = \Omega(n!)$.          (c)  $\sqrt{n + 4} - \sqrt{n} = O(1)$.

(d)  $n \log(n^2) + (n - 1)^2 \log\left(\dfrac{n}{2}\right) = \Theta\left(n^2 \log n\right)$. [7]

**7.17**   Let $f(n) = 2^{(n+10)}$. Does $f(n) = O(2^n)$? Answer by YES or NO. If yes, use the Big-Oh definition to show that $f(n)$ is $O(2^n)$. If no, use the proof by contradiction to show that $f(n)$ is not $O(2^n)$.

**7.18**   Choose the correct answer for each of the following multiple-choice questions/items.

(a) Consider the selection statement given in Algorithm 7.2. If $S_1$ and $S_2$ are $O(f_1(n))$ and $O(f_2(n))$, respectively, and the condition is "2 == $\sqrt{2}$", which can tell which branch of the selection statement is taken, then the running time of the selection statement is

   (i)  $O(f_1(n))$.                          (iii)  $O(f_2(n))$.
   (ii) $O(1 + f_1(n))$.                     (iv)  $O(\max\{f_1(n), f_2(n)\})$.

(b) Consider the while statement given in Algorithm 7.6. If the statement(s) is (are) $O(f(n))$, and the condition is "1 != $\sqrt{1}$", which is known to be true or false right from the start, then the running time of the while statement is

   (i)  $O(1)$.                                (iii)  $O(1 + f(n))$.
   (ii) $O(f(n))$.                             (iv)  None of the above.

(c) Given the program snippet in Algorithm 7.36. What is the time complexity of this program?

   (i)  $O((n^2/2) \log(n))$.                (iii)  $O(n^2 \log(n))$.
   (ii) $O(n \log^2(n))$.                     (iv)  $O((n^2/2) \log^2(n))$.

---
**Algorithm 7.36:** The algorithm of Exercise 7.18 (c)

```
1: for (i = n; i ≥ 1; i/ = 2) do
2:      for (j = 1; j ≤ log(i); j + +) do
3:          something O(n)
4:      end
5: end
```
---

[7]Hint: For (b), you may use the equation: $n \log(n^2) + (n-1)^2 \log(n/2) = 2n \log n + (n-1)^2 (\log n - \log 2)$.

---

**Algorithm 7.37:** The algorithm of Exercise 7.18 (d)

---

1: something $O(n + m)$
2: **while** $(n > 0)$ **do**
3: $\quad$ something $O(n)$
4: $\quad$ $n - -$
5: **end**
6: **while** $(m > 0)$ **do**
7: $\quad$ something $O(m)$
8: $\quad$ $m/ = 2$
9: **end**

---

(*d*) Given the program snippet in Algorithm 7.37. What is the time complexity of this program?

$\quad$ (*i*) $O(n^2 + m \log(m))$.  $\qquad\qquad$ (*iii*) $O(n^2 + m^2)$.

$\quad$ (*ii*) $O(n^2 + \log(m))$.  $\qquad\qquad$ (*iv*) $O(n^2 + m/2)$.

**7.19** Consider the fragment in Algorithm 7.26. Note that $f(n)$ is a function call. Give a simple and tight Big-Oh upper bound on the running time of Algorithm 7.26, as a function of $n$, on each of the assumptions below. Justify your answers with complete evidences.

(*a*) the running time of $f(n)$ is $O(n)$, and the value of $f(n)$ is $n$.

(*b*) the running time of $f(n)$ is $O(n)$, and the value of $f(n)$ is $n!$.

**7.20** Suppose that line (17) of Algorithm 7.27 was replaced by:

$$\text{for } (i = 1; i \leq \mathtt{cat}(n, n); i + +).$$

What would the running time of "`main`" be then? Justify your answer with complete evidence.

**7.21** Algorithm 7.38 is called a stochastic[8] interior-point algorithm and is used for solving a class of two-stage stochastic linear optimization problems of $K$ scenarios and with $n$ variables in the first stage and $m$ variables in the second stage. The input includes a matrix $A$ and a barrier function $f$, and the output is an $\epsilon$-optimal solution to the given stochastic linear program. No matter how the algorithm works and no matter how the mathematical notations therein look like and how they work, the graph structure (flowchart) of the algorithm can be visualized. Draw the graph structure for Algorithm 7.38 by completing the subgraph diagram shown in Figure 7.24.

---

[8]Deterministic problems do not consider any uncertainties in input data, whereas stochastic problems model the uncertainties in data with appropriate probability distributions.

---

**Algorithm 7.38:** The algorithm of Exercise 7.21

---

1: initialize $\epsilon > 0, \gamma \in (0,1), \theta > 0, \beta > 0, \boldsymbol{x}^{(0)}, \mu^{(0)} > 0, \boldsymbol{\lambda}^{(0)}$

2: set $\boldsymbol{x} \triangleq \boldsymbol{x}^{(0)}, \mu \triangleq \mu^{(0)}, \boldsymbol{\lambda} = \boldsymbol{\lambda}^{(0)}$

3: **while** $(\mu \geq \epsilon)$ **do**

4:     **for** $(k = 1; k < K + 1; k + +)$ **do**

5:         find $\left(\boldsymbol{y}^{(k)}, \boldsymbol{z}^{(k)}, \boldsymbol{s}^{(k)}\right)$

6:         choose a scaling element $p$

7:         compute $\left(\boldsymbol{y}_p^{(k)}, \boldsymbol{s}_p^{(k)}\right)$ by scaling $\left(\boldsymbol{y}^{(k)}, \boldsymbol{s}^{(k)}\right)$ with $p$

8:     **end**

9:     compute $\boldsymbol{g} = \nabla f(\mu, \boldsymbol{x}) - A^{\mathsf{T}} \boldsymbol{\lambda}$

10:     compute $\Delta \boldsymbol{x} = -\left(\left(\nabla^2 f\right)^{-1} - \left(\nabla^2 f\right)^{-1} A^{\mathsf{T}} \left(A \left(\nabla^2 f\right)^{-1} A^{\mathsf{T}}\right)^{-1} A \left(\nabla^2 f\right)^{-1}\right) (\boldsymbol{g})$

11:     compute $\Delta \boldsymbol{\lambda} = \left(A \left(\nabla^2 f(\mu, \boldsymbol{x})\right)^{-1} A^{\mathsf{T}}\right)^{-1} A \left(\nabla^2 f(\mu, \boldsymbol{x})\right)^{-1} (\boldsymbol{g})$

12:     compute $\delta(\mu, \boldsymbol{x}) = \sqrt{\dfrac{1}{\mu}(\Delta \boldsymbol{x})^{\mathsf{T}} \nabla^2 f(\mu, \boldsymbol{x})(\Delta \boldsymbol{x})}$

13:     **while** $(\delta > \beta)$ **do**

14:         set $\boldsymbol{x} \triangleq \boldsymbol{x} + \theta \Delta \boldsymbol{x}$

15:         set $\boldsymbol{\lambda} \triangleq \boldsymbol{\lambda} + \theta \Delta \boldsymbol{\lambda}$

16:         **for** $(k = 1; k < K + 1; k + +)$ **do**

17:             find $\left(\boldsymbol{y}^{(k)}, \boldsymbol{z}^{(k)}, \boldsymbol{s}^{(k)}\right)$

18:             choose a scaling element $p$

19:             compute $\left(\boldsymbol{y}_p^{(k)}, \boldsymbol{s}_p^{(k)}\right)$ by scaling $\left(\boldsymbol{y}^{(k)}, \boldsymbol{s}^{(k)}\right)$ with $p$

20:         **end**

21:         compute $\boldsymbol{g} = \nabla f(\mu, \boldsymbol{x}) - A^{\mathsf{T}} \boldsymbol{\lambda}$

22:         compute $\Delta \boldsymbol{x} = -\left(\left(\nabla^2 f\right)^{-1} - \left(\nabla^2 f\right)^{-1} A^{\mathsf{T}} \left(A \left(\nabla^2 f\right)^{-1} A^{\mathsf{T}}\right)^{-1} A \left(\nabla^2 f\right)^{-1}\right) (\boldsymbol{g})$

23:         compute $\Delta \boldsymbol{\lambda} = \left(A \left(\nabla^2 f(\mu, \boldsymbol{x})\right)^{-1} A^{\mathsf{T}}\right)^{-1} A \left(\nabla^2 f(\mu, \boldsymbol{x})\right)^{-1} (\boldsymbol{g})$

24:         compute $\delta(\mu, \boldsymbol{x}) = \sqrt{\dfrac{1}{\mu}(\Delta \boldsymbol{x})^{\mathsf{T}} \nabla^2 f(\mu, \boldsymbol{x})(\Delta \boldsymbol{x})}$

25:     **end**

26:     set $\mu \triangleq \gamma \mu$

27: **end**

28: **for** $(k = 1; k < K + 1; k + +)$ **do**

29:     apply inverse scaling to $\left(\boldsymbol{y}_p^{(k)}, \boldsymbol{s}_p^{(k)}\right)$ to compute $\left(\boldsymbol{y}^{(k)}, \boldsymbol{s}^{(k)}\right)$

30: **end**

---

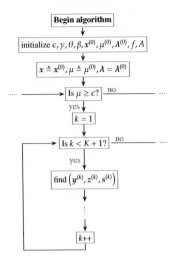

Figure 7.24: A subgraph structure of Algorithm 7.38.

**7.22**    Consider Algorithm 7.38. Suppose that the running time of the block in lines (5)-(7) is $O(m)$, and the running time of the block in lines (9)-(12) is $O(n)$. Similarly, suppose that the running time of the block in lines (17)-(19) is $O(m)$, that of the block in lines (21)-(24) is $O(n)$, and that of the statement in line (29) is $O(m)$. Suppose also that all other statements (including those of checking the conditions of the while loops) are of constant times $O(1)$. Let also $N_{in}$ be the number of times we go around the inner while loop of line (13). In view of the graph structure obtained in Exercise 7.21, determine the running time of the algorithm as a whole in each of these cases:

(a) $N_{in} = O(n + Km)$ and $\gamma \in (0, 1)$ is an arbitrarily chosen constant.

(b) $N_{in} = O(1)$ and $\gamma = 1 - \sigma/\sqrt{n + Km}$, where $\sigma > 0$.

In each case, give your answer as a function of $n, m, K, \mu^{(0)}, \epsilon$ using Big-oh, and write a paragraph (or more) that justifies your determination.

# CHAPTER 8

---

# ARRAY AND NUMERIC ALGORITHMS

## Contents

© 2022 by Baha Alzalg | Kindle Direct Publishing, Washington, United States 2022/9
B. Alzalg, *Combinatorial and Algorithmic Mathematics: From Foundation to Optimization*,
DOI 10.5281/zenodo.7110973

In this chapter, we present and analyze some standard array and numeric algorithms, such as array multiplication algorithms, array searching algorithms, array sorting algorithms, and Newton's method algorithm for solving linear and nonlinear systems. We also describe and analyze the integer Euclidean algorithm (or Euclid's algorithm), which is one of the oldest and simplest number theoretic algorithms.

# 8.1. Array multiplication algorithms

Vectors and matrices are arrays of real numbers. In this section, we will take a look at some simple matrix computations. Namely: Multiplying a matrix by a vector, and multiplying a matrix by a matrix.

### Matrix-vector multiplication

Consider a matrix with $n$ rows and $m$ columns, that is, an $n \times m$ matrix:

$$
A = \begin{bmatrix}
a_{11} & a_{12} & a_{13} & \cdots & a_{1m} \\
a_{21} & a_{22} & a_{23} & \cdots & a_{2m} \\
\vdots & \vdots & \vdots & \ddots & \vdots \\
a_{n1} & a_{n2} & a_{n3} & \cdots & a_{nm}
\end{bmatrix}.
$$

We assume that the entries of $A$ are real. Given an $m$-tuple of real numbers:

$$
x = \begin{bmatrix}
x_1 \\
x_2 \\
\vdots \\
x_m
\end{bmatrix},
$$

we can multiply $A$ by $x$ to get a product $b = Ax$, where $b$ is an $n$-tuple. The $i$th component of $b$ is obtained by taking the dot product of $i$th row of $A$ with $x$. That is, its $i$th component is given by

$$
b_i = \sum_{j=1}^{m} a_{ij} x_j. \tag{8.1}
$$

**Example 8.1** An example of matrix-vector multiplication with $n \times m = 3 \times 2$ is

$$
\begin{bmatrix}
1 & 2 \\
3 & 4 \\
5 & 6
\end{bmatrix}
\begin{bmatrix}
7 \\
8
\end{bmatrix}
=
\begin{bmatrix}
1 \times 7 + 2 \times 8 \\
3 \times 7 + 4 \times 8 \\
5 \times 7 + 6 \times 8
\end{bmatrix}
=
\begin{bmatrix}
23 \\
53 \\
77
\end{bmatrix}.
$$

---

**Algorithm 8.1:** Matrix-vector multiplication (row-oriented)

---

1: b=0
2: **for** $(i = 1; i \leq n; i++)$ **do**
3:     **for** $(j = 1; j \leq m; j++)$ **do**
4:       $b_i = b_i + a_{ij}x_j$
5:     **end**
6: **end**

---

A computer code to perform matrix-vector multiplication looks something like the one in Algorithm 8.1. Note that in Example 8.1 we have with $n = 3$ and $m = 2$ is

$$\begin{bmatrix} 23 \\ 53 \\ 77 \end{bmatrix} = 7 \begin{bmatrix} 1 \\ 3 \\ 5 \end{bmatrix} + 8 \begin{bmatrix} 2 \\ 4 \\ 6 \end{bmatrix}.$$

In general, if $b = Ax$, then $b$ is a linear combination of the columns of $A$. So, if $a_j$ denotes the $j$th column of $A$, then we have

$$b = \sum_{j=1}^{m} a_j x_j.$$

Expressing these operations as a code, we have the one in Algorithm 8.2.

If each vector operation is performed by a loop, the code becomes the one in Algorithm 8.3.

Note that Algorithms 8.1 and 8.3 are identical, except that the loops are interchanged. Algorithm 8.1 accesses $A$ by rows, while Algorithm 8.3 accesses $A$ by columns.

Now, we want to know how long it will take to complete the task of Algorithm 8.3. Note that each execution of the inner loop involves two flops. Replace each loop by a summation sign $\Sigma$. Since the inner loop is executed for $i = 1, \ldots, n$, and there are two flops per pass, the total number of flops performed on each execution of the inner loop is $\sum_{i=1}^{n} 2$. Since the outer loop runs from $j = 1$ to $j = m$, the total number of flops is

$$\sum_{j=1}^{m} \sum_{i=1}^{n} 2 = \sum_{j=1}^{m} 2n = 2nm.$$

---

**Algorithm 8.2:** Writing $b = \sum_{j=1}^{m} a_j x_j$ in a code

---

1: $b = 0$
2: **for** $(j = 1; j \leq m; j++)$ **do**
3:     $b = b + a_j x_j$
4: **end**

---

---

**Algorithm 8.3:** Matrix-vector multiplication (column-oriented)

---

1: $b = 0$
2: **for** $(j = 1; j \leq m; j + +)$ **do**
3:     **for** $(i = 1; i \leq n; i + +)$ **do**
4:         $b_i = b_i + a_{ij} x_j$
5:     **end**
6: **end**

---

## Matrix-matrix multiplication

If $A$ is an $n \times m$ matrix, and $X$ is an $m \times l$, we can form the product $B = AX$, which is $n \times l$. The $(i, j)$ entry of $B$ is the dot product of the $i$th row of $A$ with the $j$th column of $X$. That is, the $(i, j)$ entry of $B$ is

$$b_{ij} = \sum_{k=1}^{m} a_{ik} x_{kj}. \tag{8.2}$$

A computer code to perform matrix-matrix multiplication looks something like the one in Algorithm 8.4.

Note that if $l = 1$, the matrix-matrix multiplication represented in (8.2) reduces to the vector-matrix multiplication represented in (8.1), and Algorithm 8.4 reduces to Algorithm 8.1.

Now, we want to know how long it will take to complete the task of Algorithm 8.4. Note that each execution of the innermost loop involves two flops. Replacing each loop by a summation sign $\Sigma$, the total number of flops is

$$\sum_{i=1}^{n} \sum_{j=1}^{l} \sum_{k=1}^{m} 2 = 2nml.$$

In the important case when all of the matrices are square of dimension $n \times n$. We end this section by summarizing the above computational complexities in terms of $n$, where $n$ is the size of the square matrix; see Corollary 8.1.

---

**Algorithm 8.4:** Matrix-matrix multiplication

---

1: B=0
2: **for** $(i = 1; i \leq n; i + +)$ **do**
3:     **for** $(j = 1; j \leq l; j + +)$ **do**
4:         **for** $(k = 1; k \leq m; k + +)$ **do**
5:             $b_{ij} = b_{ij} + a_{ik} x_{kj}$
6:         **end**
7:     **end**
8: **end**

---

> **Corollary 8.1** *The time complexities of the above matrix algorithms are as follow:*
>
> (i) *The matrix-vector multiplication algorithms are quadratic, running in $O(n^2)$ time.*
>
> (ii) *The matrix-matrix multiplication algorithm is cubic, running in $O(n^3)$ time.*

## 8.2. Array searching algorithms

Searching algorithms are a class of algorithms that finds an element within an array or list. In this section, we introduce and analyze two well-known searching algorithms. Namely: Linear search and binary search.

### Linear search

Linear Search, also-called Sequential Search, is a linear-time searching algorithm that sequentially checks each element of the list to be searched until a target is found.

The Linear Search is stated in Algorithm 8.5, in which we search an array $A[0 : n - 1]$ to find a value $x$ that is expected to be present in the array. If the value $x$ is not present in the array, the algorithm terminates unsuccessfully.

To arrive at the Big-Oh of the running time for Algorithm 8.5, we depict a flowchart of this algorithm with running time complexity in Figure 8.1.

It is clear that the while loop of lines (3)-(5) in Algorithm 8.5 may be executed as many as $n$-times, but no more. From the summation rule, the running time of Algorithm 8.5 is $O(n)$, because that is the maximum of the time for the assignment of line (2), the time for the while loop in lines (3)-(5), and the time for the selection statement in lines (6)-(11).

---
**Algorithm 8.5:** Linear search algorithm

---
**Input:** An array $A[0 : n - 1]$ and a sought value $x$
**Output:** Index $i$ such that $x = A[i]$, otherwise $x$ is not found
1: `linear-search`(int $A[\ ]$, int $n$, int $x$)
2: $i = 0$
3: **while** $(x\ != A[i]\ \&\&\ i < n)$ **do**
4: $\quad | \quad i + +;$
5: **end**
6: **if** $(i < n)$ **then**
7: $\quad |$ `printf`("$x$ is present at index" $i$)
8: **end**
9: **else**
10: $\quad |$ `printf`("$x$ is not present in the array")
11: **end**

---

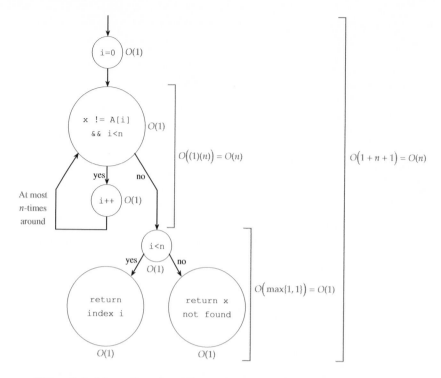

Figure 8.1: Linear Search graph structure with running time complexity.

The running time found here (which is $O(n)$) is the worst-case performance of Algorithm 8.5. What are its best- and average-case performances? This question is left as an exercise for the reader (see Exercise 8.1 $(c)$).

## Binary search

This section introduces another searching algorithm, called Binary Search. It is worth to be mentioned that Binary Search (and Merge Sort that we studied in Algorithm 8.9) are well-known examples of a class of algorithms called divide and conquer algorithms. In this class of algorithms, we divide the problem into "simpler" versions of itself, conquer each problem using the same process (usually recursively), and combine the results of the "simpler" versions to form the final solution.

Binary Search, also known as half-interval search, or logarithmic search, is a searching algorithm that finds the position of a target within a sorted array $A[low : high]$. The binary search algorithm is stated in Algorithm 8.6. For more illustration of the binary search algorithm, in Table 8.1 we show in detail how to apply Algorithm 8.6 to an 8-element array. We first target the value $x = 7$, which is present in the array at index 7. Then, we also target $x = 6$ which is not found in the array.

---

**Algorithm 8.6:** Binary search algorithm

---

**Input:** An array $A$[low:high] of length $n$ and a sought value $x$
**Output:** Index $i$ such that $x = A[i]$, otherwise $x$ is not found

1: binary-search(int $A$[ ], int low, int high, int $x$)
2: **if** (low > high) **then**
3: | printf("$x$ is not present in array")                    // constant time $c_1$
4: **end**
5: **else**
6: | mid = $\lfloor$(low+high)/2$\rfloor$                         // constant time $c_2$
7: | **if** ($x = A$[mid]) **then**
8: | | printf("$x$ is present at index" mid)                  // constant time $c_3$
9: | **end**
10: | **if** ($x < A$[mid]) **then**
11: | | binary-search($A$,low, mid-1, $x$)     // same problem of size $n/2$
12: | **end**
13: | **if** ($x > A$[mid]) **then**
14: | | binary-search($A$, mid+1, high, $x$)   // same problem of size $n/2$
15: | **end**
16: **end**

---

Let $T(n)$ be the running time of Algorithm 8.6 on an array of size $n$. From Algorithm 8.6, it follows that the recurrence formula for this recursive program is

$$
\begin{aligned}
T(1) &= c, \\
T(n) &= c + T(n/2),
\end{aligned} \tag{8.3}
$$

where $c$ might be taken equal to the constant time $c_1 + c_2 + c_3$. From Example 5.4, we have

$$T(n) = ck + T(n/2^k) = c \log n + T(1) = c \log n + c.$$

Thus, the recurrence formula (8.3) of Algorithm 8.6 describes the logarithmic function $T(n) = c \log n + c$. As a result, the binary search algorithm is logarithmic, running in $O(\log n)$ time. This is the worst-case performance of Algorithm 8.6. What are its best- and average-case performances? This question is left as an exercise for the reader (see Exercise 8.1 (e)).

We end this section with the following corollary which summarizes the computational complexity (worst behavior) in terms of the size of the list ($n$).

---

**Corollary 8.2** *The (worst-case) time complexities of the above searching algorithms are as follow:*

(i) *The linear search algorithm is linear, running in $O(n)$ time.*

(ii) *The binary search algorithm is logarithmic, running in $O(\log n)$ time.*

**Input:** $A = [1\ 2\ 3\ 4\ 5\ 7\ 9\ 11]$,
  $x = 7$.

Iteration #1:

| low | | | | | | high | |
|---|---|---|---|---|---|---|---|
| ↓ | | mid | | | | ↓ | |
| 1 | 2 | 3 | 4 | 5 | 6 | 7 | 8 |
| 1 | 2 | 3 | 4 | 5 | 7 | 9 | 11 |

Iteration #2:
As $x = 7 > A[\text{mid}] = 4$, we get

| | | | low | | high | | |
|---|---|---|---|---|---|---|---|
| | | | ↓ | mid | ↓ | | |
| | | | 5 | 6 | 7 | 8 | |
| 1 | 2 | 3 | 4 | 5 | 7 | 9 | 11 |

Iteration #3:
As $x = 7 = A[\text{mid}]$, we stop!

**Output:** $x$ is present in A at index 6.

**Input:** $A = [1\ 2\ 3\ 4\ 5\ 7\ 9\ 11]$,
  $x = 6$.

Iteration #1:

| low | | | | | | high | |
|---|---|---|---|---|---|---|---|
| ↓ | | mid | | | | ↓ | |
| 1 | 2 | 3 | 4 | 5 | 6 | 7 | 8 |
| 1 | 2 | 3 | 4 | 5 | 7 | 9 | 11 |

Iteration #2:
As $x = 6 > A[\text{mid}] = 4$, we get

| | | | low | | high | | |
|---|---|---|---|---|---|---|---|
| | | | ↓ | mid | ↓ | | |
| | | | 5 | 6 | 7 | 8 | |
| 1 | 2 | 3 | 4 | 5 | 7 | 9 | 11 |

Iteration #3:
As $x = 6 < A[\text{mid}] = 7$, we get

low mid high

↘ ↓ ↙

5

| 1 | 2 | 3 | 4 | 5 | 7 | 9 | 11 |
|---|---|---|---|---|---|---|---|

Iteration #4:
As $x = 6 > A[\text{mid}] = 5$, we get

high low

↓  ↓

5  6

| 1 | 2 | 3 | 4 | 5 | 7 | 9 | 11 |
|---|---|---|---|---|---|---|---|

Iteration #5:
As $low > high$, we stop!

**Output:** $x$ is not present in the array $A$.

Table 8.1: A concrete implementation of the binary search algorithm.

# 8.3.  Array sorting algorithms

Sorting algorithms are algorithms that put elements of an array or list in a certain order.  In this section, we introduce and analyze well-known sorting algorithms.  Namely: Insertion sort, selection sort, and merge sort.

### Insertion sort

Insertion Sort is a sorting algorithm similar to how most people arrange a hand of playing cards:

- Start with one card in your hand.

- Pick the next card (to be sorted), make room for it by shifting sorted items, and insert it to the correct location.

- Repeat the previous step for all cards.

Figure 8.2 illustrates the insertion sort algorithm by applying the above steps to sort the array [5 2 4 6 1 3]. Note that, at each iteration, the array is divided into two sub-arrays. A formal description of the insertion sort method is given in Algorithm 8.7.

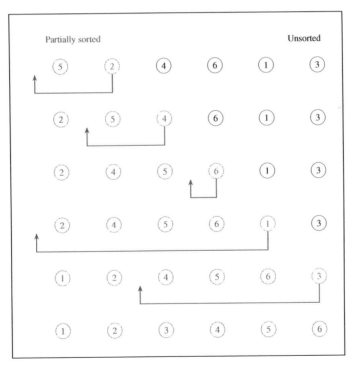

Figure 8.2: A concrete implementation of the insertion sort algorithm.

---

**Algorithm 8.7:** Insertion sort algorithm

---

**Input:** An integer array $A[0 : n - 1]$
**Output:** Array $A$ sorted in ascending order
1: `insertion-sort`(int $A[\ ]$, int $n$)
2: **for** $(i = 1; i < n; i + +)$ **do**                    // Cost $= c_1$, # times $= n$
3:         $next = A[i]$                                     // Cost $= c_2$, # times $= n - 1$
4:         $j = i - 1$                                       // Cost $= c_3$, # times $= n - 1$
5:         **while** $(j > 0\ \&\&\ A[j] > next)$ **do**      // Cost $= c_4$, # times $= \sum_{i=1}^{n-1} t_i$
6:                 $A[j + 1] = A[j]$                          // Cost $= c_5$, # times $= \sum_{i=1}^{n-1}(t_i - 1)$
7:                 $j = j - 1$                                // Cost $= c_6$, # times $= \sum_{i=1}^{n-1}(t_i - 1)$
8:         **end**
9:         $A[j + 1] = next$                                 // Cost $= c_7$, # times $= n - 1$
10: **end**

---

Note that "next" in line (3) is the item to be sorted, line (6) shifts sorted items to make room for "next", and line (9) inserts "next" to the correct location.

In the comments in gray in Algorithm 8.7, $t_i$ denotes the number of times the while statement is executed at iteration $i$ for $i = 1, 2, \ldots, n - 1$. Given this, the total cost is

$$
\begin{aligned}
f(n) \quad = \quad & c_1 n + c_2(n - 1) + c_3(n - 1) + c_4 \sum_{i=1}^{n-1} t_i \\
& + c_5 \sum_{i=1}^{n-1}(t_i - 1) + c_6 \sum_{i=1}^{n-1}(t_i - 1) + c_7(n - 1).
\end{aligned}
\tag{8.4}
$$

We now present best- and worst-case analyses for the insertion sort method:

- Best-case analysis: The array is already sorted. So, $A[j] \leq next$ upon the first time the while-loop test is run, hence $t_i = 1$ for each $i = 1, 2, \ldots, n - 1$. It follows that

$$
f(n) = \underbrace{(c_1 + c_2 + c_3 + c_4 + c_7)}_{\bar{c}\ (\text{say})} n - \underbrace{(c_2 + c_3 + c_4 + c_7)}_{\hat{c}\ (\text{say})} = \bar{c}n + \hat{c}.
$$

Thus, when best-case analysis is considered, the rate of growth for the insertion sort method is $n$.

- Worst-case analysis: The array is sorted in reverse order. So, we always have $A[j] > next$ in while-loop test. Then we have to compare "next" with all elements to the left of the $i^{\text{th}}$-position. That is, we compare with $i - 1$ element. So, $t_i = i$ for $i = 1, 2, \ldots, n - 1$.

Using (8.4) and applying the arithmetic series formula (see Corollary 2.1), it follows that

$$
\begin{aligned}
f(n) &= c_1 n + c_2(n-1) + c_3(n-1) + c_7(n-1) \\
&\quad + c_4 \sum_{i=1}^{n-1} t_i + c_5 \sum_{i=1}^{n-1} (t_i - 1) + c_6 \sum_{i=1}^{n-1} (t_i - 1) \\
&= c_1 n + c_2(n-1) + c_3(n-1) + c_7(n-1) \\
&\quad + c_4 \left( \frac{(n-1)n}{2} \right) + c_5 \left( \frac{(n-2)(n-1)}{2} \right) + c_6 \left( \frac{(n-2)(n-1)}{2} \right) \\
&= \bar{c} n^2 + \hat{c} n + \tilde{c},
\end{aligned}
$$

for some constants $\bar{c}, \hat{c}$ and $\tilde{c}$. Thus, when worst-case analysis is considered, the rate of growth of the insertion sort algorithm is $n^2$.

## Selection sort

Selection Sort is a sorting algorithm that divides the input list into two parts: a sorted sublist of items which is built up on the right-hand side and a sublist of the remaining unsorted items on the left-hand side. Initially, the sorted sublist is empty and the unsorted sublist is the entire input list. The algorithm proceeds by finding the smallest element in the unsorted sublist, swapping it with the leftmost unsorted element, and moving the sublist boundaries one element to the right.

For more illustration of this method, we apply the selection sort method to sort the 5-element list $(11, 25, 12, 22, 64)$. Table 8.2 shows all steps in this method for sorting the given list. The Selection Sort is formally stated in Algorithm 8.8.

To arrive at the Big-Oh of the running time for Algorithm 8.8, we represent the structure of this algorithm and the running time complexity by the tree structure shown in Figure 8.3.

| Unsorted sublist | Least element in the unsorted list | Sorted sublist |
|---|---|---|
| $(11, 25, 12, 22, 64)$ | 11 | ( ) |
| $(25, 12, 22, 64)$ | 12 | $(11)$ |
| $(25, 22, 64)$ | 22 | $(11, 12)$ |
| $(25, 64)$ | 25 | $(11, 12, 22)$ |
| $(64)$ | 64 | $(11, 12, 22, 25)$ |
| ( ) | | $(11, 12, 22, 25, 64)$ |

Table 8.2: A concrete implementation of the selection sort algorithm.

---

**Algorithm 8.8:** Selection sort algorithm

---

**Input:** An integer array $A[0 : n - 1]$

**Output:** Array $A$ sorted in ascending order

1: `selection-sort`(int $A[\ ]$, int $n$)
2: **for** $(i = 0; i < n - 1; i + +)$ **do**
3:      small $= i$
4:      **for** $(j = i + 1; j < n; j + +)$ **do**
5:          **if** $(A[j] < A[small])$ **then**
6:              small $= j$
7:          **end**
8:      **end**
9:      swap $= A[\text{small}]$
10:      $A[\text{small}] = A[i]$
11:      $A[i] = $ swap
12: **end**

---

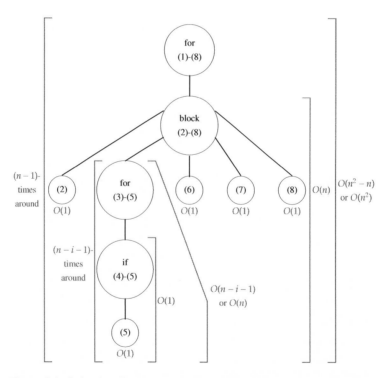

Figure 8.3: Selection Sort tree structure with running time complexity.

In closing, the running time of Selection Sort is $O(n^2)$. This is the worst-case performance of Algorithm 8.8. What are its best- and average-case performances? This question is left as an exercise for the reader (see Exercise 8.1 (c)).

## Merge sort

Merge Sort is a sorting algorithm in which the input is an unsorted array $A[first : last]$, and the output is the array $A$ sorted in ascending order. The idea behind the merge sort method is dividing the unsorted $n$-element list (or array) into $n$ sorted sublists, each containing one element (a list of one element is considered a sorted array), then repeatedly merge sublists to produce new sorted sublists until there is only one sublist remaining. This will be the sorted array. The merge sort algorithm is stated in Algorithm 8.9.

---
**Algorithm 8.9:** Merge sort algorithm

---
**Input:** An integer array $A$[first:last] of length $n$
**Output:** Array $A$ sorted in ascending order
1: `merge-sort`(int $A$[ ], int first, int last)
2: **if** (first $<$ last) **then**
3:      mid $= \lfloor$(first+last)/2$\rfloor$           // Constant time $c$
4:      `merge-sort`($A$, first, mid)       // Same problem of size $n/2$
5:      `merge-sort`($A$, mid+1, last)    // Same problem of size $n/2$
6:      `merge`($A$, first, mid, last)      // A call to "merge" takes time $n$
7: **end**

---

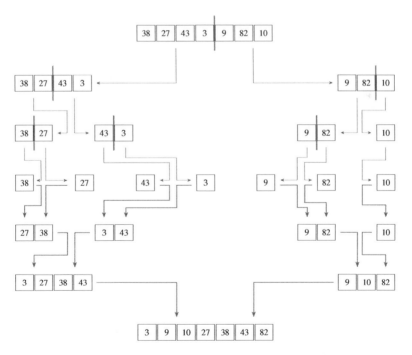

Figure 8.4: A concrete implementation of the merge sort algorithm.

Note that Algorithm 8.9 uses the "merge" function, which is an inbuilt function in C++STL[1] that merges two sorted lists into a single sorted list.

For more illustration of the merge sort algorithm, in Figure 8.4 we show in detail how to apply Algorithm 8.9 to a 7-element array.

Let $T(n)$ be the running time of Algorithm 8.9 on an array of size $n$. From Algorithm 8.9, by dropping the constant time $c$ in favor of the large time $n$, it follows that the recurrence formula for this recursive program is

$$T(n) = n + 2T(n/2). \tag{8.5}$$

From Example 5.5, we have

$$T(n) = nk + T(n/2^k) = n \log n + nT(1).$$

Thus, the recurrence formula (8.5) of Algorithm 8.9 describes the linearithmic function $T(n) = n(\log n + \text{constant time})$. As a result, the merge sort algorithm is linearithmic, running in $O(n \log n)$ time. This is the worst-case performance of Algorithm 8.9. What are its best- and average-case performances? This question is left as an exercise for the reader (see Exercise 8.1 $(e)$).

We end this section with the following corollary which summarizes the computational complexity (worst behavior) in terms of the size of the list $(n)$.

---

**Corollary 8.3** *The (worst-case) time complexities of the above sorting algorithms are as follow:*

($i$) *The insertion sort algorithm is quadratic, running in $O(n^2)$ time.*

($ii$) *The selection sort algorithm is quadratic, running in $O(n^2)$ time.*

($iii$) *The merge sort algorithm is linearithmic, running in $O(n \log n)$ time.*

---

Besides the above sorting algorithms, there are many other sorting algorithms that are not presented in this book. This includes heap sort, quicksort and counting sort (see [CLRS01]). In practical implementations a few algorithms predominate. Insertion sort is widely used for small data sets, while for large data sets an asymptotically efficient sort is used, primarily merge sort, heap sort, or quicksort.

## 8.4. Euclid's algorithm

The greatest common divisor of two integers is the largest number that divides them both without a remainder. The Euclidean algorithm or Euclid's algorithm is an efficient method for computing the greatest common divisor of two integers. The algorithm was defined by Euclid around 300 BC in Book VII of his Elements.

---

[1]C++STL stands for the C++ Standard Template Library.

---

**Algorithm 8.10:** Euclid's Algorithm (the division-based version)

---

1: $\text{gcd}(a, b)$
2: **if** $b = 0$ **then**
3:    |    **return** $a$
4: **end**
5: **else**
6:    |    **return** $\text{gcd}(b, a \bmod b)$
7: **end**

---

There are two versions of Euclid's algorithm: The division-based version which recursively calls a function while the larger number is not divisible by the smaller, and the subtraction-based version which recursively subtracts the smaller number from the larger.

The greatest common divisor of integers $a, b \in \mathbb{N}$ is written as $\text{gcd}(a, b)$. We write $a \bmod n$ to denote the unique integer $b$ such that $0 \leq b < n$ and $a \equiv b \pmod{n}$ (i.e., the remainder of $a$ when divided by $n$). For example, $39 \equiv 3 \pmod{12}$ because $38 - 3 = 36$, which is a multiple of 12. It is known that $\text{gcd}(a, b) = b$ when $b$ divides $a$. In addition, if $a$ is not divisible by $b$, the reader can show that $\text{gcd}(a, b) = \text{gcd}(b, a \bmod b)$ (see Exercise 8.5). This leads us to the correctness of the following algorithm.

For example, the $\text{gcd}(1377, 594)$ is calculated from the equivalent $\text{gcd}(594, 1377 \bmod 594) = \text{gcd}(594, 189)$. The latter greatest common divisor is calculated from the $\text{gcd}(189, 462 \bmod 189) = \text{gcd}(189, 27)$, which in turn is calculated from the $\text{gcd}(21, 189 \bmod 27) = \text{gcd}(27, 0) = 27$. Thus, $\text{gcd}(1377, 594) = 27$.

Note that, in Algorithm 8.10, if negative inputs are allowed, then the mod function may return negative values. In this case, the instruction "return $a$" must be replaced with "return $\max(a, -a)$".

Contrary to the recursive version stated in Algorithm 8.10, which works with two arbitrary integers as input, the subtraction-based version works with two positive integers and stops when they are both identical. This version of Euclid's Algorithm is stated in Algorithm 8.11.

---

**Algorithm 8.11:** Euclid's Algorithm (the subtraction-based version)

---

1: $\text{gcd}(a, b)$
2: **while** $(a \neq b)$ **do**
3:    |    **if** $(a > b)$ **then**
4:    |     |    $a \triangleq a - b$
5:    |    **end**
6:    |    **else**
7:    |     |    $b \triangleq b - a$
8:    |    **end**
9: **end**
10: **return** $a$

---

For illustration, the reader can use Algorithm 8.11 to check that $\gcd(1377, 594) =$ 27. We now estimate the worst-case time complexity of Algorithm 8.11 based on the sum, $n = a + b$, of $a$ and $b$. If we exclude our base case, this sum will decrease at each recursive step. Since the smallest possible deduction for each step is 1 (for example, $\gcd(x, 1)$), and since all positive integers are guaranteed to have a common divisor of 1, the time complexity is $O(n)$. Therefore, the number of iterations in Algorithm 8.11 can be linear based on the sum of $a$ and $b$.

We now estimate the worst-case time complexity of Algorithm 8.10 based on the minimum, $n = \min(a, b)$, of $a$ and $b$. Note that the number of iterations Algorithm 8.10 will take is maximized when the two inputs are consecutive Fibonacci numbers. As introduced in Section 3.1, the Fibonacci numbers $f_0, f_1, f_2, \ldots$, are defined by the recurrence relation

$$f_n = f_{n-1} + f_{n-2}, \text{ for } n = 2, 3, 4, \ldots, \quad \text{where } f_0 = 0 \text{ and } f_1 = 1.$$

Using this recurrence formula, we have $f_2 = 1, f_3 = 2, f_4 = 3, f_5 = 5, f_6 = 8, f_7 = 13$, etc. More specifically, if $a > b > 0$ and $\gcd(a, b)$ requires $n \geq 1$ steps, then we can verify that the smallest possible values of $a$ and $b$ are $f_{n+2}$ and $f_{n+1}$, respectively (see Exercise 8.6). Therefore,

$$b \geq f_{n+1}.$$

Exercise 5.6 uses the generating function method to solve the Fibonacci recurrence $f_n = f_{n-1} + f_{n-2}$. One can find (see the solution to Exercise 5.6) that

$$f_n = \frac{1}{\sqrt{5}}\left(\alpha^n - (1 - \alpha)^n\right),$$

where, for $n \geq 0$, $\alpha \triangleq (1 + \sqrt{5})/2$ is the 'golden ratio'. A more simplified version of the above formula is the following:

$$f_n \approx \frac{\alpha^n}{\sqrt{5}}.$$

It follows that

$$b \geq f_{n+1} \approx \frac{1}{\sqrt{5}}\alpha^{(n+1)} = \frac{3 + \sqrt{5}}{2\sqrt{5}}\alpha^{(n-1)} \geq \alpha^{(n-1)}.$$

Solving for $n$, we get

$$n \leq \log_\alpha b + 1 = O(\log_\alpha n).$$

Thus, the number of iterations in Algorithm 8.10 can be logarithmic based on the minimum of $a$ and $b$.

## 8.5. Newton's method algorithm

Newton's method (also called Newton-Raphson method) is one of the root-finding algorithms. Given a function $f$, the goal of a root-finding algorithm is to find a root of $f$, or solution of the equation $f(x) = 0$. The desired value $r$ that satisfies $f(r) = 0$ is also called a zero of the function $f$.

Let $f : [a, b] \longrightarrow \mathbb{R}$ be a twice continuously differentiable function. Let $r_0 \in [a, b]$ be an approximation to a root $r$ of $f$ such that $f'(r_0) \neq 0$ and $|r - r_0|$ is small. Consider the following Taylor expansion for $f(x)$ about $r_0$ evaluated at $x = r$.

$$f(r_0) + (r - r_0)f'(r_0) + \frac{(r - r_0)^2}{2}f''(r_0) + \frac{(r - r_0)^3}{6}f'''(r_0) + \cdots .$$

Newton's method can be derived by assuming that the terms involving $(r - r_0)^2, (r - r_0)^3, \ldots$ are tiny because $|r - r_0|$ is small. So

$$0 \approx f(r_0) + (r - r_0)f'(r_0).$$

After solving for $r$, we immediately have

$$r \approx r_0 - \frac{f(r_0)}{f'(r_0)} \equiv r_1.$$

Figure 8.5 shows the geometry of Newton's method. The tangent line to the curve $y = f(x)$, namely $y = f(r_0) + (x - r_0)f'(r_0)$, intersects the $x$-axis at $r_1$ and has the slope of $f'(r_0)$. Initializing at the approximation $r_0$, the $x$-intercept of the tangent line to the graph of $f$ at $(r_0, f(r_0))$, say $r_1$, becomes a better approximation to $r$. Re-initializing at the approximation $r_1$, the $x$-intercept of the tangent line to the graph of $f$ at $(r_1, f(r_1))$, say becomes even a better approximation than $r_1$ to $r$, and so on.

Accordingly, Newton's method initializes with an initial point $r_0$ and generates a sequence of points $\{r_k\}_{k=0}^{\infty}$ using

$$r_{k+1} = r_k - \frac{f(r_k)}{f'(r_k)}, \text{ for } k = 0, 1, 2, \ldots . \tag{8.6}$$

The following example is from Burden and Faires [BF10].

**Example 8.2** Approximate a root of the function $f(x) = \cos x - x$ using Newton's Method and starting with $r_0 = \pi/4 \approx 0.7853981635$.

**Solution** Equation (8.6) for our setting becomes

$$r_{k+1} = r_k - \frac{\cos r_k - r_k}{-\sin r_k - 1},$$

for $k = 0, 1, 2, \ldots$. This gives the approximations in Table 8.3. ∎

| $k$ | $r_{k+1}$ |
|---|---|
| 0 | 0.7395361337 |
| 1 | 0.7390851781 |
| 2 | 0.7390851332 |
| 3 | 0.7390851332 |

Table 8.3: Numerical results of Example 8.2.

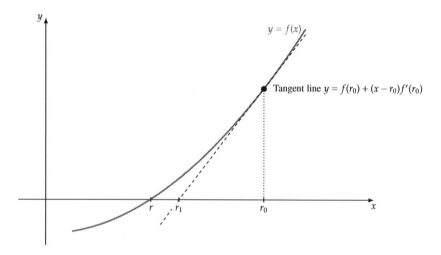

Figure 8.5: The geometry of Newton's method.

---

**Algorithm 8.12:** Newton's method algorithm

**Input:** Initial guess $r_0$, maximum number of iterations $N$, tolerance TOL
**Output:** Approximate solution $r$ to $f(x) = 0$ or message of failure

1: set $k = 0$
2: **while** $(k \leq N)$ **do**
3:      compute $r_{k+1}$ using (8.6)
4:      **if** $|r_{k+1} - r_k| <$ TOL **then**
5:          printf(*"r equals"* $r_{k+1}$)               // The procedure is successful
6:          **break**
7:      **end**
8:      set $k + +$
9:      set $r_k = r_{k+1}$
10: **end**
11: printf(*"Fails after N iterations"*)               // The procedure is unsuccessful

---

Newton's method is formally stated in Algorithm 8.12.

We give, without proof, the following convergence and complexity result in this context. For a proof, see, for example, Cheney and Kincaid [CK07].

**Theorem 8.1** *Let $f : [a,b] \longrightarrow \mathbb{R}$ be a twice continuously differentiable function. If $r \in (a,b)$ is such that $f(r) = 0$ and $f'(r) \neq 0$, then there exists $\delta > 0$ such that Newton's method generates a sequence $\{r_k\}_{k=1}^{\infty}$ converging quadratically to $r$ for any initial approximation satisfying $|r - r_0| < \delta$.*

## Newton's method for nonlinear systems

Newton's method can be also used to find roots of nonlinear equations. In general, given a system of equations when the number of unknowns equals the number of equations. We are interested to find solutions to a system of the form:

$$
\begin{aligned}
f_1(x_1; x_2; \ldots; x_n) &= 0, \\
f_2(x_1; x_2; \ldots; x_n) &= 0, \\
&\vdots \\
f_n(x_1; x_2; \ldots; x_n) &= 0,
\end{aligned}
\tag{8.7}
$$

where $f_i : \mathbb{R}^n \longrightarrow \mathbb{R}$ for $i = 1, 2, \ldots, n$. In vector form, System (8.7) can be written as

$$
f(x) = \mathbf{0},
\tag{8.8}
$$

where $f \triangleq (f_1; f_2; \ldots, f_n) : \mathbb{R}^n \longrightarrow \mathbb{R}^n$ and $x \triangleq (x_1; x_2; \ldots; x_n) \in \mathbb{R}^n$.

Newton's method can be extended to the nonlinear system (8.8) using

$$
r^{k+1} = r^k - [\mathcal{J}_x f]^{-1}|_{x=r^k} f(r^k)
$$

where $\mathcal{J}_x f$ is the $n \times n$ Jacobian matrix defined as

$$
\mathcal{J}_x f \triangleq 
\begin{bmatrix}
\frac{\partial f_1}{\partial x_1}(x) & \frac{\partial f_1}{\partial x_2}(x) & \cdots & \frac{\partial f_1}{\partial x_n}(x) \\
\frac{\partial f_2}{\partial x_1}(x) & \frac{\partial f_2}{\partial x_2}(x) & \cdots & \frac{\partial f_2}{\partial x_n}(x) \\
\vdots & \vdots & \ddots & \vdots \\
\frac{\partial f_n}{\partial x_1}(x) & \frac{\partial f_n}{\partial x_2}(x) & \cdots & \frac{\partial f_n}{\partial x_n}(x)
\end{bmatrix}
=
\begin{bmatrix}
(\nabla_x f_1(x))^\top \\
(\nabla_x f_2(x))^\top \\
\vdots \\
(\nabla_x f_n(x))^\top
\end{bmatrix},
$$

and, for a function $g : \mathbb{R}^n \to \mathbb{R}$, $\nabla g(x)$ is the gradient vector (the vector of the first derivatives of $g$) defined as

$$
\nabla_x g(x) \triangleq 
\begin{bmatrix}
\frac{\partial g}{\partial x_1}(x) \\
\frac{\partial g}{\partial x_2}(x) \\
\vdots \\
\frac{\partial g}{\partial x_n}(x)
\end{bmatrix}.
\tag{8.9}
$$

Note that the method presumes that the Jacobian $\mathcal{J}_x f$ is nonsingular at each $r^k$, so its inverse exists. The two-step procedure in the following workflow, followed by an example, will teach us to solve System (8.8).

---

**Workflow 8.1** *We solve System* (8.8) *by following two steps:*

(i) *Solve* $\mathcal{J}_x f \, \Delta r = -f(x)$ *for* $\Delta r$.

(ii) *Let* $r^{k+1} = r^k + \Delta r$.

---

The vector $\Delta r$ defined in Workflow 8.1 is called a Newton direction.

**Example 8.3** Let us perform one iteration of Newton's method to approximate the solution of the following set of equations with a starting guess $r_0 = (1/2, 1/2)^{\mathsf{T}}$.

$$
\begin{aligned}
x_1^3 + x_2 &= 1, \\
-x_1 + x_2^3 &= -1.
\end{aligned}
$$

Letting

$$
f(x) = \begin{bmatrix} f_1(x_1, x_2) \\ f_2(x_1, x_3) \end{bmatrix} = \begin{bmatrix} x_1^3 + x_2 - 1 \\ -x_1 + x_2^3 + 1 \end{bmatrix},
$$

we have

$$
\mathcal{J}_x f = \begin{bmatrix} 2x_1^2 & 1 \\ -1 & 3x_2^2 \end{bmatrix}.
$$

Then

$$
f(r_0) = \begin{bmatrix} -3/8 \\ 5/8 \end{bmatrix}, \mathcal{J}_x f(r_0) = \begin{bmatrix} 3/4 & 1 \\ -1 & 3/4 \end{bmatrix}, \text{ and hence } [\mathcal{J}_x f(r_0)]^{-1} = \frac{16}{25} \begin{bmatrix} 3/4 & -1 \\ 1 & 3/4 \end{bmatrix}.
$$

Performing one iteration of Newton's method, we get the following Newton direction

$$
\Delta r = -[\mathcal{J}_x f(r_0)]^{-1} f(r_0) = -\frac{16}{25} \begin{bmatrix} 3/4 & -1 \\ 1 & 3/4 \end{bmatrix} \begin{bmatrix} -3/8 \\ 5/8 \end{bmatrix} = \begin{bmatrix} 29/50 \\ 22/50 \end{bmatrix}.
$$

As a result, we have

$$
r^1 = r^0 + \Delta r = \begin{bmatrix} 1/2 \\ 1/2 \end{bmatrix} + \begin{bmatrix} 29/50 \\ -3/50 \end{bmatrix} = \begin{bmatrix} 27/25 \\ 11/25 \end{bmatrix},
$$

which is the approximation after the first iterate. Note that, with a sharp eye, the reader can see that there is one and only one solution to $f(x) = 0$, namely $(x_1, x_2) = (1, 0)$. ∎

## Newton's method for optimization

In this part, Newton's method is exploited to optimize twice differentiable convex functions. Consider the function $g : \mathbb{R}^n \longrightarrow \mathbb{R}$, which is a real-valued function of $n$ independent variables. We say that $g$ is differentiable on a domain $D \subseteq \mathbb{R}^n$ if $\nabla_x g$ exists for all $x \in D$, where $\nabla_x g$ is the gradient vector defined in (8.9). Similarly, $g$ is twice differentiable on $D$ if $\nabla^2_{xx} g$ exists for all $x \in D$, where $\nabla^2_{xx} g$ is the Hessian matrix (the matrix of the second derivatives of $g$) defined as

$$
\nabla^2_{xx} g \triangleq
\begin{bmatrix}
\frac{\partial^2 g}{\partial x_1^2}(x) & \frac{\partial^2 g}{\partial x_1 \partial x_2}(x) & \cdots & \frac{\partial^2 g}{\partial x_1 \partial x_n}(x) \\
\frac{\partial^2 g}{\partial x_2 \partial x_1}(x) & \frac{\partial^2 g}{\partial x_2^2}(x) & \cdots & \frac{\partial^2 g}{\partial x_2 \partial x_n}(x) \\
\vdots & \vdots & \ddots & \vdots \\
\frac{\partial^2 g}{\partial x_n \partial x_1}(x) & \frac{\partial^2 g}{\partial x_n \partial x_2}(x) & \cdots & \frac{\partial^2 g}{\partial x_n^2}(x)
\end{bmatrix}.
\tag{8.10}
$$

Note that the Hessian of a function is the Jacobian of its gradient. That is

$$
\nabla^2_{xx} g = \mathcal{J}_x \nabla_x g.
$$

In addition, $g$ is called twice continuously differentiable on $D$ if $\nabla^2_{xx} g$ is continuous on $D$. As a matter of fact, if a function is a twice continuously differentiable function, then its Hessian matrix is symmetric (positive semidefinite). Moreover, if a function is a twice continuously differentiable convex function, then its Hessian matrix is positive semidefinite (see Definition 3.3).

Assume that we are interested in minimizing a twice continuously differentiable real valued function $g : \mathbb{R}^n \longrightarrow \mathbb{R}$. That is, we are interested in a problem of the form:

$$
\min_{x \in \mathbb{R}^n} g(x).
\tag{8.11}
$$

In this case, we want to its gradient to zero. That is, we need to find the roots of the system of nonlinear equations

$$
\nabla_x g(x) = \mathbf{0}.
\tag{8.12}
$$

Note that, in (8.12), the number of unknowns equals the number of equations. Applying Newton's method here involves the Jacobian matrix of $\nabla_x g$, which is its Hessian matrix defined in (8.10).

Algorithm 8.13 is used to minimize the function $g$. In Algorithm 8.13, the vector $\Delta x$ is called a Newton direction or Newton step, $x^k$ is called the Newton iterate, and the parameter $\theta_k$ is called the step-size. Also, in Algorithm 8.13, $\|\cdot\|$ denotes the Euclidean norm.

Example 8.4, which is from [Fre04], performs one iteration of Algorithm 8.13 to get an approximation of the optimal solution of a minimization problem.

---

**Algorithm 8.13:** Newton's algorithm for unconstrained optimization

---

**Input:** Initial guess $x^0$, maximum number of iterations $N$, tolerance TOL
**Output:** Approximate optimal solution to Problem (8.11)

1: set $k = 0$
2: **while** $(k \leq N)$ **do**
3:   compute $\Delta x = -[\nabla^2_{xx} g(x^k)]^{-1} \nabla_x g(x^k)$
4:   choose step-size $\theta_k$
5:   set $x^{k+1} = x^k + \theta_k \Delta x$
6:   **if** $\|x_{k+1} - x_k\| <$ TOL **then**
7:    printf(*"Approximate optimal solution is"* $x^{k+1}$)
8:    **break**
9:   **end**
10:   set $k + +$
11:   set $x^{k+1} = x^k$
12: **end**
13: printf(*"The method fails after N iterations"*)

---

**Example 8.4** With a starting guess $x^0 = (0.85; 0.05)$, let us perform one iteration of Algorithm 8.13 to approximate the optimal solution of the problem:

$$\min_{(x_1; x_2) \in D} - \ln(1 - x_1 - x_2) - \ln x_1 - \ln x_2$$

where $D = \{(x_1; x_2) \in \mathbb{R}^2 : x_1 > 0, x_2 > 0, x_1 + x_2 < 1\}$. Letting

$$g(x) = -\ln(1 - x_1 - x_2) - \ln x_1 - \ln x_2,$$

we have

$$\nabla_x g(x) = \begin{bmatrix} \dfrac{1}{1 - x_1 - x_2} - \dfrac{1}{x_1} \\ \dfrac{1}{1 - x_1 - x_2} - \dfrac{1}{x_2} \end{bmatrix},$$

and

$$\nabla^2_{xx} g(x) = \begin{bmatrix} \dfrac{1}{(1 - x_1 - x_2)^2} + \dfrac{1}{x_1^2} & \dfrac{1}{(1 - x_1 - x_2)^2} \\ \dfrac{1}{(1 - x_1 - x_2)^2} & \dfrac{1}{(1 - x_1 - x_2)^2} + \dfrac{1}{x_2^2} \end{bmatrix}.$$

Then

$$\nabla_x g(x^0) \approx \begin{bmatrix} 8.824 \\ -10 \end{bmatrix}, \text{ and } \nabla^2_{xx} g(x^0) \approx \begin{bmatrix} 100.2 & 100 \\ 100 & 500 \end{bmatrix}.$$

Performing one iteration of Newton's method, we get the following Newton direction

$$
\begin{aligned}
\Delta x &= -[\nabla_{xx}^2 g(x^k)]^{-1} \nabla_x g(x^k) \\
&\approx -\frac{1}{40100}
\begin{bmatrix} 500 & -100 \\ -100 & 100.2 \end{bmatrix}
\begin{bmatrix} 8.824 \\ -10 \end{bmatrix} \\
&= \begin{bmatrix} -0.012468 & 0.002493 \\ 0.002493 & -0.002498 \end{bmatrix}
\begin{bmatrix} 8.824 \\ -10 \end{bmatrix}
= \begin{bmatrix} -0.134947 \\ 0.046978 \end{bmatrix}.
\end{aligned}
$$

As a result, we have

$$
x^1 = x^0 + \Delta x \approx \begin{bmatrix} 0.85 \\ 0.05 \end{bmatrix} + \begin{bmatrix} -0.134947 \\ 0.046978 \end{bmatrix} = \begin{bmatrix} 0.715053 \\ 0.096978 \end{bmatrix},
$$

which is the approximation after the first iterate. The step size was chosen to be $\theta_1 = 1$.

As an exercise, the reader is encouraged to perform more iterations to find $x^k$ for $k = 2, 3, 4, 5$. The approximate optimal solution after the fifth iteration is $x^5 = (0.333338; 0.333259)$. It is not hard to check that the optimal solution $(x_1, x_2) = (1/3, 1/3)$. ∎

The Newton direction (or generally, any search direction) is called a decent direction if $g(x + \theta \Delta x) < g(x)$ for all sufficiently small values of $\theta$. In general, without some further assumptions, there is no guarantee of decent directions in Algorithm 8.13. We have the following theorem.

> **Theorem 8.2** *If $g : \mathbb{R}^n \longrightarrow \mathbb{R}$ is a twice continuously differentiable strictly convex real valued function, then the Newton direction $\Delta x$ is descent.*

**Proof** If $g$ is strictly convex, then its Hessian matrix, $\nabla_{xx}^2 g(x)$, is positive definite (the proof for this fact is left as an exercise for the reader). It follows that

$$
(\nabla_x g(x))^\top \Delta x = -(\nabla_x g(x))^\top [\nabla_{xx}^2 g(x)]^{-1} \nabla_x g(x) < 0,
$$

where the inequality follows from the fact that $[\nabla_{xx}^2 g(x)]^{-1}$ is positive definite (see Exercise 3.7).

Now, $g$ can be approximated by its first-order Taylor expansion at $x$:

$$
g(x + \theta \Delta x) \approx g(x) + \theta (\nabla_x g(x))^\top \Delta x < g(x).
$$

Thus, $\Delta x$ is a descent direction. ∎

It can be shown that Algorithm 8.13 has a quadratic convergence under some certain circumstances (see for example [Fre04]).

# EXERCISES

**8.1**    Choose the correct answer for each of the following multiple-choice questions/items.

(*a*) Let $n$ be a positive integer. The total number of flops when we multiply an $n^2 \times n$ matrix with an $n$-tuple of real numbers is

    (*i*) $2n$.        (*ii*) $2n^2$.        (*iii*) $2n^3$.        (*iv*) $2n^4$.

(*b*) Let $n$ be a positive integer. If $A$ and $B$ are $n \times n^2$ and $n^2 \times n^3$ matrices, respectively, then the total number of flops after multiplying $A$ with $B$ is

    (*i*) $2n^3$.        (*ii*) $2n^4$.        (*iii*) $2n^5$.        (*iv*) $2n^6$.

(*c*) The best-case time complexities of Linear Search and Selection Sort are respectively

    (*i*) $O(1)$ and $O(1)$.        (*iii*) $O(1)$ and $O(n^2)$.

    (*ii*) $O(1)$ and $O(n)$.        (*iv*) $O(n)$ and $O(n^2)$.

(*d*) If the function `main` in Algorithm 8.14 calls not only the function `karger` but also the function `selection-sort` that we analyzed in Algorithm 8.8. What is the worst-case time complexity of the new resulting program?

    (*i*) $O(mn + n^2)$.        (*iii*) $O(mn + m^2)$.

    (*ii*) $O(m^2 + n^2)$.        (*iv*) $O(mn)$.

(*e*) The best-case time complexities of Binary Search and Merge Sort are respectively

    (*i*) $O(1)$ and $O(1)$.        (*iii*) $O(1)$ and $O(n \log n)$.

    (*ii*) $O(1)$ and $O(n)$.        (*iv*) $O(\log n)$ and $O(n \log n)$.

(*f*) Which of the following sorting algorithms, if any, is an example of an algorithm for which the best- and worst-case time complexities differ?

    (*i*) Merge Sort.        (*iv*) None of the above because, in all sorting algorithms, the best- and worst-case time complexities always coincide.

    (*ii*) Insertion Sort.

    (*iii*) Selection Sort.

(*g*) Which of the following, if any, is an example of a searching algorithm for which the best- and worst-case time complexities coincide?

  (*i*) Linear Search.

  (*ii*) Selection Sort.

  (*iii*) No such algorithm exists because the best- and worst-case time complexities always differ.

  (*iv*) None of the above.

(*h*) Using Algorithm 8.10, we find that gcd(55, 99) =

  (*i*) 9.          (*ii*) 10.          (*iii*) 11.          (*iv*) 12.

(*i*) Using Algorithm 8.11, we find that gcd(840, 3220) =

  (*i*) 120.          (*ii*) 140.          (*iii*) 160.          (*iv*) 180.

**8.2**    If we multiply two $n \times n$ matrices together, we require $2n^3$ flops. In 1969, Strassen [Str69] presented a method that can do the job in $O(n^s)$ flops, where $s = \log_2 7 \approx 2.81$. Since $2.81 < 3$, Strassen's method will be faster than Algorithm 8.4 if $n$ is sufficiently large. Tests have shown that even for $n$ as small as 100 or so, Strassen's method can be faster. However, since 2.81 is only slightly less than 3, $n$ has to be made quite large before Strassen's method wins by a large margin.

Even "faster" methods have been found. For example, Coppersmith and Winograd (see, for example, Higham [Hig02]) presented a method to multiply two $n \times n$ matrices in about $O(n^{2.376})$ flops. However, the computing world found that this algorithm does not beat Strassen's method. How can you justify this?

**8.3**    Given the program in Algorithm 8.14. What is the worst-case time complexity of this program?

**8.4**    What is the worst-case time complexity of the program shown in Algorithm 8.15? Justify your answer.

**8.5**    Prove that $\gcd(a, b) = \gcd(a \bmod b, b)$ for any $a, b \in \mathbb{N}$.

**8.6**    Consider Algorithm 8.11. Assume that $a > b > 0$ and that $\gcd(a, b)$ requires $n \geq 1$ steps. Show that the smallest possible values of $a$ and $b$ are $f_{n+2}$ and $f_{n+1}$, respectively, where $\{f_n\}_{n=0}^{\infty}$ is the Fibonacci sequence defined as

$$f_n = f_{n-1} + f_{n-2}, \text{ for } n = 2, 3, 4, \ldots, \text{ with } f_0 = 0 \text{ and } f_1 = 1.$$

**8.7**    Write a code to find a root of the function $f(x) = \frac{1}{2}x^2 + x + 1 - e^x$ on $[0, 4]$. Apply Newton's method starting with $x_0 = 1$.

---

**Algorithm 8.14:** The algorithm of Exercise 8.3

1: # include <stdio. $h$ >
2: int karger(*int x, int m*)
3: int linear-search(*int A[ ], int n, int x*)
    // *This line is left intentionally blank*
4: main( ) **begin**
5:     int $A[0 : n - 1], m$
6:     $a$ = karger(*0, m*)
7:     printf(*"Value a: \ n"*)
8:     **return** $a$
9: **end**
    // *This line is left intentionally blank*
10: karger(*int x, int m*) **begin**
11:     **for** $(i = 1; i \leq m; i + +)$ **do**
12:         $x$ += linear-search($A[ ], n, m$)
13:     **end**
14:     **return** $x$
15: **end**

---

**Algorithm 8.15:** The algorithm of Exercise 8.4

1: # include <stdio. $h$ >
2: int dinic(*int x, int m*)
3: int binary-search(*int A[ ], int low, int high, int x*)
4: int merge-sort(*int A[ ], int first, int last*)
    // *This line is left intentionally blank*
5: main( ) **begin**
6:     int $A[0 : n - 1], m$
7:     $A[n - 1]$ = dinic(*0, m*)
8:     $B$ = merge-sort($A[ ], 0, n - 1$)
9:     printf(*"Sorted array: \ n"*)
10:     **return** $B$
11: **end**
    // *This line is left intentionally blank*
12: dinic(*int x, int m*) **begin**
13:     **for** $(i = 1; i \leq m; i* = 2)$ **do**
14:         $x$ += binary-search($A[ ], 0, n - 1, m$)
15:     **end**
16:     **return** $x$
17: **end**

# CHAPTER 9

# ELEMENTARY COMBINATORIAL ALGORITHMS

## Contents

© 2022 by Baha Alzalg | Kindle Direct Publishing, Washington, United States 2022/9
B. Alzalg, *Combinatorial and Algorithmic Mathematics: From Foundation to Optimization*,
DOI 10.5281/zenodo.7110975

In this chapter, we present elementary algorithms for searching and exploring a graph. Our focus is on the following searching algorithms: The breadth-first search and the depth-first search. Before this, we need to learn how to represent a graph in a way that we can use such a representation as an input for the searching algorithms.

## 9.1. Graph representations

There are two standard ways to represent a graph: The adjacency list and the adjacency matrix. They both can be used to directed and undirected graphs.

### The adjacency list representation

An adjacency list is a collection of unordered lists used to represent a (directed or undirected) graph. We write an array of $|V|$ lists. Each list describes the set of neighbors of a vertex in the graph. The list that describes the set of neighbors of a vertex $u$ in a graph $G = (V, E)$ is denoted by Adj[$u$] and is defined to contain all vertices $v$ that are adjacent $u$. That is,

$$\text{Adj}[u] \triangleq \{v \in V : (u, v) \in E\}.$$

The following example illustrates this.

**Example 9.1**   Use adjacency lists to represent each of the following graphs.

(*a*) The directed graph:

(*b*) The undirected graph:

(*c*) The undirected graph:

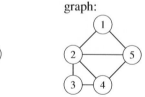
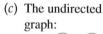
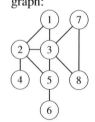

Solution   (*a*) Note that Adj[1] = {3}, Adj[2] = {1, 3}, and Adj[3] = {2}. So, the adjacency lists that represent the graph are:

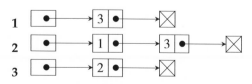

(*b*) Note that Adj[1] = {2, 5}, Adj[2] = {1, 5, 3, 4}, Adj[3] = {2, 4}, Adj[4] = {2, 5, 3}, and Adj[5] = {4, 1, 2}. So, the adjacency list representation for the graph is:

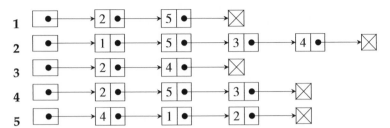

(*c*) This part is left as an exercise for the reader (see Exercise 9.3 (*a*)).

∎

Note that, for an undirected graph, in the adjacency list there are two representations of each edge in the graph.

## The adjacency matrix representation

An adjacency matrix is a square binary matrix[1] used to represent a (directed or undirected) graph. We assume that the vertices of the given graph $G = (V, E)$ are numbered $1, 2, \ldots, |V|$. The representation consists of a $(0, 1)$-matrix $A_{|V| \times |V|}$, where the entry of the *i*th row and *j*th column in the matrix is labeled $a_{ij}$ and is given by

$$a_{ij} \triangleq \begin{cases} 1, & \text{if } (i, j) \in E, \\ 0, & \text{otherwise.} \end{cases}$$

The following example illustrates this.

**Example 9.2** Use adjacency matrices to represent each of the following graphs.

(*a*) The undirected graph:

(*b*) The directed graph:

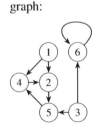

(*c*) The undirected graph:

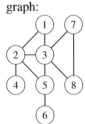

**Solution** The adjacency matrices that represent the graphs are as follow:

---

[1] A binary matrix is a matrix with entries from the set {0, 1}.

(a)                                             (b)

|   | 1 | 2 | 3 | 4 | 5 |
|---|---|---|---|---|---|
| **1** | 0 | 1 | 0 | 0 | 1 |
| **2** | 1 | 0 | 1 | 1 | 1 |
| **3** | 0 | 1 | 0 | 1 | 0 |
| **4** | 0 | 1 | 1 | 0 | 1 |
| **5** | 1 | 1 | 0 | 1 | 0 |

|   | 1 | 2 | 3 | 4 | 5 | 6 |
|---|---|---|---|---|---|---|
| **1** | 0 | 1 | 0 | 1 | 0 | 0 |
| **2** | 0 | 0 | 0 | 0 | 1 | 0 |
| **3** | 0 | 0 | 0 | 0 | 1 | 1 |
| **4** | 0 | 1 | 0 | 0 | 0 | 0 |
| **5** | 0 | 0 | 0 | 1 | 0 | 0 |
| **6** | 0 | 0 | 0 | 0 | 0 | 1 |

(c) This part is left as an exercise for the reader (see Exercise 9.3 (b)).

∎

Note that if the graph is simple, the adjacency matrix has zeros on its diagonal. Note also that, for an undirected graph, in the adjacency matrix there are two representations of each edge in the graph. In fact, if the graph is undirected, the adjacency matrix is symmetric.

We end this section with Table 9.1, which shows some differences between the adjacency list representation and the adjacency matrix representation.

|   | Adjacency list | Adjacency matrix |
|---|---|---|
| Memory required: | $O(V + E)$ | $O(V^2)$ |
| Preferred when: | $G$ is sparse | $G$ is dense |
| Time to list all vertices adjacent to $u \in V$: | $O(\deg(u))$ | $O(V)$ |
| Time to determine if an edge $(u, v) \in E$: | $O(\deg(u))$ | $O(1)$ |
| Is it a quick way to list the vertices adjacent to a given vertex? | Yes | No |
| Is it a quick way to determine if there is an edge between two vertices? | No | Yes |

Table 9.1: Comparison between the adjacency list and adjacency matrix representations of a given graph $G = (V, E)$.

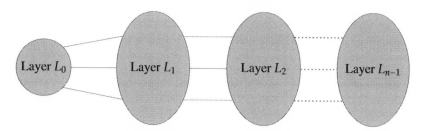

Figure 9.1: The breadth-first search scheme.

## 9.2. Breadth-first search algorithm

Breadth-first search is a method for exploring a graph. In this section, we describe and analyze this method. In the next section, we will describe some of its applications.

In a breadth-first search, we explore outward from a source vertex, say $s$, of a graph $G = (V, E)$ in all possible directions, adding vertices one layer (or level) at a time. Breadth-first search is used for both directed and undirected graphs. Figure 9.1 shows the breadth-first search scheme, where $L_0$ represents the zeroth move (which only includes the a source vertex $S$), $L_1$ represents the first move, $L_2$ represents the second move, etc.

In Figure 9.1, $L_0 \triangleq \{s\}$, $L_1 \triangleq \text{Adj}[s]$ (i.e., the neighbors of $L_0$), $L_2$ contains all vertices that do not belong $L_0$ or $L_1$ and that have an edge to a vertex in $L_1$, and $L_{i+1}$ contains all vertices that do not belong to an earlier layer, which is needed to avoid duplication, and that have an edge to a vertex in $L_i$. We have the following remark.

**Remark 9.1** *For each $i$, the layer $L_i$ consists of all vertices at distance exactly $i$ from the source vertex $s$.*

The following example explains the steps outlined in the breadth-first search method.

**Example 9.3** For each of the following graphs, use the breadth-first search method to determine the smallest number of layers (or hopes) among the vertices starting from vertex $s$.

(*a*) The undirected graph in the left-hand side of Figure 9.2 (take $s$ to be the vertex 1).

(*b*) The directed graph in the right-hand side of in Figure 9.2.

Solution   (*a*) In this part, the breadth-first search method is explained move-by-move as shown in Figure 9.3. We find that the smallest number of layers among the vertices starting from the vertex $s$ (vertex 1) equals 4.

(*b*) As shown in Figure 9.4, the smallest number of layers among the vertices starting from the vertex $s$ equals 6.

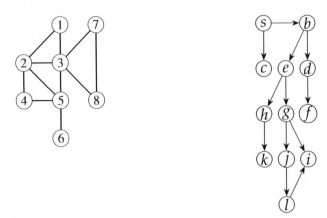

Figure 9.2: The graphs of Example 9.3.

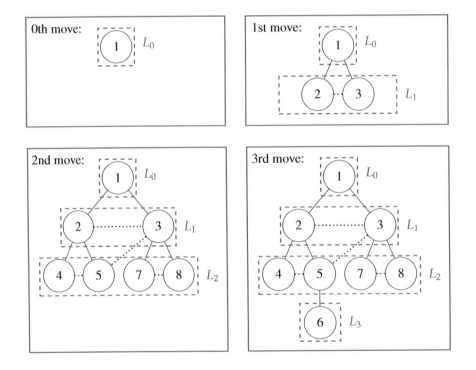

Figure 9.3: Illustrating the progress of breadth-first search on the graph of Example 9.3 (*a*).

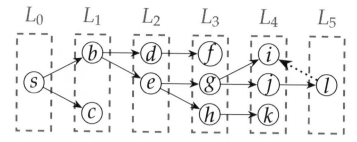

Figure 9.4: Breadth-first search on the digraph of Example 9.3 ($b$).

---

**Algorithm 9.1:** Breadth-first search algorithm

---

1:  BFS($s$, Adj)
2:  **begin**
3:       level = {$s$ : 0}
4:       parent = {$s$ : None}
5:       $i = 1$
6:       frontier = [$s$]
7:       **while** frontier **do**
8:           next = [ ]
9:           **for** $u$ in frontier **do**
10:              **for** $v$ in Adj[$u$] **do**
11:                  **if** $v$ not in level **then**
12:                      level[$v$] = $i$
13:                      parent[$v$] = $u$
14:                      next.append($v$)
15:                  **end**
16:              **end**
17:          **end**
18:          frontier = next
19:          $i+ = 1$
20:      **end**
21:      **return** parent
22: **end**

---

A version of a breadth-first search algorithm is presented in Algorithm 9.1.

Theorem 9.1 gives the running time of Algorithm 9.1 on an input graph $G = (V, E)$ given in its adjacency lists.

---

**Theorem 9.1** *Algorithm* 9.1 *runs in* $O(V + E)$ *time.*

---

**Proof** Note that, on an input graph $G = (V, E)$ given in its adjacency lists, we have at most $|V|$ layers (lists) and each vertex occurs on at most one layer and is captured at most once. So, the while-loop and the first for-loop in Algorithm 9.1 are executed at most $|V|$ times. Then we have $O(V)$. The second for-loop in Algorithm 9.1 runs

$|E_{adj}|$ times, where $E_{adj}$ is the set of edges that are incident to current vertex. The reason is that every vertex captured at most once and we examine $(u, v)$ only when $u$ is captured. Note that every edge examined at most once if $G$ is directed, and at most twice if $G$ is undirected. Therefore, from Theorem 4.1 and 4.14, $|V| \times O(E_{adj})$ is $O(E)$. The other lines in Algorithm 9.1 take constant time to execute. Thus, the total running time is $|V| \times (O(1) + O(E_{adj}))$ which turns out to $O(V + E)$. This completes the proof. ∎

From Theorem 9.1, we conclude that the breadth-first search algorithm runs in time linear in size of the adjacency list representation of $G$.

## 9.3. Applications of breadth-first search

The breadth-first search has some applications, and there are some $O(V + E)$-time algorithms based on the breadth-first search algorithm for the following problems: Computing a spanning tree (forest) of a graph, computing a path with the minimum number of edges between start vertex and assigned vertex or reporting that no such path exists, testing bipartiteness of an input graph, testing whether a graph is connected, and finding the connected component that contains a given vertex. Below we describe these applications.

### Computing spanning trees (forests)

For a given graph $G = (V, E)$, the breadth-first search algorithm directly produces a tree rooted at the source vertex. If the given graph $G$ is disconnected, then the breadth-first search produces a spanning forest of $G$. The tree (forest) produced by performing a breadth-first search is called breadth-first tree (forest), and its edges are called breadth-first edges.

For the given graph shown to the left of Figure 9.5, we compute a spanning tree by drawing the produced breadth-first tree. As shown to the right of Figure 9.5, the breadth-first edges are colored blue and the vertices are enumerated in alphabetical order starting at $a$ (the src vertex), where each vertex is also super-indexed by $(i)$ to mean that it occurs on the layer $L_i$ for some $i$. As breadth-first search does not define the order of the neighbors, we assume that in this graph the children of each vertex are ordered from top to bottom.

Figure 9.5 leads us to deduce the following remark.

> **Remark 9.2** *If $(u, v)$ is a breadth-first edge, then the levels of $u$ and $v$ differ by at most 1.*

### Computing shortest paths

Finding shortest paths is one of the applications of the breadth-first search algorithm. The shortest path problem is a well-known problem in graph theory and it has nu-

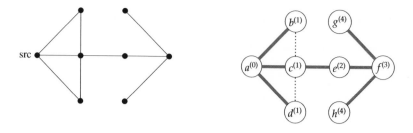

Figure 9.5: A graph and its breadth-first tree.

merous applications in the real world. It involves finding a path from a source vertex to a destination vertex which has the minimum number of edges (i.e., with the least length) among all such paths, or report that no such path exists. The following remark gives evidence when such existence is affirmed.

> **Remark 9.3** *There is a path from a (source) vertex s to a vertex t if and only if t appears in some layer while performing the breadth-first search algorithm.*

The procedure in following workflow, followed by an example, will show us how to find the shortest path(s) in an input graph $G$ given by its adjacency representation.

> **Workflow 9.1** *We find the shortest path(s) from a source vertex to a destination vertex t in an input graph G by following two steps:*
>
> (i) *Run a breadth-first search starting at the vertex s.*
>
> (ii) *Construct a path whose edges are all from breadth-first tree edges by moving backward from vertex t to its predecessors along the tree edges until we reach vertex s.*

From Workflow 9.1, we conclude that

$$(s, \ldots, \text{parent}[\text{parent}[t]], \text{parent}[t], t)$$

is a shortest path from vertex $s$ to vertex $t$. We also conclude that the length of this path equals level[$t$]. That is, if vertex $t$ occurs on a layer $L_k$ (i.e., level $k$), then the length of the above shortest path is $k$.

■ **Example 9.4** As shown Figure 9.6, by performing the breadth-first search algorithm, the layers obtained among the vertices of the graph in Figure 9.6 (*i*), starting from vertex $s$, are shown in Figure 9.6 (*ii*). The smallest number of such layers is 4. The graphs in Figure 9.6 (*iii*)-(*v*) visually show step-by-step how to find the shortest path from vertex $s$ to vertex $v$. It is clear that parent[$v$] = $c$, parent[parent[$v$]] = parent[$c$] = $x$, and parent[$x$] = $s$. So, ($s, x, c, v$) is the shortest path from $s$ to $v$. Because vertex $v$ occurs in the layer $L_3$, the length of the above shortest path is 3. The graph in Figure 9.6 (*vi*) shows that $(s, x, c, f)$ is a shortest path from vertex $s$ to vertex $f$.

Note that the shortest path may not be unique (in some cases may not even exist (see Remark 9.3)). This can be seen in Figure 9.6 (*vi*) and (*viii*). In fact, Figure 9.6 (*vii*) shows another breadth-first tree for the graph shown in Figure 9.6 (*i*), and accordingly Figure 9.6 (*viii*) shows another shortest path from vertex $s$ to vertex $f$, namely the path $(s, x, d, f)$, which is different from that shown in Figure 9.6 (*vi*).

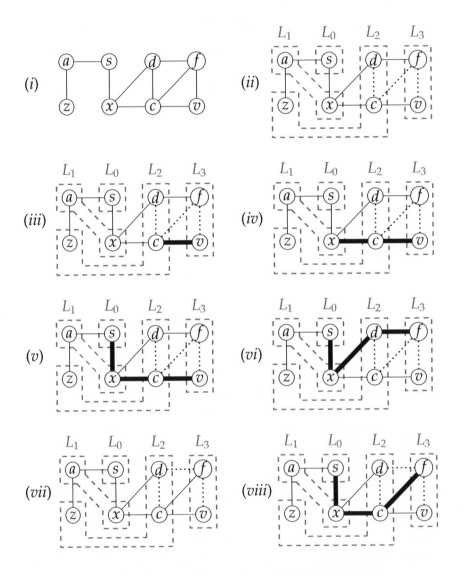

Figure 9.6: Finding shortest paths by running breadth-first searches.

### Testing bipartiteness

Breadth-first search can be used to test bipartiteness. The algorithm starts the search at any vertex and gives alternating labels to the vertices visited during the search. That is, give label 0 to the starting vertex, 1 to all its neighbors, 0 to those neighbors' neighbors, and so on. If at any step a vertex has (visited) neighbors with the same label as itself, then the graph is not bipartite. If the search ends without such a situation occurring, then the graph is bipartite.

The following workflow to test bipartiteness uses the concept of graph coloring (see Theorem 4.11) and breadth-first search.

**Workflow 9.2** *We test the bipartiteness of an input graph G by following five steps:*

(i) *Assign a color (say red) to a random source vertex.*

(ii) *Assign all the neighbors of the above vertex another color (say blue).*

(iii) *Take one neighbour at a time to assign all the neighbour's neighbors the color red.*

(iv) *Continue in this manner till all the vertices have been assigned a color.*

(v) *If at any stage, we find a neighbour which has been assigned the same color as that of the current vertex, stop the process. The graph cannot be colored using two colors. Thus the graph is not bipartite.*

We visually show the steps of Workflow 9.2 in Figure 9.7.

Finding connected components is another application of breadth-first search that is not covered in this section. In the next two sections, we present depth-first search as an alternative to breadth-first search. Since finding connected components is an application of both breadth- and depth-first searches and it can be described in both methods by the same scheme, finding connected components will be discussed together with the applications of depth-first search.

## 9.4. Depth-first search algorithm

In the previous two sections, we have presented the breadth-first search method for exploring a graph. In this section, we present depth-first search as another method for exploring a graph. In the next section, we will describe some of its applications.

Recall that the breadth-first search explores outward from a source vertex in all possible directions, adding vertices one layer at a time. The depth-first search differs from the breadth-first search in that we sequentially visit vertices until we reach a "dead end" and then we backtrack; see Figure 9.8.

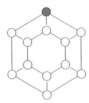

(*i*) We test the bipartiteness of this graph. Assign the source vertex a dark color.

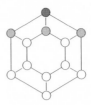

(*ii*) Assign neighbors of the source vertex another color a light color.

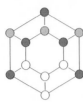

(*iii*) Assign all neighbors of the vertices colored in (*ii*) a dark color.

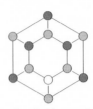

(*iv*) Assign all neighbors of the vertices colored in (*iii*) a light color.

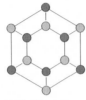

(*v*) Repeat until all vertices are colored, or a conflicting assignment occurs.

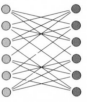

(*vi*) Since no conflicting evidence was found, the graph is bipartite.

Figure 9.7: Illustrating the progress of breadth-first search on testing bipartiteness of a graph.

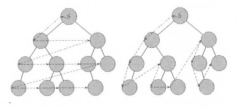

Figure 9.8: The operation of breadth-first search (left) versus that of depth-first search (right) on an undirected graph.

Like breadth-first search, depth-first search is used for both directed and undirected graphs.

As depth-first search does not define the order of the neighbors, we assume that the children of each vertex are ordered from top to bottom if those child vertices are placed to the left or right of the parent vertex as illustrated in the digraph shown to the right. We also assume that the children of each vertex are ordered from left to right if those child vertices are placed under or above the parent vertex as illustrated in the graph shown to the right of Figure 9.8. In general, we assume that the children of a vertex are ordered from top to bottom while moving in the neighborhood from left to right.

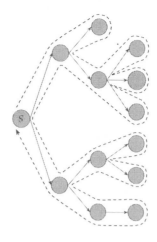

Figure 9.9: The depth-first search scheme.

In a depth-first search, we follow a path starting from a source vertex, say $s$, of a graph $G = (V, E)$ until we get stuck, then we backtrack until we reach unexplored neighbor, being careful not to repeat a vertex. See Figure 9.9.

The following example explains the steps outlined in the depth-first search method.

**Example 9.5** Show the progress of the depth-first search method on each of the following graphs starting from vertex $s$.

(a) The undirected graph:

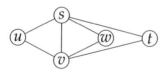

(b) The directed graph:

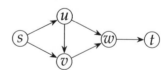

**Solution** (a) The progress of the depth-first search method on the graph given in item (a) is visualized in Figure 9.10 (i)-(iv).

(b) The progress of the depth-first search method on the graph given in item (b) is visualized in Figure 9.10 (v)-(viii).

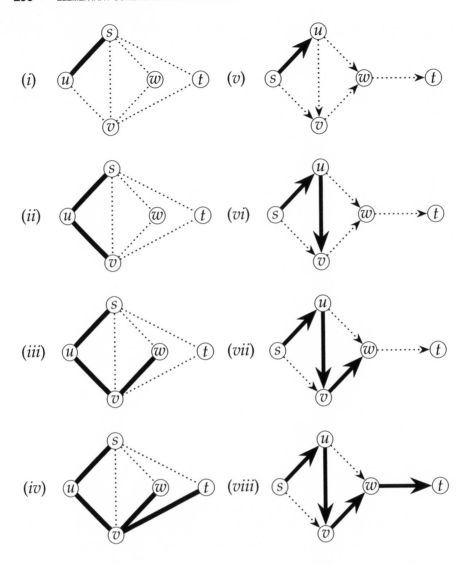

Figure 9.10: Visualization of the progress of the depth-first search method on the graphs of Example 9.5.

A version of a depth-first search algorithm is presented in Algorithm 9.2. Theorem 9.2 gives the running time of Algorithm 9.2 on an input graph $G = (V, E)$ given in its adjacency lists. This theorem concludes that the depth-first search algorithm runs in time linear in size of the adjacency list representation of $G$.

---

**Algorithm 9.2:** Depth-first search algorithm

---

```
1: begin
2:      parent = {s : None}
3:      DFS-visit(V, Adj, s)
4:      begin
5:          for v in Adj[s] do
6:              if v not in parent then
7:                  parent[v] = s
8:                  DFS-visit(V, Adj, v)
9:              end
10:         end
11:     end
12:     DFS(V, Adj)
13:     begin
14:         parent = { }
15:         for s in V do
16:             if s not in parent then
17:                 parent[s] = None
18:                 DFS-visit(V, Adj, s)
19:             end
20:         end
21:     end
22: end
```

---

**Theorem 9.2** *Algorithm* 9.2 *runs in* $O(V + E)$ *time.*

**Proof** Note that the function DFS-Visit in Algorithm 9.2 calls itself once for each vertex in $V$ since each vertex is added to the resulting tree at most once. The for-loop in DFS-Visit is executed a total of $|E|$ times for a directed graph or $2|E|$ times for an undirected graph since each edge is explored once. The for-loop in the function DFS in Algorithm 9.2 adds $O(V)$ time. Therefore, the total running time is $O(V + E)$. This completes the proof.  ∎

## 9.5. Applications of depth-first search

The depth-first search has some applications, and there are some $O(V + E)$-time algorithms based on the depth-first search algorithm for the following problems: Computing a spanning tree (forest) of a graph, detecting a cycle in a graph or reporting that no such cycle exists, testing whether a graph is connected, and finding the connected component that contains a given vertex. Below we describe some of these applications.

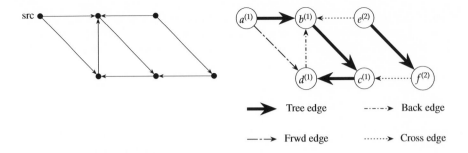

Figure 9.11: A graph and its depth-first forest.

## Computing spanning trees (forests)

For a given graph $G = (V, E)$, the depth-first search algorithm directly produces a tree rooted at the source vertex. If the given graph $G$ is disconnected, then the depth-first search produces a spanning forest of $G$. The tree (forest) produced by performing a depth-first search is called a depth-first tree (forest).

For a directed graph $G$, we define the following four edge types in terms of its depth-first tree (forest):

- Tree edges: A tree edge is an edge of a depth-first tree (forest) of $G$.

- Back edges: A back edge is a non-tree edge connecting a vertex to one of its ancestors (but not parents) in a depth-first tree (forest) of $G$. Note that a self-loop is a back edge.

- Forward edges: A forward edge is a non-tree edge connecting a vertex to one of its descendants (but not children) in a depth-first tree (forest) of $G$.

- Cross edge: A cross edge is a non-tree edge connecting some pair of vertices that have no ancestor-descendent relationship in a depth-first tree (forest) of $G$.

For an undirected graph $G$, there are no forward edges because they become back edges when we move in the opposite direction, and there are no cross edges because every edge of an undirected graph $G$ must connect an ancestor vertex to a descendant vertex. This leads us to the following remark.

**Remark 9.4** *For an undirected graph G, there are tree edges and back edges only in terms of its depth-first tree (forest). No forward or cross edges.*

For the given digraph shown to the left of Figure 9.11, we compute a spanning forest by drawing the produced depth-first forest. The computed spanning forest consists of two trees with two different roots. As shown to the right of Figure 9.11, the tree edges are colored blue and the vertices are enumerated in alphabetical order starting at $a$ (the src vertex), where each vertex is also super-indexed either by (1) to mean that it is located on the first tree that has root $a$, or by (2) to mean that it is located on the second tree that has root $e$. Note that $(d, b)$ is a back edge, $(a, d)$ is a forward edge, and $(g, e)$ and $(d, c)$ are cross edges.

## Detecting cycles

Detecting cycles is one of the applications of the depth-first search algorithm. The following remark gives evidence when the existence a cycle in an input graph is affirmed.

> **Theorem 9.3** *A graph G has a cycle if and only if there exists a back edge in G after performing any depth-first search.*

**Proof**  First, we show that if a graph $G = (V, E)$ has a cycle then $G$ has a back edge. Running a depth-first search on $G$, we can sort the vertices in decreasing order in terms of its appearance in the depth-first tree starting from source vertex. Note that tree edges, forward edges and cross edges all point forward in this ordering. However, back edges point backward to an earlier vertex in this ordering. Since by assumption these vertices form a cycle, there must be at least one back edge. Otherwise, there is no way to go from a later vertex in the ordering to an earlier vertex, which contradicts the assumption of the existence of a cycle.

Conversely, assume that $(u, v) \in E$ is a back edge. Then by definition of back edges there exists a path $P$ from $v$ to $u$ in the depth-first tree. Let $C = P + \{(u, v)\}$, then $C$ is a cycle in $G$. The proof is complete.  ∎

Recall that a directed acyclic graph (DAG) is a digraph that has no directed cycles. Theorem 9.3 leads immediately us to the following corollary.

> **Corollary 9.1** *A digraph is a DAG if and only if it has no back edges after performing any depth-first search.*

The procedure in following workflow shows us how to detect and report a cycle in an input graph $G$.

> **Workflow 9.3** *We detect and report a cycle in an input graph G by following four steps:*
>
> (i) *Run a depth-first search.*
>
> (ii) *If at any stage, we find a back edge $(u, v)$, stop the depth-first search process. A cycle is detected in the graph.*
>
> (iii) *Find the (unique) path P from v to u in the produced depth-first tree.*
>
> (iv) *Add the edge $(u, v)$ to the path P to localize the detected cycle.*

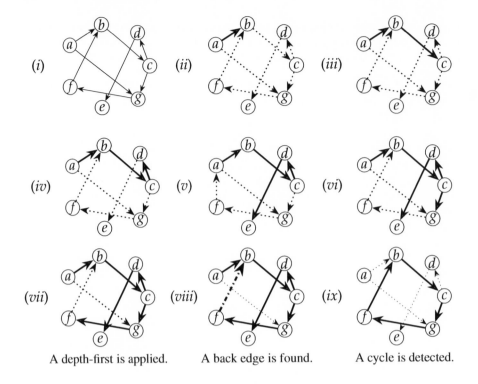

A depth-first is applied.　　A back edge is found.　　A cycle is detected.

Figure 9.12: Detecting a cycle by running a depth-first search.

We visually show the steps of Workflow 9.3 in Figure 9.12. After running a depth-first search on the graph shown in Figure 9.12 (*i*), the back edge $(f, b)$ is found, as illustrated in Figure 9.12 (*viii*), and the cycle $(b, c, g, f)$ is detected, as localized in Figure 9.12 (*ix*).

### Finding connected components

Depth- and breadth-first searches can be used to test if a graph is connected as follows: Starting from a random source vertex, if on termination of algorithm, all vertices are visited, then the graph is connected, otherwise it is disconnected. The following remark proves the correctness of this assertion.

> **Remark 9.5** *Depth- and breadth-first searches visit all vertices that are reachable from source vertex s.*

**Proof** We prove this remark by induction: Depth- and breadth-first searches will clearly reach all vertices at level 0 (just $s$ itself). This is the base case. Consider the vertices at minimum distance $i$ from vertex $s$. Call these vertices "level $i$" vertices. If depth- or breadth-first search successfully reaches all vertices at level $i$, then they must reach all vertices at level $i + 1$, since each vertex at distance $i + 1$ from $s$ must be connected to some vertex at distance $i$ from $s$. This is the inductive step. The proof is complete.                                                                              ∎

Depth- and breadth-first searches can also be used to find the connected components in an input graph (i.e., finding all vertices within each connected component).

In order to find a connected component of a graph, we pick a random source vertex and start doing a depth- or breadth-first search from that vertex. All the vertices we can reach from that vertex compose a single connected component. To find all the connected components, then, we need to go through every vertex, finding their connected components one at a time by searching the graph. Note however that we do not need to start a search from a vertex $v$ if we have already found it to be part of a previous connected component. Hence, if we keep track of what vertices we have already encountered, we will only need to perform one search for each connected component.

The connected component, say $R$, containing a start vertex, say $s$, is obtained by performing then code in Algorithm 9.3. When searching from a particular vertex $s$, we will clearly never reach any vertices outside the connected component with depth- or breadth-first search. Consequently, by Remark 9.5, Algorithm 9.3 will find each connected component correctly. That is, upon termination, $R$ will consist of vertices that are reachable from vertex $s$.

There are other applications of depth-first search such as testing bipartiteness and topological sorting. We decided to not include testing bipartiteness in this section because it was previously described as an application of breadth-first search. Topological sorting algorithm will presented in the next section as an application of depth-first search.

---

**Algorithm 9.3:** Finding connected components

---
1: $R = \{s\}$
2: **while** (there is an edge $(u, v)$ where $u \in R$ and $v \notin R$) **do**
3:    |    add $v$ to $R$
4: **end**
5: **return** $R$

---

## 9.6. Topological sort

A topological ordering or sorting of an $n$-vertex directed graph $G = (V, E)$ is an ordering of its vertices as $v_1, v_2, \ldots, v_n$ so that for every edge $(v_i, v_j) \in E$ we have $i < j$. Less informally, a topological ordering arranges the vertices of a directed graph along a horizontal line so that all edges point from left to right. See Figure 9.13.

Figure 9.13: A directed graph (left) and its topological ordering (right).

Formally, we have the following definition.

---

**Definition 9.1** *A topological ordering of a directed graph $G = (V, E)$ is a total order "$\prec$" such that $u \prec v$ for every edge $(u, v) \in E$.*

---

Topological sorting can be used to schedule tasks under precedence constraints. Assume that we have a set of tasks to do, but certain tasks have to be performed before other tasks. Only one task can be performed at a time and each task must be completed before the next task begins. Such a relationship can be represented as a directed graph and topological sorting can be used to schedule tasks under precedence constraints. We are going to determine if an order exists such that this set of tasks can be completed under the given constraints. Such an order is indeed a topological sort of a digraph $G$, where the vertices of $G$ are the tasks and the edges of $G$ are the precedence constraints. Edge $(v_i, v_{i+1})$, for instance, means that task $v_i$ must be completed be before task $v_{i+1}$ can be started. See for example Figure 9.14.

Topological sort is often useful in scheduling jobs in their proper sequence, for which the output of job $v_i$ is needed to determine the input of job $v_{i+1}$. Topological sort is also useful in selecting course prerequisites of a target program, for which the course $v_i$ must be taken before course $v_{i+1}$.

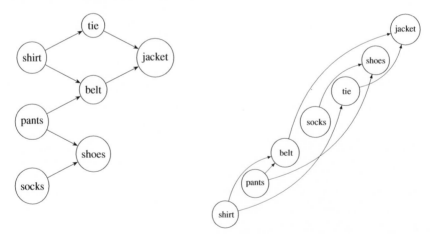

Figure 9.14: The order of dressing up different clothes (i.e., which clothes goes first and which follows second, etc.) is represented by a digraph (left), and its precedence constraints are identified by a topological sort (right).

Note that the graphs in Figures 9.13 and 9.14 are DAGs (recall that a DAG is a directed acyclic graph). We have the following theorem.

**Theorem 9.4** *A digraph has a topological order if and only if it is a DAG.*

**Proof**    First, we show that if a digraph $G$ has a topological order, then $G$ is a DAG. Suppose, on the contrary, that $G$ has a directed cycle, then a topological ordering of $G$ is clearly impossible because the right-most vertex of the cycle would have an edge pointing to the left. This contradicts the fact that $G$ has a topological order.

Conversely, assume that $G = (V, E)$ is a DAG. We now show that $G$ has a topological order. Consider an arbitrary ordering "<" of $G$. If $v < u$ for any edge $(u, v) \in E$, then $G$ contains a directed path from $v$ to $u$, and therefore contains a directed cycle through the edge $(u, v)$. Equivalently, if $G$ is acyclic, then $u < v$ for any edge $(u, v) \in E$. This means that any DAG has a topological order. The proof is complete. ∎

The three-step procedure in the following workflow, followed by an example, will learn us how to compute a topological ordering of a DAG.

**Workflow 9.4** *We compute a topological ordering of a DAG $G = (V, E)$ by following three steps:*

(*i*) *Find a vertex $v \in V$ with no incoming edges and order it first.*

(*ii*) *Delete $v$ from $G$.*

(*iii*) *Recursively compute a topological ordering of $G - \{v\}$ and append this order after $v$.*

**Example 9.6**    Compute a topological ordering for each of the following DAGs.

(*a*) The connected digraph:

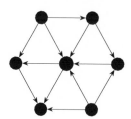

(*b*) The disconnected digraph:

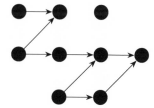

**Solution**    (*a*) The progress of Workflow 9.4 steps to compute a topological ordering for the given DAG is visualized in Figure 9.15.

(*b*) A topological ordering for the given DAG is computed and shown in Figure 9.16. ∎

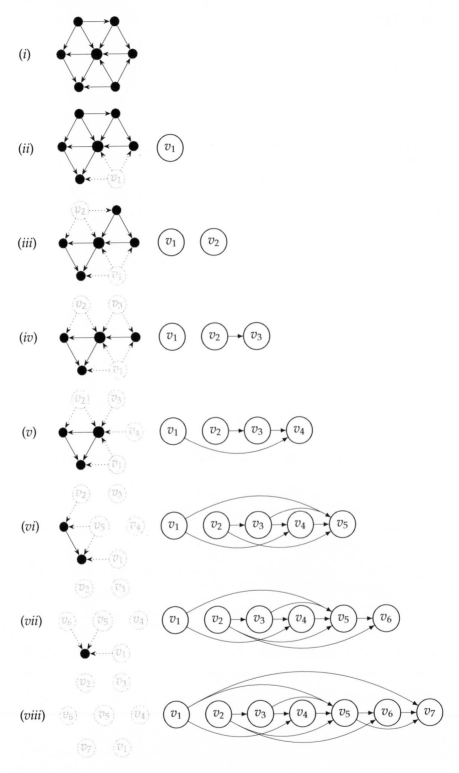

Figure 9.15: Visualization of the progress of Workflow 9.4 steps to compute a topological ordering for the DAG in Example 9.6 (*a*).

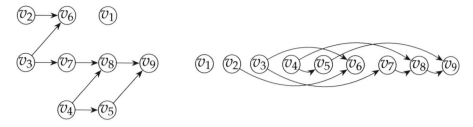

Figure 9.16: The DAG in Example 9.6 (*b*) and its topological ordering (shown to the left).

It is not hard to find that the topological sort has finishing time of the reverse of depth-first search. We have shown in Theorem 9.2 that the depth-first search runs in $O(V + E)$ time. Hence we can conclude the following corollary.

**Corollary 9.2** *The topological sort runs in* $O(V + E)$ *time.*

# EXERCISES

**9.1**    Choose the correct answer for each of the following multiple-choice questions/items.

(*a*) Which one of the following statements is false?

   (*i*) The adjacency list is preferred when the graph is a cycle while the adjacency matrix is preferred when the graph is complete.

   (*ii*) The adjacency list is faster than the adjacency matrix for listing the vertices adjacent to a given vertex.

   (*iii*) The adjacency matrix is faster than the adjacency list for determining if there is an edge between two vertices.

   (*iv*) The adjacency matrix is preferred when the graph is sparse while the adjacency list is preferred when the graph is dense.

(*b*) If we use a breadth-first search for testing bipartiteness of an input graph, we stop the process when

   (*i*) all vertices are visited.

   (*ii*) all vertices are colored red or blue, or a conflicting color assignment occurs.

   (*iii*) a back edge is found.

   (*iv*) it exceeds 10 minutes.

(*c*) If we run a depth-first search on a directed graph, and then remove all of the back edges found, the resulting graph is

    (*i*) a tree.                   (*iii*) acyclic.

    (*ii*) cyclic.                (*iv*) bipartite.

(*d*) Assume in some general connected graph *G* we run the breadth-first search and the depth-first search on G starting from the same vertex *s* and we find that the resulting trees are the same, then this necessarily means that *G* is

    (*i*) a cycle.                (*iii*) a complete graph.

    (*ii*) a tree.               (*iv*) a complete bipartite graph.

(*e*) Suppose that *G* is a graph with $n$ vertices and at most $n \log n$ edges, represented in its adjacency list representation. Then the depth-first search on *G* runs in:

    (*i*) $O(n)$ time.            (*iii*) $O(n \log n)$ time.

    (*ii*) $O(\log n)$ time.       (*iv*) $O(n/\log n)$ time.

(*f*) A person wants to visit some places. He starts from a vertex and then wants to visit every place connected to this vertex and so on. What algorithm he should use?

    (*i*) Breadth-first search.     (*iii*) Topological sorting.

    (*ii*) Depth-first search.      (*iv*) Any of the above.

(*g*) Consider the undirected graph $G = (V, E)$, where $V = \{m, n, o, p, q, r\}$ and $E = \{(m, n), (n, q), (q, m), (n, o), (o, p), (p, q), (m, r)\}$. If we run the breadth-first search on *G* starting at any vertex, which one of the following is a possible order for visiting the vertices?

    (*i*) $m, n, o, p, q, r$.       (*iii*) $r, m, q, p, o, n$.

    (*ii*) $n, q, m, p, o, r$.       (*iv*) $q, m, n, p, r, o$.

(*h*) Suppose that *G* is a graph with $n$ vertices and at most $n^2 \log n$ edges, represented in its adjacency list representation. Then the breadth-first search on *G* runs in:

    (*i*) $O(n)$ time.            (*iii*) $O(n \log n)$ time.

    (*ii*) $O(\log n)$ time.       (*iv*) $O(n^2 \log n)$ time.

(*i*) If we use a depth-first search for detecting cycles in an input graph, we stop the process when

    (*i*) all vertices are visited.

    (*ii*) all vertices are colored red or blue, or a conflicting color assignment occurs.

    (*iii*) a back edge is found.

    (*iv*) it exceeds 10 minutes.

(*j*) Three students were given a DAG, $G = (V, E)$, where $V = \{2, 3, 5, 7, 8, 9, 10, 11\}$ and $E = \{(5, 11), (7, 8), (7, 11), (3, 8), (3, 10), (8, 9), (11, 2)(11, 9), (11, 10)\}$. Each one of these students run the topological sorting on $G$ twice and obtained a result of the following. Which student made a mistake?

    (*i*) The first student who got the sorts: 5, 7, 3, 11, 8, 2, 9, 10 and 3, 5, 7, 8, 11, 2, 9, 10.

    (*ii*) The second student who got the sorts: 3, 5, 7, 8, 11, 2, 9, 10 and 5, 7, 3, 8, 11, 10, 9, 2.

    (*iii*) The third student who got the sorts: 7, 5, 11, 3, 10, 8, 9, 2 and 5, 3, 11, 2, 7, 8, 9, 10.

    (*iv*) No student made a mistake.

**9.2**     Read each of the following statements, and decide whether it is true or false.

(*a*) For undirected graphs, in both the adjacency list and the adjacency matrix, there are two representations of each edge in the graph.

(*b*) Breadth- and depth-first searches can both be used for testing the connectedness of a graph.

(*c*) A graph $G$ is acyclic if and only if there does not exist a back edge in $G$ after performing any depth-first search.

(*d*) We can find a topological sorting for any digraph, even if it contains a dicycle.

(*e*) If a topological sort exists for the vertices in a graph, then a depth-first search on the graph will produce no back edges.

(*f*) For directed graphs, in both the adjacency list and the adjacency matrix, there are two representations of each edge in the graph.

(*g*) Breadth- and depth-first searches can both be used for finding connected components.

(*h*) Given two vertices of a graph G, the shortest path in G between them is not always unique.

(*i*) For an undirected graph $G$, there are tree edges and forward edges only in terms of its depth-first tree (forest). No back or cross edges.

(*j*) A digraph $G$ has a topological order if and only if there exists a back edge in $G$ after performing any depth-first search.

**9.3**    Represent the graph in Example 9.1 (*c*) using:

(*a*) Adjacency list representation.          (*b*) Adjacency matrix representation.

**9.4**    Consider the following two graphs; one of them is undirected and the other is directed. Let $v_1$ be the source vertex in each.

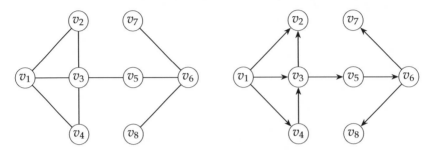

(*a*) We run a breadth-first search on the undirected graph and obtained the breadth-first tree in Figure 9.5. Use this to find the shortest path from vertex $v_1$ to vertex $v_8$. Is this shortest path unique?

(*b*) Use breadth-first search to test the undirected graph for bipartiteness.

(*c*) Run a depth-first search on the directed graph to compute a depth-first tree. Classify the edges as tree edges, back edges, forward edges, and cross edges.

(*d*) Does the directed graph has a directed cycle? Why or why not? Link your answer to the depth-first search completed in item (*c*).

(*e*) Compute a topological ordering for the directed graph.

**9.5**    Consider the following the directed graph, and let $s$ be its source vertex.

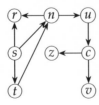

(*a*) Run a breadth-first search on the graph to obtain the breadth-first tree.

(*b*) Use breadth-first search to test the undirected version of the graph for bipartiteness.

(c) Run a depth-first search on the graph to compute a depth-first tree.

(d) Does the graph has a directed cycle? Why or why not? Link your answer to the depth-first search completed in item (c).

(e) Compute a topological ordering for the graph.

**9.6**  Answer items (a) through (e) in Exercise 9.5 for the following graph, which has s as its source vertex.

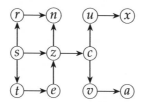

**9.7**  Prove or disapprove: A depth-first search of a directed graph always produces the same number of tree edges (i.e., independent of the order in which the vertices are provided and independent of the order of the adjacency lists).

**Part IV**

# OPTIMIZATION

# CHAPTER 10

# LINEAR PROGRAMMING

## Contents

© 2022 by Baha Alzalg | Kindle Direct Publishing, Washington, United States 2022/9
B. Alzalg, *Combinatorial and Algorithmic Mathematics: From Foundation to Optimization*,
DOI 10.5281/zenodo.7110995

Since its conception in the 1940s in connection with planning activities of the military, linear programming (also called linear optimization) has come into wide use in industry and many other fields.

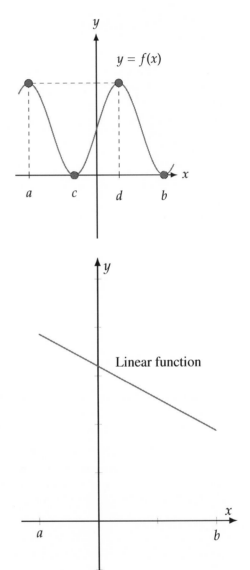

Let $f : [a, b] \to \mathbb{R}$ be a nonlinear continuous function. In Calculus, if we want to minimize/maximize the function $f(x)$, we must take the derivative, and then find the critical points. We also check the endpoints, if there are any. We can justify our maxima or minima either by the first derivative test, or the second derivative test. In the graph shown to the right, $a$ and $b$ are endpoints of the function $f(x)$, and $c$ and $d$ are their critical numbers ($f'(x) = 0$ when $x = c, d$). The function $f$ has maximum values at $x = a, d$, and has minimum values at $x = c, b$.

Now, instead of optimizing a nonlinear function on $[a, b]$, consider a linear function on $[a, b]$. In this case, we need to check only the endpoints as in the figure shown to the right. Linear optimization studies generalizations of this easy (linear) case to higher dimensions. More specifically, instead of optimizing a linear function of only one variable, say $cx$, on the closed interval $[a, b]$, we optimize a linear function of a finite number of variables, say $c^\mathsf{T}x = c_1 x_1 + c_2 x_2 + \cdots + c_n x_n$, on polytopes, which are generalizations of polygons from $\mathbb{R}^2$ to $\mathbb{R}^n$, where the set $\mathbb{R}^n$ consists of all $n$-tuples of real numbers, $\mathbb{R}$. This study is "easy" to understand because of linearity, but it is "difficult" to carry out because of high dimensionality.

In this chapter, we introduce linear programming, the graphical method, and study the linear programming duality and geometry. The reference [BT97, NW88], for example, is a good source for information relative to this topic. Then we study the most common linear programming algorithm: the simplex method. We also study an interior-point method as one of the non-simplex methods. The reference [BT97, NW88], for example, is a good source for information relative to this topic.

## 10.1. Linear programming formulation and examples

In this section, we will see that applications of linear programming touch a vast range of real-world areas. First, we present the general form of a linear programming problem.

### General form linear programs

A linear programming (LP) problem is the problem of minimizing a linear cost function subject to linear equality and inequality constraints. We have the following example.

**Example 10.1**  The following is a linear programming problem.

$$
\begin{array}{rrrrrrrrl}
\text{minimize} & 4x_1 & - & x_2 & + & 3x_3 & & & \\
\text{subject to} & x_1 & + & x_2 & & & + & x_4 & \le\ 7, \\
& & & 2x_2 & - & x_3 & & & =\ 6, \\
& & & & & x_3 & + & x_4 & \ge\ 4, \\
& x_1 & & & & & & & \ge\ 0, \\
& & & & & x_3 & & & \le\ 0.
\end{array}
$$

Here, $x_1, x_2, x_3$ and $x_4$ are the decision variables whose values are to be chosen to minimize the linear cost function $4x_1 - x_2 + 3x_3$ subject to linear equality and inequality constraints. ∎

Generally speaking, assume that we are given a cost vector $c = (c_1; c_2; \ldots; c_n)^\mathsf{T}$ and we minimize a linear cost function $c^\mathsf{T} x = \sum_{i=1}^{n} c_i x_i$ over all $n$th-dimensional vectors $x = (x_1; x_2; \ldots; x_n)^\mathsf{T}$ subject to linear equality and inequality constraints. Then we are interested in a problem of the form:

$$
\begin{array}{rll}
\min & c^\mathsf{T} x & \\
\text{s.t.} & a_i^\mathsf{T} x \ge b_i, & i = 1, 2, \ldots, m_1, \\
& a_j^\mathsf{T} x \ge b_j, & j = 1, 2, \ldots, m_2, \\
& a_k^\mathsf{T} x = b_k, & k = 1, 2, \ldots, m_3, \\
& x_p \ge 0, & p = 1, 2, \ldots, m_4, \\
& x_q \le 0, & q = 1, 2, \ldots, m_5.
\end{array}
\tag{10.1}
$$

Problem (10.1) is said to be the general form LP. We have the following definition.

**Definition 10.1** *Consider the minimization problem* (10.1). *Then:*

(a) *The variables* $x_1, x_2, \ldots, x_n$ *are called decision variables;*

(b) *A vector $x$ satisfying all of the constraints is called a feasible solution;*

(c) *The set of all feasible solutions is called the feasible set or feasible region;*

(d) *If $x_i \geq 0$ or $x_i \leq 0$, then $x_i$ is called a restricted variable, otherwise it is called a free or unrestricted variable (urs);*

(e) *The function $c^\mathsf{T} x$ is called the objective function or cost function;*

(f) *A feasible solution $x^\star$ that minimizes the objective function (that is, $c^\mathsf{T} x^\star \leq c^\mathsf{T} x$ for any feasible solution $x$) is called an optimal solution;*

(g) *The value of $c^\mathsf{T} x^\star$, corresponding to an optimal solution $x^\star$, is called the optimal cost or optimal value;*

(h) *If the optimal cost is $-\infty$, we say that the minimization problem is unbounded.*

**Example 10.2** Consider the following nonlinear minimization problem.

$$
\begin{aligned}
\min \quad & 2x_1 + |x_2| \\
\text{s.t.} \quad & 5x_1 + 7x_2 \leq 3, \\
& |x_1| + x_2 \leq 4, \\
& x_1, x_2 \text{ urs.}
\end{aligned}
$$

Using the fact that $|t| = \max\{t, -t\}$ for $t \in \mathbb{R}$, this problem can be expressed as

$$
\begin{aligned}
\min \quad & 2x_1 + y \\
\text{s.t.} \quad & 5x_1 + 7x_2 \leq 3, \\
& x_1 + x_2 \leq 4, \\
& -x_1 + x_2 \leq 4, \\
& y - x_2 \geq 0, \\
& y + x_2 \geq 0, \\
& x_1, x_2 \text{ urs,}
\end{aligned}
$$

which is an LP problem.                                                                          ▦

   Note that there is no need to study maximization problems separately, because maximizing $c^\mathsf{T} x$ subject to some constraints is equivalent to minimizing $(-c)^\mathsf{T} x$ subject to the same constraints.

## Examples of linear programming problems

This part presents some examples of linear programming problems and allows the reader gain to some familiarity with the art of constructing mathematical optimization models.

The procedure given in the following workflow, followed by some examples, will teach us to formulate linear optimization models.

---

**Workflow 10.1** *There are three steps involved in the formation of linear programming problem:*

   *(i) Identify the decision variables of interest to the decision maker and express them as $x_1, x_2, x_3, \ldots$*

  *(ii) Ascertain the objective function in terms of the decision variables. This would be a cost in case of minimization problem or a profit in case of maximization problem.*

 *(iii) Ascertain the constraints representing the maximum availability or minimum commitment or equality.*

---

■ **Example 10.3**  A firm is engaged in producing two products $P_1$ and $P_2$. Each unit of product $P_1$ requires 2 kg of raw material and 4 labour hours for processing, and each unit of product $P_2$ requires 5 kg of raw material and 3 hours of labour of the same type. Every week the firm has the availability of 50 kg of raw material and 60 labour hours. One unit of product $P_1$ sold earns profit of JD 20 and one unit of product $P_2$ sold gives JD 30 as profit. Formulate this problem as LP problem that can be used to maximize the total profit.

**Solution**  The given data can be summarized in the following table.

| Product | Row material | Labour hours | Profit |
|---------|-------------|--------------|--------|
| $P_1$ | 2 kg | 4 hrs | JD 20 |
| $P_2$ | 5 kg | 3 hrs | JD 30 |
| Restrictions | 50 kg | 60 hrs | |

The first step it to identify the decision variables. Let $x_i$ denote the number of units that should be produced from product $P_i$ per week, $i = 1, 2$.

The second step it to find out the objective function. The objective is to maximize the total profit. So, our objective function is $z = 20x_1 + 30x_2$.

The third step it to come to know the constraints. In this example, the constraints are:

■ A restriction on the row material. This can be formulated as $2x_1 + 5x_2 \leq 50$.

■ A restriction on the labour hours. This can be formulated as $4x_1 + 3x_2 \leq 60$.

- Non-negativity constraints. This can be formulated as $x_1 \geq 0$ and $x_2 \geq 0$.

As a result, this problem can be formulated as the following LP model.

$$
\begin{aligned}
\max \quad & 20x_1 + 30x_2 \\
\text{s.t.} \quad & 2x_1 + 5x_2 \leq 50, \\
& 4x_1 + 3x_2 \leq 60, \\
& x_1, x_2 \geq 0.
\end{aligned}
$$

∎

**Example 10.4**  A corn chip company is divided into two departments which put out two types of corn chips: extra larges and really smalls. The company makes a profit of $200 per kilobag of extra larges and $150 per kilobag of really smalls (a kilobag is 1000 bags). Each department has separate regulations concerning the number of bags produced per day. The company's goal is to maximize its profit with these regulations.

(*a*)  Identify the decision variables.

(*b*)  Write the objective function expressing $z$ in terms of $x$ and $y$.

(*c*)  Write inequalities expressing the following constraints:

 (*i*)  No more than 20 kilobags of extra larges and no more than 30 kilobags of really smalls can be put out per day.

 (*ii*)  No more than a total of 45 kilobags can be manufactured each day.

 (*iii*)  The number of extra larges must be no less than $\frac{3}{4}$ the number of really smalls produced per day.

 (*iv*)  More than 300 hours of labor must be used each day to meet union requirements. It takes 10 hours to make a kilobag of extra larges and 15 hours to make a kilobag of really smalls.

**Solution**  (*a*)  The decision variables are:

$x$: The number of extra large corn chips produced per day;

$y$: The number of really small corn chips produced per day.

(*b*)  $z = 200x + 150y$.

(*c*)   (*i*)  $x \leq 20$, $y \leq 30$.        (*iii*)  $x \geq \frac{3}{4}y$.

   (*ii*)  $x + y \leq 45$.        (*iv*)  $10x + 15y > 300$.

∎

**Example 10.5**  Vitamins A and B are found in two different foods $F_1$ and $F_2$. One unit of food $F_1$ contains 2 units of vitamin A and 5 units of vitamin B. One unit of food $F_2$ contains 4 units of vitamin A and 2 units of vitamin B. One unit of food $F_1$ and that of food $F_2$ cost JD 10 and JD 12.50, respectively. The minimum daily requirement (for a person) of vitamin A and B is 40 and 50 units, respectively. Formulate this problem as LP problem that can be used to minimize the total cost.

**Solution**  The given data can be summarized in the following table.

| Food/Vitamin | A | B | Cost |
|---|---|---|---|
| $F_1$ | 2 units | 5 units | JD 10 |
| $F_2$ | 4 units | 2 units | JD 12.5 |
| Restrictions | 40 units | 50 units | |

Let $x_i$ be the number of units that should be daily produced from food $F_i$ for a person, $i = 1, 2$. This problem can be formulated as the following LP model.

$$\begin{aligned} \min \quad & 10x_1 + 12.5x_2 \\ \text{s.t.} \quad & 2x_1 + 4x_2 \geq 40, \\ & 5x_1 + 2x_2 \geq 50, \\ & x_1, x_2 \geq 0. \end{aligned}$$

**Example 10.6**  A marketing manager wishes to allocate his annual advertising budget of JD 20,000 in two media A and B. The unit cost a message in media A is JD 1,000 and that in media B is JD 1,500. Media A is a monthly magazine and not more than one insertion is desired in the issue. At least five messages should appear in media B. The expected effective audience for one message in media A is 40,000 people and that in Media B is 50,000 people. Formulate this as an LP problem which can be used to maximize the total audience that might be reached after advertising.

**Solution**  The given data can be summarized in the following table.

| Media | Media A | Media B | Restrictions |
|---|---|---|---|
| Audience | 40,000 people | 50,000 people | |
| One message cost | JD 1,000 | JD 1,500 | JD 20,000 |
| Number of messages | at most 1 | at least 5 | |

Let $x_1$ and $x_2$ be the number of messages that should appear in media A and B, respectively. This problem can be formulated as the following LP model.

$$
\begin{aligned}
\max \quad & 40,000x_1 + 50,000x_2 \\
\text{s.t.} \quad & 1,000x_1 + 1,500x_2 \leq 20,000, \\
& x_1 \leq 1, \\
& x_2 \geq 5, \\
& x_1, x_2 \geq 0.
\end{aligned}
$$

$\blacksquare$

**Example 10.7**   A farmer is engaged in breeding sheep. The sheep are fed on various products grown on the farm. Because of the need to ensure nutrient constituents, it is necessary to buy additional one or two products, which we shall call A and B. The nutrient constituents (vitamins and protein) in each of the product are given below:

| Nutrient Constituents | Nutrient in product A | Nutrient in product B | Minimum requirement of nutrient constituents |
|---|---|---|---|
| X | 36 | 6 | 108 |
| Y | 3 | 12 | 36 |
| Z | 20 | 10 | 100 |

Product A costs JD 20 per unit and Product B cost JD 40 per unit. Write an LP problem that can be used to minimize the total cost.

**Solution**   Let $x_1$ and $x_2$ be the number of units that must be purchased from products A and B, respectively. This problem can be formulated as the following LP model.

$$
\begin{aligned}
\min \quad & 20x_1 + 40x_2 \\
\text{s.t.} \quad & 36x_1 + 6x_2 \geq 108, \\
& 3x_1 + 12x_2 \geq 36, \\
& 20x_1 + 10x_2 \geq 100, \\
& x_1, x_2 \geq 0.
\end{aligned}
$$

$\blacksquare$

**Example 10.8**   A University hospital needs your help with scheduling nurses in their intensive care unit. For this problem, we will assume the same schedule is repeated every day and the requirements for nurses are the same every day. The work day is broken up into four shifts: 12AM-6AM, 6AM-12PM, 12PM-6PM, and 6PM-

12AM. Nurses work two of these shifts every day. Nurses who work two consecutive shifts are paid \$20 per hour. Nurses who work a "split schedule" (e.g., 12AM-6AM and 12PM-6PM) are paid \$25 per hour. (The shifts 6PM-12AM and 12AM-6AM are considered consecutive). The table below indicates the number of nurses that are required during each of the shifts.

| Shift | Number required |
|-------|-----------------|
| 12AM-6AM | 5 |
| 6AM-12PM | 12 |
| 12PM-6PM | 7 |
| 6PM-12AM | 10 |

Formulate a linear program that can help the University Hospital determine how to schedule the nurses to meet the requirements throughout the day and minimize the total cost.

**Solution**   The decision variables are:

$x_1$: The number of nurses that work from 12AM-12PM;

$x_2$: The number of nurses that work from 6AM-6PM;

$x_3$: The number of nurses that work from 12PM-12AM;

$x_4$: The number of nurses that work from 6PM-6AM;

$x_5$: The number of nurses that work from 12AM-6AM and 12PM-6PM;

$x_6$: The number of nurses that work from 6AM-12PM and 6PM-12AM.

Minimizing the total cost, we obtain the following LP problem.

$$
\begin{aligned}
\min \quad & 20x_1 + 20x_2 + 20x_3 + 20x_4 + 25x_5 + 25x_6 \\
\text{s.t.} \quad & x_1 + x_4 + x_5 \geq 5, \\
& x_1 + x_2 + x_6 \geq 12, \\
& x_2 + x_3 + x_5 \geq 7, \\
& x_3 + x_4 + x_6 \geq 10, \\
& x_1, \ x_2, \ x_3, \ x_4, \ x_5, \ x_6 \geq 0.
\end{aligned}
$$

# 10.2. The graphical method

In this section, we discuss the graphical method for linear optimization problems of two variables. We will also visually demonstrate different LP cases which may result in different types of solutions.

We start by presenting the following workflow of six steps to find the extremum (maximum or minimum) solution graphically.

---

**Workflow 10.2** *The following steps involved in solving 2-dimensional LP problems graphically:*

 (*i*)  *Graph constraint equations on a rectangular coordinate plane.*

 (*ii*)  *Determine the valid side of each constraint equation.*

 (*iii*)  *Isolate and identify the feasible region.*

 (*iv*)  *Determine the direction of improvement.*

 (*v*)  *Locate the extreme corner.*

 (*vi*)  *Find the optimum solution and the corresponding optimal value.*

---

As a direct application of the above steps, we have the following example.

**Example 10.9**   Use the graphical method to solve the following LP problem.

$$\min z = 2x + 5y$$
$$\text{s.t.} \quad 3x + 2y \le 6,$$
$$-x + 2y \le 4,$$
$$x + y \ge 1,$$
$$x, y \ge 0.$$

**Solution**   Following the steps in Workflow 10.2, we obtain the graphical solution visualized in Figure 10.1. Note that the given objective function $z = 2x + 5y$ is perpendicular to the vector $c = (2, 5)$ for any given scalar $z$. For simplicity, we represent this using the vector $c$ in Figure 10.1. Furthermore, decreasing $z$ corresponds to moving the line $z = 2x + 5y$ in the direction of $-c$. Therefore, to minimize $z$, we move the line $2x + 5y = z$ as much as possible in the direction of $-c$, as long as we do not leave the feasible region. From Figure 10.1, we find that the unique optimal solution is $x = (1, 0)$ and the optimal value is $z = 2 \times 1 + 5 \times 0 = 2$.   ∎

For a system of linear equations $Ax = b$, we have three possibilities: The system has a unique solution, it has infinitely many solutions, or it is inconsistent. For an LP, we have the corresponding three possibilities, but we have one more possibility in addition. A LP problem may have:

- A unique/finite optimal solution;

- An unbounded solution;

- Alternative (multiple or infinite number of) optimal solutions;

- An infeasible solution.

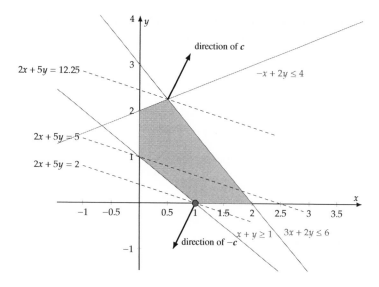

Figure 10.1: Graphical solution of the LP problem in Example 10.9.

In the context of graphical method, it is easy to visualize these four different cases. We have the following examples.

**Example 10.10**  Use the graphical method to solve the following LP problems.

(a) $\max z = 13x_1 + 23x_2$
 s.t.
$$x_1 + 3x_2 \leq 96,$$
$$x_1 + x_2 \leq 40,$$
$$7x_1 + 4x_2 \leq 238,$$
$$x_1, x_2 \geq 0.$$

(b) $\max z = x_1 + x_2$
 s.t.
$$x_1 + 3x_2 \leq 96,$$
$$x_1 + x_2 \leq 40,$$
$$7x_1 + 4x_2 \leq 238,$$
$$x_1, x_2 \geq 0.$$

**Solution**  (a) The graphical representation of the given LP problem is shown in Figure 10.2, with the feasible region shaded in gray. From the graph, we find that the maximum value for $z$ is 800 at $x = (12, 28)$. So, this LP problem has a unique optimal solution.

(b) The graphical representation of the given LP problem is shown in Figure 10.3, with the feasible region shaded in gray. Note that the $z$-line hits the entire line segment between the points $(12, 28)$ and $(26, 14)$. From the graph, we find that the maximum value for $z$ is 40, and that every point in the line segment between $(12, 28)$ and $(26, 14)$ is an optimal solution. So, this LP problem has alternative optimal solutions.

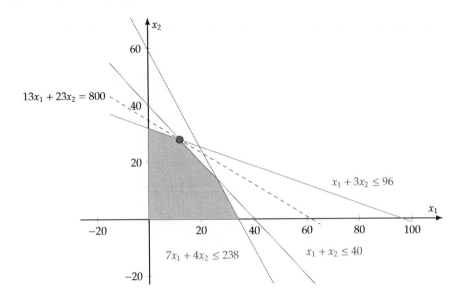

Figure 10.2: Graphical solution of the optimization problem in Example 10.10 (a).

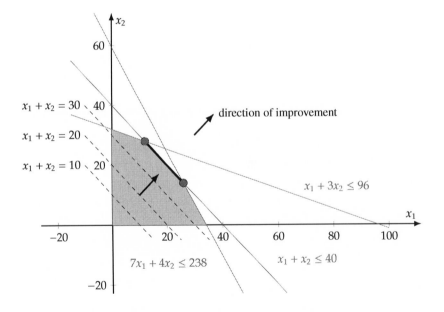

Figure 10.3: Graphical solution of the optimization problem in Example 10.10 (b).

**Example 10.11**  Use the graphical method to solve the following LP problems.

(a)  min $z = 3x_1 + x_2$

    s.t.    $5x_1 + x_2 \geq 42,$

            $2x_1 + x_2 \geq 30,$

                  $x_1, x_2 \geq 0.$

(b)  max $z = 3x_1 + x_2$

    s.t.    $5x_1 + x_2 \geq 42,$

            $2x_1 + x_2 \geq 30,$

                  $x_1, x_2 \geq 0.$

**Solution**  (a) The graphical representation of the given LP problem is shown in Figure 10.4, with the feasible region shaded in gray. From the graph, we find that the minimum value for $z$ is 34 at $x = (4, 22)$.

(b) The graphical representation of the given LP problem is shown in Figure 10.5, with the feasible region shaded in gray. Note that the $z$-line can be pushed to the top right corner of the feasible region without limit. From the graph, we find that there is no infinite optimal $z$-value. So, this LP problem is unbounded.

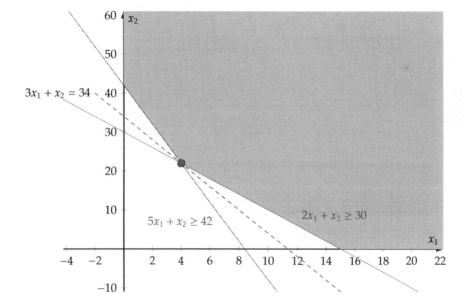

Figure 10.4: Graphical solution of the optimization problem in Example 10.11 (a).

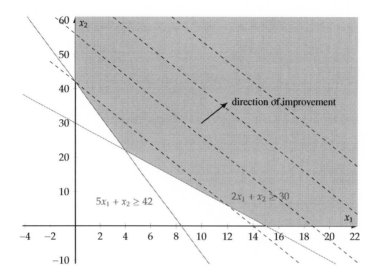

Figure 10.5: Graphical solution of the optimization problem in Example 10.11 (b).

**Example 10.12**   Use the graphical method to solve the following LP problems.

(a)  max $z = 13x_1 + 23x_2$
    s.t.
$$x_1 + 3x_2 \le 96,$$
$$x_1 + x_2 \ge 30,$$
$$7x_1 + 4x_2 \le 238,$$
$$x_1, x_2 \ge 0.$$

(b)  max $z = 13x_1 + 23x_2$
    s.t.
$$x_1 + 3x_2 \ge 96,$$
$$x_1 + x_2 \le 30,$$
$$7x_1 + 4x_2 \ge 238,$$
$$x_1, x_2 \ge 0.$$

**Solution**   (a) The graphical representation of the given LP problem is shown in Figure 10.6, with the feasible region shaded in gray. From the graph, we find that the maximum value for $z$ is 839.52 at $x = (19.41, 25.53)$. So, this LP problem has a unique optimal solution.

(b) The graphical representation of the given LP problem is shown in Figure 10.7. Note that there are no feasible solutions, i.e., there are no points satisfying all constraints. Therefore, the feasible region is empty, and the LP problem is infeasible.

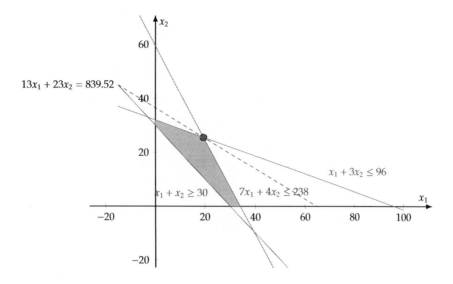

Figure 10.6: Graphical solution of the optimization problem in Example 10.12 (a).

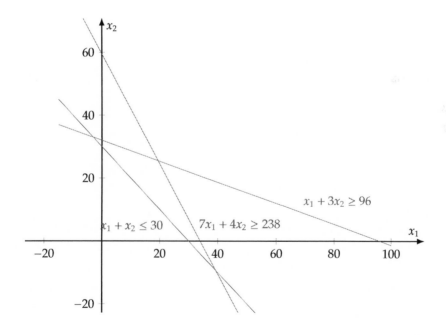

Figure 10.7: Graphical solution of the optimization problem in Example 10.12 (b).

| LP problem | LP case | Feasible region type |
|---|---|---|
| The LP in Example 10.10 (a) | Unique optimal solution | Bounded |
| The LP in Example 10.10 (b) | Alternative optimal solutions | Bounded |
| The LP in Example 10.11 (a) | Unique optimal solution | Unbounded |
| The LP in Example 10.11 (b) | Unbounded solution | Unbounded |
| The LP in Example 10.12 (a) | Unique optimal solution | Bounded |
| The LP in Example 10.12 (b) | Infeasible solution | Empty |

Table 10.1: The answer of Example 10.13.

**Example 10.13**   For the LP problems given in Examples 10.10–10.12, indicate which case the LP belongs to (that is, if the LP has a unique optimal solution, has many optimal solutions, is unbounded, or is infeasible), and which type the feasible region is found (that is, if the feasible region is bounded, unbounded, or empty).

**Solution**   The answer is given in Table 10.1.   ∎

**Example 10.14**   Consider the following LP problem.

$$
\begin{aligned}
\max \quad & y \\
\text{s.t.} \quad -x + \ & y \le 1, \\
3x + \ & 2y \le 12, \\
2x + \ & 3y \ge 12, \\
x, \quad & y \ge 0.
\end{aligned}
$$

(a) Sketch the feasible region of this LP and solve it using the graphical method.

(b) Generally speaking, if (some of) the variables are restricted to be integer-valued, then the underlying optimization problem is called an integer (a mixed-integer) program. In this example, assume that $x$ and $y$ are restricted to be integer-valued. Sketch its feasible region and solve it graphically.

**Solution**   (a) The graphical representation of the LP is shown in Figure 10.8, with the feasible region shaded in gray. We find that the optimal solution is 3 at $x = (2, 3)$.

(b) Introducing the condition $x, y \in \mathbb{Z}$ changes the feasible region, which is now indicated by the blue bullet shown in Figure 10.8. The optimal solution remains the same.

∎

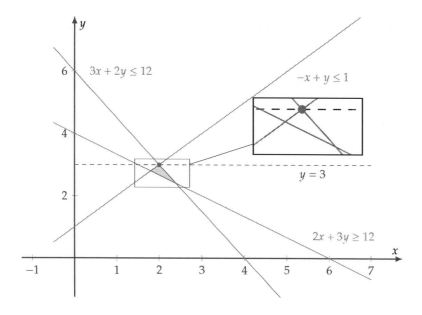

Figure 10.8: Graphical solution of the optimization problem in Example 10.14.

## 10.3. Standard form linear programs

Recall that the general form LP is:

$$
\begin{aligned}
\min \quad & c^\mathsf{T} x \\
\text{s.t.} \quad & a_i^\mathsf{T} x \geq b_i, \quad i = 1, 2, \ldots, m_1, \\
& a_j^\mathsf{T} x \geq b_j, \quad j = 1, 2, \ldots, m_2, \\
& a_k^\mathsf{T} x = b_k, \quad k = 1, 2, \ldots, m_3, \\
& x_p \geq 0, \qquad p = 1, 2, \ldots, m_4, \\
& x_q \leq 0, \qquad q = 1, 2, \ldots, m_5.
\end{aligned}
\tag{10.2}
$$

where $c, x \in \mathbb{R}^n$.

Recall also that there is no need to study maximization problems separately, because maximizing $c^\mathsf{T} x$ subject to some constraints is equivalent to minimizing $-c^\mathsf{T} x$ subject to the same constraints. In addition, because

- $a_i^\mathsf{T} x = b_i$ is equivalent to $a_i^\mathsf{T} x \leq b_i$ and $a_i^\mathsf{T} x \geq b_i$;

- $a_i^\mathsf{T} x \leq b_i$ can be written as $(-a_i)^\mathsf{T} x \geq -b_i$;

- $x_i \geq 0$ and $x_i \leq 0$ are special cases of $u^\mathsf{T} x \geq 0$ and $(-u)^\mathsf{T} x \geq 0$, respectively, where $u$ is a unit vector in $\mathbb{R}^n$,

Problem (10.2) can be expressed exclusively in terms of inequality constraints of the form $a_i^T x \geq b_i$. As a result, Problem (10.2) can be formulated in vector form as

$$
\begin{aligned}
\min \quad & c^T x \\
\text{s.t.} \quad & a_i^T x \geq b_i, \ i = 1, 2, \ldots, m,
\end{aligned}
\tag{10.3}
$$

or, more compactly, in matrix form as

$$
\begin{aligned}
\min \quad & c^T x \\
\text{s.t.} \quad & Ax \geq b,
\end{aligned}
\tag{10.4}
$$

where $A \in \mathbb{R}^{m \times n}$ is the matrix whose rows are the row vectors $a_1^T, a_2^T, \ldots, a_m^T$ and $b = (b_1, b_2, \ldots, b_m)^T$. We have the following example.

**Example 10.15** The LP problem in Example 10.1 can be written as

$$
\begin{aligned}
\min \quad & 2x_1 \ - \ x_2 \ + \ 4x_3 \\
\text{s.t.} \quad & -x_1 \ - \ x_2 \ \ \ \ \ \ \ \ \ \ - \ x_4 \geq -2, \\
& \ \ \ \ \ \ \ \ \ 3x_2 \ - \ x_3 \ \ \ \ \ \ \ \ \geq \ \ 5, \\
& \ \ \ \ \ - \ 3x_2 \ + \ x_3 \ \ \ \ \ \ \ \ \geq -5, \\
& \ \ \ \ \ \ \ \ \ \ \ \ \ \ \ \ \ \ x_3 \ + \ x_4 \geq \ \ 3, \\
& \ \ x_1 \ \ \ \ \ \ \ \ \ \ \ \ \ \ \ \ \ \ \ \ \ \ \ \ \geq \ \ 0, \\
& \ \ \ \ \ \ \ \ \ \ \ \ \ \ - \ x_3 \ \ \ \ \ \ \ \ \ \geq \ \ 0.
\end{aligned}
$$

This can be also written in the matrix form (10.4) with

$$
c = \begin{bmatrix} 2 \\ -1 \\ 4 \\ 0 \end{bmatrix}, \ A = \begin{bmatrix} -1 & -1 & 0 & -1 \\ 0 & 3 & -1 & 0 \\ 0 & -3 & 1 & 0 \\ 0 & 0 & 1 & 1 \\ 1 & 0 & 0 & 0 \\ 0 & 0 & -1 & 0 \end{bmatrix}, \ \text{and} \ b = \begin{bmatrix} -2 \\ 5 \\ -5 \\ 3 \\ 0 \\ 0 \end{bmatrix}.
$$

An LP problem of the form

$$
\begin{aligned}
\min \quad & c^T x \\
\text{s.t.} \quad & A x = b, \\
& x \geq 0
\end{aligned}
\tag{10.5}
$$

is said to be the standard form LP problem.

We can convert an LP problem to the standard form by eliminating of free variables and eliminating of inequality constraints as detailed in the following workflow.

---

**Workflow 10.3** *We can convert a linear programming problem to the standard form by following three steps:*

(i) *Elimination of free variables: We replace each unrestricted variable $x_i$ with $x_i^+ - x_i^-$, where $x_i^+, x_i^- \geq 0$.*

(ii) *Elimination of "$\leq$" constraints: We replace $\sum_{j=1}^{n} a_{ij}x_j \leq b_i$ with $\sum_{j=1}^{n} a_{ij}x_j + s_i = b_i$, where $s_i \geq 0$ is called a slack variable.*

(iii) *Elimination of "$\geq$" constraints: We replace $\sum_{j=1}^{n} a_{ij}x_j \geq b_i$ with $\sum_{j=1}^{n} a_{ij}x_j - e_i = b_i$, where $e_i \geq 0$ is called an excess variable.*

---

We have the following example.

**Example 10.16** The LP problem

$$\begin{aligned} \min \quad & 3x_1 + 7x_2 \\ \text{s.t.} \quad & x_1 + x_2 \geq 3, \\ & 5x_1 + 3x_2 = 19, \\ & x_1 \geq 0, \end{aligned}$$

is equivalent to the standard form problem

$$\begin{aligned} \min \quad & 3x_1 + 7x_2^+ - 7x_2^- \\ \text{s.t.} \quad & x_1 + x_2^+ - x_2^- - x_3 = 3, \quad (\text{letting } x_3 = s_3) \\ & 5x_1 + 3x_2^+ - 3x_2^- = 19, \\ & x_1, x_2^+, x_2^-, x_3 \geq 0. \end{aligned}$$

For instance, given the feasible solution $(x_1, x_2) = (2, 3)$ to the original problem, we obtain the feasible solution $(x_1, x_2^+, x_2^-, x_3) = (2, 3, 0, 2)$ to the standard form problem. In Exercise 10.14, we seek the feasible solution $(x_1, x_2)$ for the original problem given the feasible solution $(x_1, x_2^+, x_2^-, x_3) = (4, 0, 1/3, 2/3)$ to the standard form problem.

# 10.4. Duality in linear programming

Linear programming duality studies the relationships between pairs of linear programs and their solutions.

The linear programming problem in the primal standard form is defined as

$$
\begin{aligned}
\min \quad & c^\mathsf{T} x \\
\text{s.t.} \quad & A x = b, \qquad\qquad \text{(P|LP)} \\
& x \geq 0,
\end{aligned}
$$

where $A \in \mathbb{R}^{m \times n}$, $b \in \mathbb{R}^m$ and $c \in \mathbb{R}^n$ constitute given data, and $x \in \mathbb{R}^n$ is called the primal decision variable.

The linear programming problem in the dual standard form is the dual of (P|LP), which is defined as

$$
\begin{aligned}
\max \quad & b^\mathsf{T} y \\
\text{s.t.} \quad & A^\mathsf{T} y \leq c, \qquad\qquad \text{(D|LP)} \\
& y \quad \text{urs,}
\end{aligned}
$$

where $y \in \mathbb{R}^m$ is called the dual decision variable.

### Lagrangian duality and LP duality

Problem (D|LP) can be derived from (P|LP) through the usual Lagrangian approach. The optimization problems are classified into two classes: Constrained optimization problems and unconstrained optimization problems. This classification is based on whether or not we have constraints on the variables. The Lagrangian approach is a technique by which a constrained optimization problem becomes an unconstrained optimization problem by adding Lagrangian multipliers for the equality constraints. The Lagrangian function is a function that combines the objective function being optimized with functions penalizing constraint violations linearly.

The Lagrangian function for (P|LP) is defined as

$$
\mathcal{L}(x, \lambda, \nu) \triangleq c^\mathsf{T} x - \lambda^\mathsf{T}(A x - b) - \nu^\mathsf{T} x.
$$

The vectors $\lambda$ and $\nu$ are called Lagrangian multipliers. The dual of (P|LP) has the objective function

$$
q(\lambda, \nu) \triangleq \inf_x \mathcal{L}(x, \lambda, \nu) = \lambda^\mathsf{T} b + \inf_x (c - A^\mathsf{T}\lambda - \nu)^\mathsf{T} x.
$$

The dual problem is obtained by maximizing $q(\lambda, \nu)$ subject to $\nu \geq 0$.

If $c - A^\mathsf{T}\lambda - \nu \neq 0$, the infimum is clearly $-\infty$. So we can exclude $\lambda$ for which $c - A^\mathsf{T}\lambda - \nu \neq 0$. When $c - A^\mathsf{T}\lambda - \nu = 0$, the dual objective function is simply $\lambda^\mathsf{T} b$. Hence, we can write the dual problem as follows:

$$
\begin{aligned}
\max \quad & b^\mathsf{T} \lambda \\
\text{s.t.} \quad & A^\mathsf{T}\lambda + \nu = c, \qquad\qquad \text{(10.6)} \\
& \nu \geq 0.
\end{aligned}
$$

| PRIMAL | MINIMUM | MAXIMUM | DUAL |
|:------:|:-------:|:-------:|:----:|
| C | $\geq b$ | $\geq 0$ | V |
| N | $\leq b$ | $\leq 0$ | A |
| S | $= b$ | urs | R |
| V | $\geq 0$ | $\leq c$ | C |
| A | $\leq 0$ | $\geq c$ | N |
| R | urs | $= c$ | S |

Table 10.2: Correspondence rules between primal and dual linear programs.

Replacing $\lambda$ and $\nu$ in (10.6) by $x$ and $z$, respectively, we get (D|LP).

In general, linear programs can be written in a variety of forms different from the standard forms (P|LP) and (D|LP). However, if we consider linear programs in other forms, we can use Table 10.2, which is a summary of the correspondence rules between primal and dual linear programs.

In light of Table 10.2, we have the following remark.

---

**Remark 10.1** *The following are three typical pairs of primal and dual linear programming problems:*

$$(P|LP) \quad \begin{aligned} \min \quad & c^\mathsf{T}x \\ \text{s.t.} \quad & Ax = b, \\ & x \geq 0; \end{aligned} \qquad\qquad \begin{aligned} \max \quad & b^\mathsf{T}y \\ \text{s.t.} \quad & A^\mathsf{T}y \leq c, \end{aligned} \quad (D|LP)$$

$$(\overline{P|LP}) \quad \begin{aligned} \min \quad & c^\mathsf{T}x \\ \text{s.t.} \quad & Ax \geq b; \end{aligned} \qquad\qquad \begin{aligned} \max \quad & b^\mathsf{T}y \\ \text{s.t.} \quad & A^\mathsf{T}y = c, \\ & y \geq 0, \end{aligned} \quad (\overline{D|LP})$$

$$(\widehat{P|LP}) \quad \begin{aligned} \min \quad & c^\mathsf{T}x \\ \text{s.t.} \quad & Ax \geq b, \\ & x \geq 0; \end{aligned} \qquad\qquad \begin{aligned} \max \quad & b^\mathsf{T}y \\ \text{s.t.} \quad & A^\mathsf{T}y \leq c, \\ & y \geq 0. \end{aligned} \quad (\widehat{D|LP})$$

---

The dual of the dual is the primal (see Proposition 10.1), so it does not matter which problem is called the primal.

**Example 10.17** The following is a pair of primal-dual linear programs.

$$
\begin{array}{llll}
\max & 5x_1 + 4x_2 - 3x_3 & \qquad \min & 4y_1 + 5y_2 \\
\text{s.t.} & x_1 \qquad\;\; - 5x_3 \;\geq\; 4, & \qquad \text{s.t.} & y_1 + 3y_2 \;\geq\; 5, \\
& 3x_1 + x_2 + 2x_3 \;\leq\; 5, & & y_2 \;=\; 4, \\
& x_1 \geq 0, x_2 \text{ urs}, x_3 \geq 0; & & -5y_1 + 2y_2 \;\geq\; -3, \\
& & & y_1 \leq 0, y_2 \geq 0.
\end{array}
$$

If we take the dual of the dual, we get

$$
\begin{array}{ll}
\max & 5z_1 + 4z_2 - 3z_3 \\
\text{s.t.} & z_1 \qquad\;\; - 5z_3 \;\geq\; 4, \\
& 3z_1 + z_2 + 2z_3 \;\leq\; 5, \\
& z_1 \geq 0, z_2 \text{ urs}, z_3 \geq 0,
\end{array}
$$

which is the primal problem.   ◼

The proof of the following proposition is left as an exercise for the reader.

> **Proposition 10.1** *The dual of the dual is the primal.*

### The duality theorem

The duality theorem is a very powerful theoretical tool that is very useful in applications because it leads to an interesting class of optimization algorithms. In this part, we state and prove the weak and strong duality theorems for the primal-dual pair (P|LP) and (D|LP). All the results in this part are stated for the pair (P|LP) and (D|LP), but we indicate that all the results established in this section are satisfied for any primal-dual pair, including the pair $\overline{\text{(P|LP)}}$ and $\overline{\text{(D|LP)}}$ as well as the pair $\widehat{\text{(P|LP)}}$ and $\widehat{\text{(D|LP)}}$ outlined in Remark 10.1. We have the following definition.

> **Definition 10.2**
>
> (*a*) *An optimization problem is called feasible if it has at least one feasible point, and infeasible otherwise.*
>
> (*b*) *An optimization problem is called unbounded if it is feasible and has unbounded optimal value. More specifically, an minimization (maximization) problem is called unbounded if it is feasible and has the optimal cost* $-\infty$ *(optimal cost* $+\infty$*).*

We state the weak duality property in Theorem 10.1.

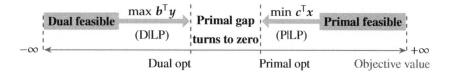

Figure 10.9: The duality gap between the primal and dual LP problems.

---

**Theorem 10.1 (Weak duality in LP)** *Consider the primal-dual pair (P|LP) and (D|LP). Let (P|LP) and (D|LP) be both feasible. If $x$ is a feasible solution to (P|LP) and $y$ is a feasible solution to (D|LP), then $b^{\mathsf{T}}y \leq c^{\mathsf{T}}x$.*

---

**Proof** Note that, in (D|LP), the constraint $A^{\mathsf{T}}y \leq c$ can be written as $A^{\mathsf{T}}y + s = c$ with $s \geq 0$. It follows that $c^{\mathsf{T}}x - b^{\mathsf{T}}y = (A^{\mathsf{T}}y + s)^{\mathsf{T}}x - b^{\mathsf{T}}y = y^{\mathsf{T}}Ax + s^{\mathsf{T}}x - y^{\mathsf{T}}b = y^{\mathsf{T}}(Ax - b) + s^{\mathsf{T}}x = x^{\mathsf{T}}s \geq 0$, where the last equality follows from the constraint $Ax = b$ stated in (P|LP), and the inequality follows from the fact that $x \geq 0$ and $s \geq 0$. The proof is complete. ∎

The following corollary is now easy to obtain.

---

**Corollary 10.1** *Consider the primal-dual pair (P|LP) and (D|LP).*

*(a) If (P|LP) is unbounded, then (D|LP) is infeasible.*

*(b) If (D|LP) is unbounded, then (P|LP) is infeasible.*

---

**Proof** If we prove item (a), item (b) immediately follows by a symmetrical argument. Suppose, in the contrary, that Problem (P|LP) is feasible, with the optimal cost $-\infty$, and that Problem (D|LP) is also feasible. Let $w$ be the optimal cost in (D|LP). By weak duality, we have $w \leq -\infty$. That is, $w \leq r$ for all $r \in \mathbb{R}$, which is impossible. This means that (D|LP) cannot have a feasible solution. This proves item (a), and hence completes the proof. ∎

In Figure 10.9, we show visually how the duality gap between the primal and dual LP problems turns to zero. That is, the difference $c^{\mathsf{T}}x - b^{\mathsf{T}}y$ becomes zero when $x$ is an optimal solution to (P|LP) and $y$ is an optimal solution to (D|LP). This is the essence of the strong duality property, which is stated below in Theorem 10.2.

---

**Theorem 10.2 (Strong duality in LP)** *Consider the primal-dual pair (P|LP) and (D|LP). Assume that (P|LP) and (D|LP) are both feasible. If one of (P|LP) or (D|LP) has a finite optimal solution, so does the other, and their optimal values are equal.*

---

**Proof** Let $\bar{x}$ and $\bar{y}$ be feasible solutions to Problems (P|LP) and (D|LP), respectively. From the weak duality (Theorem 10.1), we have that $b^{\mathsf{T}}\bar{y} \leq c^{\mathsf{T}}\bar{x}$. This means that both (P|LP) and (D|LP) are bounded. Let $z$ and $w$ be the optimal values of (P|LP) and

(D|LP), respectively. Using the weak duality again, we have $w \leq z$. To prove that $w = z$, suppose, in the contrary, that $w < z$, then there is no $\boldsymbol{y}$ satisfying

$$A^{\mathsf{T}}\boldsymbol{y} \leq \boldsymbol{c} \text{ and } \boldsymbol{b}^{\mathsf{T}}\boldsymbol{y} \geq z,$$

or equivalently

$$\begin{bmatrix} A^{\mathsf{T}} \\ -\boldsymbol{b}^{\mathsf{T}} \end{bmatrix} \boldsymbol{y} \leq \begin{bmatrix} \boldsymbol{c} \\ -z \end{bmatrix}. \tag{10.7}$$

Letting

$$\hat{A} \triangleq \begin{bmatrix} A \vdots -\boldsymbol{b} \end{bmatrix}, \text{ and } \hat{c} \triangleq \begin{bmatrix} \boldsymbol{c} \\ -z \end{bmatrix},$$

we can rewrite (10.7) as $\hat{A}^{\mathsf{T}}\boldsymbol{y} \leq \hat{c}$. Using Farkas' lemma (Version II; see Theorem 3.16), there exists a vector $\hat{\boldsymbol{x}}$ satisfying

$$\hat{A}\hat{\boldsymbol{x}} = \boldsymbol{0}, \ \hat{c}^{\mathsf{T}}\hat{\boldsymbol{x}} < 0, \text{ and } \hat{\boldsymbol{x}} \geq \boldsymbol{0}. \tag{10.8}$$

Note that the vector $\hat{\boldsymbol{x}}$ can be written as $\hat{\boldsymbol{x}} \triangleq (\boldsymbol{x}^{\mathsf{T}}, \alpha)^{\mathsf{T}}$ with $\alpha \neq 0$. This rewrites (10.8) as

$$\begin{bmatrix} A \vdots -\boldsymbol{b} \end{bmatrix} \begin{bmatrix} \boldsymbol{x} \\ \alpha \end{bmatrix} = \boldsymbol{0}, \ \begin{bmatrix} \boldsymbol{c} \\ -z \end{bmatrix}^{\mathsf{T}} \begin{bmatrix} \boldsymbol{x} \\ \alpha \end{bmatrix} < 0, \text{ and } \begin{bmatrix} \boldsymbol{x} \\ \alpha \end{bmatrix} \geq \boldsymbol{0}. \tag{10.9}$$

To see that $\alpha \neq 0$, suppose in the contrary that $\alpha = 0$. Then, from (10.9), we have $A\boldsymbol{x} = \boldsymbol{0}, \boldsymbol{c}^{\mathsf{T}}\boldsymbol{x} < 0$, and $\boldsymbol{x} \geq \boldsymbol{0}$. Using Farkas' lemma (Version II) again, there does not exist a vector $\boldsymbol{y}$ satisfying $A^{\mathsf{T}}\boldsymbol{y} \leq \boldsymbol{c}$. This means that Problem (D|LP) is infeasible, a contradiction.

Note that $\frac{1}{\alpha}\boldsymbol{x} \geq \boldsymbol{0}$, and that $A\boldsymbol{x} - \alpha\boldsymbol{c} = \boldsymbol{0}$, or equivalently $A(\frac{1}{\alpha}\boldsymbol{x}) = \boldsymbol{c}$. This means that the vector $\frac{1}{\alpha}\boldsymbol{x}$ is feasible for (P|LP). But, from (10.9), $\boldsymbol{c}^{\mathsf{T}}\boldsymbol{x} - \alpha z < 0$, so $\boldsymbol{c}^{\mathsf{T}}(\frac{1}{\alpha}\boldsymbol{x}) < z$. This contradicts the fact that $z$ is the optimal value of (D|LP). Thus, $w = z$. The proof is complete. ∎

The following example, which is taken from [NW88], is a direct application of Theorem 10.2.

**Example 10.18** Consider the following primal-dual pair of problems.

$$
\begin{array}{llll}
\min & 7x_1 + 2x_2 & \max & 4y_1 + 20y_2 - 7y_3 \\
\text{s.t.} & -x_1 + 2x_2 \leq 4, & \text{s.t.} & -y_1 + 5y_2 - 2y_3 \geq 7, \\
& 5x_1 + x_2 \leq 20, & & 2y_1 + y_2 - 2y_3 \geq 2, \\
& -2x_1 - 2x_2 \leq -7, & & y_1, y_2, y_3 \geq 0. \\
& x_1, x_2 \leq 0;
\end{array}
$$

Let $\boldsymbol{x}^{\star} \triangleq (\frac{36}{11}, \frac{40}{11})^{\mathsf{T}}$ and $\boldsymbol{y}^{\star} \triangleq (\frac{3}{11}, \frac{16}{11}, 0)^{\mathsf{T}}$. One can easily see that $\boldsymbol{x}^{\star}$ and $\boldsymbol{y}^{\star}$ are feasible in the primal and dual problems, respectively. One can also easily see that

$b^\top y^\star = 30\frac{2}{11}$ and $c^\top x^\star = 30\frac{2}{11}$. Based on the strong duality property (Theorem 10.2), since $b^\top y^\star = c^\top x^\star$, we conclude that $x^\star$ and $y^\star$ are optimal in the primal and dual problems, respectively, and their optimal value is $30\frac{2}{11}$. ∎

It is natural question to ask: Can Problems (P|LP) and (D|LP) be both infeasible? The following example answers this question positively.

**Example 10.19**  The following primal-dual pair of problems are both infeasible.

$$
\begin{array}{ll}
\min & x_1 + 2x_2 \\
\text{s.t.} & x_1 + x_2 = 2, \\
& 3x_1 + 3x_2 = 4;
\end{array}
\qquad
\begin{array}{ll}
\max & 2y_1 + 4y_2 \\
\text{s.t.} & y_1 + 3y_2 = 1, \\
& y_1 + 3y_2 = 2.
\end{array}
$$

∎

It is not hard now to establish the following corollary.

---

**Corollary 10.2**  *Consider the primal-dual pair (P|LP) and (D|LP).*

*(a) If (P|LP) is infeasible, then (D|LP) is either infeasible or unbounded.*

*(b) If (D|LP) is infeasible, then (P|LP) is either infeasible or unbounded.*

---

**Proof**  Note that the possibility that Problems (P|LP) and (D|LP) could be both infeasible has been grounded in Example 10.19. To prove item (a), it remains to show that if (P|LP) is infeasible and (D|LP) is feasible, then (D|LP) must be unbounded. Assume that (P|LP) is infeasible and let $\bar{y}$ be a feasible solution for (D|LP). Due to the infeasibility of (P|LP), there does not exist $x$ satisfying $Ax = b$ and $x \geq 0$. Using Farkas' lemma (Version I); see Theorem 3.15, there is a vector $\hat{y}$ satisfying $A^\top \hat{y} \geq 0$ and $b^\top \hat{y} < 0$. Due to the feasibility of $\bar{y}$ in (D|LP), we have $A^\top \bar{y} \leq c$. Define $y_\alpha \triangleq \bar{y} - \alpha\hat{y}$ for $\alpha \geq 0$. Then

$$A^\top y_\alpha = A^\top(\bar{y} - \alpha\hat{y}) = A^\top \bar{y} - \alpha A^\top \hat{y} \leq c - \alpha A^\top \hat{y} \leq c.$$

This means that $y_\alpha$ is feasible in (D|LP). Note that, because $b^\top \hat{y} < 0$, we have

$$b^\top y_\alpha = b^\top(\bar{y} - \alpha\hat{y}) = b^\top \bar{y} - \alpha b^\top \hat{y} \longrightarrow b^\top \bar{y} + \infty = \infty,$$

as $\alpha \longrightarrow \infty$, which implies that Problem (D|LP) is unbounded. This proves item (a). Item (b) can be proven by following a symmetrical argument of that of item (a) and using Farkas' lemma (Version II). We leave the proof of this part as an exercise for the reader (see Exercise 10.13). The proof is complete. ∎

Corollary 10.3 is now obvious. See also Table 10.3.

| | | (P\|LP) Finite optimum | (P\|LP) Unbounded | (P\|LP) Infeasible |
|---|---|:---:|:---:|:---:|
| (D\|LP) | Finite optimum | ✓ | ✗ | ✗ |
| (D\|LP) | Unbounded | ✗ | ✗ | ✓ |
| (D\|LP) | Infeasible | ✗ | ✓ | ✓ |

Table 10.3: Possibilities for the primal and the dual linear programs.

---

**Corollary 10.3** *There are only four possibilities for the primal-dual pair (P\|LP) and (D\|LP). Namely:*

(*a*)  *Both (P\|LP) and (D\|LP) are feasible and their optimal values are finite and equal.*

(*b*)  *(P\|LP) is infeasible and (D\|LP) is unbounded.*

(*c*)  *(P\|LP) is unbounded and (D\|LP) is infeasible.*

(*d*)  *Both (P\|LP) and (D\|LP) are infeasible.*

---

### Complementary slackness

Complementary slackness is another important property of primal-dual pairs. We have the following definition.

---

**Definition 10.3** *A slack variable is a variable that is added to an inequality constraint to transform it into an equality. Likewise, an excess (also called surplus or negative slack) variable is a variable that is subtracted to an inequality constraint to transform it into an equality.*

---

Consider the pair $(\widehat{\text{P}\|\text{LP}})$ and $(\widehat{\text{D}\|\text{LP}})$ outlined in Remark 10.1.

$$
(\widehat{\text{P}\|\text{LP}}) \quad
\begin{array}{lrcl}
\min & c^\mathsf{T}x & & \\
\text{s.t.} & Ax & \geq & b, \\
& x & \geq & 0;
\end{array}
\qquad
\begin{array}{lrcl}
\max & b^\mathsf{T}y & & \\
\text{s.t.} & A^\mathsf{T}y & \leq & c, \\
& y & \geq & 0.
\end{array}
\quad (\widehat{\text{D}\|\text{LP}})
$$

Let $s \triangleq c - A^\mathsf{T}y \geq 0$ be the vector of slack variables of $(\widehat{\text{P}\|\text{LP}})$, and $e \triangleq Ax - b \geq 0$ be the vector of excess variables of $(\widehat{\text{D}\|\text{LP}})$. Then the pair $(\widehat{\text{P}\|\text{LP}})$ and $(\widehat{\text{D}\|\text{LP}})$ can be written as

$$
\begin{array}{lrcl}
\min & c^\mathsf{T}x & & \\
\text{s.t.} & Ax - e & = & b, \\
& x, e & \geq & 0;
\end{array}
\qquad
\begin{array}{lrcl}
\max & b^\mathsf{T}y & & \\
\text{s.t.} & A^\mathsf{T}y + s & = & c, \\
& y, s & \geq & 0.
\end{array}
$$

The complementary slackness conditions for linear programming are provided in the following theorem.

**Theorem 10.3 (Complementary slackness)** *Consider the primal-dual pair* $(\widehat{P|LP})$ *and* $(\widehat{D|LP})$. *If* $x^\star$ *is an optimal solution to* $(\widehat{P|LP})$ *and* $y^\star$ *is an optimal solution to* $(\widehat{D|LP})$, *then* $x_i^\star s_i^\star = 0$ *for all i, and* $y_j^\star e_j^\star = 0$ *for all j, where* $e^\star \triangleq Ax^\star - b$ *and* $s^\star \triangleq c - A^\mathsf{T}y^\star$.

**Proof** Note that

$$
\begin{aligned}
c^\mathsf{T}x^\star &= (A^\mathsf{T}y^\star + s^\star)^\mathsf{T}x^\star \\
&= y^{\star\mathsf{T}}Ax^\star + s^{\star\mathsf{T}}x^\star \\
&= y^{\star\mathsf{T}}(b + e^\star) + s^{\star\mathsf{T}}x^\star = b^\mathsf{T}y^\star + y^{\star\mathsf{T}}e^\star + s^{\star\mathsf{T}}x^\star.
\end{aligned}
$$

Note also that the strong duality property (Theorem 10.2) implies that $c^\mathsf{T}x^\star = b^\mathsf{T}y^\star$. It follow that $y^{\star\mathsf{T}}e^\star + s^{\star\mathsf{T}}x^\star = 0$. Because $x^\star, e^\star, y^\star$ and $s^\star$ are all nonnegative vectors, we have $x_i^\star s_i^\star = 0$ for all $i$, and $y_j^\star e_j^\star = 0$ for all $j$. The proof is complete. ∎

**Example 10.20 (Example 10.18 revisited)** To see how the complementary slackness conditions hold for the primal-dual pair in Example 10.18, note that the slack and excess variables are $s^\star \triangleq (0, 0, 6\frac{9}{11})^\mathsf{T}$ and $e^\star \triangleq (0, 0)^\mathsf{T}$, respectively. Clearly, $x_i^\star s_i^\star = 0$ for $i = 1, 2$, and $y_j^\star e_j^\star = 0$ for $j = 1, 2, 3$. ∎

## 10.5. Geometry of linear programming

The graphical method for linear optimization problems tells us that an optimal solution to a linear programming lies at a "corner" of a polyhedron. In this section, we define a vertex, an extreme point, and a basic feasible solution of a given nonempty polyhedron.

### Extreme points, vertices, and basic feasible solutions

The definitions of a vertex, an extreme point, and a basic feasible solution are three different ways of defining the concept of a corner of a polyhedron. The first two definitions are geometric.

**Definition 10.4** *Let P be a nonempty polyhedron. A vector* $x \in P$ *is called a vertex of P if there is some c such that* $c^\mathsf{T}x < c^\mathsf{T}y$ *for all* $y \in P$ *different from* $x$.

In light of Definition 10.4, we observe that $x$ is a vertex of a polyhedron $P$ if it is the optimal solution of some linear program with $P$ as the feasible region. In Figure

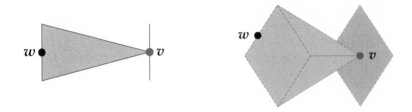

Figure 10.10: Vertices versus ($v$'s) nonvertices ($w$'s).

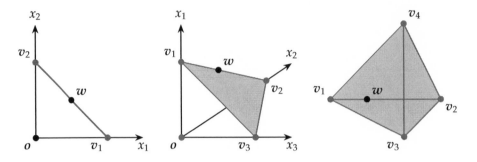

Figure 10.11: Extreme points ($v_i$'s) versus nonextreme points ($w$'s).

10.10, we show two polyhedra. In each polyhedron, the hyperplane $\{y : c^\mathsf{T} y = c^\mathsf{T} v\}$ on the right-hand side touches $P$ as a single point and the point $v$ is a vertex. In contrast, the point $w$ is not a vertex because there is no hyperplane that meets $P$ only at $w$.

> **Definition 10.5** *Let $P$ be a nonempty polyhedron. A vector $x \in P$ is called an extreme point of $P$ if there are no $y, z \in P$ and a scalar $\lambda \in (0, 1)$ such that $x = \lambda y + (1 - \lambda)z$.*

In Figure 10.11, we show three polyhedra. In each polyhedron, the vectors $v_i$'s are extreme points, and the vector $w$ is not an extreme point because $w$ is a convex combination of $v_1$ and $v_2$.

Definitions 10.4 and 10.5 are geometric, and hence intuitive. We need an equivalent definition that is algebraic, so that we can do computations. Before this, we need an intermediate concept for connecting geometry and algebra.

> **Definition 10.6** *If a vertex $x^\star$ satisfies an inequality $a^\mathsf{T} x \geq b$ (or $a^\mathsf{T} x \leq b$) as an equality, i.e., $a^\mathsf{T} x^\star = b$, then we say that this inequality is active or binding at $x^\star$.*

If $P \subset \mathbb{R}^n$ is a polyhedron defined by linear equality and inequality constraints, then $x^\star \in \mathbb{R}^n$ may or may not be feasible with respect the constraints. Now, if

$x^\star \in \mathbb{R}^n$ is feasible (i.e., $x^\star \in P$; satisfying all the constraints), then from Definition 10.6 all the equality constraints are active at $x^\star$.

We have the following example to more illustrate Definition 10.6.

**Example 10.21** The polyhedron shown in the middle of Figure 10.11 is expressed as

$$P = \{(x_1, x_2, x_3) : x_1 + x_2 + x_3 = 1, x_1, x_2, x_3 \geq 0\}. \qquad (10.10)$$

There are three constraints that are binding at each of the points $v_1, v_2$ and $v_3$. Namely, the constraints $x_1 + x_2 + x_3 = 1$, $x_2 = 0$ and $x_3 = 0$ are active at $v_1$, the constraints $x_1 + x_2 + x_3 = 1$, $x_1 = 0$ and $x_3 = 0$ are active at $v_2$, and the constraints $x_1 + x_2 + x_3 = 1$, $x_1 = 0$ and $x_2 = 0$ are active at $v_3$. Also, at the point $w$, there are two constraints that are binding, which are $x_1 + x_2 + x_3 = 1$ and $x_3 = 0$. ∎

If there are $n$ constraints that are binding at a vector $x^\star \in \mathbb{R}^n$, then $x^\star$ satisfies a system of $n$ linear equations in $n$ unknowns. In view of Theorem 3.13, this system has a unique solution if and only if these $n$ equations are linearly independent.

Now, we are ready to introduce the algebraic definition of a corner point.

---

**Definition 10.7** *Let P be a polyhedron defined by linear equality and inequality constraints, and $x^\star \in \mathbb{R}^n$. We say that the vector*

*(a)* $x^\star$ *is a basic solution if the following two statements hold:*

  *(i) All equality constraints are active.*

  *(ii) Out of the constraints that are active at $x^\star$, there are n of them that are linearly independent.*

*(b)* $x^\star$ *is a basic feasible solution if it is a basic solution, and satisfies all of the constraints (i.e., $x^\star \in P$).*

---

The following two examples illustrate Definition 10.7.

**Example 10.22** In the polyhedron shown in the middle of Figure 10.11, which is represented in (10.10), we observe that the points $v_i$'s are basic feasible solutions. The point $o$ does not satisfy the equality constraint $x_1 + x_2 + x_3 = 1$, and hence it is not a basic solution. The point $w$ is feasible, but not basic according to Definition 10.7. However, if the equality constraint $x_1 + x_2 + x_3 = 1$ is replaced with the inequality constraints $x_1 + x_2 + x_3 \leq 1$, then $o$ becomes a basic feasible solution; see Figure 10.12. ∎

**Example 10.23** In Figure 10.13, the points $a, b, c, d, e, f$ and $g$ are all basic solutions because at each one of them, we can find two linearly independent constraints that are binding. The points $a, b, d, e$ and $f$ are basic feasible solutions because they satisfy all of the constraints. ∎

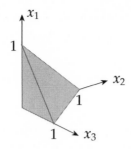

Figure 10.12: The polyhedron given in Example 10.22 with 4 corners.

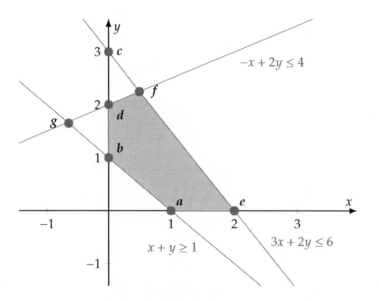

Figure 10.13: Basic solutions and basic feasible solutions.

We give, without proof, the following result in this context. For a proof, see, for example, Bertsimas and Tsitsiklis [BT97].

---

**Theorem 10.4** *Let $x^\star$ be a point in a nonempty polyhedron P. Then the following are equivalent:*

(a) $x^\star$ *is a vertex.*

(b) $x^\star$ *is an extreme point.*

(c) $x^\star$ *is a basic feasible solution.*

---

**Definition 10.8** *Two distinct basic solutions to a set of linear constraints in $\mathbb{R}^n$ are called adjacent if there are $n - 1$ linearly independent constraints that are binding at both of them.*

As an example, in Figure 10.13, the points $a$ and $g$ are adjacent to the point $b$, and the points $d$ and $e$ are adjacent to $f$.

In the subsequent development, we will see that we find an optimal corner point of a linear programming problem by moving from one basic feasible solution to an adjacent basic feasible solution that improves the objective function value, and so on, repeating this step until we cannot go to an adjacent basic feasible solution that improves the objective function value.

Let $n$ and $m$ be positive integers such that $m \leq n$. Let also $b \in \mathbb{R}^m$ and $A \in \mathbb{R}^{m \times n}$ with rank$(A) = m$ (i.e., $A$ has a full-row rank). The set $P = \{x \in \mathbb{R}^n : Ax = b, x \geq 0\}$ is called a polyhedron in standard form. Note that the number of equality constraints in $P$ is $m$.

## Finding basic feasible solutions

The question that arises now is, how to find basic solutions of polyhedra in standard form? The system $Ax = b$ gives m linearly independent constraints as rank$(A) = m$. Consequently, we need $n - m$ more binding constraint from $x \geq 0$ (this is $n$ nonnegativity constraints: $x_1 \geq 0, x_2 \geq 0, \ldots, x_n \geq 0$). Which $n - m$ (out of those $n$) constraints to select for our purpose? We cannot choose any $(n - m)$ $x_i$'s. Theorem 10.5 helps in this task. Before this theorem we give some definitions.

As a matter of notation, we use ";" for adjoining vectors and matrices in a column, and use "," or ":" for adjoining them in a row.

We write $A$ as $A = [a_1 : a_2 : \cdots : a_n]$ where $a_j$ is the $j$th column of $A$. Since rank$(A) = m$, there exists an invertible matrix

$$A_B \triangleq [a_{B_1} : a_{B_2} : \cdots : a_{B_m}] \in \mathbb{R}^{m \times m}. \tag{10.11}$$

Let $B \triangleq \{B_1, B_2, \ldots, B_m\}$ and $N \triangleq \{1, 2, \ldots, n\} - B$. We can permute the columns of $A$ so that $A = [A_B : A_N]$. We can write the system $Ax = b$ as $A_B x_B + A_N x_N = b$ where $x = (x_B; x_N)$ (equivalently, $x^\mathsf{T} = (x_B^\mathsf{T}, x_N^\mathsf{T})$).

**Definition 10.9** *The $m \times m$ nonsingular matrix $A_B$ is called a basis matrix. The vector $x_B$ is called a basic solution (also called the vector of basic variables). The vector $x_N$ is called a nonbasic solution (also called the vector of nonbasic variables).*

We are now ready to state the following theorem which will be given without proof. For a proof, see, for example, Bertsimas and Tsitsiklis [BT97].

> **Theorem 10.5** *Let $b \in \mathbb{R}^m$ and $A \in \mathbb{R}^{m \times n}$ have linearly independent rows. Consider the constraints $Ax = b$ and $x \geq 0$. A vector $x \in \mathbb{R}^n$ is a basic solution if and only if we have*
>
> *(a)  The columns of $A_B$ are linearly independent.*
>
> *(b)  $x_N = 0$.*

Since $A_B$ is nonsingular, we can solve the system of $m$ linear equations $Ax = b$ for $x_B$. The solution is given by $x_N = 0$ and $x_B = A_B^{-1}b$. The three-step procedure in the following workflow, followed by an example, will teach us to construct such basic solutions.

> **Workflow 10.4** *We construct all basic solutions to a standard form polyhedron by following three steps:*
>
> *(i)  Choose $m$ linearly independent columns $a_{B(1)}, a_{B(2)}, \ldots, a_{B(m)}$.*
>
> *(ii)  Set $x_N = 0$.*
>
> *(iii)  Calculate $x_B = A_B^{-1}b$. If $x_B \geq 0$, then the vector $x = (x_B; x_N)$ is a basic feasible solution. Otherwise, $x = (x_B; x_N)$ is a basic solution.*

It is clear that the maximum number of basic feasible solutions is $\binom{n}{m}$. Note that, generally, not all of $\binom{n}{m}$ choices of $m$ columns may produce a basis (i.e., a nonsingular matrix $A_B$). Hence, the number of basic solutions may be smaller than $\binom{n}{m}$. Note also that not all of these $\binom{n}{m}$ bases may lead to basic feasible solutions.

In Example 10.24, we have that $n = 5$ and $m = 3$, and that each of $\binom{5}{3} = 10$ choices produces a basic solution.

**Example 10.24**   Consider the linear system

$$
\begin{aligned}
x_1 &+& x_2 &\geq& 2, \\
3x_1 &+& x_2 &\geq& 4, \\
3x_1 &+& 2x_2 &\leq& 10, \\
x_1, && x_2 &\geq& 0.
\end{aligned}
\tag{10.12}
$$

The resulting polyhedron is shown in Figure 10.14. In the standard form, we have

$$
\begin{aligned}
x_1 &+& x_2 &-& x_3 &&&&&=& 2, \\
3x_1 &+& x_2 &&&-& x_4 &&&=& 4, \\
3x_1 &+& 2x_2 &&&&&+& x_5 &=& 10, \\
x_1, && x_2, && x_3, && x_4, && x_5 &\geq& 0.
\end{aligned}
$$

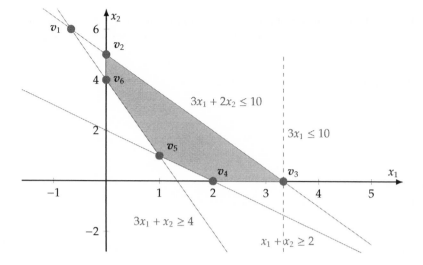

Figure 10.14: The polyhedron in Example 10.24.

Consequently, the following arrays draw the resulting polyhedron.

$$A = \begin{bmatrix} 1 & 1 & -1 & 0 & 0 \\ 3 & 1 & 0 & -1 & 0 \\ 3 & 2 & 0 & 0 & 1 \end{bmatrix}, \text{ and } b = \begin{bmatrix} 2 \\ 4 \\ 10 \end{bmatrix}.$$

The columns of $A$ are

$$a_1 = \begin{bmatrix} 1 \\ 3 \\ 3 \end{bmatrix}, a_2 = \begin{bmatrix} 1 \\ 1 \\ 2 \end{bmatrix}, a_3 = \begin{bmatrix} -1 \\ 0 \\ 0 \end{bmatrix}, a_4 = \begin{bmatrix} 0 \\ -1 \\ 0 \end{bmatrix}, \text{ and } a_5 = \begin{bmatrix} 0 \\ 0 \\ 1 \end{bmatrix}.$$

Choosing $B = \{1, 2, 3\}$ (hence $N = \{4, 5\}$) gives

$$A_B = \begin{bmatrix} 1 & 1 & -1 \\ 3 & 1 & 0 \\ 3 & 2 & 0 \end{bmatrix}, \text{ and } \det(B) = -3 \neq 0 \text{ (hence } B \text{ is invertible).}$$

Let $x_N = (x_4; x_5) = (0; 0)$. Finding $A_B^{-1}$ and calculating $x_B = A_B^{-1}b$, we get

$$x_B = \begin{bmatrix} x_1 \\ x_2 \\ x_3 \end{bmatrix} = \begin{bmatrix} -2/3 \\ 6 \\ 10/3 \end{bmatrix}.$$

| Vertex | $B$ | $\det(A_B)$ | The variable $x$ | Basic feasible solution? |
|--------|-----|-------------|------------------|--------------------------|
| $v_1$ | $\{1,2,3\}$ | -3 | $(-2/3;6;10/3;0;0)$ | ✗ |
| $v_2$ | $\{2,3,4\}$ | 2 | $(0;5;3;1;0)$ | ✓ |
| $v_3$ | $\{1,3,4\}$ | 3 | $(10/3;0;4/3;6;0)$ | ✓ |
| $v_4$ | $\{1,4,5\}$ | -1 | $(2;0;0;2;4)$ | ✓ |
| $v_5$ | $\{1,2,5\}$ | -2 | $(1;1;0;0;5)$ | ✓ |
| $v_6$ | $\{2,3,5\}$ | 1 | $(0;4;2;0;2)$ | ✓ |

Table 10.4: Correspondences between the basic feasible solutions in the standard form polyhedron and the vertices visualized in Figure 10.14.

This point is a basic solution, but it is not a basic feasible solution because not all entries are nonnegative. This point corresponds to the vertex $v_1 = (-2/3; 6)$.

Choosing $B = \{2,3,4\}$ (hence $N = \{1,5\}$) gives

$$A_B = \begin{bmatrix} 1 & -1 & 0 \\ 1 & 0 & -1 \\ 2 & 0 & 0 \end{bmatrix}, \text{ and } \det(B) = 2 \neq 0 \text{ (hence } B \text{ is invertible).}$$

Let $x_N = (x_1; x_5) = (0; 0)$. Finding $A_B^{-1}$ and calculating $x_B = A_B^{-1}b$, we get

$$x_B = \begin{bmatrix} x_2 \\ x_3 \\ x_4 \end{bmatrix} = \begin{bmatrix} 5 \\ 3 \\ 1 \end{bmatrix}.$$

Thus $x = (0; 5; 3; 1; 0)$. This point is a basic feasible solution because all the entries are nonnegative. This point corresponds to the vertex $v_2 = (0; 5)$.

In Table 10.4, we summarize the correspondences between the basic feasible solutions in the standard form polyhedron and the vertices visualized in Figure 10.14.

∎

We indicate that we can find the basic feasible solution in the polyhedron standard form corresponding to each corner point by inspection. In other words, we do not necessarily have to enumerate all $\binom{n}{m}$ choices for bases. For instance, in Example 10.24, at vertex $v_5$, the constraint $x_1 + x_2 \geq 2$ and $3x_1 + x_2 \geq 4$ are binding, while $3x_1 + 2x_2 \leq 10$ is not active. Hence $x_3 = x_4 = 0$ and $x_5 > 0$ in the corresponding basic feasible solutions from the standard for polyhedron. Also, observe that both $x_1$ and $x_2$ are strictly positive. Hence $x_B = (x_1; x_2; x_5)$ is the corresponding basis.

Similarly, at vertex $v_2$, the constraint $x_1 + x_2 \geq 2$ and $3x_1 + x_2 \geq 4$ are not binding, while $3x_1 + 2x_2 \leq 10$ and $x_1 \geq 0$ are active. Hence $x_3, x_4 > 0$ but $x_5 = 0$ in the corresponding basic feasible solutions from the standard for polyhedron. Also, observe that $x_5$ is strictly positive. Hence $x_B = (x_2; x_3; x_4)$ is the corresponding basis.

***Degeneracy*** At a basic solution, we must have $n$ linearly independent active constraints. Because no more than $n$ constraints can be linearly independent at a point in the $n$th-dimensional space, there is still a possibility that more than $n$ active constraints exist at a basic solution. Such a basic solution is called a degenerate.

> **Definition 10.10** *A basic solution $x \in \mathbb{R}^n$ is called degenerate if more than n of the constraints are active at $x$. In a nonempty polyhedron in standard form with $A \in \mathbb{R}^{m \times n}$, $x$ is a degenerate basic solution if more than $n - m$ of the components of $x$ are zero.*

**Example 10.25 (Example 10.24 revisited)** Adding the constraint $x_1 \le 10/3$ to System 10.12 produces three active constraints at $v_3$ (see Figure 10.14). Hence, $v_3$ is a degenerate basic feasible solution. In the standard form, we would have $x_1 + x_6 = 10/3$, where $x_6$ is the slack variable for the constraint $x_1 \le 10/3$. In this case, we have $n = 6, m = 4, x_2 = x_5 = x_6 = 0$, hence more than $n - m = 2$ of the components of $x$ are zero. ∎

Degeneracy may not be much of a problem in small instances. But when we solve large instances of a linear program, it could create inefficiencies. A typical algorithm tries to move from one basic feasible solution to an adjacent basic feasible solution such that the objective function value improves, or at least does not become worse. It could happen that we cycle through several degenerate basic feasible solutions before ultimately jumping to a vertex that actually strictly improves the objective function value. However, degeneracy depends very much on how we represent the polyhedron. For instance, in Example 10.25, we could throw out the constraint $x_1 \le 10/3$ without changing the polyhedron, and remove the degeneracy at $v_3$. Also, if permitted, we could avoid degeneracy by perturbing some of the constraint a tiny bit, i.e., replacing the constraint $x_1 \le 10/3$ by $x_1 \le 10/3 + 0.001$. But whether we can do so will depend very much on the specific application in the problem.

## Pointedness

A polyhedron is pointed if it contains no lines (a line is a straight one-dimensional figure formed when two points are connected with minimum distance between them, and both the ends extended to infinity). Figure 10.15 shows two polyhedra, one of them (namely, $P_1$) is pointed but the other (namely, $P_2$) is non-pointed. Note that every nonempty polyhedron subset of a pointed polyhedron is pointed.

A good question to ask: Is every nonempty polyhedron pointed? We give the following theorem without proof. For a proof, see, for example, Bertsimas and Tsitsiklis [BT97].

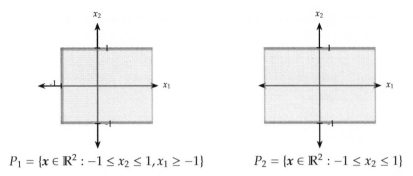

$$P_1 = \{x \in \mathbb{R}^2 : -1 \leq x_2 \leq 1, x_1 \geq -1\} \qquad P_2 = \{x \in \mathbb{R}^2 : -1 \leq x_2 \leq 1\}$$

Figure 10.15: Pointed polyhedron (left) versus non-pointed polyhedron (right).

**Theorem 10.6** *Assume that the polyhedron* $P = \{x \in \mathbb{R}^n : a_i^{\mathsf{T}} x \geq b_i, i = 1, \ldots, m\}$ *is nonempty. Then the following are equivalent:*

(a) *The polyhedron P is pointed.*

(b) *The polyhedron P has at least one extreme point.*

(c) *There exist n vectors out of the family* $\bar{a}_1, \ldots, a_m$, *which are linearly independent.*

Note that, from Theorem 10.6, a bounded polyhedron is pointed. Similarly, the nonnegative orthant cone $\mathbb{R}_+^n \triangleq \{x \in \mathbb{R}^n : x \geq 0\}$ is pointed. Since any standard form polyhedron is a subset of the nonnegative orthant cone, it is pointed too. The following two corollaries are now immediate.

**Corollary 10.4** *Every nonempty bounded polyhedron has at least one basic feasible solution.*

**Corollary 10.5** *Every nonempty polyhedron in standard form has at least one basic feasible solution.*

Note also that, from Theorem 10.6, every nonempty polyhedron $P = \{x \in \mathbb{R}^n : Ax \geq b\}$, with $A \in \mathbb{R}^{m \times n}$ and $m < n$, cannot have any basic feasible solution.

## Optimality

In the above part, we have established the conditions for the existence of extreme points. In this part, we will see that if a nonempty polyhedron $P$ has no corner poins, then the linear programming problem of minimizing a linear objective function over $P$ cannot have a unique optimal solution. The following theorem presents the contrapositive of this statement.

> **Theorem 10.7** *Consider the linear programming problem over a polyhedron P. If P has at least one extreme point and there exists an optimal solution, then there exists an extreme point of P which is optimal.*

**Proof** Let $P = \{x \in \mathbb{R}^n : Ax \geq b\}$, and $v$ be the optimal value of the cost $c^T x$ which we have assumed to be attained. Then $P_{\text{opt}} \triangleq \{x \in \mathbb{R}^n : Ax \geq b, c^T x = v\}$ contains all optimal solutions in $P$. By assumption, $P_{\text{opt}}$ is a nonempty polyhedron. From Theorem 10.6, $P$ is pointed. Since $P_{\text{opt}} \subset P$, $P_{\text{opt}}$ is pointed too. Using Theorem 10.6 again, $P_{\text{opt}}$ has an extreme point, say $x^\star$. Since $x^\star \in P_{\text{opt}}$, we have $c^T x^\star = v$, i.e., $x^\star$ is optimal. To complete the proof, it remains to show that $x^\star$ is an extreme point of $P$.

Suppose, in the contrary, that $x^\star$ is not an extreme point of $P$. Then, there exist $y, z \in P$ and a scalar $\lambda \in (0, 1)$ such that $x^\star = \lambda y + (1 - \lambda)z$. Consequently, $v = c^T x^\star = \lambda c^T y + (1 - \lambda)c^T z$. Furthermore, since $v$ is the optimal cost, $c^T y \geq v$ and $c^T z \geq v$. It follows that $c^T y = c^T z = v$, and therefore $y, z \in P_{\text{opt}}$. But this contradicts the fact that $x^\star$ is an extreme point of $P_{\text{opt}}$. Thus, $x^\star$ is an extreme point of $P$. The proof is complete. ∎

A more general result than that in Theorem 10.7 is stated in the following theorem, which will be given without proof. For a proof, see, for example, Bertsimas and Tsitsiklis [BT97].

> **Theorem 10.8** *Consider the linear programming problem over a polyhedron P. If P has at least one extreme point, then either the optimal cost is equal to $-\infty$, or there exists an extreme point of P which is optimal.*

Theorems 10.7 and 10.8 cover the case of polyhedra with the condition that there is at least one extreme point. Now, what about polyhedra without this condition? In fact, any linear programming problem over a polyhedron (with or without extreme points) can be transformed into an equivalent problem in standard form on which Theorem 10.8 can be applied (see Corollary 10.5). From this observation one concludes the following corollary.

> **Corollary 10.6** *Consider the linear programming problem over a polyhedron P. Then either the optimal cost is equal to $-\infty$, or there exists an optimal solution.*

Generally, Corollary 10.6 does not hold for nonlinear programming problems. For example, the nonlinear optimization problem

$$\min \quad 1/x$$
$$\text{s.t.} \quad x \geq 1,$$

has no optimal solution, but the optimal cost is not $-\infty$.

# 10.6. The simplex method

We have introduced linear optimization problems and studied its geometry and duality. Now, we are ready to introduce the simplex method. The word "simplex" is a general term of LP feasible region. Simplex method is used to solve LPs with any number of variables and constraints. The idea behind this method is to move from one basic feasible solution to an adjacent basic feasible solution so that the objective function value improves.

## The simplex method

We first describe the simplex method for the maximization LP.

**Simplex method for maximization**    The six-step procedure in Workflow 10.5, followed by Example 10.26, will teach us to apply the simplex method for solving the maximization problem. First, we need the following definition.

---

**Definition 10.11** *Consider the standard form LP:*

$$\begin{aligned} \max \quad & z = c^\mathsf{T} x \\ s.t. \quad & Ax = b, \\ & x \geq 0. \end{aligned} \tag{10.13}$$

*The corresponding canonical form is*

$$\begin{aligned} z - c^\mathsf{T} x &= 0, \\ Ax &= b. \end{aligned}$$

---

For example, the canonical form corresponding to the standard form LP:

$$\begin{aligned} \max \quad & z = 2x_1 + 3x_2 \\ s.t. \quad & x_1 + 2x_2 = 4, \\ & 2x_1 + x_2 + x_3 = 8, \\ & x_1, x_2, x_3 \geq 0, \end{aligned}$$

is the system

$$\begin{aligned} z - 2x_1 - 3x_2 &= 0, \\ x_1 + 2x_2 &= 4, \\ 2x_1 + x_2 + x_3 &= 8. \end{aligned}$$

---

**Workflow 10.5 (The simplex method)** *We solve a maximization LP problem by following five steps:*

(i) *Write the given LP in the standard form.*

(ii) *Convert the standard form to a canonical form.*

(iii) *Find a basic feasible solution for the canonical form.*

(iv) *If the current basic feasible solution is optimal, stop. If not, find which basic variable must become nonbasic and which nonbasic variable must become basic, and apply elementary row operations in order to move to an adjacent basic feasible solution with a higher value for the objective function.*

(v) *Go to Step (iv).*

---

More details on how to find which basic (respectively, nonbasic) variable must become nonbasic (respectively, basic), and on how to check the current basic feasible solution for optimality can be found in Remarks 10.2-10.4 below.

**Example 10.26** Use the simplex method to solve the following LP.

$$
\begin{aligned}
\max \quad & z = 2x_1 + 3x_2 \\
\text{s.t.} \quad & x_1 + 2x_2 && \leq \quad 6, \\
& 2x_1 + x_2 && \leq \quad 8, \\
& x_1, x_2 && \geq \quad 0.
\end{aligned}
\tag{10.14}
$$

**Solution** We apply the steps in Workflow 10.5. Problem (10.18) in standard form is written as

$$
\begin{aligned}
\max \quad & z = 2x_1 + 3x_2 \\
\text{s.t.} \quad & x_1 + 2x_2 + s_1 && = \quad 6, \\
& 2x_1 + x_2 + s_2 && = \quad 8, \\
& x_1, x_2, s_1, s_2 && \geq \quad 0.
\end{aligned}
\tag{10.15}
$$

Problem (10.15) in the canonical form is written as

$$
\begin{aligned}
z \;-\; 2x_1 \;-\; 3x_2 && && = \; 0, \\
x_1 \;+\; 2x_2 \;+\; s_1 && && = \; 6, \\
2x_1 \;+\; x_2 && +\; s_2 && = \; 8.
\end{aligned}
\tag{10.16}
$$

The canonical variables, which correspond to the unit columns, are the variables $z, s_1$ and $s_2$. Let BV denote the set of the basic variables. We select BV $= \{z, s_1, s_2\}$. Generally, we have $|BV| = m + 1$ ($m = 2$ in this example), and we choose $z \in BV$ always. Therefore, the BV contains the variable $z$ plus $m$ canonical variables. Let NBV denote the set of the nonbasic variables. Then NBV $= \{x_1, x_2\}$.

Fix $x_1 = x_2 = 0$. System (10.16) now reads $z = 0$, $s_1 = 6$ and $s_2 = 8$, which are a basic feasible solution.

The task now is to check if the current basic feasible solution is optimal. Note that the current basic feasible solution is optimal if we cannot improve the value of $z$ by increasing the value of any nonbasic variable (from zero). Here, $z = 2x_1 + 3x_2 = 0$ as $x_1 = x_2 = 0$ (NBV = $\{x_1, x_2\}$). Increasing $x_1$ from 0 to 1 increases $z$ from 0 to 2, while increasing $x_2$ from 0 to 1 increases $z$ from 0 to 3 (we are increasing one variable at a time, while keeping the other nonbasic variable fixed at zero). It is more beneficial to increase $x_2$ here than $x_1$. In general, we pick the nonbasic variable with the largest positive coefficient in the expression for $z$ to enter the basis. In the canonical form, we pick the nonbasic variable with the most negative coefficient in row-0 to enter the basis. The following remark summarizes this discussion. We will move forward in solving Example 10.26 after this remark.

---

**Remark 10.2 (Criterion for choosing the entering variable in maximization)**
*The entering variable in a maximization LP problem is the nonbasic variable having the most negative coefficient in the $z$-row.*

---

In light of Remark 10.2, we call $x_2$ the entering variable in this step.

We have identified an entering variable. Now, we find a new adjacent basic feasible solution by identifying also a leaving variable. We cannot increase $x_2$, the entering variable, without limits. When $x_2$ increases, $s_1$ or $s_2$ may decrease, and we need to make sure they stay nonnegative, so that we stay feasible.

From row 1 and row 2, with $x_1 = 0$, we get

$$\text{Row 1:} \quad 2x_2 + s_1 = 6, \quad \text{which implies} \quad s_1 = 6 - 2x_2 \geq 0,$$
$$\text{Row 2:} \quad x_2 + s_2 = 8, \quad \text{which implies} \quad s_2 = 8 - x_2 \geq 0.$$

Note that $s_1$ and $s_2$ need to be nonnegative for feasibility. To keep $s_1 \geq 0$, we cannot increase $x_2$ beyond 6/2=3. To keep $s_2 \geq 0$, we cannot increase $x_2$ beyond 8/1=8.

Thus, we let $x_2 = 3$ which makes $s_1 = 0$. In this step, $x_2$ is called the entering variable, and $s_1$ is called the leaving variable.

The test in the following remark summarizes the above discussion. Applying this test guarantees that the basic solution remains feasible. We will move forward in solving Example 10.26 after this remark.

---

**Remark 10.3 (Minimum ratio test for choosing the leaving variable)** *For each constraint row that has a positive coefficient[a] for the entering variable, we compute the ratio:*

$$\frac{\text{right-hand side of row}}{\text{coefficient of entering variable in row}}.$$

*Among all these ratios, the nonbasic variable with smallest nonnegative ratio is the leaving variable.*

---
[a]We do not consider the row(s) with negative coefficients.

---

Note that the smallest among all the ratios computed in Remark 10.3 is the largest value that the entering variable can take. Going back to Example 10.26, the ratios are:

$$\text{Row 1:} \quad \tfrac{6}{2} = 3; \quad \leftarrow \quad \text{The winner!}$$
$$\text{Row 2:} \quad \tfrac{8}{1} = 8.$$

Therefore, $s_1$ leaves the basis, i.e., it becomes nonbasic, and the entering variable $x_2$ takes its place.

We use elementary row operations in order to make the entering variable basic in the row that the minimum ratio test meets the requirement outlined in Remark 10.3.

$$
\begin{aligned}
R_0: &\quad z &-\ 2x_1 &-\ 3x_2 & & & & & = &\ 0, \\
R_1: & & x_1 &+\ 2x_2 &+\ s_1 & & & & = &\ 6, \\
R_2: & & 2x_1 &+\ x_2 & & &+\ s_2 & & = &\ 8; \\[4pt]
\tfrac{3}{2}R_1 + R_0 \to R_0: &\quad z &-\ \tfrac{1}{2}x_1 & &+\ \tfrac{3}{2}s_1 & & & & = &\ 9, \\
\tfrac{1}{2}R_1 \to R_1: & & \tfrac{1}{2}x_1 &+\ x_2 &+\ \tfrac{1}{2}s_1 & & & & = &\ 3, \\
\tfrac{-1}{2}R_1 + R_2 \to R_2: & & \tfrac{3}{2}x_1 & &-\ \tfrac{1}{2}s_1 &+\ s_2 & & & = &\ 5; \\[4pt]
R_0 + \tfrac{1}{3}R_2 \to R_0: &\quad z & & &+\ \tfrac{4}{3}s_1 &+\ \tfrac{1}{3}s_2 & & & = &\ \tfrac{32}{3}, \\
R_1 - \tfrac{1}{3}R_2 \to R_1: & & &x_2 &+\ \tfrac{2}{3}s_1 &-\ \tfrac{1}{3}s_2 & & & = &\ \tfrac{4}{3}, \\
\tfrac{2}{3}R_2 \to R_2: & & x_1 & &-\ \tfrac{1}{3}s_1 &+\ \tfrac{2}{3}s_2 & & & = &\ \tfrac{10}{3}.
\end{aligned}
$$

We will finish solving Example 10.26 after the following remark.

> **Remark 10.4 (Criterion for optimality in maximization)** *In a maximization LP problem, the optimum is reached at the iteration where all the z-row coefficient of the non-basic variables are nonnegative.*

Note that in our example we cannot improve the value of $z$ anymore (by making $s_1$ or $s_2$ basic). Hence, we have an optimal solution. The optimal solution is $(x_1, x_2) = (10/3, 4/3)$ with the optimal value $z = 32/3$. ∎

## The full tableau method

The full tableau method is a more convenient way to perform the computations needed by the simplex method.

*Simplex tableau for maximization*   If we have a maximization problem, the structure of the simplex tableau is as follows:

| $z$ | $x$ | rhs | |
|---|---|---|---|
| 1 | $c^{\mathsf{T}} - c_B^{\mathsf{T}} A_B^{-1} A$ | $-c_B^{\mathsf{T}} A_B^{-1} b$ | |
| 0 | $A_B^{-1} A$ | $A_B^{-1} b$ | $= x_B$ |

Here $A_B$ is defined in (10.11) and $c_B$ is the cost vector corresponding to the basic variables. We keep maintaining and updating the above table till we reach the optimality. Example 10.27 resolves Example 10.26 using the simplex tableau method.

◼ **Example 10.27 (Example 10.26 revisited)** Use the simplex tableau method to solve the following maximization problem.

$$
\begin{array}{rlr}
\max & z = 2x_1 + 3x_2 \\
\text{s.t.} & x_1 + 2x_2 & \leq \ 6, \\
& 2x_1 + x_2 & \leq \ 8, \\
& x_1, x_2 & \geq \ 0.
\end{array}
$$

**Solution**  After introducing slack variables, we obtain the standard form problem given in (10.15). Note that $x = (0, 0, 6, 8)$ is a basic feasible solution. Hence, we have the following initial tableau:

| $z$ | $x_1$ | $x_2$ | $s_1$ | $s_2$ | rhs | |
|---|---|---|---|---|---|---|
| 1 | $-2$ | $-3$ | 0 | 0 | 0 | |
| 0 | 1 | ②  | 1 | 0 | 6 | $= s_1$ |
| 0 | 2 | 1 | 0 | 1 | 8 | $= s_2$ |

Since we are maximizing the objective function, we select a nonbasic variable with the greatest positive reduced cost to be the one that enters the basis. Indicating the pivot element with a circled number, we obtain the following tableau:

| $z$ | $x_1$ | $x_2$ | $s_1$ | $s_2$ | rhs | |
|---|---|---|---|---|---|---|
| 1 | $-1/2$ | 0 | 3/2 | 0 | 9 | |
| 0 | 1/2 | 1 | 1/2 | 0 | 3 | $= x_2$ |
| 0 | ③/2 | 0 | $-1/2$ | 1 | 5 | $= s_2$ |

Note that we brought $x_2$ into the basis and $s_1$ exited. We then bring $x_1$ into the basis; $s_2$ exits and we obtain the following tableau:

| $z$ | $x_1$ | $x_2$ | $s_1$ | $s_2$ | rhs | |
|---|---|---|---|---|---|---|
| 1 | 0 | 0 | 4/3 | 1/3 | 32/3 | $= z$ |
| 0 | 0 | 1 | 2/3 | $-1/3$ | 4/3 | $= x_2$ |
| 0 | 1 | 0 | $-1/3$ | 2/3 | 10/3 | $= x_1$ |

The reduced costs in the zeroth row of the tableau are all nonnegative, so the current basic feasible solution is optimal. In terms of the original variables $x_1$ and $x_2$, this solution is $x = (10/3, 4/3)$. The optimal value is $z = 32/3$.

The above series of tableaux can be combined in one single table as follows:

| EROs | $z$ | $x_1$ | $x_2$ | $s_1$ | $s_2$ | rhs | MR |
|---|---|---|---|---|---|---|---|
| $R_0:$ | 1 | $-2$ | $-3$ | 0 | 0 | 0 | |
| $R_1:$ | 0 | 1 | (2) | 1 | 0 | 6 | 6/2 |
| $R_2:$ | 0 | 2 | 1 | 0 | 1 | 8 | 8/1 |
| $R_0 + \frac{3}{2}R_1 \rightarrow R_0:$ | 1 | $-1/2$ | 0 | 3/2 | 0 | 9 | |
| $\frac{1}{2}R_1 \rightarrow R_1:$ | 0 | 1/2 | 1 | 1/2 | 0 | 3 | 3/0.5 |
| $\frac{-1}{2}R_1 + R_2 \rightarrow R_2:$ | 0 | (3/2) | 0 | $-1/2$ | 1 | 5 | 5/1.5 |
| $R_0 + \frac{1}{3}R_2 \rightarrow R_0:$ | 1 | 0 | 0 | 4/3 | 1/3 | 32/3 | |
| $R_1 - \frac{1}{3}R_2 \rightarrow R_1:$ | 0 | 0 | 1 | 2/3 | $-1/3$ | 4/3 | Optimal |
| $\frac{2}{3}R_2 \rightarrow R_2:$ | 0 | 1 | 0 | $-1/3$ | 2/3 | 10/3 | tableau! |

**Example 10.28** Use the simplex method to solve the following LP.

$$\max \quad z = 2x_1 - x_2 + x_3$$

$$\text{s.t.} \quad \begin{aligned} 3x_1 + x_2 + x_3 &\leq 60, &\text{(adding } s_1) \\ x_1 - x_2 + 2x_3 &\leq 10, &\text{(adding } s_2) \\ x_1 + x_2 - x_3 &\leq 20, &\text{(adding } s_3) \\ x_1, x_2, x_3 &\geq 0. \end{aligned}$$

(10.17)

**Solution** We have the following tableaux:

| EROs | $z$ | $x_1$ | $x_2$ | $x_3$ | $s_1$ | $s_2$ | $s_3$ | rhs | MR |
|---|---|---|---|---|---|---|---|---|---|
| $R_0:$ | 1 | $-2$ | 1 | $-1$ | 0 | 0 | 0 | 0 | |
| $R_1:$ | 0 | 3 | 1 | 1 | 1 | 0 | 0 | 60 | 60/3 |
| $R_2:$ | 0 | (1) | $-1$ | 2 | 0 | 1 | 0 | 10 | 10/1 |
| $R_3:$ | 0 | 1 | 1 | $-1$ | 0 | 0 | 1 | 20 | 20/1 |
| $R_0 + 2R_2 \rightarrow R_0:$ | 1 | 0 | $-1$ | 3 | 0 | 2 | 0 | 20 | |
| $R_1 - 3R_2 \rightarrow R_1:$ | 0 | 0 | 4 | $-5$ | 1 | $-3$ | 0 | 30 | 30/4 |
| $R_2 \rightarrow R_2:$ | 0 | 1 | $-1$ | 2 | 0 | 1 | 0 | 10 | |
| $-R_2 + R_3 \rightarrow R_3:$ | 0 | 0 | (2) | $-3$ | 0 | $-1$ | 1 | 10 | 10/2 |
| $R_0 + \frac{1}{2}R_3 \rightarrow R_0:$ | 1 | 0 | 0 | 3/2 | 0 | 3/2 | 1/2 | 25 | |
| $R_1 - 2R_3 \rightarrow R_1:$ | 0 | 0 | 0 | 1 | 1 | $-1$ | $-2$ | 10 | Optimal |
| $R_2 + \frac{1}{2}R_3 \rightarrow R_2:$ | 0 | 1 | 0 | 1/2 | 0 | 1/2 | 1/2 | 15 | tableau! |
| $\frac{1}{2}R_3 \rightarrow R_3:$ | 0 | 0 | 1 | $-3/2$ | 0 | $-1/2$ | 1/2 | 5 | |

The optimal solution is given by $(x_1, x_2, x_3) = (15, 5, 0)$, and the optimal value is $z = 25$.

***Detecting the existence of alternative optimal solutions***   The simplex method can tell if alternative optimal solutions (i.e., infinitely many solutions) exist. The following remark tells us when we have alternative optimal solutions.

> **Remark 10.5** *If the coefficient of a nonbasic variable in the zeroth row of the tableau is zero, then the linear programming problem has alternative optimal solutions.*

■   **Example 10.29** Use the simplex tableau method to solve the following maximization problem.

$$\begin{aligned} \max \quad & z = 4x_1 + x_2 \\ \text{s.t.} \quad & 8x_1 + 2x_2 \leq 16, \quad \text{(adding } s_1) \\ & 5x_1 + 2x_2 \leq 12, \quad \text{(adding } s_2) \\ & x_1, x_2 \geq 0. \end{aligned}$$

**Solution**   We have the following tableaux:

| EROs | z | $x_1$ | $x_2$ | $s_1$ | $s_2$ | rhs | MR |
|---|---|---|---|---|---|---|---|
| $R_0$ : | 1 | −4 | −1 | 0 | 0 | 0 | |
| $R_1$ : | 0 | ⑧ | 2 | 1 | 0 | 16 | $16/8 = 2$ |
| $R_2$ : | 0 | 5 | 2 | 0 | 1 | 12 | $12/5 = 2.4$ |
| $R_0 + \frac{1}{2}R_1 \to R_0$ : | 1 | 0 | 0 | 1/2 | 0 | 8 | |
| $\frac{1}{8}R_1 \to R_1$ : | 0 | 1 | 1/4 | 1/8 | 0 | 2 | This tableau |
| $\frac{-5}{8}R_1 + R_2 \to R_2$ : | 0 | 0 | ③/④ | −5/8 | 1 | 2 | is optimal! |
| $R_0 \to R_0$ : | 1 | 0 | 0 | 1/2 | 0 | 8 | |
| $R_1 - \frac{1}{3}R_2 \to R_1$ : | 0 | 1 | 0 | 1/3 | −1/3 | 4/3 | This tableau |
| $\frac{4}{3}R_2 \to R_2$ : | 0 | 0 | 1 | −5/6 | 4/3 | 8/3 | is also optimal! |

In view of Remark 10.5, we have alternative optimal solutions. An optimal solution is given by $(x_1, x_2) = (2, 0)$. Another optimal solution is given by $(x_1, x_2) = (4/3, 8/3)$. The optimal value is $z = 8$. As an exercise for the reader, use the graphical method to reach the same conclusion.   ■

***Detecting unboundedness***   The simplex method can be used to detect the unboundedness. The following remark tells us when we have an unbounded problem.

> **Remark 10.6** *If there is no candidate for the minimum ratio test, then the linear programming problem is unbounded.*

**Example 10.30**  Use the simplex tableau method to solve the following maximization problem.

$$\begin{aligned}
\max \quad & z = 2x_2 \\
\text{s.t.} \quad & x_1 - x_2 \leq 4, \quad (\text{adding } s_1) \\
& -x_1 + x_2 \leq 1, \quad (\text{adding } s_2) \\
& x_1, x_2 \geq 0.
\end{aligned}$$

**Solution**  We have the following tableaux:

| EROs | $z$ | $x_1$ | $x_2$ | $s_1$ | $s_2$ | rhs | MR |
|---|---|---|---|---|---|---|---|
| $R_0:$ | 1 | 0 | $-2$ | 0 | 0 | 0 | |
| $R_1:$ | 0 | 1 | $-1$ | 1 | 0 | 4 | $16/8 = 2$ |
| $R_2:$ | 0 | $-1$ | ①  | 0 | 1 | 1 | $12/5 = 2.4$ |
| $R_0 - 2R_2 \to R_0:$ | 1 | $-2$ | 0 | 0 | 2 | 2 | |
| $R_1 + R_2 \to R_1:$ | 0 | 0 | 0 | 1 | 1 | 5 | The LP is |
| $R_2 \to R_2:$ | 0 | $-1$ | 1 | 0 | 1 | 1 | unbounded! |

Note that there is no candidate for the minimum ratio test. In view of Remark 10.6, we have an unbounded linear programming problem. As an exercise for the reader, use the graphical method to reach the same conclusion. ∎

**Breaking ties**  In what follows, we learn how to break ties in selection of a nonbasic variable. The following remark tells us how to break ties for entering or leaving variable if any.

> **Remark 10.7**  *If there are ties for entering or leaving, we can break them arbitrarily.*

**Example 10.31**  Use the simplex tableau method to solve the following maximization problem.

$$\begin{aligned}
\max \quad & z = x_1 + x_2 \\
\text{s.t.} \quad & x_1 + x_2 + x_3 \leq 1, \quad (\text{adding } s_1) \\
& x_1 + 2x_3 \leq 1, \quad (\text{adding } s_2) \\
& x_1, x_2, x_3 \geq 0.
\end{aligned}$$

**Solution**  We have the following tableaux:

| EROs | z | $x_1$ | $x_2$ | $x_3$ | $s_1$ | $s_2$ | rhs | MR |
|---|---|---|---|---|---|---|---|---|
| $R_0$ : | 1 | $-1$ | $-1$ | 0 | 0 | 0 | 0 | |
| $R_1$ : | 0 | ① | 1 | 1 | 1 | 0 | 1 | 1 (A candidate) |
| $R_2$ : | 0 | 1 | 0 | 2 | 0 | 1 | 1 | 1 (Another candidate!) |
| $R_0 - 2R_2 \rightarrow R_0$ : | 1 | 0 | 0 | 1 | 1 | 0 | 1 | Alternative |
| $R_1 + R_2 \rightarrow R_1$ : | 0 | 1 | 1 | 1 | 1 | 0 | 1 | optimal |
| $R_2 \rightarrow R_2$ : | 0 | 0 | $-1$ | 1 | $-1$ | 1 | 0 | solutions! |

We have alternative optimal solutions. An optimal solution is given by $(x_1, x_2, x_3) = (1, 0, 0)$. The optimal value is $z = 1$.  ∎

***Simplex tableau for minimization***  So far, we have studied the simplex method for solving linear maximization problems. This method can also be applied to solve linear minimization problems. The rules for entering and optimality for minimization LP problems are opposite to those for maximization LP problems, as in the following remarks.

---

**Remark 10.8 (Criterion for choosing the entering variable in minimization)**
*The entering variable in a minimization LP problem is the nonbasic variable having the most positive coefficient in the z-row.*

---

---

**Remark 10.9 (Criterion for optimality in minimization)** *In a minimization LP problem, the optimum is reached at the iteration where all the z-row coefficient of the non-basic variables are nonpositive.*

---

We have the following examples.

**Example 10.32** Use the simplex tableau method to solve the following minimization problem.

$$\begin{aligned}
\min \quad & z = -x_1 - x_2 \\
\text{s.t.} \quad & x_1 - x_2 && \leq \ 1, && (\text{adding } s_1) \\
& x_1 + x_2 && \leq \ 2, && (\text{adding } s_2) \\
& x_1, x_2 && \geq \ 0.
\end{aligned}$$

**Solution** We have the following tableaux:

| EROs | z | $x_1$ | $x_2$ | $s_1$ | $s_2$ | rhs | MR |
|---|---|---|---|---|---|---|---|
| $R_0$ : | 1 | 1 | 1 | 0 | 0 | 0 | |
| $R_1$ : | 0 | ①  | −1 | 1 | 0 | 1 | $1/1 = 1$ |
| $R_2$ : | 0 | 1 | 1 | 0 | 1 | 2 | $2/1 = 2$ |
| $R_0 - R_1 \rightarrow R_0$ : | 1 | 0 | 2 | −1 | 0 | −1 | |
| $R_1 \rightarrow R_1$ : | 0 | 1 | −1 | 1 | 0 | 1 | |
| $-R_1 + R_2 \rightarrow R_2$ : | 0 | 0 | ②  | −1 | 1 | 1 | $1/2 = 0.5$ |
| $R_0 - R_1 \rightarrow R_0$ : | 1 | 0 | 0 | 0 | −1 | −2 | |
| $R_1 \rightarrow R_1$ : | 0 | 1 | 0 | 1/2 | 1/2 | 3/2 | *Optimal* |
| $-R_1 + R_2 \rightarrow R_2$ : | 0 | 0 | 1 | −1/2 | 1/2 | 1/2 | *tableau!* |

The reduced costs in the zeroth row of the tableau are all nonpositive, so the current basic feasible solution is optimal. In terms of the original variables $x_1$ and $x_2$, this solution is $x = (3/2, 1/2)$. The optimal value is $z = -2$. In addition, since the coefficient of the nonbasic variable $s_1$ in the zeroth row of the last tableau is zero, we have alternative optimal solutions. ∎

**Example 10.33** Use the simplex tableau method to solve the following minimization problem.

$$\min \quad z = -x_1 - x_2$$
$$\text{s.t.} \quad 2x_1 + x_2 \leq 4, \quad (\text{adding } s_1)$$
$$\qquad 3x_1 + 5x_2 \leq 15, \quad (\text{adding } s_2)$$
$$\qquad x_1, x_2 \geq 0.$$

**Solution** We have the following tableaux:

| EROs | z | $x_1$ | $x_2$ | $s_1$ | $s_2$ | rhs | MR |
|---|---|---|---|---|---|---|---|
| $R_0$ : | 1 | 1 | 1 | 0 | 0 | 0 | |
| $R_1$ : | 0 | ②  | 1 | 1 | 0 | 4 | $4/2 = 2$ |
| $R_2$ : | 0 | 3 | 5 | 0 | 1 | 15 | $15/3 = 5$ |
| $R_0 - \frac{1}{2}R_1 \rightarrow R_0$ : | 1 | 0 | 1/2 | −1/2 | 0 | −2 | |
| $\frac{1}{2}R_1 \rightarrow R_1$ : | 0 | 1 | 1/2 | 1/2 | 0 | 2 | $2/0.5 = 4$ |
| $-\frac{3}{2}R_1 + R_2 \rightarrow R_2$ : | 0 | 0 | ⑦⁄₂  | −3/2 | 1 | 9 | $9/3.5 \approx 2.57$ |
| $R_0 - \frac{1}{7}R_2 \rightarrow R_0$ : | 1 | 0 | 0 | −2/7 | −1/7 | −23/7 | |
| $R_1 - \frac{1}{7}R_2 \rightarrow R_1$ : | 0 | 1 | 0 | 5/7 | −1/7 | 5/7 | *Optimal* |
| $\frac{2}{7}R_2 \rightarrow R_2$ : | 0 | 0 | 1 | −3/7 | 2/7 | 18/7 | *tableau!* |

The optimal solution is $x = (5/7, 18/7)$. The optimal value is $z = -23/7$. ∎

***Problems with nonpositive variables and/or free variables***   So far, we have studied the simplex method for solving linear optimization problems with nonnegative variables.

Handling problems with nonpositive variables is easy. The idea is to define a new nonnegative variable that equals the negative of the original variable. That is, if for some $j$, $x_j \leq 0$, we replace $x_j$ with $-x'_j$ and add $x'_j \geq 0$.

Handling problems with unrestricted-in-sign (or free) variables is also easy. The idea is to define two new nonnegative variables whose difference equals the original variable. That is, if for some $j$, $x_j$ is unrestricted-in-sign, we replace $x_j$ with $x'_j - x''_j$ and add $x'_j, x''_j \geq 0$. In this case, at most one of $x'_j$ and $x''_j$ can be basic in one tableau, not both. We have the following example.

**Example 10.34** Use the simplex tableau method to solve the following maximization problem.

$$
\begin{aligned}
\max \quad & z = 2x_1 + x_2 \\
\text{s.t.} \quad & 3x_1 + x_2 \leq 6, \\
& x_1 + x_2 \leq 4, \\
& x_1 \geq 0.
\end{aligned}
$$

**Solution**   Note that the variable $x_2$ is unrestricted-in-sign. An equivalent problem is

$$
\begin{aligned}
\max \quad & z = 2x_1 + x'_2 - x''_2 \\
\text{s.t.} \quad & 3x_1 + x'_2 - x''_2 \leq 6, \quad \text{(adding } s_1\text{)} \\
& x_1 + x'_2 - x''_2 \leq 4, \quad \text{(adding } s_2\text{)} \\
& x_1, x'_2, x''_2 \geq 0.
\end{aligned}
$$

We then have the following tableaux:

| EROs | $z$ | $x_1$ | $x'_2$ | $x''_2$ | $s_1$ | $s_2$ | rhs | MR |
|---|---|---|---|---|---|---|---|---|
| $R_0:$ | 1 | $-2$ | $-1$ | 1 | 0 | 0 | 0 | |
| $R_1:$ | 0 | ③ | 1 | $-1$ | 1 | 0 | 6 | $6/3 = 2$ |
| $R_2:$ | 0 | 1 | 1 | $-1$ | 0 | 1 | 4 | $4/1 = 4$ |
| $R_0 + \frac{2}{3}R_1 \to R_0:$ | 1 | 0 | $-1/3$ | 1/3 | 2/3 | 0 | 4 | |
| $\frac{1}{3}R_1 \to R_1:$ | 0 | 1 | 1/3 | $-1/3$ | 1/3 | 0 | 2 | $2/(1/3) = 6$ |
| $-\frac{1}{3}R_1 + R_2 \to R_2:$ | 0 | 0 | ②/③ | $-2/3$ | $-1/3$ | 1 | 2 | $2/(2/3) = 3$ |
| $R_0 + \frac{1}{2}R_2 \to R_0:$ | 1 | 0 | 0 | 0 | 1/2 | 1/2 | 5 | |
| $R_1 - \frac{1}{2}R_2 \to R_1:$ | 0 | 1 | 0 | 0 | 1/2 | $-1/2$ | 1 | Optimal |
| $\frac{3}{2}R_2 \to R_2:$ | 0 | 0 | 1 | $-1$ | $-1/2$ | 3/2 | 3 | tableau! |

An optimal solution is given by $x_1 = 1$ and $x_2 = x_2' - x_2'' = 3 - 0 = 3$. The optimal value is $z = 5$. We point out that the columns corresponding to $x_2'$ and $x_2''$ are always identical but with opposite signs. ∎

## The big-M method

So far, we have studied the simplex method for solving linear optimization problems with "≤" constraints. The question that arises now is how to handle maximization and minimization problems involving "≥" or "=" constraints. A common approach of handling such problems is to use the so-called big-M method. In other words, the big-M method extends the simplex algorithm to problems that contain "greater-than" and/or "equal" constraints.

***Problems with "greater-than" and/or "equal" constraints*** With only "≤" constraints, we get an obvious starting basic feasible solution (namely the slack variables). Now, how to get a starting basic feasible solution when we have "≥" and/or "=" constraints? The idea behind the big-M method is to add artificial variables for each "≥" and "=" constraints and follows the steps in the following workflow.

---

**Workflow 10.6 (The big-M method)** *We solve problems with "greater-than" and/or "equal" constraints by following six steps:*

  (*i*) *Modify constraints as needed so that all the right-hand side values are nonnegative.*

 (*ii*) *Add an artificial variable, say $a_i$, for constraint $i$ if it is a "≥" or "=" constraint. Then add the nonnegativity constraint $a_i \geq 0$.*

(*iii*) *Add $\pm M a_i$ to the objective function, where $M$ is a big positive number, as follows:*

  ▪ *For a maximization LP problem, add $-M a_i$.*

  ▪ *For a minimization LP problem, add $+M a_i$.*

 (*iv*) *Convert the resulting LP into the standard form by adding slack/excess variables.*

  (*v*) *Convert the LP into the canonical form and make the coefficient of $a_i$ in the zeroth row zero by using elementary row operations involving $M$.*

 (*vi*) *Operate Steps (iii)-(vi) in Workflow* 10.5.

---

As a direct application of Workflow 10.6, we have the following example.

■ **Example 10.35** Use the simplex tableau method to solve the following minimization problem.

$$\begin{aligned} \min \quad & z = 2x_1 + 3x_2 \\ \text{s.t.} \quad & 2x_1 + x_2 && \geq && 4, \\ & x_1 - x_2 && \geq && -1, \\ & x_1, x_2 && \geq && 0. \end{aligned}$$

**Solution** We start by operating Steps (i)-(v) in Workflow 10.6. First, we modify the constraints so that all the right-hand side values are nonnegative, and get

$$\begin{aligned} \min \quad & z = 2x_1 + 3x_2 \\ \text{s.t.} \quad & 2x_1 + x_2 && \geq && 4, \\ & -x_1 + x_2 && \leq && 1, \\ & x_1, x_2 && \geq && 0. \end{aligned}$$

Then, we add an artificial variable $a_i$ for constraint $i$ if it is a "$\geq$" or "$=$" constraint. Then add $a_i \geq 0$. We also add $Ma_i$ to the objective function, where $M$ is a big positive number. This yields

$$\begin{aligned} \min \quad & z = 2x_1 + 3x_2 + Ma_1 \\ \text{s.t.} \quad & 2x_1 + x_2 + a_1 && \geq && 4, \\ & -x_1 + x_2 && \leq && 1, \\ & x_1, x_2, a_1 && \geq && 0. \end{aligned}$$

Next, we convert the resulting LP into the standard form to get

$$\begin{aligned} \min \quad & z = 2x_1 + 3x_2 + Ma_1 \\ \text{s.t.} \quad & 2x_1 + x_2 + a_1 - e_1 && = && 4, \\ & -x_1 + x_2 + s_1 && = && 1, \\ & x_1, x_2, s_1, e_1, a_1 && \geq && 0. \end{aligned}$$

Now, we convert the LP into the canonical form, use elementary row operations to make the coefficient of $a_i$ in the zeroth row zero, and then proceed by operating Steps (iii)-(vi) in Workflow 10.5. This can be seen in the following tableaux:

| EROs | $z$ | $x_1$ | $x_2'$ | $e_1$ | $s_1$ | $a_1$ | rhs | |
|---|---|---|---|---|---|---|---|---|
| $R_0$ : | 1 | $-2$ | $-3$ | 0 | 0 | $-M$ | 0 | Not in |
| $R_1$ : | 0 | 2 | 1 | $-1$ | 0 | 1 | 4 | canonical |
| $R_2$ : | 0 | $-1$ | 1 | 0 | 1 | 0 | 1 | form |
| $R_0 + MR_1 \rightarrow R_0$ : | 1 | $2M-2$ | $M-3$ | $-M$ | 0 | 0 | $4M$ | In |
| $R_1 \rightarrow R_1$ : | 0 | 2 | 1 | $-1$ | 0 | 1 | 4 | canonical |
| $R_2 \rightarrow R_2$ : | 0 | $-1$ | 1 | 0 | 1 | 0 | 1 | form |
| $R_0 + (1-M)R_1 \rightarrow R_0$ : | 1 | 0 | $-2$ | $-1$ | 0 | $-M+1$ | 4 | |
| $\frac{1}{2}R_1 \rightarrow R_1$ : | 0 | 1 | 1/2 | $-1/2$ | 0 | 1/2 | 2 | Optimal |
| $\frac{1}{2}R_1 + R_2 \rightarrow R_2$ : | 0 | 0 | 3/2 | $-1/2$ | 1 | 1/2 | 3 | tableau! |

The optimal solution is given by $(x_1, x_2) = (2, 0)$. The optimal value is given by $z = 4$.

$\blacksquare$

**Example 10.36** Use the simplex tableau method to solve the following minimization problem.

$$
\begin{aligned}
\min \quad & z = 2x_1 - 3x_2 \\
\text{s.t.} \quad & x_1 + 3x_2 \leq 9, \\
& 2x_1 + 5x_2 \geq -6, \\
& x_2 \geq 1, \\
& x_2 \geq 0.
\end{aligned}
$$

**Solution** Note that the variable $x_1$ is unrestricted-in-sign. Considering an equivalent problem with equality constraints and nonnegative variables only, we are interested in a problem of the form

$$
\begin{aligned}
\min \quad & z = 2(x_1' - x_1'') - 3x_2 + Ma_1 \\
\text{s.t.} \quad & (x_1' - x_1'') + 3x_2 + s_1 = 9, \\
& -2(x_1' - x_1'') - 5x_2 + s_2 = 6, \\
& x_2 - e_1 + a_1 = 1, \\
& x_1', x_1'', x_2, s_1, s_2, e_1, a_1 \geq 0.
\end{aligned}
$$

We then have the following tableaux:

| EROs | z | $x_1'$ | $x_1''$ | $x_2$ | $s_1$ | $s_2$ | $e_1$ | $a_1$ | rhs | |
|---|---|---|---|---|---|---|---|---|---|---|
| $R_0$ : | 1 | −2 | 2 | 3 | 0 | 0 | 0 | −M | 0 | |
| $R_1$ : | 0 | 1 | −1 | 3 | 1 | 0 | 0 | 0 | 9 | |
| $R_2$ : | 0 | −2 | 2 | −5 | 0 | 1 | 0 | 0 | 6 | |
| $R_3$ : | 0 | 0 | 0 | 1 | 0 | 0 | −1 | 1 | 1 | |
| $R_0 + MR_3 \to R_0$ : | 1 | −2 | 2 | $M+3$ | 0 | 0 | −M | 0 | M | |
| $R_1 \to R_1$ : | 0 | 1 | −1 | 3 | 1 | 0 | 0 | 0 | 9 | |
| $R_2 \to R_2$ : | 0 | −2 | 2 | −5 | 0 | 1 | 0 | 0 | 6 | |
| $R_3 \to R_3$ : | 0 | 0 | 0 | 1 | 0 | 0 | −1 | 1 | 1 | |
| $R_0 - (M+3)R_3 \to R_0$ : | 1 | −2 | 2 | 0 | 0 | 0 | 3 | $-M-3$ | −3 | |
| $R_1 - 3R_3 \to R_1$ : | 0 | 1 | −1 | 0 | 1 | 0 | 3 | −3 | 6 | |
| $R_2 + 5R_3 \to R_2$ : | 0 | −2 | 2 | 0 | 0 | 1 | −5 | 5 | 11 | |
| $R_3 \to R_3$ : | 0 | 0 | 0 | 1 | 0 | 0 | −1 | 1 | 1 | |
| $R_0 - R_1 \to R_0$ : | 1 | −3 | 3 | 0 | −1 | 0 | 0 | −M | −9 | |
| $\frac{1}{3}R_1 \to R_1$ : | 0 | 1/3 | −1/3 | 0 | 1/3 | 0 | 1 | −1 | 2 | |
| $\frac{5}{3}R_1 + R_2 \to R_2$ : | 0 | −1/3 | 1/3 | 0 | 5/3 | 1 | 0 | 0 | 21 | |
| $\frac{1}{3}R_1 + R_3 \to R_3$ : | 0 | 1/3 | −1/3 | 1 | 1/3 | 0 | 0 | 0 | 3 | |
| $R_0 - 9R_2 \to R_0$ : | 1 | 0 | 0 | 0 | −16 | −9 | 0 | −M | −198 | O |
| $R_1 + R_2 \to R_1$ : | 0 | 0 | 0 | 0 | 2 | 3 | 1 | −1 | 23 | P |
| $3R_2 \to R_2$ : | 0 | −1 | 1 | 0 | 5 | 3 | 0 | 0 | 63 | T |
| $R_2 + R_3 \to R_3$ : | 0 | 0 | 0 | 1 | 2 | 1 | 0 | 0 | 24 | M |

The optimal solution is given by $x_1 = x_1' - x_1'' = 0 - 63 = -63$, $x_2 = 24$ and $e_1 = 23$, hence $(x_1, x_2) = (-63, 24)$. The optimal value is $z = -198$. ∎

**Detecting infeasibility**   The big-M method can be used to detect the infeasibility. The following remark tells us when we have an infeasible problem.

> **Remark 10.10** *If any artificial variable is basic in the optimal tableau, i.e.* $a_i > 0$ *for some i, then the linear programming problem is infeasible.*

As a direct application, we have the following example.

**Example 10.37** Use the simplex tableau method to solve the following minimization problem.

$$\begin{aligned}
\min \quad & z = 3x_1 \\
\text{s.t.} \quad & 2x_1 + x_2 \geq 6, \\
& 3x_1 + 2x_2 = 4, \\
& x_1, x_2 \geq 0.
\end{aligned}$$

**Solution**   Considering an equivalent problem with equality constraints and nonnegative variables only, we are interested in a problem of the form

$$
\begin{array}{lrcl}
\min & z = 3x_1 + Ma_1 + Ma_2 & & \\
\text{s.t.} & 2x_1 + x_2 + a_1 - e_1 & = & 6, \\
& 3x_1 + 2x_2 + a_2 & = & 4, \\
& x_1, x_2, a_1, a_2, e_1 & \geq & 0.
\end{array}
$$

We then have the following tableaux:

| EROs | $z$ | $x_1$ | $x_2'$ | $e_1$ | $s_1$ | $a_1$ | rhs | |
|---|---|---|---|---|---|---|---|---|
| $R_0:$ | 1 | 3 | 0 | 0 | $-M$ | $-M$ | 0 | |
| $R_1:$ | 0 | 2 | 1 | $-1$ | 1 | 0 | 6 | |
| $R_2:$ | 0 | 3 | 2 | 0 | 0 | 1 | 4 | |
| $R_0 + MR_1 + MR_2 \rightsquigarrow R_0:$ | 1 | $5M-3$ | $3M$ | $-M$ | 0 | 0 | $10M$ | |
| $R_1 \rightarrow R_1:$ | 0 | 2 | 1 | $-1$ | 1 | 0 | 6 | |
| $R_2 \rightarrow R_2:$ | 0 | 3 | 2 | 0 | 0 | 1 | 4 | |
| $R_0 + (1 - \frac{5}{3}M)R_2 \rightarrow R_0:$ | 1 | 0 | $-M/3 + 2$ | $-M$ | 0 | $-5M/3 + 1$ | $10M/3 + 4$ | O |
| $R_1 - \frac{2}{3}R_2 \rightarrow R_1:$ | 0 | 0 | $-1/3$ | $-1$ | 1 | $-2/3$ | $10/3$ | P |
| $\frac{1}{3}R_2 \rightarrow R_2:$ | 0 | 1 | $2/3$ | 0 | 0 | $1/3$ | $4/3$ | T |

The last tableau is optimal. Since $a_1 = 10/3 > 0$, the problem is infeasible. As an exercise for the reader, use the graphical method to reach the same conclusion. ∎

**_Summary of the simplex method steps_**   We now summarize the above description of the simplex method.

---

**Workflow 10.7 (Overview of the simplex method)**   *We solve a linear optimization problem by operating the following steps:*

(*i*) *Modify constraints as needed so that all the right-hand side values are nonnegative.*

(*ii*) *Add an artificial variable, say $a_i$, for constraint i if it is a "$\geq$" or "$=$" constraint. Then add the nonnegativity constraint $a_i \geq 0$.*

(*iii*) *Add $\pm Ma_i$ to the objective function, where M is a big positive number.*

(*iv*) *Convert the resulting LP into the standard form by adding slack/excess variables.*

(*v*) *Convert the LP into the canonical form and make the coefficient of $a_i$ in the zeroth row zero by using elementary row operations involving M.*

(*vi*) *Find a basic feasible solution for the canonical form.*

---

> (*vii*) *If the current basic feasible solution is optimal, stop. If not, move to an adjacent basic feasible solution with a higher value for the objective function by applying elementary row operations and noting that*
>
> - *If the coefficient of a nonbasic variable in the zeroth row of the tableau is zero, then the LP has alternative optimal solutions.*
>
> - *If there is no candidate for the minimum ratio test, then the LP is unbounded.*
>
> - *If any artificial variable is basic in the optimal tableau, i.e. $a_i > 0$ for some i, then the LP is infeasible.*
>
> (*viii*) *Go to Step (vii).*

We end this part with the following theorem which gives the amount of computation per iteration (worst behavior) in terms of the size of the coefficient matrix. We give Theorem 10.9, without proof. For a proof, see, for example, Bertsimas and Tsitsiklis [BT97, Section 3].

> **Theorem 10.9** *Assume that the matrix A in Problem* (10.13) *is $m \times n$. The number of arithmetic operations in each iteration of the simplex tableau algorithm solving Problem* (10.13) *is $O(mn)$.*

Note that the estimate of the computational complexity in Theorem 10.9 refers to a single iteration. This complexity estimate is for both the worst-case time and the best-case time.

### Anticycling

The simplex method can be indeed cycling. In order to prevent cycling, there are two anticycling rules under which the simplex method is guaranteed to terminate. Namely: The lexicographic rule and Bland's rule (after Robert Bland who discovered it in 1976). In this part, we only discuss Bland's rule leaving the lexicographic rule probably for a future edition of the book. However, there are a number of good references for learning anticycling rules, see for example [BT97, Section 3.4].

Recall that the ordinary rule for choosing the entering and leaving variables are:

- Choose $j$ with the most negative $c_j$ as the entering column.

- Break ties by picking the variable with the smallest index to enter.

Considering the above rule, we may get back to the starting tableau after some iterations in some linear programming problems, as in the following example due to [BT97].

**Example 10.38** Consider the following LP problem.

$$\max \quad z = \tfrac{3}{4}x_1 - 20x_2 + \tfrac{1}{2}x_3 - 6x_4$$

$$\begin{aligned}
\text{s.t.} \quad &\tfrac{1}{4}x_1 - 8x_2 - x_3 + 9x_4 && \leq && 0, && (\text{adding } x_5) \\
&\tfrac{1}{2}x_1 - 12x_2 - \tfrac{1}{2}x_3 + 3x_4 && \leq && 0, && (\text{adding } x_6) && (10.18)\\
&x_3 + 6x_4 && \leq && 1, && (\text{adding } x_7) \\
&x_1, x_2, x_3, x_4 && \geq && 0.
\end{aligned}$$

If we use the simplex method to solve Problem 10.18 with the ordinary rule, we obtain the following tableaux:

| EROs | z | $x_1$ | $x_2$ | $x_3$ | $x_4$ | $x_5$ | $x_6$ | $x_7$ | rhs |
|---|---|---|---|---|---|---|---|---|---|
| $R_0:$ | 1 | $-3/4$ | 20 | $-1/2$ | 6 | 0 | 0 | 0 | 0 |
| $R_1:$ | 0 | (1/4) | $-8$ | $-1$ | 9 | 1 | 0 | 0 | 0 |
| $R_2:$ | 0 | 1/2 | $-12$ | $-1/2$ | 3 | 0 | 1 | 0 | 0 |
| $R_3:$ | 0 | 0 | 0 | 1 | 6 | 0 | 0 | 1 | 1 |
| $R_0 + 3R_1 \to R_0:$ | 1 | 0 | $-4$ | $-7/2$ | 33 | 3 | 0 | 0 | 0 |
| $4R_1 \to R_1:$ | 0 | 1 | $-32$ | $-4$ | 36 | 4 | 0 | 0 | 0 |
| $-2R_1 + R_2 \to R_2:$ | 0 | 0 | (4) | 3/2 | $-15$ | $-2$ | 1 | 0 | 0 |
| $R_3 \to R_3:$ | 0 | 0 | 0 | 1 | 0 | 0 | 0 | 1 | 1 |
| $R_0 + R_2 \to R_0:$ | 1 | 0 | 0 | $-2$ | 18 | 1 | 1 | 0 | 0 |
| $R_1 + 8R_2 \to R_1:$ | 0 | 1 | 0 | (8) | $-84$ | $-12$ | 8 | 0 | 0 |
| $\tfrac{1}{4}R_2 \to R_2:$ | 0 | 0 | 1 | 3/8 | $-15/4$ | $-1/2$ | 1/4 | 0 | 0 |
| $R_3 \to R_3:$ | 0 | 0 | 0 | 1 | 0 | 0 | 0 | 1 | 1 |
| $R_0 + \tfrac{1}{4}R_1 \to R_0:$ | 1 | 1/4 | 0 | 0 | $-3$ | $-2$ | 3 | 0 | 0 |
| $\tfrac{1}{8}R_1 \to R_1:$ | 0 | 1/8 | 0 | 1 | $-21/2$ | $-3/2$ | 1 | 0 | 0 |
| $-\tfrac{3}{64}R_1 + R_2 \to R_2:$ | 0 | $-3/64$ | 1 | 0 | (3/16) | 1/16 | $-1/8$ | 0 | 0 |
| $-\tfrac{1}{8}R_1 + R_3 \to R_3:$ | 0 | $-1/8$ | 0 | 0 | 21/2 | 3/2 | $-1$ | 1 | 1 |
| $R_0 + 16R_2 \to R_0:$ | 1 | $-1/2$ | 16 | 0 | 0 | $-1$ | 1 | 0 | 0 |
| $R_1 + 56R_2 \to R_1:$ | 0 | $-5/2$ | 56 | 1 | 0 | (2) | $-6$ | 0 | 0 |
| $\tfrac{16}{3}R_2 \to R_2:$ | 0 | $-1/4$ | 16/3 | 0 | 1 | 1/3 | $-2/3$ | 0 | 0 |
| $-56R_2 + R_3 \to R_3:$ | 0 | 5/2 | $-56$ | 0 | 0 | $-2$ | 6 | 1 | 1 |
| $R_0 + \tfrac{1}{2}R_1 \to R_0:$ | 1 | $-7/4$ | 44 | 1/2 | 0 | 0 | $-2$ | 0 | 0 |
| $\tfrac{1}{2}R_1 \to R_1:$ | 0 | $-5/4$ | 28 | 1/2 | 0 | 1 | $-3$ | 0 | 0 |
| $-\tfrac{1}{6}R_1 + R_2 \to R_2:$ | 0 | 1/6 | $-4$ | $-1/6$ | 1 | 0 | (1/3) | 0 | 0 |
| $R_1 + R_3 \to R_3:$ | 0 | 0 | 0 | 1 | 0 | 0 | 0 | 1 | 1 |
| $R_0 + 6R_2 \to R_0:$ | 1 | $-3/4$ | 20 | $-1/2$ | 6 | 0 | 0 | 0 | 0 |
| $R_1 + 9R_2 \to R_1:$ | 0 | 1/4 | $-8$ | $-1$ | 9 | 1 | 0 | 0 | 0 |
| $3R_2 \to R_2:$ | 0 | 1/2 | $-12$ | $-1/2$ | 3 | 0 | 1 | 0 | 0 |
| $R_3 \to R_3:$ | 0 | 0 | 0 | 1 | 6 | 0 | 0 | 1 | 1 |

Note that the ending tableau is identical to the starting tableau. This means that the simplex method is cycling here! ■

Example 10.38 will be revisited in order to avoid cycling after discussing Bland's rule.

**Bland's rule** Bland's rule is one of the algorithmic refinements of the simplex method to avoid cycling.

---

**Remark 10.11 (Bland's rule)** *Bland's rule under which the simplex method for linear optimization terminates is as follows:*

- *Choose the smallest $j$ with $c_j < 0$ as the entering column.*

- *Break ties by picking the variable with the smallest index to enter.*

---

**Example 10.39 (Example 10.38 revisited)** If we use the simplex method to solve Problem 10.18 with Bland's rule, we obtain the following tableaux:

| EROs | $z$ | $x_1$ | $x_2$ | $x_3$ | $x_4$ | $x_5$ | $x_6$ | $x_7$ | rhs |
|---|---|---|---|---|---|---|---|---|---|
| $R_0$ : | 1 | $-3/4$ | 20 | $-1/2$ | 6 | 0 | 0 | 0 | 0 |
| $R_1$ : | 0 | (1/4) | $-8$ | $-1$ | 9 | 1 | 0 | 0 | 0 |
| $R_2$ : | 0 | 1/2 | $-12$ | $-1/2$ | 3 | 0 | 1 | 0 | 0 |
| $R_3$ : | 0 | 0 | 0 | 1 | 6 | 0 | 0 | 1 | 1 |
| $R_0 + 3R_1 \to R_0$ : | 1 | 0 | $-4$ | $-7/2$ | 33 | 3 | 0 | 0 | 0 |
| $4R_1 \to R_1$ : | 0 | 1 | $-32$ | $-4$ | 36 | 4 | 0 | 0 | 0 |
| $-2R_1 + R_2 \to R_2$ : | 0 | 0 | (4) | 3/2 | $-15$ | $-2$ | 1 | 0 | 0 |
| $R_3 \to R_3$ : | 0 | 0 | 0 | 1 | 0 | 0 | 0 | 1 | 1 |
| $R_0 + R_2 \to R_0$ : | 1 | 0 | 0 | $-2$ | 18 | 1 | 1 | 0 | 0 |
| $R_1 + 8R_2 \to R_1$ : | 0 | 1 | 0 | (8) | $-84$ | $-12$ | 8 | 0 | 0 |
| $\frac{1}{4}R_2 \to R_2$ : | 0 | 0 | 1 | 3/8 | $-15/4$ | $-1/2$ | 1/4 | 0 | 0 |
| $R_3 \to R_3$ : | 0 | 0 | 0 | 1 | 0 | 0 | 0 | 1 | 1 |
| $R_0 + \frac{1}{4}R_1 \to R_0$ : | 1 | 1/4 | 0 | 0 | $-3$ | $-2$ | 3 | 0 | 0 |
| $\frac{1}{8}R_1 \to R_1$ : | 0 | 1/8 | 0 | 1 | $-21/2$ | $-3/2$ | 1 | 0 | 0 |
| $-\frac{3}{64}R_1 + R_2 \to R_2$ : | 0 | $-3/64$ | 1 | 0 | (3/16) | 1/16 | $-1/8$ | 0 | 0 |
| $-\frac{1}{8}R_1 + R_3 \to R_3$ : | 0 | $-1/8$ | 0 | 0 | 21/2 | 3/2 | $-1$ | 1 | 1 |
| $R_0 + 16R_2 \to R_0$ : | 1 | $-1/2$ | 16 | 0 | 0 | $-1$ | 1 | 0 | 0 |
| $R_1 + 56R_2 \to R_1$ : | 0 | $-5/2$ | 56 | 1 | 0 | (2) | $-6$ | 0 | 0 |
| $\frac{16}{3}R_2 \to R_2$ : | 0 | $-1/4$ | 16/3 | 0 | 1 | 1/3 | $-2/3$ | 0 | 0 |
| $-56R_2 + R_3 \to R_3$ : | 0 | 5/2 | $-56$ | 0 | 0 | $-2$ | 6 | 1 | 1 |
| $R_0 + \frac{1}{5}R_3 \to R_0$ : | 1 | 0 | 24/5 | 0 | 0 | $-7/5$ | 11/5 | 1/5 | 1/5 |
| $R_1 + R_3 \to R_1$ : | 0 | 0 | 0 | 1 | 0 | 0 | 0 | 1 | 1 |
| $R_2 + \frac{1}{10}R_3 \to R_2$ : | 0 | 0 | $-4/15$ | 0 | 1 | (2/15) | $-11/15$ | 1/10 | 1/10 |
| $\frac{2}{5}R_3 \to R_3$ : | 0 | 1 | $-112/5$ | 0 | 0 | $-4/5$ | 12/5 | 2/5 | 2/5 |
| $R_0 + \frac{21}{2}R_2 \to R_0$ : | 1 | 0 | 2 | 0 | 21/2 | 0 | 3/2 | 5/4 | 5/4 |
| $R_1 \to R_1$ : | 0 | 0 | 0 | 1 | 0 | 0 | 0 | 1 | 1 |
| $\frac{15}{2}R_2 \to R_2$ : | 0 | 0 | $-2$ | 0 | 15/2 | 1 | $-1/2$ | 3/4 | 3/4 |
| $6R_2 + R_3 \to R_3$ : | 0 | 1 | $-24$ | 0 | 6 | 0 | 2 | 1 | 1 |

Since we have applied Bland's rule, the simplex method has terminated. The last tableau is optimal, the optimal solution is $x = (1; 0; 1; 0)$, and the optimal value is $z = 5/4$.    ∎

The question that remains now in this context is: How to prevent cycling when we solve linear maximization problems? One answer stems from the following remark.

---

**Remark 10.12** *If you start with a minimization problem, say* $\min f(x)$ *subject to* $x \in S$, *where* $f : \mathbb{R}^n \longrightarrow \mathbb{R}$ *is a function and S is a set, then an equivalent maximization problem is* $\max -f(x)$ *subject to* $x \in S$. *Similarly, if you start with a maximization problem, say* $\max f(x)$ *subject to* $x \in S$, *then an equivalent minimization problem is* $\min -f(x)$ *subject to* $x \in S$.

---

In light of Remark 10.12, to prevent cycling when we maximize $c^\mathsf{T}x$ subject to some constraints, we use the simplex method and apply Bland's rule to minimize $-c^\mathsf{T}x$ subject to the same constraints. We have the same optimal solution, but the optimal value of the maximization problem equals that of the minimization problem multiplied by -1.

## Optimal values of the dual variables

Recall that, from Theorem 10.2, If the primal and dual problems are both feasible and one of them has a finite optimal solution, so does the other, and their optimal values are equal. The question that arises now is: How do you find the optimal solution of a dual problem from the simplex tableau of the primal problem? We answer this question by providing the following remark.

---

**Remark 10.13** *If we are given the simplex tableau of a primal maximization problem, then*

$$Optimal\ y_i = \begin{cases} Coefficient\ of\ s_i\ in\ R_0, & if\ the\ ith\ constraint\ is\ ``\leq "; \\ -(Coefficient\ of\ e_i\ in\ R_0), & if\ the\ ith\ constraint\ is\ ``\geq "; \\ (Coefficient\ of\ a_i\ in\ R_0) - M^a, & if\ the\ ith\ constraint\ is\ ``= ". \end{cases}$$

*If we are given the simplex tableau of a primal minimization problem, then*

$$Optimal\ y_i = \begin{cases} Coefficient\ of\ s_i\ in\ R_0, & if\ the\ ith\ constraint\ is\ ``\leq "; \\ -(Coefficient\ of\ e_i\ in\ R_0), & if\ the\ ith\ constraint\ is\ ``\geq "; \\ (Coefficient\ of\ a_i\ in\ R_0) + M, & if\ the\ ith\ constraint\ is\ ``= ". \end{cases}$$

---

$^a$This also holds when the $i$th constraint is "≥".

The following example illustrates how Remark 10.13 is used.

**Example 10.40**  The dual problem of the maximization LP problem

$$\begin{aligned}
\text{max} \quad & z = 30x_1 + 100x_2 \\
\text{s.t.} \quad & x_1 + x_2 \le 7, \\
& 4x_1 + 10x_2 \le 40, \\
& 10x_1 \ge 30, \\
& x_1 \ge 0, x_2 \ge 0,
\end{aligned} \quad (10.19)$$

is the minimization LP problem

$$\begin{aligned}
\text{min} \quad & w = 7y_1 + 40y_2 + 30y_3 \\
\text{s.t.} \quad & y_1 + 4y_2 + 10y_3 \ge 30, \\
& y_1 + 10y_2 \ge 100, \\
& y_1 \ge 0, y_2 \ge 0, y_3 \le 0.
\end{aligned} \quad (10.20)$$

Solving Problem (10.19) by the simplex tableau method, we obtain

| EROs | $z$ | $x_1$ | $x_2$ | $s_1$ | $s_2$ | $e_3$ | $a_3$ | rhs | |
|---|---|---|---|---|---|---|---|---|---|
| $R_0:$ | 1 | $-30$ | $-100$ | 0 | 0 | 0 | $M$ | 0 | |
| $R_1:$ | 0 | 1 | 1 | 1 | 0 | 0 | 0 | 7 | |
| $R_2:$ | 0 | 4 | 10 | 0 | 1 | 0 | 0 | 40 | |
| $R_3:$ | 0 | 10 | 0 | 0 | 0 | $-1$ | 1 | 30 | |
| $R_0 - MR_3 \to R_0:$ | 1 | $-30 - 10M$ | $-100$ | 0 | 0 | $M$ | 0 | $-30M$ | |
| $R_1 \to R_1:$ | 0 | 1 | 1 | 1 | 0 | 0 | 0 | 7 | |
| $R_2 \to R_2:$ | 0 | 4 | 10 | 0 | 1 | 0 | 0 | 40 | |
| $R_3 \to R_3:$ | 0 | 10 | 0 | 0 | 0 | $-1$ | 1 | 30 | |
| $R_0 + (3+M)R_3 \to R_0:$ | 1 | 0 | $-100$ | 0 | 0 | $-3$ | $M+3$ | 90 | |
| $R_1 - \frac{1}{10}R_3 \to R_1:$ | 0 | 0 | 1 | 1 | 0 | 1/10 | $-1/10$ | 4 | |
| $R_2 - \frac{2}{5}R_3 \to R_2:$ | 0 | 0 | 10 | 0 | 1 | 3/5 | $-3/5$ | 28 | |
| $\frac{1}{10}R_3 \to R_3:$ | 0 | 1 | 0 | 0 | 0 | $-1/10$ | 1/10 | 3 | |
| $R_0 + 10R_1 \to R_0:$ | 1 | 0 | 0 | 0 | 10 | 1 | $M-1$ | 370 | O |
| $R_1 \to R_1:$ | 0 | 0 | 0 | 1 | $-1/10$ | 3/50 | $-3/50$ | 1.2 | P |
| $R_2 \to R_2:$ | 0 | 0 | 1 | 0 | $-1/10$ | 1/25 | $-1/25$ | 2.8 | T |
| $R_3 \to R_3:$ | 0 | 1 | 0 | 0 | 0 | $-1/10$ | 1/10 | 3 | M |

The last tableau is optimal. The optimal value is $z = 370$, the primal optimal solution is $x_1 = 3$ and $x_2 = 2.8$. According to Remark 10.13, the dual optimal solution is:

$$\begin{aligned}
y_1 &= \text{Coefficient of } s_1 \text{ in } R_0 & &= 0; \\
y_2 &= \text{Coefficient of } s_2 \text{ in } R_0 & &= 10; \\
y_3 &= -(\text{Coefficient of } e_3 \text{ in } R_0) & &= -1, \\
\text{or } y_3 &= (\text{Coefficient of } a_3 \text{ in } R_0) - M & &= (M-1) - M = -1.
\end{aligned}$$

To check this, note that the dual optimal value is $w = 7y_1 + 10y_2 + 30y_3 = 370$, which is exactly the primal optimal value.  ∎

## Complexity

Like any algorithmic method, the computational complexity of the simplex method is determined by the following two factors: (a) The computational complexity of each iteration. (b) The total number of iterations.

The following theorem is known to hold (see Section 3.3 in [BT97]). It indicates that the amount of computation in each iteration of the full tableau method is propositional to the size of the tableau.

> **Theorem 10.10** *The number of arithmetic operations in each iteration of the full tableau method is $O(mn)$.*

Practically, the advantage of the simplex method is that it has been observed that the method takes only $O(m)$ iterations to find an optimal solution. Theoretically, the disadvantage of the method is that the above observation is not true for every linear programming problem. In fact, there are a family of problems for which an exponential number of iterations is required [BT97, Section 3.7]. The reason for this is that the number of extreme points of the feasible set can increase exponentially with the number of variables and constraints.

## 10.7.  A homogeneous interior-point method

Interior-point methods [NN94] are one of the efficient methods developed to solve linear and nonlinear programming problems. Contrary to the simplex method, interior-point methods reach a best solution by traversing the interior of the feasible region. There are several Interior-point algorithms for LPs; see for example [BT97, Chapter 9] and the references contained therein.

Homogeneous self-dual algorithms are one of the interior-point methods used for solving linear and nonlinear programs. In this section, we present a homogeneous interior-point algorithm for solving (P|LP) and (D|LP) introduced in Section 10.4. The material of this section has appeared in [Tuc57, YTM94].

We define the following feasibility sets for the primal-dual pair (P|LP) and (D|LP).

$$
\begin{aligned}
\mathcal{F}_{\text{P|LP}} &\triangleq \{x \in \mathbb{R}^n : Ax = b,\ x \geq 0\}, \\
\mathcal{F}_{\text{D|LP}} &\triangleq \left\{(y, s) \in \mathbb{R}^m \times \mathbb{R}^n : A^\mathsf{T} y + s = c,\ s \geq 0\right\}, \\
\mathcal{F}_{\text{P|LP}}^{\circ} &\triangleq \{x \in \mathbb{R}^n : Ax = b,\ x > 0\}, \\
\mathcal{F}_{\text{D|LP}}^{\circ} &\triangleq \left\{(y, s) \in \mathbb{R}^m \times \mathbb{R}^n : A^\mathsf{T} y + s = c,\ s > 0\right\}, \\
\mathcal{F}_{\text{LP}}^{\circ} &\triangleq \mathcal{F}_{\text{P|LP}}^{\circ} \times \mathcal{F}_{\text{D|LP}}^{\circ}.
\end{aligned}
$$

We also make the following assumptions about the primal-dual pair (P|LP) and (D|LP).

> **Assumption 10.1** *The m rows of the matrix A are linearly independent.*

> **Assumption 10.2** *The set $\mathcal{F}^{\circ}_{LP}$ is nonempty.*

Assumption 10.1 is for convenience. Assumption 10.2 requires that Problem (P|LP) and its dual (D|LP) have strictly feasible solutions, which guarantees strong duality for the linear programming problem.

The following primal-dual LP model provides sufficient conditions (but not always necessary) for an optimal solution of (P|LP) and (D|LP).

$$
\begin{aligned}
Ax &= b, \\
A^\mathsf{T}y + s &= c, \\
x^\mathsf{T}s &= 0, \\
x, s &\geq 0.
\end{aligned}
\tag{10.21}
$$

The homogeneous LP model for the pair (P|LP) and (D|LP) is as follows:

$$
\begin{array}{rrrrcl}
Ax & & -b\tau & & = & 0, \\
& -A^\mathsf{T}y & -s & +c\tau & = & 0, \\
-c^\mathsf{T}x & +b^\mathsf{T}y & & -\kappa & = & 0, \\
x & & & & \geq & 0, \\
& s & & & \geq & 0, \\
& & \tau & & \geq & 0, \\
& & & \kappa & \geq & 0.
\end{array}
\tag{10.22}
$$

The first two equations in (10.22), with $\tau = 1$, represent primal and dual feasibility (with $x, s \geq 0$) and reversed weak duality. So they, together with the third equation after forcing $\kappa = 0$, define primal and dual optimal solutions. Note that homogenizing $\tau$ (i.e., making it a variable) adds the required variable dual to the third equation, introducing the artificial variable $\kappa$ achieves feasibility, and adding the third equation in (10.22) achieves self-duality.

One can show that $x^\mathsf{T}s + \tau\kappa = 0$ (see Exercise 10.22). The next theorem relates (10.21) to (10.22), and it is easily proved. Here, as defined previously, $\mathbb{R}^n_+ = \{x \in \mathbb{R}^n : x \geq 0\}$ is the nonnegative orthant cone.

> **Theorem 10.11** *The primal-dual LP model* (10.21) *has a solution if and only if the homogeneous LP model* (10.22) *has a solution*
>
> $$(x^\star, y^\star, s^\star, \tau^\star, \kappa^\star) \in \mathbb{R}_+^n \times \mathbb{R}^m \times \mathbb{R}_+^n \times \mathbb{R}_+ \times \mathbb{R}_+$$
>
> *such that* $\tau^\star > 0$ *and* $\kappa^\star = 0$.

The main step at each iteration of the homogeneous interior-point algorithm for solving (P|LP) and (D|LP) is the computation of the search direction $(\Delta x, \Delta y, \Delta s)$ from the Newton equations defined by the system:

$$
\begin{aligned}
A\,\Delta x && && -b\,\Delta\tau && &&= \eta r_p, \\
&& -\,A^\mathsf{T}\,\Delta y && -\,\Delta s && +c\,\Delta\tau && &&= \eta r_d, \\
-c^\mathsf{T}\,\Delta x && +\,b^\mathsf{T}\,\Delta y && && && -\,\Delta\kappa &&= \eta r_g, \\
&& && && \kappa\,\Delta\tau && +\,\tau\,\Delta\kappa &&= \gamma\mu - \tau\kappa, \\
S\Delta x && && +\,X\Delta s && && &&= \gamma\mu\mathbf{1} - Xs,
\end{aligned}
\tag{10.23}
$$

where $\mathbf{1}$ is a vector of ones with an appropriate dimension. $\eta$ and $\gamma$ are two parameters, $X \triangleq \operatorname{Diag}(x) \in \mathbb{R}^{n \times n}$ is the diagonal matrix with the vector $x \in \mathbb{R}^n$ on its diagonal, $S \triangleq \operatorname{Diag}(s) \in \mathbb{R}^{n \times n}$ is the diagonal matrix with the vector $s \in \mathbb{R}^n$ on its diagonal, and

$$
\begin{aligned}
r_p &\triangleq b\tau - Ax, \\
r_d &\triangleq A^\mathsf{T}y + s - \tau c, \\
r_g &\triangleq c^\mathsf{T}x - b^\mathsf{T}y + \kappa, \\
\mu &\triangleq \tfrac{1}{n+1}(x^\mathsf{T}s + \tau\kappa).
\end{aligned}
$$

We state the generic homogeneous algorithm for solving the pair (P|LP) and (D|LP) in Algorithm 10.1.

---

**Algorithm 10.1:** Generic homogeneous self-dual algorithm for LP

---

**Input:** Data in Problems (P|LP) and (D|LP) $(x, y, s, \tau, \kappa) \triangleq (\mathbf{1}, 0, \mathbf{1}, 1, 1)$
**Output:** An approximate optimal solution to Problem (P|LP)

1: **while** a stopping criterion is not satisfied **do**
2:     choose $\eta, \gamma$
3:     compute the solution $(\Delta x, \Delta y, \Delta s, \Delta\tau, \Delta\kappa)$ of the linear system (10.23)
4:     compute a step length $\theta$ so that
    $x + \theta\Delta x > 0$
    $s + \theta\Delta s > 0$
    $\tau + \theta\Delta\tau > 0$
    $\kappa + \theta\Delta\kappa > 0$
5:     set the new iterate according to
    $(x, y, s, \tau, \kappa) \triangleq (x, y, s, \tau, \kappa) + \theta(\Delta x, \Delta y, \Delta s, \Delta\tau, \Delta\kappa)$
6: **end**

---

The following theorem is known to hold (see [YTM94]). It gives the computational complexity (worst behavior) of Algorithm 10.1 in terms of the dimension of the decision variable ($n$).

---

**Theorem 10.12**  *Let $\epsilon_0 > 0$ be the residual error at a starting point, and $\epsilon > 0$ be a given tolerance. Under Assumptions 10.2 and 10.1, if the pair (P|LP) and (D|LP) has a solution $(x^\star, y^\star, s^\star)$, then Algorithm 10.1 finds an $\epsilon$-approximate solution (i.e., a solution with residual error less than or equal to $\epsilon$) in at most*

$$O\left( \sqrt{n} \ln \left( \mathbf{1}^{\mathsf{T}} (x^\star + s^\star) \left( \frac{\epsilon_0}{\epsilon} \right) \right) \right)$$

*iterations.*

---

Theoretically, the advantage of this interior-point method is maintaining the iteration complexity of $O(\sqrt{n} \ln(L))$, where $L$ is the data length of the underlying LP. Practically, the disadvantage of this method is the doubled dimension of the system of equations, which must be solved at each iteration.

# EXERCISES

**10.1**  Choose the correct answer for each of the following multiple-choice questions/items.

(*a*)  All linear programming problems may be solved using graphical method.

   (*i*) True.                           (*ii*) False.

(*b*)  XYZ Inc. produces two types of paper towels, called regular and super-soaker. Marketing has imposed a constraint that the total monthly production of regular should be no more than twice the monthly production of super-soakers. Letting $x_1$ be the number of units of regular produced per month and $x_2$ represent the number of units of super-soaker produced per month, the appropriate constraint(s) will be

   (*i*) $2x_1 \le x_2$.        (*iii*) $x_1 \le 0.5x_2$.        (*v*) $x_1 - 0.5x_2 \ge 0$.
   (*ii*) $x_1 \le 2x_2$.       (*iv*) $x_1 - x_2 \le 0$.

(*c*)  Problem A is a given formulation of a linear program with an optimal solution. Problem B is a formulation obtained by multiplying the objective function of Problem A by a positive constant and leaving all other things unchanged. Problems A and B will have

   (*i*) the same optimal solution and same objective function value.

(*ii*) the same optimal solution but different objective function values.

(*iii*) different optimal solutions but same objective function value.

(*iv*) different optimal solutions and different objective function values.

(*d*) Consider the following linear programming problem:

$$
\begin{array}{rrcl}
\max & 12x + 10y & & \\
\text{s.t.} & 4x + 3y & \leq & 480, \\
& 2x + 3y & \leq & 360, \\
& x, y & \geq & 0.
\end{array}
$$

Which of the following points $(x, y)$ could be a feasible corner point?

(*i*) $(40, 48)$.        (*iii*) $(180, 120)$.        (*v*) None of the above.

(*ii*) $(120, 0)$.        (*iv*) $(30, 36)$.

(*e*) XYZ Inc. produces two types of printers, called regular and high-speed. Regular uses 2 units of recycled plastic per unit and high-speed uses 1 unit of recycled plastic per unit of production. The total amount of recycled plastic available per month is 5,000. A critical machine is needed to manufacture the printers and each unit of regular requires 5 units of time in this machine and each unit of high-speed requires 3 units of time in this machine. The total time available in this machine per month is 10000 units. Letting $x_1$ be the number of units of regular produced per month and $x_2$ represent the number of units of high-speed produced per month, the appropriate constraint(s) will be:

(*i*) $2x_1 + x_2 = 5000$.        (*iii*) $5x_1 + 3x_2 \leq 10000$.        (*v*) (b) and (c).

(*ii*) $2x_1 + x_2 \leq 5000$.        (*iv*) (a) and (c).

(*f*) Problem A is a given formulation of a linear program with an optimal solution and its constraint 1 is $\leq$ type. Problem B is a formulation obtained from Problem A by replacing the $\leq$ constraint by an equality constraint and leaving all other things unchanged. Problems A and B will have

(*i*) the same optimal solution and same objective function value.

(*ii*) the same optimal solution but different objective function values.

(*iii*) different optimal solutions but same objective function value.

(*iv*) same or different solution profile depending on the role of the constraints in the solutions.

(g) Consider the following linear programming problem:

$$\begin{array}{lrcl} \max & 12x + 10y & & \\ \text{s.t.} & 4x + 3y & \leq & 480, \\ & 2x + 3y & \leq & 360, \\ & x, y & \geq & 0. \end{array}$$

Which of the following points $(x, y)$ is not in the feasible region?

(i) $(30, 60)$.      (iii) $(0, 110)$.      (v) None of the above.

(ii) $(105, 0)$.      (iv) $(100, 10)$.

(h) In any graphically solvable linear program, if two points are feasible, then any weighted average of the two points where weights are non-negative and add up to 1.0 will also be feasible.

(i) True.      (ii) False.

(i) In a two variable graphical linear program, if the coefficient of one of the variables in the objective function is changed (while the other remains fixed), then slope of the objective function expression will change.

(i) True.      (ii) False.

(j) XYZ Inc. produces two types of printers, called regular and high-speed. Regular uses 2 units of recycled plastic per unit and high-speed uses 1 unit of recycled plastic per unit of production. XYZ is committed to using at least 5,000 units of recycled plastic per month. A critical machine is needed to manufacture the printers and each unit of regular requires 5 units of time in this machine and each unit of high-speed requires 3 units of time in this machine. The total time available in this machine per month is 10000 units. Let $x_1$ be the number of units of regular produced per month and $x_2$ represent the number of units of high-speed produced per month. Imposing both of these constraints, and non-negativity constraints one of the feasible corner points is (assuming the first number in parenthesis is $x_1$ and the second number in the parenthesis is $x_2$):

(i) $(0, 0)$.      (iii) None exists.      (v) $(2500, 0)$.

(ii) $(2000, 0)$.      (iv) $(0, 5000)$.

(k) A point that satisfies all of a problem's constraints simultaneously is a(n):

(i) optimal solution.

(ii) corner point.

      (*iii*) intersection of the profit line and a constraint.

      (*iv*) intersection of two or more constraints.

      (*v*) None of the above.

(*l*) If a graphically solvable linear program is unbounded, then it can always be converted to a regular bounded problem by removing a constraint.

      (*i*) True.                      (*ii*) False.

(*m*) In a two variable graphical linear program, if the RHS of one of the constraints is changed (keeping all other things fixed) then the plot of the corresponding constraint will move in parallel to its old plot.

      (*i*) True.                      (*ii*) False.

(*n*) Two models of a product - Regular ($x$) and Deluxe ($y$) - are produced by a company. A linear programming model is used to determine the production schedule. The formulation is as follows:

$$
\begin{array}{llll}
\max & 50x + 60y & & \text{(maximum profit)} \\
\text{s.t.} & 8x + 10y \leq 800 & & \text{(labor hours),} \\
& x + y \leq 120 & & \text{(total units demanded),} \\
& 4x + 5y \leq 500 & & \text{(raw materials),} \\
& x, y \geq 0 & & \text{(non-negativity).}
\end{array}
$$

The optimal solution is $x = 100, y = 0$. How many units of the labor hours must be used to produce this number of units?

    (*i*) 400.                 (*iii*) 500.                 (*v*) None of the above.

    (*ii*) 200.               (*iv*) 5000.

(*o*) LP theory states that the optimal solution to any problem will lie at:

      (*i*) the origin.

      (*ii*) a corner point of the feasible region.

      (*iii*) the highest point of the feasible region.

      (*iv*) the lowest point in the feasible region.

      (*v*) none of the above.

(*p*) Dual of a linear programming problem with maximize objective function, all $\leq$ constraints and non-negative variables has minimize objective function, all $\geq$ constraints and non-negative decision variables.

    (*i*) True.                            (*ii*) False.

(*q*) The two objective functions (max $5x + 7y$, and min $-15x - 21y$) will produce the same solution to a linear programming problem.

    (*i*) True.                             (*ii*) False.

(*r*) In order for a linear programming problem to have a unique solution, the solution must exist

    (*i*) at the intersection of the non-negativity constraints.

    (*ii*) at the intersection of the objective function and a constraint.

    (*iii*) at the intersection of two or more constraints.

    (*iv*) none of the above.

(*s*) If a minimization problem has an objective function of $2x_1 + 5x_2$ , which of the following corner points is the optimal solution?

    (*i*) $(0,2)$.                 (*iii*) $(3,3)$.              (*v*) $(2,0)$.

    (*ii*) $(0,3)$.                 (*iv*) $(1,1)$.

(*t*) In a linear programming problem, when the objective function is parallel to one of the constraints, then

    (*i*) the solution is not optimal.

    (*ii*) multiple optimal solutions may exist.

    (*iii*) a single corner point solution exists.

    (*iv*) no feasible solution exists.

    (*v*) none of the above.

(*u*) A linear programming problem cannot have

    (*i*) no optimal solutions.

    (*ii*) exactly two optimal solution.

    (*iii*) as many optimal solutions as there are decision variables.

    (*iv*) an infinite number of optimal solutions.

    (*v*) none of the above.

**10.2**    A house wife wishes to mix two types of food $F_1$ and $F_2$ in such a way that the vitamin contents of the mixture contain at least 8 units of vitamin $A$ and 11 units of vitamin $B$. Food $F_1$ costs \$60/Kg and Food $F_2$ costs \$80/kg. Food $F_1$ contains 3 units/kg of vitamin $A$ and 5 units/kg of vitamin $B$ while Food $F_2$ contains 4 units/kg of vitamin $A$ and 2 units/kg of vitamin $B$. Formulate this problem as a linear programming problem to minimize the cost of the mixtures.

**10.3**    A baker has 30 oz of flour and 5 packages of yeast. Baking a loaf of bread requires 5 oz of flour and 1 package of yeast. Each loaf of bread can be sold for 30 cents. The baker may purchase additional flour at 4 cents/oz or sell leftover flour at the same price. Formulate an LP to help the baker maximize profits (revenues – costs).

**10.4**    An agriculturist has a farm with 126 acres. He produces Radish, Onion and Potato. Whatever the product is fully sold in the market, he gets USD 5 for radish per kg, USD 4 for Onion per kg and USD 5 for Potato per kg. The average yield is 1,500 kg of Radish per acre, 1800 kg of Onion per acre and 1200 kg of Potato per acre. To produce 100 kg of Radish, 100 kg of Onion, and 80 kg of Potato, a sum of USD 12.5 has to be used for water. The labour requirement for each acre to raise the crop is 6 man-days for Radish and Potato each and 5 man-days for Onion. A total of 500 man-days of labour at a rate of USD 40 per-man day are available. Formulate this as a linear programming model to maximize the agriculturist's total profit.

**10.5**    Use the graphical method to solve the following LP problem.

$$\min z = 15x_1 + 10x_2$$
$$\text{s.t.} \quad 0.25x_1 + \quad x_2 \le 65,$$
$$1.25x_1 + 0.5x_2 \le 90,$$
$$x_1 + \quad x_2 \le 85,$$
$$x_1, x_2 \ge 0.$$

**10.6**    A firm makes two products X and Y and has a total production capacity of 9 tons per day, X and Y requiring the same production capacity. The firm has a permanent contract to supply at least 2 tons of X and at least 3 tons of Y per day to another company. Each ton of X requires 20 machine hours production time and each ton of Y requires 50 machine hours production time. The daily maximum possible number of machine hours is 360. All the firm's output can be sold and the profit made is JD 80 per tons of X and JD 120 per tons of Y.

(*a*) Formulate this as a linear programming problem that can be used to maximize the total profit.

(*b*) Solve this optimization problem graphically.

**10.7**   A small paint company manufactures two types of paint, $P_1$ and $P_2$, from two raw materials, $M_1$ and $M_2$. The following table provides the basic data of the problem.

| Tons of raw material per ton of paints produced | | | |
|---|---|---|---|
| | $P_1$ | $P_2$ | Availability |
| $M_1$ | 6 | 4 | 24 |
| $M_2$ | 1 | 2 | 6 |
| Profit per ton (in $) | 500 | 400 | |

A market survey restricts the maximum daily demand of $P_2$ to 2 tons, and the daily demand for $P_1$ cannot exceed that of $P_2$ by more than 1.

(*a*)  Write an LP formulation for the problem.

(*b*)  Solve the LP model obtained in item (a) using the graphical method.

(*c*)  If the number of tons to be produced for $P_2$ is restricted to be integer-valued, the problem obtained in item (a) is called a mixed integer program. Sketch its feasible region and solve it graphically.

(*d*)  If the number of tons to be produced for $P_1$ and $P_2$ are both restricted to be integer-valued, the problem obtained in item (a) becomes a pure integer program. Sketch its feasible region and solve it graphically.

**10.8**   Use the graphical method to solve the following optimization problems.

(*a*) min   $5x + 7y$  
     s.t.  $x + 3y \geq 6$,  
         $5x + 2y \geq 10$,  
         $y \leq 4$,  
       $x, \quad y \geq 0$.

(*b*) max   $5x + 4y$  
     s.t.  $6x + 4y \leq 24$,  
         $x + 2y \leq 6$,  
         $-x + y \leq 1$,  
         $y = 2$,  
       $x, \quad y \geq 0$.

(*c*) max   $x_2$  
     s.t.  $-x_1 + x_2 \leq 1$,  
         $3x_1 + 2x_2 \leq 12$,  
         $2x_1 + 3x_2 \geq 12$,  
       $x_1, \quad x_2 \geq 0$,  
       $x_1, \quad x_2 \in \mathbb{Z}$.

**10.9**   Consider the following LP problem.

$$\max \quad z = 5x_1 + 4x_2$$
$$\text{s.t.} \quad 3x_1 + 2x_2 \leq 12,$$
$$x_1 + 2x_2 \leq 6,$$
$$-x_1 + x_2 \leq 1,$$
$$x_2 \leq 2,$$
$$x_1, \quad x_2 \geq 0.$$

Sketch the feasible region and solve it graphically for each of the following cases:

(a) The variable $x_2$ is restricted to be integer-valued; in this case the problem becomes a mixed integer program.

(b) The variables $x_1$ and $x_2$ are both restricted to be integer-valued; in this case the problem becomes a pure integer program.

**10.10**    Transform the following LP into the standard form.

$$\min z = 2x_1 - 4x_2 + 5x_3 - 30$$
$$\text{s.t.} \quad 3x_1 + 2x_2 - x_3 \geq 10,$$
$$-2x_1 \qquad + 4x_3 \leq 35,$$
$$4x_1 - x_2 \qquad \leq 20,$$
$$x_1 \leq 6, \quad x_2 \leq 8, \quad x_3 \leq 10.$$

**10.11**    Consider the following primal-dual pair of problems.

$$\min \quad 13x_1 + 10x_2 + 6x_3 \qquad\qquad \max \quad 8y_1 + 3y_2$$
$$\text{s.t.} \quad 5x_1 + x_2 + 3x_3 = 8, \qquad\qquad \text{s.t.} \quad 5y_1 + 3y_2 \leq 13,$$
$$3x_1 + x_2 = 3, \qquad\qquad\qquad y_1 + y_2 \leq 10,$$
$$x_1, x_2, x_3 \geq 0; \qquad\qquad\qquad\qquad 3y_1 \leq 6.$$

Show that $x^\star \triangleq (1, 0, 1)^\mathsf{T}$ and $y^\star \triangleq (2, 1)^\mathsf{T}$ are optimal in the primal and dual problems, respectively, and find the corresponding optimal values.

**10.12**    In Example 10.19, we give a pair of problems with the property that the primal and dual problems are both infeasible. Give an example of another pair with this property.

**10.13**    Prove item (b) in Corollary 10.2.

**10.14**    Find the feasible solution $(x_1, x_2)$ for the original LP problem in Example 10.16 given the feasible solution $(x_1, x_2^+, x_2^-, x_3) = (4, 0, 1/3, 2/3)$ to the same problem in the standard form.

**10.15**    Choose the correct answer for each of the following multiple-choice questions/items.

(a) A two variable linear programming problem cannot be solved by the simplex method.

(i) True.                    (ii) False.

(*b*) If, when we are using a simplex table to solve a maximization problem, we find that the ratios for determining the pivot row are all negative, then we know that the solution is:

(*i*) unbounded.     (*iii*) degenerate.     (*v*) none of the above.

(*ii*) infeasible     (*iv*) optimal.

(*c*) In converting a greater-than-or-equal constraint for use in a simplex table, we must add:

(*i*) an artificial variable.

(*ii*) a slack variable.

(*iii*) a slack and an artificial variable.

(*iv*) an excess and an artificial variable.

(*v*) a slack and an excess variable.

(*d*) For a minimization problem using a simplex table, we know we have reached the optimal solution when the row $R_0$

(*i*) has no numbers in it.

(*ii*) has no positive numbers in it.

(*iii*) has no negative numbers in it.

(*iv*) has no nonzero numbers in it.

(*v*) none of the above.

(*e*) A feasible solution requires that all artificial variables are

(*i*) greater than zero.

(*ii*) less than zero.

(*iii*) equal to zero.

(*iv*) there are no special requirements on artificial variables; they may take on any value.

(*v*) none of the above.

(*f*) If the right-hand side of a constraint is changed, the feasible region will not be affected and will remain the same.

(*i*) True.     (*ii*) False.

(*g*) With Bland's rule, the simplex algorithm solves feasible linear minimization problems without cycling when

  (*i*) we choose the rightmost nonbasic column with a negative cost to select the entering variable.

 (*ii*) we choose the rightmost nonbasic column with a negative cost to select the leaving variable.

(*iii*) we choose the leftmost nonbasic column with a negative cost to select the entering variable.

(*iv*) we choose the leftmost nonbasic column with a negative cost to select the leaving variable.

**10.16** Use the simplex tableau method to solve the following maximization problems.

(*a*)

$$\max \quad z = x_1 + 1.5x_2$$
$$\text{s.t.} \quad 2x_1 + 4x_2 \le 12,$$
$$3x_1 + 2x_2 \le 10,$$
$$x_1, x_2 \ge 0.$$

(*c*)

$$\max \quad z = 2x_1 - x_2 + x_3$$
$$\text{s.t.} \quad 3x_1 + x_2 + x_3 \le 6,$$
$$x_1 + x_2 + 2x_3 \le 1,$$
$$x_1 + x_2 - x_3 \le 2,$$
$$x_1, x_2, x_3 \ge 0.$$

(*b*)

$$\max \quad z = 3x_1 + 5x_2 + 4x_3$$
$$\text{s.t.} \quad 2x_1 + 3x_2 \le 8,$$
$$2x_2 + 5x_3 \le 10,$$
$$3x_1 + 2x_2 + 4x_3 \le 15,$$
$$x_1, x_2, x_3 \ge 0.$$

(*d*)

$$\max \quad z = 60x_1 + 30x_2 + 20x_3$$
$$\text{s.t.} \quad 8x_1 + 6x_2 + x_3 \le 48,$$
$$4x_1 + 2x_2 + 1.5x_3 \le 20,$$
$$2x_1 + 1.5x_2 + 0.5x_3 \le 8,$$
$$x_2 \le 5,$$
$$x_1, x_2, x_3 \ge 0.$$

**10.17** Consider the maximization problem presented by the following tableau. The parameters $a$ and $b$ are unknown.

| $x_1$ | $x_2$ | $s_1$ | $s_2$ | $s_3$ | rhs |
|---|---|---|---|---|---|
| 0 | 0 | 17 | $-3 + 2a$ | 0 | 10 |
| 1 | 0 | 3 | $-1$ | 0 | 2 |
| 0 | 1 | 4 | $a$ | 0 | 2 |
| 0 | 0 | 1 | $b$ | 1 | 6 |

For each of the following cases, explicitly discuss how many optimal solutions, if any, there are to the linear program. (If the LP is unbounded state that).

(*a*) $a = -2$ and $b = 0$.      (*b*) $a = 2$ and $b = -1$.      (*c*) $a = 3/2$ and $b = 1$.

**10.18**    Consider the following tableau of the simplex method for a maximization LP problem

| $z$ | $x_1$ | $x_2$ | $x_3$ | $x_4$ | $x_5$ | $x_6$ | rhs |
|---|---|---|---|---|---|---|---|
| 1 | 0 | 0 | 0 | $c_1$ | $c_2$ | $c_3$ | $z^\star$ |
| 0 | 0 | $-2$ | $a_3$ | $a_5$ | $-1$ | 0 | 0 |
| 0 | 1 | $a_1$ | 0 | $-3$ | 0 | $a_7$ | 2 |
| 0 | 0 | $a_2$ | $a_4$ | $-4$ | $a_6$ | $a_8$ | $b$ |

(*a*) There have to be three basic variables. Find them and give conditions on (all or some of) the unknowns $c_1, c_2, c_3, a_1, a_2, \ldots, a_8$ that make these variables basic.

(*b*) Give a condition on $b$ that makes the LP feasible and conditions on $c_1, c_2$ and $c_3$ that make the LP optimal.

(*c*) Do we have alternative optimal solutions? Justify your answer.

**10.19**    Consider the following optimization problem:

$$\max \quad z = 5x_1 - x_2$$
$$\text{s.t.} \qquad x_1 - 3x_2 \le 1,$$
$$x_1 - 4x_2 \le 3,$$
$$x_1, \quad x_2 \ge 0.$$

Use simplex algorithm to show that this LP is an unbounded LP problem.

**10.20**    Consider the following LP problem.

$$\min \quad z = 5x_1 + 3x_2 - 2x_3$$
$$\text{s.t.} \qquad x_1 + x_2 + x_3 \ge 4,$$
$$2x_1 + 3x_2 - x_3 \ge 9,$$
$$x_2 + 3x_3 \le 5,$$
$$x_1, \quad x_2, \quad x_3 \ge 0.$$

(*a*) Write down the corresponding dual LP problem.

(*b*) Suppose that the simplex method has been applied directly to the primal problem, and the resulting optimal tableau is:

| $z$ | $x_1$ | $x_2$ | $x_3$ | $e_1$ | $a_1$ | $e_2$ | $a_2$ | $s_3$ | rhs |
|---|---|---|---|---|---|---|---|---|---|
| 1 | $-2.5$ | 0 | 0 | 0 | $-M$ | $-1.25$ | $1.25 - M$ | $-0.75$ | 7.5 |
| 0 | $-0.5$ | 0 | 1 | 0 | 0 | 0.25 | $-0.25$ | 0.75 | 1.5 |
| 0 | 0.5 | 1 | 0 | 0 | 0 | $-0.25$ | 0.25 | 0.25 | 3.5 |
| 0 | $-1$ | 0 | 0 | 1 | $-1$ | 0 | 0 | 1 | 1 |

(*i*) Deduce the optimal solution to the primal problem and the optimal value.

(*ii*) Deduce the optimal solution to the corresponding dual problem.

**10.21** This exercise involves the implementation of the revised simplex method and the tableau simplex method in Octave/Matlab (or another program of your choice), and comparing the performances of your programs with that of standard optimization software. You should test these programs on a set of LP problems for which you need to generate the data randomly. You will also submit a well structured solution.

(*a*) Write an Octave function that solves a linear program (LP) in standard form using the revised simplex method. The function should take as input the constraint matrix $A$, the right hand-side vector $\mathbf{b}$, and the cost vector $\mathbf{c}$, and output an optimal solution vector $\mathbf{x}$ and the optimal cost, or indicate that the LP is unbounded or infeasible. It should also output the number of simplex pivots or iterations used.

The function should have the flexibility in terms of the rules used for choosing the entering and leaving variables. As far as choosing the entering variable is concerned, the function should provide the choice to implement the following options.

(*i*) After calculating all reduced costs, choose the variable with the smallest, i.e., most negative reduced cost to enter the basis. This should be the default option.

(*ii*) Calculate the reduced costs one at a time, and choose the variable that first gives a negative reduced cost to enter. In this option, you must not calculate all reduced costs.

For choosing the leaving variable, the function should provide the following option: Smallest index rule: From among the candidates, the variable $x_j$ with the smallest $j$ leaves. This should be the default option.

(*b*) Write a Octave function that solves an LP in standard form using the tableau simplex method. The function should take as input the constraint matrix $A$, the right hand-side vector $\mathbf{b}$, and the cost vector $\mathbf{c}$, and output an optimal solution vector $\mathbf{x}$ and the optimal cost, or indicate that the LP is unbounded or infeasible. It should also output the number of simplex pivots or iterations used.

The function should have the flexibility in terms of the rules used for choosing the entering and leaving variables. As far as choosing the entering variable is concerned, the function should provide the choice to implement the following options:

(*i*)  After calculating all reduced costs, choose the variable with the smallest, i.e., most negative reduced cost to enter the basis. This should be the default option.

(*ii*)  After calculating all reduced costs, choose the variable with the smallest index with a negative reduced cost to enter the basis.

For choosing the leaving variable, the function should provide the following options.

- Smallest index rule: From among the candidates, the variable $x_j$ with the smallest $j$ leaves. This should be the default option.

- Lexicographic rule: The leaving variable corresponds to the lexicographically smallest row, after scaling.

**10.22**    Use $(10.22)$ to show that $x^\mathsf{T}s + \tau\kappa = 0$.

# CHAPTER 11

# SECOND-ORDER CONE PROGRAMMING

## Contents

© 2022 by Baha Alzalg | Kindle Direct Publishing, Washington, United States 2022/9
B. Alzalg, *Combinatorial and Algorithmic Mathematics: From Foundation to Optimization*,
DOI 10.5281/zenodo.7111001

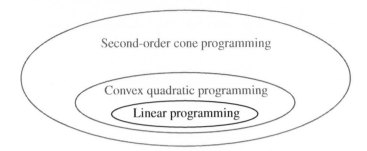

Figure 11.1: A Venn diagram of different classes of optimization problems.

In Chapter 10, we have studied linear programming. In modern convex optimization, the class of optimization that is an immediate enlargement of linear programming is not quadratic programming, but rather the so-called second-order cone programming. Second-order cone programming (SOCP for short) problems, which include linear programming problems and quadratic programming problems as special cases, are a class of convex optimization problems in which the variable is not a vector whose each of its components is required to be nonnegative, but rather a block vector whose each of its subvectors is required to reside in a second-order cone (see Definition 11.1).

Figure 11.1 shows graphical relationships among different classes of optimization problems. So in an SOCP problem, we optimize a linear function over the intersection of an affine linear manifold[1] with the Cartesian product of second-order cones. This chapter is devoted to studying SOCP problems. We also refer to Alizadeh and Goldfarb [AG03] for an excellent survey paper on this topic.

## 11.1. The second-order cone and its algebraic structure

This section aims to introduce algebraic tools needed to study the SOCP problems. We start this by introducing notations that will be used throughout this chapter. As in earlier chapters, we use "," for adjoining scalars, vectors and matrices in a row, and use ";" for adjoining them in a column. For example, a vector $x \in \mathbb{R}^n$ can be written as $x = (x_1; x_2; \ldots; x_n)$. For each vector $x \in \mathbb{R}^n$ whose first entry is indexed with 0, we write $\widetilde{x}$ for the subvector consisting of entries 1 through $n - 1$; therefore $x = (x_0; \widetilde{x}) \in \mathbb{R} \times \mathbb{R}^{n-1}$. We let $\mathbb{E}^n$ denote the $n$th-dimensional real Euclidean space $\mathbb{R} \times \mathbb{R}^{n-1}$ whose elements $x$ are indexed from 0.

---

[1] A manifold, in mathematics, can be viewed as a generalization and abstraction of the notion of a curved surface.

**Definition 11.1 (The second-order cone)** *The nth-dimensional second-order cone is defined as*

$$\mathbb{E}_+^n \triangleq \{x \in \mathbb{E}^n : x_0 \geq \|\widetilde{x}\|\},$$

*where $\|\cdot\|$ denotes the Euclidean norm. The interior of this cone is the set int $\mathbb{E}_+^n \triangleq \{x \in \mathbb{E}^n : x_0 > \|\widetilde{x}\|\}$.*

The graph to the right shows the 3rd-dimensional second-order cone $\mathbb{E}_+^3$. The cone $\mathbb{E}_+^n$ is closed, pointed (i.e., it does not contain a pair of opposite nonzero vectors) and convex with nonempty interior in $\mathbb{R}^n$. The cone $\mathbb{E}_+^n$ is one of the well-known examples of of the so-called symmetric cones (or self-scaled cones, see [SA03, NT98] for definitions).

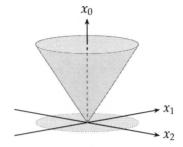

It is known that the space $\mathbb{E}^n$ under the bilinear map $\circ : \mathbb{E}^n \times \mathbb{E}^n \longrightarrow \mathbb{E}^n$ defined as

$$x \circ y \triangleq \begin{bmatrix} x^\mathsf{T} y \\ x_0 \bar{y} + y_0 \bar{x} \end{bmatrix}$$

forms a Euclidean Jordan algebra (see [Far94, SA03] for definitions) equipped with the standard inner product $\langle x, y \rangle \triangleq x^\mathsf{T} y$.

The spectral factorization of a given element is a factorization of this element into eigenvectors together with eigenvalues.

**Property 11.1 (Spectral decomposition in $\mathbb{E}^n$)** *Any $x \in \mathbb{E}^n$ can be expressed in exactly one way as a product:*

$$x = \underbrace{(x_0 + \|\widetilde{x}\|)}_{\lambda_1(x)} \underbrace{\left(\frac{1}{2}\right) \begin{bmatrix} 1 \\ \dfrac{\widetilde{x}}{\|\widetilde{x}\|} \end{bmatrix}}_{c_1(x)} + \underbrace{(x_0 - \|\widetilde{x}\|)}_{\lambda_2(x)} \underbrace{\begin{bmatrix} 1 \\ -\dfrac{\widetilde{x}}{\|\widetilde{x}\|} \end{bmatrix}}_{c_2(x)}. \tag{11.1}$$

The following definition is based on Property 11.1.

**Definition 11.2 (Eigenvalues and eigenvectors)** *The decomposition in (11.1) is called the spectral decomposition of $x \in \mathbb{E}^n$. The values $\lambda_{1,2}(x) \triangleq x_0 \pm \|\widetilde{x}\|$ are called the eigenvalues of $x$, and the vectors $c_{1,2}(x) \triangleq (1; \pm\widetilde{x}/\|\widetilde{x}\|)$ are called the eigenvectors of $x$.*

Having eigenvalues, we can define spectral notions such as trace and determinant.

> **Definition 11.3 (Trace and determinant)** *The trace and determinant of $x \in \mathbb{E}^n$ are defined as*
>
> $$trace(x) \triangleq \lambda_1(x) + \lambda_2(x) = 2x_0 \quad and \quad \det(x) \triangleq \lambda_1(x)\lambda_2(x) = x_0^2 - \|\widetilde{x}\|^2.$$

Note that $x \in \mathbb{E}_+^n$ ($x \in$ int $\mathbb{E}_+^n$) if and only if $\det(x) \geq 0$ ($\det(x) > 0$). Note also that $x \bullet y = \frac{1}{2}\text{trace}(x \circ y) = x^\mathsf{T} y$.

We call $e_n \triangleq (1; 0) \in \mathbb{E}^n$ the identity vector of $\mathbb{E}^n$. The logarithmic barrier function is defined on int $\mathbb{E}_+^n$ as $x \longmapsto -\ln \det(x)$. This map will play an important role for our subsequent development.

One can show that the eigenvectors $c_1(x)$ and $c_2(x)$ satisfy the properties:

$$c_1(x) \circ c_2(x) = 0,$$
$$c_1(x) \circ c_1(x) = c_1(x),$$
$$c_2(x) \circ c_2(x) = c_2(x),$$
$$c_1(x) + c_2(x) = e_n.$$

Any continuous function $f : \mathbb{R} \longrightarrow \mathbb{R}$ is defined on $\mathbb{E}^n$ as

$$f(x) \triangleq f(\lambda_1(x))c_1(x) + f(\lambda_2(x))c_2(x).$$

In particular, $x^n$ for $n \geq 2$, which is defined recursively as $x^n \triangleq x \circ x^{n-1}$, can be redefined as

$$x^n \triangleq \lambda_1^n(x)c_1(x) + \lambda_2^n(x)c_2(x).$$

Note that $x^{-1} \circ x = e_n$. The vector $x$ is called invertible if $x^{-1}$ is defined, and noninvertible otherwise. Note also that every positive definite element is invertible and its inverse is also positive definite.

The Frobenius norm of $x \in \mathbb{E}^n$ is defined as

$$\|x\|_F \triangleq \sqrt{\lambda_1^2(x) + \lambda_2^2(x)} = 2\|x\|.$$

It can be shown that, for any $x, y \in \mathbb{E}^n$, we have

$$\|x \circ y\|_F \leq \|x\| \, \|y\|_F \leq \|x\|_F \, \|y\|_F \quad \text{and} \quad \|x + y\|_F^2 = \|x\|_F^2 + \|y\|_F^2 + 4x^\mathsf{T} y. \quad (11.2)$$

Let $x \in \mathbb{E}^n$. The arrow-shaped matrix is defined so that $x \circ y = \text{Arw}(x)y$ for every $y \in \mathbb{E}^n$.

> **Definition 11.4 (Arrow-shaped matrix)** *Associated with each vector $x \in \mathbb{E}^n$, the arrow-shaped matrix of $x$ is defined as*
>
> $$\text{Arw}(x) \triangleq \begin{bmatrix} x_0 & \widetilde{x}^\mathsf{T} \\ \widetilde{x} & x_0 I \end{bmatrix}.$$

Note that $x \in \mathbb{E}_+^n$ ($x \in \text{int } \mathbb{E}_+^n$) if and only if $\text{Arw}(x)$ is positive semidefinite ($\text{Arw}(x)$ is positive definite).

---

**Definition 11.5 (Quadratic representation matrix)** *Associated with each vector $x \in \mathbb{E}^n$, the quadratic representation matrix of $x$ is defined as*

$$Q_x \triangleq 2\text{Arw}^2(x) - \text{Arw}(x^2) = \begin{bmatrix} \|x\|^2 & 2x_0\widetilde{x}^\mathsf{T} \\ 2x_0\widetilde{x} & \det(x)I + 2\widetilde{x}\widetilde{x}^\mathsf{T} \end{bmatrix}.$$

---

Note that $Q_x$ is also a linear operator in $\mathbb{E}^n$. Note also that $\text{Arw}(x)e = x$, $\text{Arw}(x)x = x^2$, $\text{Arw}(e) = Q_e = I$, $\text{trace}(e) = 2$ and $\det(e) = 1$ (since each of the two eigenvalues of $e$ equals one).

We say that two vectors $x, y \in \mathbb{E}^n$ are simultaneously decomposed if they share a Jordan frame, i.e., $x = \lambda_1(x)c_1 + \cdots + \lambda_r(x)c_r$ and $y = \lambda_1(y)c_1 + \cdots + \lambda_r(y)c_r$ for a Jordan frame $\{c_1, \ldots, c_r\}$ (hence, we have $c_i(x) = c_i(y)$ for each $i = 1, \ldots, r$).

We say that two vectors $x$ and $y$ operator commute if for all $z$, we have that $x \circ (y \circ z) = y \circ (x \circ z)$.

It is known that two elements of $\mathbb{E}^n$ are simultaneously decomposed if and only if they operator commute [SA03, Theorem 27].

Table 11.1 summarizes the notions associated with the algebra of the second-order cone.

All the above notions are also used in the block sense as follows: Let $x \triangleq (x_1; x_2; \ldots; x_r)$, $y \triangleq (y_1; y_2; \ldots; y_r)$, and $x_i, y_i \in \mathbb{E}^{n_i}$ for $i = 1, 2, \ldots, r$. Then

(a) $\mathbb{E}_r^n \triangleq \mathbb{E}^{n_1} \times \mathbb{E}^{n_2} \times \cdots \times \mathbb{E}^{n_r}$;   (b) $\mathbb{E}_{r+}^n \triangleq \mathbb{E}_+^{n_1} \times \mathbb{E}_+^{n_2} \times \cdots \times \mathbb{E}_+^{n_r}$;

(c) $\text{int } \mathbb{E}_{r+}^n \triangleq \text{int } \mathbb{E}_+^{n_1} \times \text{int } \mathbb{E}_+^{n_2} \times \cdots \times \text{int } \mathbb{E}_+^{n_r}$;

(d) $x \circ y \triangleq (x_1 \circ y_1; \ldots; x_r \circ y_r)$;

(e) $x^\mathsf{T} y \triangleq \sum_{i=1}^r x_i^\mathsf{T} y_i$;   (h) $\text{Arw}(x) \triangleq \bigoplus_{i=1}^r \text{Arw}(x_i)^2$;

(f) $\det(x) \triangleq \prod_{i=1}^r \det(x_i)$;   (i) $Q_x \triangleq \bigoplus_{i=1}^r Q_{x_i}$;

(g) $\text{trace}(x) \triangleq \sum_{i=1}^r \text{trace}(x_i)$;   (j) $\|x\|_F^2 \triangleq \sum_{i=1}^r \|x_i\|_F^2$;

(k) $f(x) \triangleq (f(x_1); f(x_2); \ldots; f(x_r))$. In particular, $x^{-1} \triangleq (x_1^{-1}; x_2^{-1}; \ldots; x_r^{-1})$;

(l) $e \triangleq (e_{n_1}; e_{n_2}; \ldots; e_{n_r})$ is the identity vector of $\mathbb{E}_r^n$;

(m) The logarithmic barrier of $x \in \text{int } \mathbb{E}_{r+}^n$ is defined as $x \longmapsto -\ln \det(x)$;

(n) $x$ and $y$ operator commute iff $x_i$ and $y_i$ operator commute for each $i = 1, 2, \ldots, r$.

---

[2]The direct sum of two square matrices $A$ and $B$ is the block diagonal matrix $A \oplus B \triangleq \begin{bmatrix} A & 0 \\ 0 & B \end{bmatrix}$.

| Algebraic notions | Algebra of second-order cone |
|---|---|
| Space | $\mathbb{E}^n \triangleq \left\{ x \triangleq \begin{pmatrix} x_0 \\ \bar{x} \end{pmatrix} : x \in \mathbb{R} \times \mathbb{R}^{n-1} \right\}$ |
| Second-order cone | $\mathbb{E}^n_+ \triangleq \{ x \in \mathbb{E}^n : x_0 \geq \|\bar{x}\| \}$ |
| Dimension | $\dim(\mathbb{E}^n_+) = n$ |
| Rank | $\operatorname{rank}(\mathbb{E}^n, \circ) = 2$ |
| Inner product | $x \bullet y \triangleq x_0 y_0 + \bar{x}^\mathsf{T} \bar{y} = x^\mathsf{T} y$ |
| Bilinear map | $x \circ y \triangleq \begin{bmatrix} x^\mathsf{T} y \\ x_0 \bar{y} + y_0 \bar{x} \end{bmatrix}$ |
| Algebra | $(\mathbb{E}^n, \circ)$ |
| Identity | $e_n \triangleq \begin{bmatrix} 1 \\ \mathbf{0} \end{bmatrix}$ |
| Reflection matrix | $R \triangleq \begin{bmatrix} 1 & \mathbf{0}^\mathsf{T} \\ \mathbf{0} & -I_{n-1} \end{bmatrix}$ |
| Eigenvalues | $\lambda_1(x) \triangleq x_0 + \|\bar{x}\|^2, \quad \lambda_2(x) \triangleq x_0 - \|\bar{x}\|^2$ |
| Eigenvectors | $c_1(x) \triangleq \frac{1}{2} \begin{bmatrix} 1 \\ \frac{\bar{x}}{\|\bar{x}\|} \end{bmatrix}, \quad c_2(x) \triangleq \frac{1}{2} \begin{bmatrix} 1 \\ \frac{-\bar{x}}{\|\bar{x}\|} \end{bmatrix}$ |
| Spectral decomposition | $x = \lambda_1(x) c_2(x) + \lambda_2(x) c_2(x)$ |
| Trace | $\operatorname{trace}(x) := 2x_0$ |
| Determinant | $\det(x) := x_0^2 - \|\bar{x}\|^2$ |
| Square | $x^2 \triangleq x \circ x = \begin{bmatrix} \|x\|^2 \\ 2 x_0 \bar{x} \end{bmatrix}$ |
| Inverse | $x^{-1} \triangleq \frac{1}{\lambda_1(x)} c_2(x) + \frac{1}{\lambda_2(x)} c_2(x) = \frac{1}{\det(x)} R x$ |
| Frobenius norm | $\|x\|_F \triangleq \sqrt{2} \sqrt{x \bullet x} = \sqrt{2} \|x\|$ |
| Arrow-shaped matrix | $\operatorname{Arw}(x) \triangleq \begin{bmatrix} x_0 & \bar{x}^\mathsf{T} \\ \bar{x} & x_0 I_{n-1} \end{bmatrix}$ |
| Quadratic representation | $Q_x \triangleq \begin{bmatrix} \|x\|^2 & 2x_0 \bar{x}^\mathsf{T} \\ 2x_0 \bar{x} & \det(x) I_{n-1} + 2\bar{x}\bar{x}^\mathsf{T} \end{bmatrix}$ |
| Logarithmic barrier | $-\ln \det(x)$ |
| Gradient of log barrier | $-x^{-1}$ |
| Hessian of log barrier | $-Q_{x^{-1}} = -\frac{1}{\det^2(x)} \begin{bmatrix} \|x\|^2 & -2x_0 \bar{x}^\mathsf{T} \\ -2x_0 \bar{x} & \det(x) I_{n-1} + 2\bar{x}\bar{x}^\mathsf{T} \end{bmatrix}$ |

Table 11.1: The algebraic notions and concepts associated with the second-order cone.

Note that $x$ has $2r$ eigenvalues (including multiplicities) comprised of the union of the eigenvalues of each $x_i$ for $i = 1, 2, \ldots, r$.

We end this section by introducing some notations that will be used throughout this chapter. We write $x \geq 0$ to mean that $x \in \mathbb{E}^n_+$ (i.e., $x$ is a positive semidefinite vector). We also write $x > 0$ to mean that $x \in \text{int } \mathbb{E}^n_+$ (i.e., $x$ is a positive definite vector), and write $x \geq y$ $(x > y)$ to mean that $x - y \geq 0$ $(x - y > 0)$. Note that $x \geq 0$ $(x > 0)$ iff $\lambda_i(x) \geq 0$ $(\lambda_i(x) > 0)$ for $i = 1, 2, \ldots, r$.

We use $\mathbb{S}^n$ to denote the set of symmetric matrices of order $n$, and $\mathbb{S}^{n+}$ to denote the set of symmetric positive semidefinite matrices of order $n$. The set $\mathbb{S}^{n+}$ is a convex cone and optimization problems over this cone will be studied in Chapter 12. Throughout this and next chapters, we write $X \geq 0$ to mean that $X \in \mathbb{S}^n_+$ (i.e., $x$ is a positive semidefinite matrix). We also write $X > 0$ to mean that $X \in \text{int } \mathbb{S}^n_+$ (i.e., $X$ is a positive definite matrix), and write $X \geq Y$ $(X > Y)$ to mean that $X - Y \geq 0$ $(X - Y > 0)$.

## 11.2. Second-order cone programming formulation

In this section, we introduce the SOCP problem and formulate known class of optimization problems as SOCPs.

### Problem formulation

Let $r \geq 1$ be an integer. For each $i = 1, 2, \ldots, r$, let $m, n, n_i$ be positive integers such that $n = \sum_{i=1}^r n_i$. Let also $x, c$ and $z$ be vectors in $\mathbb{R}^n$, $y$ and $b$ be vectors in $\mathbb{R}^m$, and $A$ be a matrix in $\mathbb{R}^{m \times n}$ such that they are all conformally partitioned as

$$
\begin{aligned}
x &\triangleq (x_1; x_2; \ldots; x_r), \\
s &\triangleq (s_1; s_2; \ldots; s_r), \\
c &\triangleq (c_1; c_2; \ldots; c_r), \\
A &\triangleq (A_1, A_2, \ldots, A_r),
\end{aligned}
$$

where $x_i, s_i, c_i \in \mathbb{E}^{n_i}$ and $A_i \in \mathbb{R}^{m_i \times n_i}$ for $i = 1, 2, \ldots, r$. The SOCP problem and its dual in multi-block structures are defined as

$$
\begin{array}{llll}
& \min & c_1^\mathsf{T} x_1 + \cdots + c_r^\mathsf{T} x_r & \\
(\text{P}|\text{SOCP}_i) & \text{s.t.} & A_1 x_1 + \cdots + A_r x_r = b, \\
& & x_i \geq 0, \ i = 1, \ldots, r;
\end{array}
\qquad
\begin{array}{lll}
\max & b^\mathsf{T} y & \\
(\text{D}|\text{SOCP}_i) \quad \text{s.t.} & A_i^\mathsf{T} y + s_i = c_i, \ i = 1, \ldots, r, \\
& s_i \geq 0, \ i = 1, \ldots, r.
\end{array}
$$

The pair $(\text{P}|\text{SOCP}_i, \text{D}|\text{SOCP}_i)$ can be compactly rewritten as

$$
\begin{array}{lll}
& \min & c^\mathsf{T} x \\
(\text{P}|\text{SOCP}) & \text{s.t.} & Ax = b, \\
& & x \geq 0;
\end{array}
\qquad
\begin{array}{lll}
\max & b^\mathsf{T} y \\
(\text{D}|\text{SOCP}) \quad \text{s.t.} & A^\mathsf{T} y + s = c, \\
& s \geq 0.
\end{array}
$$

## Formulating problems as SOCPs

In this part, we formulate three general classes of optimization problems as SOCPs. We start with linear optimization.

***Linear programming***   The linear optimization in the standard form is the problem

$$
\begin{aligned}
\min \quad & c_1 x_1 + \cdots + c_r x_r \\
\text{s.t.} \quad & x_1 \boldsymbol{a}_1 + \cdots + x_r \boldsymbol{a}_r = \boldsymbol{b}, \\
& x_i \geq 0, \ i = 1, \ldots, r.
\end{aligned}
\tag{11.3}
$$

Clearly, the linear optimization problem (11.3) is Problem (P|SOCP$_i$) with $n_1 = n_2 = \cdots = n_r = 1$. In other words, when the vector $\boldsymbol{x} \in \mathbb{R}^r$ resides the following Cartesian product of second-order cones:

$$
\underbrace{\mathbb{E}_+^1 \times \mathbb{E}_+^1 \times \cdots \times \mathbb{E}_+^1}_{r-\text{times}},
$$

we have $\boldsymbol{x} \geq \boldsymbol{0}$ (as, by definition, $\mathbb{E}_+^1 \triangleq \{t \in \mathbb{R} : t \geq 0\}$), and hence the SOCP problem reduces to a linear programming problem.

***Convex quadratic programming***   In convex quadratic optimization problems, we minimize a strictly convex quadratic function subject to affine constraint functions:

$$
\begin{aligned}
\min \quad & q(\boldsymbol{x}) \triangleq \boldsymbol{x}^\mathsf{T} Q \boldsymbol{x} + \boldsymbol{c}^\mathsf{T} \boldsymbol{x} \\
\text{s.t.} \quad & A\boldsymbol{x} = \boldsymbol{b}, \\
& \boldsymbol{x} \geq \boldsymbol{0}.
\end{aligned}
\tag{11.4}
$$

Since Problem (12.6) is strictly convex, the matrix $Q$ must be a symmetric positive definite matrix (i.e., $Q = Q^\mathsf{T}$ and $Q > O$). It follows that there exists another positive definite matrix (hence nonsingular), say $Q^{\frac{1}{2}}$, whose square is $Q$. Note that

$$
q(\boldsymbol{x}) = \boldsymbol{x}^\mathsf{T} Q \boldsymbol{x} + \boldsymbol{c}^\mathsf{T} \boldsymbol{x} = \left\| Q^{\frac{1}{2}} \left( \boldsymbol{x} + \frac{1}{2} Q^{-1} \boldsymbol{c} \right) \right\|^2 - \frac{1}{4} \boldsymbol{c}^\mathsf{T} Q^{-1} \boldsymbol{c} = \left\| \bar{\boldsymbol{y}} \right\|^2 - \frac{1}{4} \boldsymbol{c}^\mathsf{T} Q^{-1} \boldsymbol{c},
$$

where $\bar{\boldsymbol{y}} \triangleq Q^{\frac{1}{2}} \left( \boldsymbol{x} + \frac{1}{2} Q^{-1} \boldsymbol{c} \right)$. Therefore, the quadratic optimization problem (12.6) can be formulated as the SOCP problem

$$
\begin{aligned}
\min \quad & y_0 \\
\text{s.t.} \quad & A\boldsymbol{x} = \boldsymbol{b}, \\
& \bar{\boldsymbol{y}} - \boldsymbol{x} = \tfrac{1}{2} Q^{-1} \boldsymbol{c}, \\
& \boldsymbol{y} \geq \boldsymbol{0}, \ \boldsymbol{x} \geq \boldsymbol{0},
\end{aligned}
\tag{11.5}
$$

where the underlying cone is the $(n + 1)$st-dimensional second-order cone $\mathbb{E}^n$. Note that Problems (12.6) and (11.5) have the same minimizer but their optimal objective values are equal up to the constant $\frac{1}{4}c^T Q^{-1} c$. We also point out that, more generally, convex quadratically constrained quadratic optimization problems can also be formulated as SOCP problems (see [AG03]).

***Rotated quadratic cone programming*** Let $n$ be a positive integer and $M$ be a nonsingular matrix of order $n - 2$. The $n$th dimensional rotated quadratic cone is defined as

$$\mathcal{K}^n \triangleq \left\{ x = (x_0; x_1; \hat{x}) \in \mathbb{R} \times \mathbb{R} \times \mathbb{R}^{n-2} : 2x_0 x_1 \geq \|\hat{x}\|^2, \ x_0 \geq 0, x_1 \geq 0 \right\}.$$

One can see that the rotated quadratic cone is obtained by rotating the second-order cone through an angle of $\pi/6$ in the $x_0 x_1$-plane. We call the constraint on $x$ that satisfies the inequality $2x_0 x_1 \geq \|\hat{x}\|^2$ the hyperbolic constraint.

In rotated quadratic cone optimization problems, a linear objective function is minimized subject to linear constraints and hyperbolic constraints. In fact, a rotated quadratic cone optimization problem can be expressed as an SOCP problem because the hyperbolic constraint is equivalent to a second-order cone constraint, and this can be easily deduced by observing that

$$
\begin{aligned}
x = (x_0; x_1; \hat{x}) \in \mathcal{K}^n \quad &\Longleftrightarrow \quad 2x_0 x_1 \geq \|\hat{x}\|^2 \\
&\Longleftrightarrow \quad 4x_0 x_1 \geq -4x_0 x_1 + 4\|\hat{x}\|^2 \\
&\Longleftrightarrow \quad (2x_0 + x_1)^2 \geq \|(2x_0 - x_1; 2\hat{x})\|^2 \qquad (11.6) \\
&\Longleftrightarrow \quad 2x_0 + x_1 \geq \|(2x_0 - x_1; 2\hat{x})\| \\
&\Longleftrightarrow \quad (2x_0 + x_1; 2x_0 - x_1; 2\hat{x}) \in \mathbb{E}^n_+.
\end{aligned}
$$

The next section contains an application with hyperbolic constraints.

## 11.3. Applications in engineering and finance

In this section, we describe three applications of SOCP in geometry and finance. Namely, Euclidean facility location problem, the portfolio optimization problem with loss risk constraints, and the optimal covering random ellipsoid problem. The material of this section has appeared in [LVBL98] (see also [Alz12]). For more applications of SOCP, we refer the reader to [LVBL98, AG03, BUS13, AA22].

### Euclidean facility location problem

In facility location problems (FLPs) we are interested in choosing a location to build a new facility or locations to build multiple new facilities so that an appropriate measure of distance from the new facilities to existing facilities is minimized. FLPs arise locating airports, regional campuses, wireless communication towers, etc.

The following are two ways of classifying FLPs: We can classify FLPs based on the number of new facilities in the following sense: If we add only one new facility then we get a problem known as a single FLP, while if we add multiple new facilities instead of adding only one, then we get more a general problem known as a multiple FLP. Another way of classification is based on the distance measure used in the model between the facilities: If we use the Euclidean distance then these problems are called Euclidean FLPs, and if we use the rectilinear distance then these problems are called rectilinear FLPs.

In single Euclidean FLP, we are given $r$ existing facilities represented by the fixed points $a_1, a_2, \ldots, a_r$ in $\mathbb{R}^n$, and we plan to place a new facility represented by $x$ so that we minimize the weighted sum of the distances between $x$ and each of the points $a_1, a_2, \cdots, a_r$. This leads us to the problem

$$\min \quad \sum_{i=1}^{r} w_i \, \|x - a_i\|$$

or, alternatively, to the problem

$$\begin{aligned} \min \quad & \sum_{i=1}^{r} w_i \, t_i \\ \text{s.t.} \quad & (t_i; x - a_i) \geq 0, \ i = 1, 2, \ldots, r, \end{aligned}$$

where $w_i$ is the weight associated with the $i$th existing facility and the new facility for $i = 1, 2, \ldots, r$. The resulting model is an SOCP model.

Assume that we need to add $m$ new facilities, namely $x_1, x_2, \ldots, x_m \in \mathbb{R}^n$, instead of adding only one. We have two cases depending on whether or not there is an interaction among the new facilities in the underlying model. If there is no interaction between the new facilities, we are just concerned in minimizing the weighted sums of the distance between each one of the new facilities on one hand and each one of the existing facilities. In other words, we solve the following SOCP model:

$$\begin{aligned} \min \quad & \sum_{j=1}^{m} \sum_{i=1}^{r} w_{ij} \, t_{ij} \\ \text{s.t.} \quad & (t_{ij}; x_j - a_i) \geq 0, \ i = 1, 2, \ldots, r, \ j = 1, 2, \ldots, m, \end{aligned}$$

where $w_{ij}$ is the weight associated with the $i$th existing facility and the $j$th new facility for $j = 1, 2, \ldots, m$ and $i = 1, 2, \ldots, r$.

If interaction exists among the new facilities, then, in addition to the above requirements, we need to minimize the sum of the Euclidean distances between each pair of the new facilities. In this case, we are interested in a model of the form:

$$\begin{aligned} \min \quad & \sum_{j=1}^{m} \sum_{i=1}^{r} w_{ij} t_{ij} + \sum_{j=1}^{m} \sum_{j'=1}^{j-1} \hat{w}_{jj'} \, \hat{t}_{jj'} \\ \text{s.t.} \quad & (t_{ij}; x_j - a_i) \geq 0, \ i = 1, 2, \ldots, r, \ j = 1, 2, \ldots, m, \\ & (\hat{t}_{js}; x_j - x_s) \geq 0, \ j = 1, 2, \ldots, m-1, \ s = j+1, j+2, \ldots, m, \end{aligned}$$

where $\hat{w}_{jj'}$ is the weight associated with the new facilities $j'$ and $j$ for $j' = 1, 2, \ldots, j-1$ and $j = 1, 2, \ldots, m$.

## Portfolio optimization with loss risk constraints

We consider the problem of maximizing the expected return subject to loss risk constraints. This application is a well-known problem from portfolio optimization.

Consider a portfolio problem with $n$ assets or stocks over a period of time. We start by letting $x_i$ denote the amount of asset $i$ held at the beginning of (and throughout) the given period, and $p_i$ will denote the price change of asset $i$ over this period. So, the vector $p \in \mathbb{R}^n$ is the price vector over the period. For simplicity, we let $p$ be Gaussian with known mean $\bar{p}$ and covariance $\Sigma$, so the return over this period is the (scalar) Gaussian random variable $r = p^\mathsf{T} x$ with mean $\bar{r} = \bar{p}^\mathsf{T} x$ and variance $\sigma = x^\mathsf{T}\Sigma x$ where $x = (x_1; x_2; \ldots; x_n)$. As pointed out in [LVBL98], the choice of portfolio variable $x$ involves the (classical, Markowitz) tradeoff between random return mean and random variance.

The optimization variables are the portfolio vectors $x \in \mathbb{R}^n$. For this portfolio vector, we take the simplest assumption $x \geq 0$ (i.e., no short position [LVBL98]) and $\sum_{i=1}^n x_i = 1$ (i.e., unit total budget [LVBL98]).

Let $\alpha$ be a given unwanted return level over the given period, and let $\beta$ be a given maximum probability over the period. Assuming the above data is given, our goal is to determine the amount of the asset $i$ (which is $x_i$), i.e., determine $x$, such that the expected return over the period is maximized subject to the following loss risk constraint: The constraint $P(r \leq \alpha) \leq \beta$ must be satisfied over the given period.

As noted in [LVBL98], the constraint $P(r \leq \alpha) \leq \beta$ is equivalent to the second-order cone constraint

$$\left(\alpha - \bar{r}; \ \Phi^{-1}(\beta)(\Sigma^{\frac{1}{2}}x)\right) \geq 0,$$

provided $\beta \leq 1/2$ (i.e.,$\Phi^{-1}(\beta) \leq 0$), where

$$\Phi(z) = \frac{1}{\sqrt{2\pi}} \int_{-\infty}^{z} e^{-t^2/2}dt$$

is the cumulative normal distribution function of a zero mean unit variance Gaussian random variable. To prove this (see also [LVBL98]), notice that the constraint $P(r \leq \alpha) \leq \beta$ can be written as

$$P\left(\frac{r - \bar{r}}{\sqrt{\sigma}} \leq \frac{\alpha - \bar{r}}{\sqrt{\sigma}}\right) \leq \beta.$$

Since $(r - \bar{r})/\sqrt{\sigma}$ is a zero mean unit variance Gaussian random variable, the probability above is simply $\Phi((\alpha - \bar{r})/\sqrt{\sigma})$, thus the constraint $P(r \leq \alpha) \leq \beta$ can be expressed as $\Phi((\alpha - \bar{r})/\sqrt{\sigma}) \leq \beta$ or $((\alpha - \bar{r})/\sqrt{\sigma}) \leq \Phi^{-1}(\beta)$, or equivalently $\bar{r} + \Phi^{-1}(\beta)\sqrt{\sigma} \geq \alpha$. Since

$$\sqrt{\sigma} = \sqrt{x^\mathsf{T}\Sigma x} = \sqrt{(\Sigma^{1/2}x)^\mathsf{T}(\Sigma^{1/2}x)} = \|\Sigma^{1/2}x\|,$$

the constraint $P(r \leq \alpha) \leq \beta$ is equivalent to the second-order cone constraint $\bar{r} + \Phi^{-1}(\beta)\|\Sigma^{1/2} x\| \geq \alpha$ or equivalently

$$\left(\alpha - \bar{r}; \; \Phi^{-1}(\beta)(\Sigma^{\frac{1}{2}}x)\right) \geq 0.$$

Our goal is to determine the amount of the asset $i$ (which is $x_i$ over the given period, i.e., determine $x$, such that the expected return over this period is maximized. This problem can be cast as an SOCP as follows:

$$\begin{aligned} \max \quad & \bar{p}^\mathsf{T} x \\ \text{s.t.} \quad & \left(\alpha - \bar{p}^\mathsf{T}x; \Phi^{-1}(\beta)(\Sigma^{1/2}x)\right) \geq 0, \\ & \mathbf{1}^\mathsf{T}x = 1, x \geq 0. \end{aligned}$$

The simple problem described above has many extensions (see [LVBL98]). One of these extensions is imposing several loss risk constraints, i.e., the constraints $P(r \leq \alpha_i) \leq \beta_i$, $i = 1, 2, \ldots, k$ (where $\beta_i \leq 1/2$, for $i = 1, 2, \ldots, k$), or equivalently

$$\left(\alpha_i - \bar{r}; \Phi^{-1}(\beta_i)(\Sigma^{1/2}x)\right) \geq 0, \text{ for } i = 1, 2, \ldots, k,$$

to be satisfied over the period. So our problem becomes

$$\begin{aligned} \max \quad & \bar{p}^\mathsf{T} x \\ \text{s.t.} \quad & \left(\alpha_i - \bar{r}; \Phi^{-1}(\beta_i)(\Sigma^{1/2}x)\right) \geq 0, \; i = 1, 2, \ldots, k, \\ & \mathbf{1}^\mathsf{T}x = 1, x \geq 0. \end{aligned}$$

### Optimal covering ellipsoid problem

Suppose we have $k$ existing ellipsoids:

$$\mathcal{E}_i \triangleq \{x \in \mathbb{R}^n : x^\mathsf{T}H_ix + 2g_i^\mathsf{T}x + v_i \leq 0\} \subset \mathbb{R}^n, i = 1, 2, \ldots, k,$$

where $H_i \in \mathbb{S}_+^n$, $g_i \in \mathbb{R}^n$ and $v_i \in \mathbb{R}$ are given data for $i = 1, 2, \ldots, k$. We need to determine a ball that contains all $k$ existing ellipsoids.

We consider the same assumptions as in [Alz12]. We assume that the cost of choosing the ball has two components:

- The cost of the center, which is proportional to the Euclidean distance to the center from the origin;

- The cost of the radius, which is proportional to the square of the radius.

The center and the radius of the ball are determined so that the total cost is minimized. In [Alz12], the author describes a concrete stochastic version of this generic application. All given data are deterministic in our description. In this concrete version. It is assumed that $n \triangleq 2$. The existing fixed ellipsoids contain targets that need to be destroyed. Fighter aircrafts take off from the origin with a planned disk of coverage that contains the fixed ellipsoids. The disk of coverage needs to be determined so that the total cost is minimized.

Our goal is to determine $\bar{x} \in \mathbb{R}^n$ and $\gamma \in \mathbb{R}$ such that the ball $\mathcal{B}$ defined by

$$\mathcal{B} \triangleq \{x \in \mathbb{R}^n : x^\mathsf{T}x - 2\bar{x}^\mathsf{T}x + \gamma \leq 0\}$$

contains the existing ellipsoids $\mathcal{E}_i$ for $i = 1, 2, \ldots, k$.

Notice that the center of the ball $\mathcal{B}$ is $\bar{x}$, its radius is $r \triangleq \sqrt{\bar{x}^\mathsf{T}\bar{x} - \gamma}$, and the distance from the origin to its center is $\sqrt{\bar{x}^\mathsf{T}\bar{x}}$.

We introduce the constraints $d_1^2 \geq \bar{x}^\mathsf{T}\bar{x}$ and $d_2 \geq r^2 = \bar{x}^\mathsf{T}\bar{x} - \gamma$. That is, $d_1$ is an upper bound on the distance between the center of the ball $\mathcal{B}$ and the origin, $\sqrt{\bar{x}^\mathsf{T}\bar{x}}$, and $d_2$ is an upper bound on square of the radius of the ball $\mathcal{B}$.

In order to proceed, we need the following lemma which is due to Sun and Freund [SF04].

---

**Lemma 11.1** *Suppose that we are given two ellipsoids $\mathcal{E}_i \subset \mathbb{R}^n$, $i = 1, 2$, defined by $\mathcal{E}_i \triangleq \{x \in \mathbb{R}^n : x^\mathsf{T}H_ix + 2g_i^\mathsf{T}x + v_i \leq 0\}$, where $H_i \in \mathbb{S}_+^n$, $g_i \in \mathbb{R}^n$ and $v_i \in \mathbb{R}$ for $i = 1, 2$, then $\mathcal{E}_1$ contains $\mathcal{E}_2$ if and only if there exists $\tau \geq 0$ such that the linear matrix inequality*

$$\begin{bmatrix} H_1 & g_1 \\ g_1^\mathsf{T} & v_1 \end{bmatrix} \leq \tau \begin{bmatrix} H_2 & g_2 \\ g_2^\mathsf{T} & v_2 \end{bmatrix}$$

*holds.*

---

In view of Lemma 11.1 and the requirement that the ball $\mathcal{B}$ contains the existing ellipsoids $\mathcal{E}_i$ for $i = 1, 2, \ldots, k$, we accordingly add the following constraints:

$$\begin{bmatrix} I & -\bar{x} \\ -\bar{x}^\mathsf{T} & \gamma \end{bmatrix} \leq \tau_i \begin{bmatrix} H_i & g_i \\ g_i^\mathsf{T} & v_i \end{bmatrix}, \quad i = 1, 2, \ldots, k,$$

or equivalently

$$M_i \geq 0, \quad \forall i = 1, \ldots, k, \quad \text{where} \quad M_i \triangleq \begin{bmatrix} \tau_i H_i - I & \tau_i g_i + \bar{x} \\ \tau_i g_i^\mathsf{T} + \bar{x}^\mathsf{T} & \tau_i v_i - \gamma \end{bmatrix},$$

for each $i = 1, \ldots, k$.

Since we are looking to minimizing $d_2$, where $d_2$ is an upper bound on square of the radius of the ball $\mathcal{B}$, we can write the constraint $d_2 \geq \bar{x}^\mathsf{T}\bar{x} - \gamma$ as $d_2 = \bar{x}^\mathsf{T}\bar{x} - \gamma$. So, the matrix $M_i$ can be then written as

$$M_i = \begin{bmatrix} \tau_i H_i - I & \tau_i g_i + \bar{x} \\ \tau_i g_i^\mathsf{T} + \bar{x}^\mathsf{T} & \tau_i v_i + d_2 - \bar{x}^\mathsf{T}\bar{x} \end{bmatrix}.$$

Now, let $H_i \triangleq \Xi_i \Lambda_i \Xi_i^\mathsf{T}$ be the spectral decomposition of $H_i$, where $\Lambda_i \triangleq \mathrm{Diag}(\lambda_{i1}; \dots ; \lambda_{ik})$, and let $\mathbf{u}_i \triangleq \Xi_i^\mathsf{T}(\tau_i \mathbf{g}_i + \bar{\mathbf{x}})$. Then, for each $i = 1, \dots, k$, we have

$$\bar{M}_i \triangleq \begin{bmatrix} \Xi_i^\mathsf{T} & \mathbf{0} \\ \mathbf{0}^\mathsf{T} & 1 \end{bmatrix} M_i \begin{bmatrix} \Xi_i & \mathbf{0} \\ \mathbf{0}^\mathsf{T} & 1 \end{bmatrix} = \begin{bmatrix} \tau_i \Lambda_i - I & \mathbf{u}_i \\ \mathbf{u}_i^\mathsf{T} & \tau_i v_i + d_2 - \bar{\mathbf{x}}^\mathsf{T} \bar{\mathbf{x}} \end{bmatrix}.$$

Consequently, $M_i \succeq 0$ if and only if $\bar{M}_i \succeq 0$ for each $i = 1, 2, \dots, k$. Now our formulation of the problem in SOCP depends on the following lemma (see also [LVBL98]).

---

**Lemma 11.2** *For each $i = 1, 2, \dots, k$, $\bar{M}_i \succeq 0$ if and only if $\tau_i \lambda_{\min}(H_i) \geq 1$ and $\bar{\mathbf{x}}^\mathsf{T} \bar{\mathbf{x}} \leq d_2 + \tau_i v_i + \mathbf{1}^\mathsf{T} \mathbf{s}_i$, where $\mathbf{s}_i = (s_{i1}; s_{i2}; \dots ; s_{in})$, $s_{ij} = u_{ij}^2 / (\tau_i \lambda_{ij} - 1)$ for all $j$ such that $\tau_i \lambda_{ij} > 1$, and $s_{ij} = 0$ otherwise.*

---

The proof of Lemma 11.2 is left as an exercise for the reader (see Exercise 11.1).

Now, since we are minimizing $d_2$ and $\tilde{d}_2$, then for all $j = 1, 2, \dots, n$, we can relax the definitions of $s_{ij}$ and $\tilde{s}_{ij}$ by replacing them by $u_{ij}^2 \leq s_{ij}(\tau_i \lambda_{ij} - 1)$ for all $i = 1, 2, \dots, k$.

When the realizations of the random ellipsoids become available, if necessary, we determine $\tilde{\lambda}$ so that the new ball $\tilde{\mathcal{B}} \triangleq \{\mathbf{x} \in \mathbb{R}^n : \mathbf{x}^\mathsf{T} \mathbf{x} - 2\bar{\mathbf{x}}^\mathsf{T} \mathbf{x} + \tilde{\lambda} \leq 0\}$ contains all the realizations of the random ellipsoid. This new ball $\tilde{\mathcal{B}}$ has the same center $\bar{\mathbf{x}}$ as $\mathcal{B}$ but a larger radius, $\tilde{r} \triangleq \sqrt{\bar{\mathbf{x}}^\mathsf{T} \bar{\mathbf{x}}}$. We note that $\tilde{r}^2 - r^2 = (\bar{\mathbf{x}}^\mathsf{T} \bar{\mathbf{x}} - \tilde{\gamma}) - (\bar{\mathbf{x}}^\mathsf{T} \bar{\mathbf{x}} - \gamma) = \gamma - \tilde{\gamma}$, and thus we introduce the constraint $0 \leq \gamma - \tilde{\gamma} \leq z$ where $z$ is an upper bound of $\tilde{r}^2 - r^2$. Let $\bar{c} > 0$ denote the cost per unit of the Euclidean distance between the center of the ball $\mathcal{B}$ and the origin; let $\alpha > 0$ be the cost per unit of the square of the radius of $\mathcal{B}$; and let $\beta > 0$ be the cost per unit increase of the square of the radius if it becomes necessary after the realizations of the random ellipsoids are available.

We now define the decision variable $\mathbf{x} \triangleq (d_1; d_2; \bar{\mathbf{x}}; \gamma; \tau)$. Then, by introducing the unit cost vector $\mathbf{c} \triangleq (\bar{c}; \alpha; \mathbf{0}; 0; \mathbf{0})$, and combining all of the above, we get the SOCP model:

$$
\begin{aligned}
\min \quad & \mathbf{c}^\mathsf{T} \mathbf{x} + \\
\text{s.t.} \quad & \mathbf{u}_i = \Xi_i^\mathsf{T}(\tau_i \mathbf{g}_i + \bar{\mathbf{x}}), i = 1, 2, \dots, k, \\
& u_{ij}^2 \leq s_{ij}(\tau_i \lambda_{ij} - 1), i = 1, 2, \dots, k, \; j = 1, 2, \dots, n, \\
& \bar{\mathbf{x}}^\mathsf{T} \bar{\mathbf{x}} \leq \sigma, \\
& \sigma \leq d_2 + \tau_i v_i - \mathbf{1}^\mathsf{T} \mathbf{s}_i, i = 1, 2, \dots, k, \\
& \tau_i \geq 1/\lambda_{\min}(H_i), i = 1, 2, \dots, k, \\
& \bar{\mathbf{x}}^\mathsf{T} \bar{\mathbf{x}} \leq d_1^2,
\end{aligned}
$$

which includes only linear and hyperbolic constraints.

# 11.4. Duality in second-order cone programming

The duality in SOCP requires the application of the Karush-Kuhn-Tucker theorem which we state in Theorem 11.1. We first need the following definition.

---

**Definition 11.6** *The Lagrangian function of the general minimization problem*

$$
\begin{aligned}
min \quad & f(x) \\
s.t. \quad & g_i(x) = 0, \ i = 1, 2, \ldots, p, \\
& h_j(x) \leq 0, \ j = 1, 2, \ldots, q,
\end{aligned}
\tag{11.7}
$$

*is defined as*

$$
\mathcal{L}(x, u, v) \triangleq f(x) + \sum_{i=1}^{p} u_i g_i(x) + \sum_{j=1}^{q} v_j h_j(x),
$$

*for $x \in \mathbb{R}^n, u \in \mathbb{R}^p_+$ and $v \in \mathbb{R}^q$.*

---

The Karush-Kuhn-Tucker conditions[3] (KKT conditions for short) are also called first-order necessary conditions for a point to be a local minimizer.

---

**Theorem 11.1 (KKT Theorem)** *Let $x$ be a local minimizer for Problem (11.7), then there exist $u \in \mathbb{R}^p$ and $v \in \mathbb{R}^q$ such that all the following KKT conditions are satisfied.*

$$
\begin{aligned}
& \nabla_x \mathcal{L}(x, u, v) = 0, && \textit{(stationary)}; \\
& u_i g_i(x) = 0, \ \forall i, && \textit{(complementary slackness)}; \\
& g_i(x) = 0, h_j(x) \leq 0, \ \forall i, \ \forall j, && \textit{(primal feasibility)}; \\
& u_i \geq 0, \ \forall i, && \textit{(dual feasibility)}.
\end{aligned}
$$

---

The following example is from [CZ13].

**Example 11.1** Consider the following problem.

$$
\begin{aligned}
min \quad & x_1 x_2 \\
s.t. \quad & x_1 + x_2 \geq 2, \\
& -x_1 + x_2 \geq 0.
\end{aligned}
$$

---

[3]The KKT conditions were derived independently by William Karush in 1939 and by Harold Kuhn and Albert Tucker in 1951.

The Lagrangian function is

$$
\begin{aligned}
\mathcal{L}(x, u, v) &= x_1 x_2 + v_1(2 - x_1 - x_2) + v_2(x_1 - x_2) \\
&= x_1 x_2 - (v_1 - v_2)x_1 - (v_1 + v_2)x_2 + 2v_2,
\end{aligned}
$$

for $v_1, v_2 \in \mathbb{R}$. It follows that

$$
\nabla_x \mathcal{L}(x, u, v) = \begin{bmatrix} x_2 - v_1 + v_2 \\ x_1 - v_1 - v_2 \end{bmatrix}.
$$

The KKT conditions are

$$
\begin{aligned}
x_2 - v_1 + v_2 &= 0, \\
x_1 - v_1 - v_2 &= 0, \\
v_1(2 - x_1 - x_2) + v_2(x_1 - x_2) &= 0, \\
x_1 - x_2 &\leq 0, \\
2 - x_1 - x_2 &\leq 0, \\
v_1, v_2 &\geq 0.
\end{aligned}
$$

Note that the points $x^\star = (1; 1)$ and $v^\star = (1; 0)$ satisfy the KKT conditions, therefore $x^\star$ is a candidate for being a minimizer. ∎

We point out that, in Example 11.1, there is no guarantee that $x^\star$ is indeed a minimizer, because the KKT conditions are, in general, only necessary. As a central result in convex programming, if Problem (11.7) is a convex program, then the KKT conditions are not only necessary, but also sufficient for optimality.

In this section, we present a glimpse of the duality theory associated with SOCP and introduce the complementarity condition as one of the optimality conditions of SOCP. The KKT conditions have been introduced because they are needed to establish some of these duality results.

Much, but not all, of the duality theory for SOCP is very similar to duality theory for linear programming. Recall that a regular cone is self-dual if it equals its dual cone (see Definition 3.22). Now we need the following lemma.

---

**Lemma 11.3** *The second-order cone $\mathbb{E}_+^n$ is self-dual.*

---

**Proof** We verify that the second-order cone $\mathbb{E}_+^n$ equals its dual cone, which is defined as

$$
(\mathbb{E}_+^n)^\star \triangleq \left\{ x \in \mathbb{E}^n : x^\mathsf{T} y \geq 0 \text{ for all } y \in \mathbb{E}_+^n \right\}.
$$

We first show that $\mathbb{E}_+^n \subseteq (\mathbb{E}_+^n)^\star$. Let $x = (x_0; \widetilde{x}) \in \mathbb{E}_+^n$, we need to show $x \in (\mathbb{E}_+^n)^\star$. For any $y \in \mathbb{E}_+^n$, we have

$$
x^\mathsf{T} y = x_0 y_0 + \widetilde{x}^\mathsf{T} \widetilde{y} \geq \|\widetilde{x}\| \|\widetilde{y}\| + \widetilde{x}^\mathsf{T} \widetilde{y} \geq \left| \widetilde{x}^\mathsf{T} \widetilde{y} \right| + \widetilde{x}^\mathsf{T} \widetilde{y} \geq 0,
$$

where the first inequality follows from the fact that $x, y \in \mathbb{E}^n_+$, and the second inequality follows from Cauchy-Schwartz inequality. This implies that $x \in (\mathbb{E}^n_+)^\star$ and hence $\mathbb{E}^n_+ \subseteq (\mathbb{E}^n_+)^\star$.

To prove the reverse inclusion, let $y \in (\mathbb{E}^n_+)^\star$, we need to show that $y \in \mathbb{E}^n_+$, which is trivial if $\widetilde{y} = 0$. If $\widetilde{y} \neq 0$, let $x \triangleq (\|\widetilde{y}\|; -\widetilde{y}) \in \mathbb{E}^n_+$. Then, we have

$$x^\mathsf{T} y = x_0 y_0 + \widetilde{x}^\mathsf{T} \widetilde{y} = y_0 \|\widetilde{y}\| - \widetilde{y}^\mathsf{T} \widetilde{y} = y_0 \|\widetilde{y}\| - \|\widetilde{y}\|^2,$$

where we used Cauchy-Schwartz inequality, where the equality is attained, to obtain the last equality. Since $x$ belongs to the second-order cone $\mathbb{E}^n_+$ and $y$ belongs to its dual, it follows that

$$0 \leq x^\mathsf{T} y = \|\widetilde{y}\| \left( y_0 - \|\widetilde{y}\| \right).$$

As $\widetilde{y} \neq 0$, this implies that $y_0 \geq \|\widetilde{y}\|$. That is, $y \in \mathbb{E}^n_+$ and hence $(\mathbb{E}^n_+)^\star \subseteq \mathbb{E}^n_+$. The proof is complete. ∎

Now consider Problem (P|SOCP). Note that the variable $x$ is the primal variable, and the variables $y$ and $s$ are the dual variables. We call $x \in \mathbb{E}^n$ primal feasible if $Ax = b$ and $x \geq 0$. Similarly, $(s, y) \in \mathbb{E}^n \times \mathbb{R}^m$ is called dual feasible if $A^\mathsf{T} y + s = c$ and $s \geq 0$. Note that the matrix $A$ is defined to map $\mathbb{E}^n$ into $\mathbb{R}^m$, and its transpose $A^\mathsf{T}$ is defined to map $\mathbb{R}^m$ into $\mathbb{E}^n$ such that $x^\mathsf{T}(A^\mathsf{T} y) = (Ax)^\mathsf{T} y$.

We state and prove the following weak duality property.

---

**Lemma 11.4 (Weak duality in SOCP)** *If $x$ and $(y, s)$ are primal and dual feasible solutions in (P|SOCP) and (D|SOCP), respectively, then the duality gap is*

$$c^\mathsf{T} x - b^\mathsf{T} y = x^\mathsf{T} s \geq 0.$$

---

**Proof** Consider Problems (P|SOCP) and (D|SOCP). Due to their constraints, we can replace $c$ with $A^\mathsf{T} y + s$ and $b$ with $Ax$ and get

$$c^\mathsf{T} x - b^\mathsf{T} y = \left( A^\mathsf{T} y + s \right)^\mathsf{T} x - (Ax)^\mathsf{T} y = \left( A^\mathsf{T} y \right)^\mathsf{T} x + s^\mathsf{T} x - (Ax)^\mathsf{T} y = x^\mathsf{T} s.$$

Note that $x, s \in \mathbb{E}^n$. By the self duality of the second-order cone $\mathbb{E}^n_+$, it concludes that $x^\mathsf{T} s \geq 0$ and this completes the proof. ∎

The authors in [NN94] show that the strong duality property can fail in general conic optimization. However, despite of this, a slightly weaker property can be always shown in conic optimization. Now, we give conditions for such a slightly weaker property to hold in SOCP. We say that the primal problem is strictly feasible if there exists a primal feasible point $\hat{x}$ such that $\hat{x} > 0$. We make the following assumption for convenience.

---

**Assumption 11.1** *The $m$ rows of the matrix $A$ are linearly independent.*

---

Using the KKT conditions (see Theorem 11.1), we state and prove the following semi-strong duality result.

> **Lemma 11.5 (Semi-strong duality in SOCP)**  *Consider the primal–dual pair (P|SOCP) and (D|SOCP). If the primal problem is strictly feasible and solvable, then the dual problem is solvable and their optimal values are equal.*

**Proof**  By the assumption of the lemma, the primal problem is strictly feasible and solvable. So, let $x$ be an optimal solution of the primal problem where we can apply the KKT conditions. This implies that there are Lagrange multiplier vectors $y$ and $s$ such that $(x, y, s)$ satisfies

$$Ax = b,$$
$$A^\mathsf{T} y + s = c,$$
$$x^\mathsf{T} s = 0,$$
$$x, s \geq 0.$$

This means that $(y, s)$ is feasible for the dual problem. Let $(v, z)$ be any feasible solution of the dual problem, then we have that $b^\mathsf{T} v \leq c^\mathsf{T} x = s^\mathsf{T} x + b^\mathsf{T} y = b^\mathsf{T} y$, where we used the weak duality to obtain the inequality and the complementary slackness to obtain the last equality. Thus, $(y, s)$ is an optimal solution of the dual problem and $c^\mathsf{T} x = b^\mathsf{T} y$ as desired.    ∎

The following strong duality result can be obtained by applying the duality relations to our problem formulation.

> **Theorem 11.2 (Strong duality in SOCP)**  *Consider the primal–dual pair (P|SOCP) and (D|SOCP). If both the primal and dual problems have strictly feasible solutions, then they both have optimal solutions $x^\star$ and $(y^\star, s^\star)$, respectively, and*
> $$p^\star \triangleq c^\mathsf{T} x^\star = d^\star \triangleq b^\mathsf{T} y^\star (i.e., \ x^{\star\mathsf{T}} s^\star = 0).$$

The following lemma describes the complementarity condition as one of the optimality conditions of SOCP.

> **Lemma 11.6 (Complementarity condition in SOCP)**  *If $x, s \geq 0$, then*
> $$x^\mathsf{T} s = 0 \iff x \circ s = 0.$$

**Proof**  For any $x, s \in \mathbb{E}^n$ having $x, s \geq 0$, we need to show that $x^\mathsf{T} s = 0 \Leftrightarrow x \circ s = 0$, or equivalently

$$x^\mathsf{T} s = 0 \iff (i) \ x^\mathsf{T} s = x_0 s_0 + \langle \widetilde{x}, \widetilde{s} \rangle = 0, \ \text{and} \ (ii) \ x_0 \widetilde{s} + s_0 \widetilde{x} = 0.$$

The direction from right to left is very clear because the required is itself a part of the assumptions. To prove the other direction, assume that $x^\mathsf{T} s = 0$ (which is again (i) itself), it is enough to show that (ii) is satisfied. If $x_0 = 0$ or $s_0 = 0$, the result is clearly trivial. Therefore, we only consider the cases when $x_0 > 0$ and $s_0 > 0$. Using Cauchy-Schwartz inequality and the fact that $x, s \in \mathbb{E}^n$, we have

$$\langle \widetilde{x}, \widetilde{s} \rangle = \widetilde{x}^\mathsf{T} \widetilde{s} = \widetilde{x}^\mathsf{T} \widetilde{s} \geq -\|\widetilde{x}\| \|\widetilde{s}\| \geq -x_0 s_0, \tag{11.8}$$

Note that $x^\mathsf{T} s = 0$ if and only if $\langle \widetilde{x}, \widetilde{s} \rangle = -x_0 s_0$, therefore the inequalities in (11.8) are satisfied as equalities. This holds if and only if either $x = 0$ or $s = 0$, in which case (i) and (ii) trivially hold, or $x \neq 0$ and $s \neq 0$, $\widetilde{x} = -\alpha \widetilde{s}$, where $\alpha > 0$, and $x_0 = \|\widetilde{x}\| = \alpha \|\widetilde{s}\| = \alpha s_0$, that is $\widetilde{x} + \alpha \widetilde{s} = \widetilde{x} + (x_0/s_0)\widetilde{s} = 0$. This implies that $x_0 \widetilde{s} + s_0 \widetilde{x} = 0$. The proof is complete. ∎

As a result of Lemma 12.3, the complementarity slackness condition for the primal and dual SOCP problems (P|SOCP) and (D|SOCP) can be equivalently represented by the equation $x \circ s = 0$. From the above results, we get the following corollary.

---

**Corollary 11.1 (Optimality conditions in SOCP)** *Consider the primal–dual pair (P|SOCP) and (D|SOCP). Assume that both the primal and dual problems are strictly feasible, then $(x, (y, s)) \in \mathbb{E}^n \times \mathbb{R}^m \times \mathbb{E}^n$ is a pair of optimal solutions to the SOCP (P|SOCP) and (D|SOCP) if and only if*

$$
\begin{aligned}
Ax &= b, \\
A^\mathsf{T} y + s &= c, \\
x \circ s &= 0, \\
x, s &\geq 0.
\end{aligned}
\tag{11.9}
$$

---

We have established the duality relations in SOCP. The focus in the remaining part of this chapter is to solve SOCP algorithmically.

## 11.5. A primal-dual path-following algorithm

As we mentioned earlier, interior-point methods reach a best solution by traversing the interior of the feasible region. There are several Interior-point algorithms for SOCPs; see for example [LVBL98, AG03, Alz20, Alz17, Alz11, Alz14a, AP17a, ABA19, Alz14b, AP17b] and the references contained therein.

In this section, we present a primal-dual path-following algorithm for solving SOCP problems. The material presented in this section is based on, and similar to, any primal-dual path-following algorithm proposed for SOCP (see for instance [Alz17]). The general scheme of the path-following algorithms for SOCP is as follows. We associate the perturbed problems to second-order cone programming problems (P|SOCP) and (D|SOCP), then we draw a path of the centers defined by the perturbed KKT optimality conditions. After that, Newton's method is applied to treat the corresponding perturbed equations in order to obtain a descent search direction.

Let $\mu > 0$ be a barrier parameter. The perturbed primal problem corresponding to the primal problem (P|SOCP) is

$$
\begin{aligned}
&\text{min} && f_\mu(x) \triangleq c^\mathsf{T} x - \mu \ln \det(x) + r\mu \ln \mu \\
(\text{P|SOCP}_\mu)\quad &\text{s.t.} && Ax = b, \\
& && x > 0,
\end{aligned}
$$

and the perturbed dual problem corresponding to the dual problem (D|SOCP) is

$$\max \quad g_\mu(y,s) \triangleq b^\mathsf{T} y + \mu \ln \det(s) - r\mu \ln \mu$$

$$(\text{D|SOCP}_\mu) \quad \text{s.t.} \quad A^\mathsf{T} y + s = c,$$

$$s > 0.$$

Now, we define the following feasibility sets:

$$\begin{aligned}
\mathcal{F}_{\text{P|SOCP}} &\triangleq \{x \in \mathbb{E}^n : Ax = b,\ x \geq 0\}, \\
\mathcal{F}_{\text{D|SOCP}} &\triangleq \left\{(y;s) \in \mathbb{R}^m \times \mathbb{E}^n : A^\mathsf{T} y + s = c,\ s \geq 0\right\}, \\
\mathcal{F}_{\text{P|SOCP}}^{\circ} &\triangleq \{x \in \mathbb{E}^n : Ax = b,\ x > 0\}, \\
\mathcal{F}_{\text{D|SOCP}}^{\circ} &\triangleq \left\{(y;s) \in \mathbb{R}^m \times \mathbb{E}^n : A^\mathsf{T} y + s = c,\ s > 0\right\}, \\
\mathcal{F}_{\text{SOCP}}^{\circ} &\triangleq \mathcal{F}_{\text{P|SOCP}}^{\circ} \times \mathcal{F}_{\text{D|SOCP}}^{\circ}.
\end{aligned}$$

We also make the following assumption about the primal-dual pair (P|SOCP) and (D|SOCP).

> **Assumption 11.2** *The set $\mathcal{F}_{\text{SOCP}}^{\circ}$ is nonempty.*

Assumption 11.2 requires that Problem (P|SOCP$_\mu$) and its dual (D|SOCP$_\mu$) have strictly feasible solutions, which guarantees strong duality for the second-order cone programming problem. Note that the feasible region for Problems (P|SOCP$_\mu$) and (D|SOCP$_\mu$) is described implicitly by $\mathcal{F}_{\text{SOCP}}^{\circ}$. Due to the coercivity of the function $f_\mu$ on the feasible set of (P|SOCP$_\mu$), Problem (P|SOCP$_\mu$) has an optimal solution.

The following lemma proves the convergence of the optimal solution of Problem (P|SOCP$_\mu$) to the optimal solution of Problem (P|SOCP) when $\mu$ approaches zero.

> **Lemma 11.7** *Let $\bar{x}_\mu$ be an optimal primal solution of Problem (P|SOCP$_\mu$), then $\bar{x} = \lim_{\mu \to 0} \bar{x}_\mu$ is an optimal solution of Problem (P|SOCP).*

**Proof** Let $f_\mu(x) \triangleq f(x, \mu)$ and $f(x) \triangleq f(x, 0)$. Due to the coercivity of the function $f_\mu$ on the feasible set of (P|SOCP$_\mu$), Problem (P|SOCP$_\mu$) has an optimal solution, say $\bar{x}_\mu$, such that $\nabla_x f_\mu(\bar{x}_\mu) = \nabla_x f(\bar{x}_\mu, \mu) = 0$. Then, for all $x \in \mathcal{F}_{\text{P|SOCP}}^{\circ}$, we have that

$$\begin{aligned}
f(x) &\geq f(\bar{x}_\mu, \mu) + (x - \bar{x}_\mu)^\mathsf{T} \nabla_x f(\bar{x}_\mu, \mu) + (0 - \mu)\frac{\partial}{\partial \mu} f(\bar{x}_\mu, \mu) \\
&\geq f(\bar{x}_\mu, \mu) + \mu \ln \det \bar{x}_\mu - r\mu \ln \mu - r\mu \\
&\geq c^\mathsf{T} \bar{x}_\mu - \mu \ln \det \bar{x}_\mu + r\mu \ln \mu + \mu \ln \det \bar{x}_\mu - r\mu \ln \mu - r\mu \geq c^\mathsf{T} \bar{x}_\mu - r\mu.
\end{aligned}$$

Since $x$ was arbitrary in $\mathcal{F}_{\text{P|SOCP}}^{\circ}$, this implies that $\min_{x \in \mathcal{F}_{\text{P|SOCP}}^{\circ}} f(x) \geq c^\mathsf{T} \bar{x}_\mu - r\mu \geq c^\mathsf{T} \bar{x}_\mu = f(\bar{x}_\mu)$. On the other side, we have $f(\bar{x}_\mu) \geq \min_{x \in \mathcal{F}_{\text{P|SOCP}}^{\circ}} f(x)$. As $\mu$ goes to 0, it immediately follows that $f(\bar{x}) = \min_{x \in \mathcal{F}_{\text{P|SOCP}}^{\circ}} f(x)$. Thus, $\bar{x}$ is an optimal solution of Problem (P|SOCP). The proof is complete. ∎

### Newton's method and commutative directions

As we mentioned, the objective function of Problem (P|SOCP$_\mu$) is strictly convex, hence the KKT conditions are necessary and sufficient to characterize an optimal solution of Problem (P|SOCP). Consequently, the points $\bar{x}_\mu$ and $(\bar{y}_\mu, \bar{s}_\mu)$ are optimal solutions of (P|SOCP$_\mu$) and (D|SOCP$_\mu$) respectively if and only if they satisfy the perturbed nonlinear system

$$
\begin{aligned}
Ax &= b, & x &> 0, \\
A^\top y + s &= c, & s &> 0, \\
x \circ s &= \mu e, & \mu &> 0,
\end{aligned}
\tag{11.10}
$$

where $e \triangleq (e_{n_1}; e_{n_2}; \ldots; e_{n_r})$ is the identity vector of $\mathbb{E}_r^n$.

We call the set of all solutions of system (11.10), denoted by $(x_\mu; y_\mu; s_\mu)$ with $\mu > 0$, the central path. We say that a point $(x, y, s)$ is near to the central path if it belongs to the set $\mathcal{N}_{\text{SOCP}}(\mu)$, which is defined as

$$
\mathcal{N}_{\text{SOCP}}(\mu) \triangleq \left\{ (x; y; s) \in \mathcal{F}_{\text{P|SOCP}}^\circ \times \mathcal{F}_{\text{D|SOCP}}^\circ : d_{\text{SOCP}}(x, s) \leq \theta\mu, \theta \in (0, 1) \right\},
$$

where

$$
d_{\text{SOCP}}(x, s) \triangleq \|Q_{x^{1/2}} s - \mu e\|_F.
$$

Now, we can apply Newton's method to system (11.10) and obtain the following linear system

$$
\begin{aligned}
A\Delta x &= 0, \\
A^\top \Delta y + \Delta s &= 0, \\
x \circ \Delta s + \Delta x \circ s &= \sigma\mu e - x \circ s.
\end{aligned}
\tag{11.11}
$$

where $(\Delta x; \Delta y; \Delta s) \in \mathbb{E}_r^n \times \mathbb{R}^m \times \mathbb{E}_r^n$ is the search direction, $\mu = \frac{1}{r} x^\top s$ is the normalized duality gap corresponding to $(x; y; s)$, and $\sigma \in (0, 1)$ is the centering parameter.

Note that the strict second-order cone inequalities $x, s > 0$ imply that $d_{\text{SOCP}}(x, s) \leq \|x \circ s - \mu e\|_F$ with equality holds when $x$ and $s$ operator commute [SA03, Lemma 30]. In fact, it is known that many interesting properties become apparent for the analysis of interior-point methods when $x$ and $s$ operator commute.

Denote by $C(x, s)$ the set of all elements so that the scaled elements operator commute. That is,

$$
C(x, s) \triangleq \left\{ p \in \mathbb{E}_r^n : p^{-1} \text{ exists, and } Q_p x \,\&\, Q_{p^{-1}} s \text{ operator commute} \right\}.
\tag{11.12}
$$

From [SA03, Lemma 28], the equality $x \circ s = \mu e$ holds if and only if the equality $(Q_p x) \circ (Q_{p^{-1}} s) = \mu e$ holds, for any nonsingular vector $p$ in $\mathbb{E}_r^n$. Therefore, for any given nonsingular vector $p \in \mathbb{E}_r^n$, the system (11.10) is equivalent to the system

$$
\begin{aligned}
Ax &= b, & x &> 0, \\
A^T y + s &= c, & s &> 0, \\
(Q_p x) \circ (Q_{p^{-1}} s) &= \mu e, & \mu &> 0.
\end{aligned}
\tag{11.13}
$$

Let $v \in \mathbb{E}_r^n$. With respect to a nonsingular vector $p \in \mathbb{E}_r^n$, we define the scaling vectors $\overline{v}$ and $\underline{v}$ and the scaling matrix $\underline{A}$ as

$$
\overline{v} \triangleq Q_p v, \quad \underline{v} \triangleq Q_{p^{-1}} v, \quad \text{and} \quad \underline{A} \triangleq Q_p A.
\tag{11.14}
$$

The definitions in (11.14) are valid for both the single and multiple block cases.

Using this change of variables and the fact that $Q_p\left(\mathbb{E}_r^n\right) = \mathbb{E}_r^n$, we conclude that the system (11.11) is equivalent to the following Newton system

$$
\begin{aligned}
\underline{A}\,\overline{\Delta x} &= b - \underline{A}\,\overline{x}, \\
\underline{A}^T \Delta y + \underline{\Delta s} &= \underline{c} - \underline{s} - \underline{A}^T y, \\
\overline{x} \circ \underline{\Delta s} + \overline{\Delta x} \circ \underline{s} &= \sigma \mu e - \overline{x} \circ \underline{s}.
\end{aligned}
\tag{11.15}
$$

Here, the normalized duality gap is $\mu = \frac{1}{r}\overline{x}^T \underline{s} = \frac{1}{r}x^T s$. In fact,

$$
\overline{x}^T \underline{s} = \left(Q_p x\right)^T Q_{p^{-1}} s = x^T Q_p Q_{p^{-1}} s = x^T s.
\tag{11.16}
$$

Solving the scaled Newton system (11.15) yields the search direction $(\overline{\Delta x}; \Delta y; \underline{\Delta s})$. Then, we apply the inverse scaling to $(\overline{\Delta x}; \underline{\Delta s})$ to obtain the Newton direction $(\Delta x; \Delta s)$. Note that the search direction $(\overline{\Delta x}; \Delta y; \underline{\Delta s})$ belongs to the so-called the MZ family of directions (due to Monteiro [Mon97] and Zhang [Zha98b]).

Clearly, the set $C(x, s)$ defined in (11.12) is a subclass of the MZ family of search directions. Our focus is in vectors $p \in C(x, s)$. We discuss the following three choices of $p$ (see [SA03, Section 3]):

- The first one is to choose $p = x^{-1/2}$, which gives $\overline{x} = e$.

- The second one is to choose $p = s^{1/2}$, which gives $\underline{s} = e$.

- The third choice of $p$ is given by $p = (Q_{x^{1/2}}(Q_{x^{1/2}}s)^{-1/2})^{-1/2}$, which yields $Q_p^2 x = s$, and therefore $\underline{s} = Q_{p^{-1}} s = Q_p x = \overline{x}$.

The first two choices of directions are respectively called the HRVW/KSH/M direction and dual HRVW/KSH/M direction (due to Helmberg et al. [HRVW96], Monteiro [Mon97] and Kojima et al. [KSH97]). The third choice of directions is called the NT direction (due to Nesterov and Todd [NT98]).

## Path-following algorithm

The path-following algorithm for solving SOCP problem is formally stated in Algorithm 11.1.

---

**Algorithm 11.1:** Path-following algorithm for SOCP

---

**Input:** Data in Problems (P|SOCP) and (D|SOCP), $k = 0$,
$$\left(x^{(0)}; y^{(0)}; s^{(0)}\right) \in \mathcal{N}_{\text{SOCP}}\left(\mu^{(0)}\right), \epsilon > 0, \sigma^{(0)}, \theta \in (0, 1)$$
**Output:** An $\epsilon$-optimal solution to Problem (P|SOCP)

1: **while** $x^{(k)^\mathsf{T}} s^{(k)} \geq \epsilon$ **do**

2:      choose $p^{(k)} \in C\left(x^{(k)}, s^{(k)}\right)$

3:      compute $\left(\overline{x^{(k)}}; y^{(k)}; \underline{s^{(k)}}\right)$ by applying scaling to $\left(x^{(k)}; y^{(k)}; s^{(k)}\right)$

4:      set $\mu^{(k)} \triangleq \frac{1}{r}\overline{x}^{(k)^\mathsf{T}}\underline{s}^{(k)}$, $h^{(k)} \triangleq \sigma^{(k)}\mu^{(k)}e - \overline{x^{(k)}} \circ \underline{s}^{(k)}$ and $\Psi^{(k)} \triangleq \frac{1}{\mu}\underline{A}\,\overline{x^{(k)^2}}\underline{A}^\mathsf{T}$

5:      compute $\left(\overline{\Delta x^{(k)}}; \Delta y^{(k)}; \underline{\Delta s^{(k)}}\right)$ by solving the scaled system (11.15) to get
$$\left(\overline{\Delta x^{(k)}}; \Delta y^{(k)}; \underline{\Delta s^{(k)}}\right) \triangleq \left(\left(h^{(k)} - \overline{x^{(k)}} \circ \underline{\Delta s^{(k)}}\right) \circ \underline{s}^{(k)^{-1}}; -\Psi^{(k)^{-1}}\underline{A}\left(\underline{s}^{(k)^{-1}} \circ h^{(k)}\right); -\underline{A}^\mathsf{T}\Delta y\right)$$

6:      compute $\left(\Delta x^{(k)}; \Delta y^{(k)}; \Delta s^{(k)}\right)$ by applying inverse scaling to $\left(\overline{\Delta x^{(k)}}; \Delta y^{(k)}; \underline{\Delta s^{(k)}}\right)$

7:      set the new iterate according to
$$\left(x^{(k+1)}; y^{(k+1)}; s^{(k+1)}\right) \triangleq \left(x^{(k)} + \alpha^{(k)}\Delta x^{(k)}; y^{(k)} + \alpha^{(k)}\Delta y^{(k)}; s^{(k)} + \alpha^{(k)}\Delta s^{(k)}\right)$$

8:      set $k = k + 1$

9: **end**

---

Algorithm 11.1 selects a sequence of displacement steps $\{\alpha^{(k)}\}$ and centrality parameters $\{\sigma^{(k)}\}$ according to the following rule: For all $k \geq 0$, we take $\sigma^{(k)} = 1 - \delta/\sqrt{r}$, where $\delta \in [0, \sqrt{r})$. The author in [Alz17] discusses in Section 4 various selections for calculating the displacement step $\alpha^{(k)}$.

In the rest of this section, we prove that the complementary gap and the function $f_\mu$ decrease for a given displacement step. The proof of this result depends essentially on the following lemma.

---

**Lemma 11.8** *Let* $(x; y; s) \in int\, \mathbb{R}^n_{+ \times \mathbb{R}^m \times int\, \mathbb{E}^n_{r+}}$, $(\overline{x}; y; \underline{s})$ *be obtained by applying scaling to* $(x; y; s)$, *and* $\left(\overline{\Delta x}; \Delta y; \underline{\Delta s}\right)$ *be a solution of System* (11.15). *Then we have*

*(a)* $\overline{\Delta x}^\mathsf{T}\underline{\Delta s} = 0$.

*(b)* $\overline{x}^\mathsf{T}\underline{\Delta s} + \overline{\Delta x}^\mathsf{T}\underline{s} = trace(h)$, *where* $h \triangleq \sigma\mu e - \overline{x} \circ \underline{s}$ *such that* $\sigma \in (0, 1)$ *and* $\mu = \frac{1}{r}\overline{x}^\mathsf{T}\underline{s}$.

*(c)* $\overline{x}^{+^\mathsf{T}}\underline{s}^+ = (1 - \alpha(1 - \sigma/2))\,\overline{x}^\mathsf{T}\underline{s}$, $\forall \alpha \in \mathbb{R}$, *where* $\overline{x}^+ \triangleq \overline{x} + \alpha\overline{\Delta x}$ *and* $\underline{s}^+ \triangleq \underline{s} + \alpha\underline{\Delta s}$.

(d) $x^{+^\mathsf{T}}s^+ = (1 - \alpha(1 - \sigma/2))\, x^\mathsf{T}s,\ \forall \alpha \in \mathbb{R},\ where\ x^+ \triangleq x + \alpha\Delta x\ and$
$s^+ \triangleq s + \alpha\Delta s.$

**Proof** By the first two equations of System (11.15), we get

$$\overline{\Delta x}^\mathsf{T}\underline{\Delta s} = -\overline{\Delta x}^\mathsf{T}\underline{A}^\mathsf{T}\Delta y = -\left(\underline{A}\overline{\Delta x}\right)^\mathsf{T}\Delta y = 0.$$

This proves item (a).

We prove item (b) by noting that

$$
\begin{aligned}
\mathrm{trace}(h) &= \mathrm{trace}\left(\sigma\mu e - \overline{x} \circ \underline{s}\right) \\
&= \mathrm{trace}\left(\overline{x} \circ \underline{\Delta s} + \overline{\Delta x} \circ \underline{s}\right) \\
&= \mathrm{trace}\left(\overline{x} \circ \underline{\Delta s}\right) + \mathrm{trace}\left(\overline{\Delta x} \circ \underline{s}\right) = \overline{x}^\mathsf{T}\underline{\Delta s} + \overline{\Delta x}^\mathsf{T}\underline{s},
\end{aligned}
$$

where we used the last equation of System (11.15) to obtain the first equality.

Item (c) is left as an exercise for the reader (see Exercise 11.4). Item (d) follows from item (c) and the fact that $\overline{x}^\mathsf{T}\underline{s} = x^\mathsf{T}s$ (see (11.16)), and similarly that $\overline{x}^{+^\mathsf{T}}\underline{s}^+ = x^{+^\mathsf{T}}s^+$. The proof is complete. ∎

The result in the following theorem is a special case of that in [Alz17, Lemma 5.2].

---

**Theorem 11.3** *Let $(x; y; s)$ and $(x^+; y^+; s^+)$ be strictly feasible solutions of the pair of problems (P|SOCP$_\mu$) and (D|SOCP$_\mu$) with*

$$(x^+; y^+; s^+) = (x + \alpha\Delta x;\ y + \alpha\Delta y;\ s + \alpha\Delta s),$$

*where $\alpha$ is a displacement step and $(\Delta x; \Delta y; \Delta s)$ is the Newton direction. Then*

(a) $x^{+^\mathsf{T}}s^+ < \overline{x}^\mathsf{T}\underline{s}.$    (b) $f_\mu(x^+) < f_\mu(x).$

---

**Proof** Note that

$$x^{+^\mathsf{T}}s^+ = \left(1 - \alpha\left(1 - \frac{\sigma}{2}\right)\right)\overline{x}^\mathsf{T}\underline{s} < \overline{x}^\mathsf{T}\underline{s},$$

where the equality follows from item (d) of Lemma 11.8 and the strict inequality follows from $(1 - \alpha(1 - \sigma/2)) < 1$ (as $\alpha > 0$ and $\sigma \in (0, 1)$). This proves item (a).

To prove item (b), note that $f_\mu(x^+) \simeq f_\mu(x) + \nabla_x f_\mu(x)^\mathsf{T}(x^+ - x)$, and therefore $f_\mu(x^+) - f_\mu(x) \simeq \alpha\nabla_x f_\mu(x)^\mathsf{T}\Delta x$. Since $\nabla_x f_\mu(x) = -\nabla^2_{xx} f_\mu(x)\Delta x$, we have that

$$f_\mu(x^+) - f_\mu(x) \simeq -\alpha\, \Delta x^\mathsf{T}\nabla^2_{xx} f_\mu(x)\Delta x < 0,$$

where the strict inequality follows from the positive definiteness of the Hessian matrix $\nabla^2_{xx} f_\mu(x)$ (as $f_\mu$ is strictly convex). Thus, $f_\mu(x^+) < f_\mu(x)$. The proof is complete. ∎

## Complexity estimates

In this part, we analyze the complexity of the proposed path-following algorithm for SOCP. More specifically, we prove that the iteration-complexity of Algorithm 11.1 is bounded by

$$O\left(\sqrt{r}\ln\left(\epsilon^{-1}x^{(0)^{\mathsf{T}}}s^{(0)}\right)\right).$$

Our proof depends essentially on the following two lemmas.

---

**Lemma 11.9** *Let* $(x; y; s) \in \mathcal{F}_{P\backslash SOCP}^{\circ} \times \mathcal{F}_{D\backslash SOCP}^{\circ}$. *Let also* $(\overline{x}; y; \underline{s})$ *be obtained by applying scaling to* $(x; y; s)$ *with* $h = \sigma\mu e - \overline{x} \circ \underline{s}$, *and* $(\overline{\Delta x}; \Delta y; \underline{\Delta s})$ *be a solution of System* (11.15). *For any* $\alpha \in \mathbb{R}$, *we set*

$$(x(\alpha); y(\alpha); s(\alpha)) \triangleq (\overline{x}; y; \underline{s}) + \alpha(\overline{\Delta x}; \Delta y; \underline{\Delta s}),$$
$$\mu(\alpha) \triangleq \frac{1}{r}x(\alpha)^{\mathsf{T}}s(\alpha),$$
$$v(\alpha) \triangleq x(\alpha) \circ s(\alpha) - \mu(\alpha)e.$$

*Then*

$$v(\alpha) = (1 - \alpha)(\overline{x} \circ \underline{s} - \mu e) + \alpha^2 \overline{\Delta x} \circ \underline{\Delta s}. \qquad (11.17)$$

---

**Proof** See Exercise 11.5. ∎

---

**Lemma 11.10** *Let* $(x; y; s) \in \mathcal{F}_{P\backslash SOCP}^{\circ} \times \mathcal{F}_{D\backslash SOCP}^{\circ}$, *and* $(\overline{x}; y; \underline{s})$ *be obtained by applying scaling to* $(x; y; s)$ *such that* $\|\overline{x} \circ \underline{s} - \mu e\| \leq \theta\mu$, *for some* $\theta \in [0, 1)$ *and* $\mu > 0$. *Let also* $(\overline{\Delta x}; \Delta y; \underline{\Delta s})$ *be a solution of System* (11.15), $h = \sigma\mu e - \overline{x} \circ \underline{s}$, $\delta_x \triangleq \mu\|\overline{\Delta x} \circ \overline{x}^{-1}\|_F, \delta_s \triangleq \|\overline{x} \circ \underline{\Delta s}\|_F$. *Then, we have*

$$\delta_x\delta_s \leq \frac{1}{2}\left(\delta_x^2 + \delta_s^2\right) \leq \frac{\|h\|_F^2}{2(1 - \theta)^2}. \qquad (11.18)$$

---

**Proof** See Exercise 11.6. ∎

The following theorem analyzes the behavior of one iteration of Algorithm 11.1. This theorem is a special case of [Alz17, Theorem 6.1].

---

**Theorem 11.4** *Let* $\theta \in (0, 1)$ *and* $\delta \in [0, \sqrt{r})$ *be given such that*

$$\frac{\theta^2 + \delta^2}{2(1 - \theta)^2\left(1 - \frac{\delta}{\sqrt{r}}\right)} \leq \theta \leq \frac{1}{2}. \qquad (11.19)$$

*Suppose that* $(\overline{x}; y; \underline{s}) \in \mathcal{N}_{SOCP}(\mu)$ *and let* $(\overline{\Delta x}; \Delta y; \underline{\Delta s})$ *denote the solution of system* (11.15) *with* $h = \sigma\mu e - \overline{x} \circ \underline{s}$ *and* $\sigma = 1 - \delta/\sqrt{r}$. *Then, we have*

*(a)* $\overline{x}^{+\mathsf{T}}\underline{s}^+ = (1 - \delta/\sqrt{r})\overline{x}^{\mathsf{T}}\underline{s}.$

(b) $(\overline{x}^+; y^+; \underline{s}^+) = (\overline{x}; y; \underline{s}) + (\overline{\Delta x}; \Delta y; \underline{\Delta s}) \in \mathcal{N}_{SOCP}(\mu)$.

(c) $(x^+; y^+; s^+) = (x; y; s) + (\Delta x; \Delta y; \Delta s) \in \mathcal{N}_{SOCP}(\mu)$.

**Proof** Item (a) follows directly from item (c) of Lemma 11.8 with $\alpha = 1$ and $\sigma = 1 - \delta/\sqrt{r}$. We now prove item (b). Define

$$\mu^+ \triangleq \frac{1}{r} \overline{x}^{+T} \underline{s}^+ = \left(1 - \frac{\delta}{r}\right)\mu \tag{11.20}$$

and let $(\overline{x}; y; \underline{s}) \in \mathcal{N}_{SOCP}(\mu)$, we then have

$$\begin{aligned}
\left\|\sigma\mu e - \overline{x} \circ \underline{s}\right\|_F^2 &\leq \left\|(\sigma - 1)\mu e\right\|_F^2 + \left\|\mu e - \overline{x} \circ \underline{s}\right\|_F^2 \\
&\leq \left((\sigma - 1)^2 r + \theta^2\right)\mu^2 = \left(\delta^2 + \theta^2\right)\mu^2.
\end{aligned} \tag{11.21}$$

Since $\|\overline{x} \circ \underline{s} - \mu e\| \leq \theta\mu$ and $h = \sigma\mu e - \overline{x} \circ \underline{s}$, using Lemma 11.10 it follows that

$$\left\|\overline{\Delta x} \circ \overline{x}^{-1}\right\|_F \left\|\overline{x} \circ \underline{\Delta s}\right\|_F \leq \frac{\left\|\sigma\mu e - \overline{x} \circ \underline{s}\right\|_F^2}{2(1-\theta)^2\mu}. \tag{11.22}$$

Defining $v^+ \triangleq v(1) = \overline{x}^+ \circ \underline{s}^+ - \mu^+ e$ and using (11.17) with $\alpha = 1$, (11.22), (11.21), (11.19) and (11.20), we get

$$\begin{aligned}
\|v^+\|_F &= \left\|\overline{\Delta x} \circ \underline{\Delta s}\right\|_F \\
&\leq \left\|\overline{\Delta x} \circ \overline{x}^{-1}\right\|_F \left\|\overline{x} \circ \underline{\Delta s}\right\|_F \\
&\leq \frac{\left\|\sigma\mu e - \overline{x} \circ \underline{s}\right\|_F^2}{2(1-\theta)^2\mu} \\
&\leq \frac{(\delta^2 + \theta^2)\mu}{2(1-\theta)^2} \\
&\leq \theta\left(1 - \frac{\delta}{\sqrt{r}}\right)\mu = \theta\mu^+.
\end{aligned}$$

Consequently,

$$\left\|\overline{x}^+ \circ \underline{s}^+ - \mu^+ e\right\|_F \leq \theta\mu^+. \tag{11.23}$$

By using the right-hand side inequality in (11.18), and using (11.21) and (11.19), we have

$$\left\|\overline{\Delta x} \circ \overline{x}^{-1}\right\|_F \leq \frac{\left\|\sigma\mu e - \overline{x} \circ \underline{s}\right\|_F}{(1-\theta)\mu} \leq \frac{\sqrt{\delta^2 + \theta^2}}{(1-\theta)} \leq \sqrt{2\theta\left(1 - \frac{\delta}{\sqrt{r}}\right)} < 1,$$

where the strict inequality follows from $\theta \leq \frac{1}{2}$ and $0 < 1 - \delta/\sqrt{r} < 1$.

One can easily see that $\|\overline{\Delta x} \circ \overline{x}^{-1}\|_F < 1$ implies that $e + \overline{\Delta x} \circ \overline{x}^{-1} > 0$, and therefore

$$\overline{x}^+ = \overline{\Delta x} + \overline{x} = \left(e + \overline{\Delta x} \circ \overline{x}^{-1}\right) \circ \overline{x} > 0.$$

Note that, from (11.23), we have $\lambda_{\min}(\overline{x}^+ \circ \underline{s}^+) \geq (1 - \theta)\mu^+ > 0$, and therefore $\overline{x}^+ \circ \underline{s}^+ > 0$. Since $\overline{x}^+ > 0$ and $\overline{x}^+$ and $\underline{s}^+$ operator commute, we conclude that $\underline{s}^+ > 0$. Using the first equation of system (11.15), we get

$$\underline{A}\overline{x}^+ = \underline{A}\left(\overline{x} + \overline{\Delta x}\right) = \underline{A}\overline{x} + \underline{A}\overline{\Delta x} = b, \quad \text{and hence} \quad \overline{x}^+ \in \mathcal{F}_{\text{PISOCP}}^{\circ}.$$

By using the second equation of system (11.15), we get

$$\underline{A}^{\mathsf{T}}y^+ + \overline{s}^+ = \underline{A}^{\mathsf{T}}(y + \Delta y) + \left(\overline{s} + \overline{\Delta s}\right) = \underline{A}^{\mathsf{T}}y + \overline{s} + \underline{A}^{\mathsf{T}}\Delta y + \overline{\Delta s} = c,$$

and hence

$$\left(y^+; \overline{s}^+\right) \in \mathcal{F}_{\text{DISOCP}}^{\circ}.$$

Thus, in view of (11.23), we deduce that $(\overline{x}^+; y^+; \overline{s}^+) \in \mathcal{N}_{SOCP}(\mu)$. Item (b) is therefore established. Item (c) follows from item (b) and [SA03, Proposition 29]. The proof is now complete. ∎

---

**Corollary 11.2** *Let $\theta$ and $\delta$ as given in Theorem* 11.4 *and* $(x^0; y^0; s^{(0)}) \in \mathcal{N}_{SOCP}(\mu)$. *Then Algorithm* 11.1 *generates a sequence of points* $\{(x^{(k)}; y^{(k)}; s^{(k)})\} \subset \mathcal{N}_{SOCP}(\mu)$ *such that*

$$x^{(k)^{\mathsf{T}}} s^{(k)} = \left(1 - \frac{\delta}{\sqrt{r}}\right)^k x^{(0)^{\mathsf{T}}} s^{(0)}, \quad \forall k \geq 0.$$

*Moreover, given a tolerance $\epsilon > 0$, Algorithm* 11.1 *computes an iterate* $\{(x^{(k)}; y^{(k)}; s^{(k)})\}$ *satisfying* $x^{(k)^{\mathsf{T}}} s^{(k)} \leq \epsilon$ *in at most*

$$O\left(\sqrt{r} \ln\left(\frac{x^{(0)^{\mathsf{T}}} s^{(0)}}{\epsilon}\right)\right)$$

*iterations.*

---

**Proof** Looking recursively at item (a) of Theorem 11.4, for each $k$ we have that

$$x^{(k)^{\mathsf{T}}} s^{(k)} = \left(1 - \frac{\delta}{\sqrt{r}}\right)^k x^{(0)^{\mathsf{T}}} s^{(0)} \leq \epsilon.$$

By taking natural algorithm of both sides, we get

$$k \ln\left(1 - \frac{\delta}{\sqrt{r}}\right) \leq \ln\left(\frac{\epsilon}{x^{(0)^{\mathsf{T}}} s^{(0)}}\right),$$

which holds only if the inequality $k(-\delta/\sqrt{r}) \leq \ln(\epsilon/x^{(0)^{\mathsf{T}}} s^{(0)})$ holds, or equivalently, $k \geq \delta^{-1}\sqrt{r} \ln(x^{(0)^{\mathsf{T}}} s^{(0)}/\epsilon)$. The proof is complete. ∎

## 11.6. A homogeneous self-dual algorithm

In this section, we present a homogeneous self-dual algorithm for SOCP. The material of this section has appeared in [Alz14b, Section 3]. The algorithm presented in this section for SOCP generalizes the one proposed in Section 10.7 for linear programming.

The following primal-dual SOCP model provides sufficient conditions (but not always necessary) for an optimal solution of (P|SOCP) and (D|SOCP).

$$
\begin{aligned}
Ax &= b, \\
A^\mathsf{T} y + s &= c, \\
x^\mathsf{T} s &= 0, \\
x, s &\geq 0.
\end{aligned}
\tag{11.24}
$$

The homogeneous SOCP model for the pair (P|SOCP) and (D|SOCP) is as follows:

$$
\begin{aligned}
Ax & & & -b\tau & &= 0, \\
& -A^\mathsf{T} y & -s & +c\tau & &= 0, \\
-c^\mathsf{T} x & +b^\mathsf{T} y & & & -\kappa &= 0, \\
x & & & & &\geq 0, \\
& & s & & &\geq 0, \\
& & & \tau & &\geq 0, \\
& & & & \kappa &\geq 0.
\end{aligned}
\tag{11.25}
$$

The first two equations in (11.25), with $\tau = 1$, represent primal and dual feasibility (with $x, s \geq 0$) and reversed weak duality. So they, together with the third equation after forcing $\kappa = 0$, define primal and dual optimal solutions. Note that homogenizing $\tau$ (i.e., making it a variable) adds the required variable dual to the third equation, introducing the artificial variable $\kappa$ achieves feasibility, and adding the third equation in (11.25) achieves self-duality.

It can be easily shown that $x^\mathsf{T} s + \tau\kappa = 0$. The following theorem relates (11.24) to (11.25), and it is easily proved.

---

**Theorem 11.5** *The primal-dual SOCP model (11.24) has a solution if and only if the homogeneous SOCP model (11.25) has a solution*

$$
(x^\star; y^\star; s^\star; \tau^\star; \kappa^\star) \in \mathbb{E}_{r+}^n \times \mathbb{R}^m \times \mathbb{E}_{r+}^n \times \mathbb{R}_+ \times \mathbb{R}_+
$$

*such that $\tau^\star > 0$ and $\kappa^\star = 0$.*

---

The main step at each iteration of the homogeneous interior-point algorithm for solving (P|SOCP) and (D|SOCP) is the computation of the search direction $(\Delta x; \Delta y; \Delta s)$ from the symmetrized Newton equations with respect to an invertible vector $p$ (which is chosen as the function of $(x; y; s)$) defined by the system:

$$
\begin{array}{rcl}
A\,\Delta x \qquad\qquad\qquad -b\,\Delta\tau \qquad\qquad\qquad &=& \eta r_p, \\
-\ A^{\mathsf{T}}\,\Delta y \quad -\ \Delta s \ +c\,\Delta\tau \qquad\qquad &=& \eta r_d, \\
-c^{\mathsf{T}}\,\Delta x \ + \ b^{\mathsf{T}}\,\Delta y \qquad\qquad\qquad -\quad \Delta\kappa &=& \eta r_g, \\
\kappa\,\Delta\tau \ + \ \tau\,\Delta\kappa &=& \gamma\mu - \tau\kappa, \\
(Q_p\Delta x)\ \circ\ (Q_{p^{-1}}s)\ +\ (Q_p x)\ \circ\ (Q_{p^{-1}}\Delta s) &=& \gamma\mu e - (Q_p x)\circ(Q_{p^{-1}}s),
\end{array}
$$
$$(11.26)$$

where

$$
\begin{array}{llll}
r_p &\triangleq& b\tau - Ax, & \qquad r_d \ \triangleq\ A^{\mathsf{T}}y + s - \tau c, \\
r_g &\triangleq& c^{\mathsf{T}}x - b^{\mathsf{T}}y + \kappa, & \qquad \mu \ \triangleq\ \frac{1}{2r+1}(x^{\mathsf{T}}s + \tau\kappa),
\end{array}
$$

and $\eta$ and $\gamma$ are two parameters.

The vectors $x$ and $s$ may not operator commute. As we mentioned in the preceding section, many interesting properties become apparent for the analysis of interior-point methods when $x$ and $s$ operator commute. We need now to scale the primal-dual pair (P|SOCP) and (D|SOCP) so that the scaled decision variables in the resulting pair of problems (which should be equivalent to the pair (P|SOCP) and (D|SOCP)) operator commute. For this purpose, we use an effective way of scaling illustrated in (11.14). With this change of variables, the pair (P|SOCP) and (D|SOCP) becomes

$$
\begin{array}{lll lll}
& \min & \underline{c}^{\mathsf{T}}\overline{x} & & \max & b^{\mathsf{T}}y \\
\overline{(\text{P|SOCP})} & \text{s.t.} & \underline{A}\overline{x} = b, & \underline{(\text{D|SOCP})} & \text{s.t.} & \underline{A}^{\mathsf{T}}y + \underline{s} = \underline{c}, \\
& & \overline{x} \ge 0; & & & \underline{s} \ge 0.
\end{array}
$$

Note that $x^{\mathsf{T}}s = \overline{x}^{\mathsf{T}}\underline{s}$ (see (11.15)). From [AG03, Lemma 31] it can be seen that the search direction $(\Delta x; \Delta y; \Delta s)$ solves System (11.26) iff $(\overline{\Delta x}; \Delta y; \underline{\Delta s})$ solves the system:

$$
\begin{array}{rcl}
\underline{A}\,\overline{\Delta x} \qquad\qquad\qquad -b\,\Delta\tau \qquad\qquad\qquad &=& \eta\hat{r}_p, \\
-\ \underline{A}^{\mathsf{T}}\,\Delta y \quad -\ \underline{\Delta s} \ +\underline{c}\,\Delta\tau \qquad\qquad &=& \eta\hat{r}_d, \\
-\underline{c}^{\mathsf{T}}\,\overline{\Delta x} \ + \ b^{\mathsf{T}}\,\Delta y \qquad\qquad\qquad -\quad \Delta\kappa &=& \eta\hat{r}_g, \\
\kappa\,\Delta\tau \ + \ \tau\,\Delta\kappa &=& \gamma\mu - \tau\kappa, \\
\overline{\Delta x}\ \circ\ \underline{s} \qquad\quad +\qquad \overline{x}\ \circ\ \underline{\Delta s} &=& \gamma\mu e - \overline{x}\circ\underline{s},
\end{array}
$$
$$(11.27)$$

where

$$
\begin{array}{llll}
\hat{r}_p &\triangleq& b\tau - \underline{A}\overline{x}, & \qquad \hat{r}_d \ \triangleq\ \underline{A}^{\mathsf{T}}y + \underline{s} - \tau\underline{c}, \\
\hat{r}_g &\triangleq& \underline{c}^{\mathsf{T}}\overline{x} - b^{\mathsf{T}}y + \kappa, & \qquad \mu \ \triangleq\ \frac{1}{2r+1}(\overline{x}^{\mathsf{T}}\underline{s} + \tau\kappa).
\end{array}
$$

---

**Algorithm 11.2:** Generic homogeneous self-dual algorithm for SOCP

---

**Input:** Data in Problems (P|SOCP) and (D|SOCP)
$$(x; y; s; \tau; \kappa) \triangleq (e; 0; e; 1; 1)$$
**Output:** An approximate optimal solution to Problem (P|SOCP)
1: choose a scaling element $p$ and compute $(\overline{x}, \underline{s})$
2: **while** a stopping criterion is not satisfied **do**
3:    choose $\eta, \gamma$
4:    compute the solution $(\overline{\Delta x}; \Delta y; \underline{\Delta s}; \Delta\tau; \Delta\kappa)$ of the linear system (11.27)
5:    compute $(\Delta x; \Delta s)$ by applying inverse scaling to $(\overline{\Delta x}; \underline{\Delta s})$
6:    compute a step length $\theta$ so that
      $$x + \theta\Delta x > 0$$
      $$s + \theta\Delta s > 0$$
      $$\tau + \theta\Delta\tau > 0$$
      $$\kappa + \theta\Delta\kappa > 0$$
7:    set the new iterate according to
      $$(x; y; s; \tau; \kappa) \triangleq (x; y; s; \tau; \kappa) + \theta(\Delta x; \Delta y; \Delta s; \Delta\tau; \Delta\kappa)$$
8: **end**

---

For each choice of $p$, we get a different search direction. The three choices of $p$ that we discussed in the preceding section are the most common in practice. We emphasize that $(\overline{\Delta x}; \Delta y; \underline{\Delta s})$ is the result of applying Newton's method to the primal and dual feasibility and complementarity relations arising from the scaled problems (P|SOCP) and (D|SOCP). It depends on the choice of $p$, while $(\Delta x; \Delta y; \Delta s)$ results as a special case when $p = e$.

We state the generic homogeneous algorithm for solving the pair (P|SOCP) and (D|SOCP) in Algorithm 11.2.

The following theorem, which is known to hold, gives the computational complexity (worst behavior) of Algorithm 11.2 in terms of the rank of the underlying second-order cone.

---

**Theorem 11.6** *Let $\epsilon_0 > 0$ be the residual error at a starting point, and $\epsilon > 0$ be a given tolerance. Under Assumptions 11.2 and 11.1, if the pair (P|SOCP) and (D|SOCP) has a solution $(x^\star; y^\star; s^\star)$, then Algorithm 11.2 finds an $\epsilon$-approximate solution (i.e., a solution with residual error less than or equal to $\epsilon$) in at most*
$$O\left( \sqrt{2r} \ln\left( trace\,(x^\star + s^\star)\left(\frac{\epsilon_0}{\epsilon}\right)\right)\right)$$
*iterations.*

---

We point out that the result in Theorem 11.6 for SOCP is the counterpart of that in Theorem 10.12 for linear programming.

# EXERCISES

**11.1**    Prove Lemma 11.2.

**11.2**    The $p$th-order cone of dimension $n$ is defined as

$$\mathcal{P}_p^n \triangleq \left\{ x \in \mathbb{E}^n : x_0 \geq \|\widetilde{x}\|_p \right\},$$

where $\| \cdot \|_{p \geq 1}$, is the $p$-norm of $\xi$:

$$\|\xi\|_p := \left( \sum_{i=1}^m |\xi_i|^p \right)^{1/p}, \text{ for } \xi \in \mathbb{R}^m.$$

Clearly, when $p = 2$, the $p$th-order cone reduces to the second-order cone, i.e., $\mathbb{E}_+^n = \mathcal{P}_2^n$. In the proof of Lemma 11.3, we used the Cauchy-Schwartz inequality to show that the second-order cone is self-dual (i.e., $\mathcal{P}_2^{n\star} = \mathcal{P}_2^n$). More generally, use the Hölder's inequality to show that $\mathcal{P}_p^{n\star} = \mathcal{P}_q^n$ for $p \in [1, \infty]$, where $q$ is the conjugate of $p$.

**11.3**    The $n$th-dimensional elliptic cone is defined as

$$\mathcal{K}_M^n \triangleq \left\{ x = (x_0; \widetilde{x}) \in \mathbb{R} \times \mathbb{R}^{n-1} : x_0 \geq \left\| M\widetilde{x} \right\| \right\}, \tag{11.28}$$

where $M$ be a nonsingular matrix of order $n - 1$. Clearly, when $M = I_{n-1}$, the elliptic cone reduces to the second-order cone, i.e., $\mathbb{E}_+^n = \mathcal{K}_{I_{n-1}}^n$. In the proof of Lemma 11.3, we showed that the second-order cone is self-dual (i.e., $(\mathcal{K}_{I_{n-1}}^n)^\star = \mathcal{K}_{I_{n-1}}^n$). More generally, show that

$$(\mathcal{K}_M^n)^\star = \mathcal{K}_{(M^{-1})^\top}^n.$$

**11.4**    Prove item (c) in Lemma 11.8.

**11.5**    Prove Lemma 11.9.

**11.6**    Prove Lemma 11.10.

**11.7**    Implement Algorithm 11.1 and test it on the class of instances of the pair (P|SOCP) and (D|SOCP) with $n = 2m$ and

$$\begin{aligned}
c &\triangleq 10e - 2\,\mathbf{1} + 4\,\mathbf{rand}(n, 1) \in \mathbb{E}^n, \\
b &\triangleq 10e - 2\,\mathbf{1} + 4\,\mathbf{rand}(m, 1) \in \mathbb{R}^m, \\
A &\triangleq \left[ \hat{A} \vdots \mathbf{Randn}(m, n - m) \right] \in \mathbb{R}^{m \times n},
\end{aligned}$$

where, for $1 \leq i, j \leq m$,

$$
\hat{a}_{ij} \triangleq \begin{cases} 2, & \text{if } i = j - 1, \\ 100, & \text{if } i = j, \\ -2, & \text{if } i = j + 1, \\ 0, & \text{otherwise.} \end{cases}
$$

Here, **1** is a vector of ones with with an appropriate dimension, and **rand**$(\cdot, 1)$ (respectively, Randn$(\cdot, \cdot)$) is a random vector (respectively, matrix) with the indicated dimension (respectively, size). Take $x^{(0)} = e \in \mathbb{E}^n$ and $y^{(0)} = \mathbf{0} \in \mathbb{R}^m$ as your initial strictly feasible points. You may also take $\epsilon = 10^{-6}, \sigma = 0.1$ and $\rho = 0.99$.

**11.8**    In this and the following exercises, we practice writing the homogeneous model of an SOCP problem and the corresponding search direction system. The underlying problems are SOCPs, in primal and dual standard forms, with block diagonal structures. Let $r \geq 1$ be an integer. For $i = 1, 2, \dots, r$ and $k = 0, 1, \dots, K$, let $m, n_k, n_{ki}$ be positive integers such that $n_k = \sum_{i=1}^{r} n_{ik}$. Let the data

$$
\begin{aligned}
W_0 &\triangleq (W_{01}, W_{02}, \dots, W_{0r}), \text{ where } W_{0i} \in \mathbb{R}^{m_0 \times n_{0i}}; \\
c_0 &\triangleq (c_{01}; c_{02}; \dots; c_{0r}), \text{ where } c_{0i} \in \mathbb{E}^{n_{0i}}; \\
B_k &\triangleq (B_{k1}, B_{k2}, \dots, B_{kr}), \text{ where } B_{ki} \in \mathbb{R}^{m_1 \times n_{ki}}; \\
W_k &\triangleq (W_{k1}, W_{k2}, \dots, W_{kr}), \text{ where } W_{ki} \in \mathbb{R}^{m_1 \times n_{ki}}; \\
c_k &\triangleq (c_{k1}; c_{k2}; \dots; c_{kr}), \text{ where } c_{ki} \in \mathbb{E}^{n_{ki}},
\end{aligned}
$$

for $k = 1, 2, \dots, K$ be given. Consider the SOCP problem:

$$
\begin{array}{llllll}
\min & c_0^{\mathsf{T}} x_0 & + & c_1^{\mathsf{T}} x_1 & + \cdots + & c_K^{\mathsf{T}} x_K \\
\text{s.t.} & W_0 x_0 & & & & = h_0, \\
& B_1 x_0 & + & W_1 x_1 & & = h_1, \\
& \vdots & & & \ddots & \vdots \\
& B_K x_0 & + & & W_K x_K & = h_K, \\
& x_0, & & x_1, & \cdots, & x_K \geq \mathbf{0},
\end{array} \quad (11.29)
$$

where

$$
\begin{aligned}
x_0 &\triangleq (x_{01}; x_{02}; \dots; x_{0r}), \text{ where } x_{0i} \in \mathbb{E}^{n_{0i}}; \\
x_k &\triangleq (x_{k1}; x_{k2}; \dots; x_{kr}), \text{ where } x_{ki} \in \mathbb{E}^{n_{ki}},
\end{aligned}
$$

for $k = 1, 2, \ldots, K$, are the primal decision variables, and $h_0 \in \mathbb{R}^{m_0}, h_k \in \mathbb{R}^{m_1}$, $k = 1, 2, \ldots, K$, are right-hand side vectors. The dual of (11.29) is the problem:

$$
\begin{aligned}
\max \quad & h_0^{\mathsf{T}} y_0 \;+\; h_1^{\mathsf{T}} y_1 \;+\; \cdots \;+\; h_K^{\mathsf{T}} y_K \\
\text{s.t.} \quad & W_0^{\mathsf{T}} y_0 \;+\; B_1^{\mathsf{T}} y_1 \;+\; \cdots \;+\; B_K^{\mathsf{T}} y_K \;+\; s_0 \;=\; c_0, \\
& \qquad\qquad\quad W_1^{\mathsf{T}} y_1 \qquad\qquad\qquad\qquad\quad +\; s_1 \;=\; c_1, \\
& \qquad\qquad\qquad\qquad\quad \ddots \qquad\qquad\qquad\qquad\qquad \vdots \\
& \qquad\qquad\qquad\qquad\qquad\qquad\quad W_K^{\mathsf{T}} y_K \;+\; s_K \;=\; c_K, \\
& \qquad s_0, \qquad\quad s_1, \qquad \ldots, \qquad\quad s_K \qquad\quad \geq\; \mathbf{0},
\end{aligned}
\tag{11.30}
$$

where

$$
\begin{aligned}
y \;&\triangleq\; (y_0; y_1; \ldots; y_K) \in \mathbb{R}^{m_0 + K m_1}; \\
s_k \;&\triangleq\; (s_{k1}; s_{k2}; \ldots; s_{kr}), \text{ where } s_{ki} \in \mathbb{E}^{n_{ki}},
\end{aligned}
$$

for $k = 0, 1, \ldots, K$, are the dual decision variables. Write the homogeneous model for the pair (11.29) and (11.30).

**11.9** Write the search direction system corresponding to the homogeneous model obtained in Exercise 11.8.

# CHAPTER 12

# SEMIDEFINITE PROGRAMMING AND COMBINATORIAL OPTIMIZATION

## Contents

© 2022 by Baha Alzalg | Kindle Direct Publishing, Washington, United States 2022/9
B. Alzalg, *Combinatorial and Algorithmic Mathematics: From Foundation to Optimization*,
DOI 10.5281/zenodo.7111009

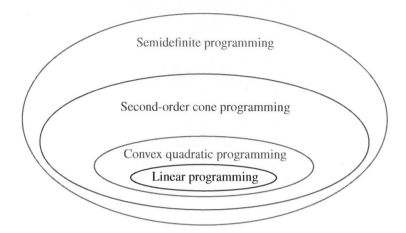

Figure 12.1: A Venn diagram of different classes of optimization problems.

In Chapters 10 and 11, we have studied linear programming and second-order cone programming, respectively. In modern convex optimization, the class of optimization that is an immediate enlargement of second-order cone optimization is the so-called semidefinite optimization. Figure 12.1 shows graphical relationships among different classes of optimization problems. Semidefinite programming (SDP for short) problems, which include linear programming problems and second-order cone programming problems as special cases, are a class of convex optimization problems in which the variable is not a vector that is required to be nonnegative, but rather a symmetric matrix that is required to be positive semidefinite (see Definition 3.3). This chapter is devoted to studying SDP problems. We also refer to Todd [Tod01] for an excellent survey paper on this topic.

## 12.1. The cone of positive semidefinite matrices

Recall that a square matrix is positive semidefinite (respectively, positive definite) if it is symmetric and all its eigenvalues are nonnegative (respectively, positive). As an alternative to the above definition, a matrix $U \in \mathbb{R}^{n \times n}$ is positive semidefinite (respectively, positive definite) if it is symmetric and $x^\mathsf{T} U x \geq 0$ for all $x \in \mathbb{R}^n$ (respectively, $x^\mathsf{T} U x > 0$ for all $x \in \mathbb{R}^n - \{0\}$). An immediate corollary here is that $xx^\mathsf{T}$ is a positive semidefinite matrix for all $x \in \mathbb{R}^n$.

This section aims to introduce tools needed to study the SDP problems. We start this by introducing notations that will be used throughout this chapter. We use $\mathbb{S}^n$ to denote the space of real symmetric $n \times n$ matrices. An identity matrix of appropriate dimension is denoted by $I$. The set of the positive semidefinite matrices in $\mathbb{S}^n$ is a convex self-dual cone (see Lemma 12.1). See Figure 12.2 which shows a 3D plot of the boundary of the cone of the $2 \times 2$ positive semidefinite matrices.

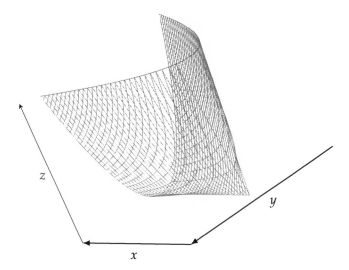

Figure 12.2: A 3D plot of the boundary of the cone of the positive semidefinite matrices in $\mathbb{S}^2$. The boundary is the continuous surface shaped by the above blue mesh.

For $U, V \in \mathbb{S}^n$, we write $U \geq 0$ ($U \succ 0$) to mean that $U$ is positive semidefinite (positive definite), and we use $U \geq V$ or $V \leq U$ to mean that $U - V \geq 0$.

The bilinear map $\circ : \mathbb{R}^{n \times n} \times \mathbb{R}^{n \times n} \to \mathbb{R}^{n \times n}$ is defined as

$$U \circ V \triangleq \frac{1}{2}(UV + VU). \tag{12.1}$$

The Frobenius inner product $\bullet : \mathbb{R}^{m \times n} \times \mathbb{R}^{mn \times n} \to \mathbb{R}$ between is defined as

$$U \bullet V \triangleq \text{trace}\,(U^\mathsf{T} V).$$

It is known that the space $\mathbb{S}^n$ under the bilinear map $\circ : \mathbb{S}^n \times \mathbb{S}^n \longrightarrow \mathbb{S}^n$ defined in (12.1) forms a Euclidean Jordan algebra (see [Far94, SA03] for definitions) equipped with the standard inner product $\langle U, V \rangle \triangleq U \bullet V = \text{trace}\,(U \circ V) = \text{trace}\,(UV)$. The Frobenius norm of a matrix $U \in \mathbb{R}^{m \times n}$ is defined as

$$\|U\|_F \triangleq U \bullet U = \sum_i \lambda_i^2(U),$$

where $\lambda_i's(U)$ are the eigenvalues of the matrix $U$.

It is known that, for any $U, V \in \mathbb{S}^n$, we have (see [HJ90])

$$\begin{aligned}
\|U + V\|_F^2 &= \|U\|_F^2 + \|V\|_F^2 + 2U \bullet V, \\
\|UV\|_F &\leq \|U\|_2\,\|V\|_F \leq \|U\|_F\,\|V\|_F, \\
\|UV\|_F &\leq \|U\|_F\,\|V\|_2 \leq \|U\|_F\,\|V\|_F,
\end{aligned} \tag{12.2}$$

where $\|U\|_2 \triangleq \max\{\|Ux\|_2 : \|x\|_2 = 1\} = \max_i |\lambda_i(U)|$ is the operator norm (or 2-norm) of the matrix $U$.

The following properties of matrices are taken from [Tod01, Section 3] and are given here without proofs. For proofs, see, for example, [HJ90, Wat91, Tod01].

> **Property 12.1 (Trace commutativity)** *If $A \in \mathbb{R}^{m \times n}$ and $B \in \mathbb{R}^{n \times m}$, then $trace(AB) = trace(BA)$.*

In particular, like any inner product, we have that $U \bullet V = V \bullet U$ for $U, V \in \mathbb{R}^{m \times n}$. The following property in $\mathbb{S}^n$ is the counterpart of Property 11.1 in $\mathbb{E}^n$.

> **Property 12.2 (Spectral decomposition in $\mathbb{S}^n$)** *Any $A \in \mathbb{S}^n$ can be expressed in exactly one way as a product $A = Q(A)\Lambda(A)Q(A)^\mathsf{T}$, where $Q(A) \in \mathbb{R}^{n \times n}$ is an orthogonal matrix whose columns are the eigenvectors of $A$, and $\Lambda(A) \in \mathbb{R}^{n \times n}$ is a diagonal matrix whose diagonal entries are the eigenvalues of $A$.*

Recall that a square matrix is called lower triangular if all the entries above the main diagonal are zero. A unit lower triangular matrix $A$ is a lower triangular matrix with $a_{ii} = 1$ for $1 \le i \le n$. Therefore, a unit lower triangular matrix has the form:

$$
\begin{bmatrix}
1 & & & & \\
\times & 1 & & \mathbf{0} & \\
\times & \times & 1 & & \\
\times & \times & \times & 1 & \\
\times & \times & \times & \times & 1
\end{bmatrix}.
$$

Unit lower triangular matrices arise in the following property about the $LDL^\mathsf{T}$ decomposition, which is a variant of the spectral decomposition.

> **Property 12.3 ($LDL^\mathsf{T}$ decomposition)** *Any $A \ge 0$ ($A > 0$) can be expressed in exactly one way as a product $A = L(A)D(A)L^\mathsf{T}(A)$, where $L(A) \in \mathbb{R}^{n \times n}$ is a unit lower triangular matrix, and $D(A) \in \mathbb{R}^{n \times n}$ is a diagonal matrix whose diagonal entries are nonnegative (positive).*

An immediate corollary of Property 12.3 is that every $A \ge 0$ ($A > 0$) has a square root $A^{1/2} \ge 0$ ($A^{1/2} > 0$). Simply, we take $A^{1/2} = L(A)D^{1/2}(A)L^\mathsf{T}(A)$. Another immediate corollary of this property is that every $A > 0$ has an inverse. Simply, we take $A^{-1} = L(A)D^{-1}(A)L^\mathsf{T}(A)$. The Cholesky decomposition is a variant of the $LDL^\mathsf{T}$ decomposition.

> **Property 12.4 (Cholesky decomposition)** *Any $A > 0$ can be expressed in exactly one way as a product $A = LL^\mathsf{T}$, where $L \in \mathbb{R}^{n \times n}$ is lower triangular and has all main diagonal entries positive.*

The following property is used to represent the quadratic forms with the Frobenius inner product.

> **Property 12.5 (Representing quadratics)** *If $A \in \mathbb{S}^n$ and $x \in \mathbb{R}^n$, then*
>
> $$x^\mathsf{T} A x = A \bullet x x^\mathsf{T}.$$

In the following property and in what follows, we say that two matrices are simultaneously diagonalized if they have spectral decompositions with the same $Q$.

> **Property 12.6 (Commutativity and symmetry)** *If $A, B \in \mathbb{S}^n$, then $A$ and $B$ commute if and only if $AB$ is symmetric, if and only if $A$ and $B$ are simultaneously diagonalized.*

The Schur complement property presented in Property 12.7 has many applications in SDP.

> **Property 12.7 (Schur complement)** *Let $A$ and $B$ be symmetric matrices and $A > 0$. Then*
>
> $$\begin{bmatrix} A & B^\mathsf{T} \\ B^\mathsf{T} & C \end{bmatrix} \geq 0 \; (\succ 0) \iff C - B^\mathsf{T} A^{-1} B \geq 0 \; (\succ 0).$$
>
> *The matrix $C - B^\mathsf{T} A^{-1} B$ is called the Schur complement of the submatrix $A$.*

Associated with $U_i \in \mathbb{S}^n (i = 1, 2, \ldots, m)$, we define the linear operator $\mathcal{U} : \mathbb{S}^n \to \mathbb{R}^m$ and its adjoint operator $\mathcal{U}^\star : \mathbb{R}^m \to \mathbb{S}^n$ as

$$\mathcal{U}X \triangleq \begin{bmatrix} U_1 \bullet X \\ U_2 \bullet X \\ \vdots \\ U_m \bullet X \end{bmatrix} \in \mathbb{R}^m, \text{ and } \mathcal{U}^\star z \triangleq \sum_{l=1}^{m} z_i U_i \in \mathbb{S}^n, \tag{12.3}$$

respectively. Note that

$$(\mathcal{U}X)^\mathsf{T} z = \sum_{l=l}^{m} (U_i \bullet X) z_i = (\sum_{l=1}^{m} z_i U_i \bullet X) = \mathcal{U}^\star z \bullet X, \tag{12.4}$$

for $X \in \mathbb{S}^n$ and $z \in \mathbb{R}^m$.

Associated with each nonsingular matrix $P \in \mathbb{R}^{n \times n}$, the symmetrization operator $\mathcal{H}_P : \mathbb{R}^{n \times n} \longrightarrow \mathbb{S}^n$ is defined as

$$\mathcal{H}_P(M) \triangleq \frac{1}{2} \left( PMP^{-1} + \left( PMP^{-1} \right)^\mathsf{T} \right) \tag{12.5}$$

for $M \in \mathbb{R}^{n \times n}$. We will use the operator $\mathcal{H}_P(\cdot)$ in order to symmetrize the optimality condition system, which is needed for applying Newton's method.

To end the notations, we point out that in the sequel $\mathrm{Diag}(x) \in \mathbb{S}^n$ is written for the diagonal matrix with the vector $x \in \mathbb{R}^n$ on its diagonal.

## 12.2.  Semidefinite programming formulation

In this section, we introduce the SDP problem and formulate known class of optimization problems as SDPs.

### Problem formulation

We will be concentrating on SDP problems in primal standard form and in the corresponding dual form. The problem in primal form can be written as

$$
\text{(P|SDP)} \quad
\begin{aligned}
\min \quad & C \bullet X \\
A_i \bullet X \ =\ & b_i, \quad i = 1, \ldots, m, \\
X \ \geq\ & 0,
\end{aligned}
$$

where the data $C$, $A_i$, $i = 1, \ldots, m$, are real symmetric $n \times n$ matrices while $b$ is a real $m$-vector and the variable $X$ is a real symmetric $n \times n$ matrix.

It is convenient to introduce a *slack matrix* $S$ and rewrite the problem as

$$
\text{(D|SDP)} \quad
\begin{aligned}
\max \quad & b^{\mathsf{T}}y \\
\textstyle\sum_{i=1}^{m} y_i A_i \ +\ & S \ =\ C, \\
S \ \geq\ & 0.
\end{aligned}
$$

Here the variable $y$ is a real $m$-vector while $S$ is a real symmetric $n \times n$ matrix.

Using the notations introduced in (12.3), we can write our problems more compactly as

$$
\text{(P|SDP)} \quad
\begin{aligned}
\min \quad & C \bullet X \\
\mathcal{A}X \ =\ & b, \\
X \ \geq\ & 0;
\end{aligned}
\qquad
\text{(D|SDP)} \quad
\begin{aligned}
\max \quad & b^{\mathsf{T}}y \\
\mathcal{A}^{\star}y \ +\ & S \ =\ C, \\
S \ \geq\ & 0.
\end{aligned}
$$

### Formulating problems as SDPs

In this part, we formulate four general classes of optimization problems as SDPs. We start with linear optimization.

*Linear programming*   It is easy to see that the following pair of primal and dual linear programming problems:

$$
\begin{aligned}
\min \quad & c^{\mathsf{T}}x \\
\text{s.t.} \quad Ax \ =\ & b, \\
x \ \geq\ & 0;
\end{aligned}
\qquad
\begin{aligned}
\max \quad & b^{\mathsf{T}}y \\
\text{s.t.} \quad A^{\mathsf{T}}y + s \ =\ & c, \\
s \ \geq\ & 0,
\end{aligned}
$$

is equivalent to the pair (P|SDP) and (D|SDP), respectively, with

$$C = \text{Diag}(c), X = \text{Diag}(x), S = \text{Diag}(s), \text{ and } A_i = \text{Diag}(a_i),$$

where $a_i$ is the $i$th row of $A$.

***Convex quadratic programming***    In convex quadratic optimization problems, we minimize a strictly convex quadratic function subject to affine constraint functions:

$$\begin{aligned} \min \quad & q(x) \triangleq x^T Q x + c^T x \\ \text{s.t.} \quad & Ax = b, \\ & x \geq 0. \end{aligned} \tag{12.6}$$

Since Problem (12.6) is strictly convex, the matrix $Q$ must be a symmetric positive definite matrix (i.e., $Q = Q^T$ and $Q > O$). Let $Q = LL^T$ (see Property 12.4). Then, using Schur complements (Property 12.7), the inequality $q(x) \leq t$ can be written as

$$\begin{bmatrix} I & L^T x \\ x^T L & t - c^T x \end{bmatrix} \geq 0.$$

Therefore, the quadratic optimization problem (12.6) can be formulated as the following SDP relaxation.

$$\begin{aligned} \min \quad & t \\ \text{s.t.} \quad & Ax = b, \\ & \begin{bmatrix} I & L^T x \\ x^T L & t - c^T x \end{bmatrix} \geq 0, \\ & x \geq 0. \end{aligned}$$

We also point out that, more generally, convex quadratically constrained quadratic optimization problems can also be formulated as SDP problems (see for example [Tod01, Example 5]).

***Second-order cone programming***    Consider the second-order cone programming problem (as introduced in Section 11.2):

$$\begin{aligned} \min \quad & c^T x \\ \text{s.t.} \quad & Ax = b, \\ & x \geq 0. \end{aligned}$$

The second-order cone constraint $x \geq 0$ is the inequality $x_0 \geq \|\widetilde{x}\|$ (i.e., $x \in \mathbb{E}_+^n$; see Definition 11.1). Using Schur complements (Property 12.7), we have that

$$x = (x_0; \widetilde{x}) \geq 0 \iff \begin{bmatrix} x_0 I & \widetilde{x} \\ \widetilde{x}^T & x_0 \end{bmatrix} \geq 0, \tag{12.7}$$

and equivalently, $\mathrm{Arw}(x) \succeq 0$, where $\mathrm{Arw}(x)$ is the arrow-shaped matrix introduced in Definition 11.4.

***Rotated quadratic cone programming***    Rotated quadratic cone programs were introduced in Section 11.2 in the context of second-order cone optimization. Let $n$ be a positive integer and $M$ be a nonsingular matrix of order $n - 2$. Recall that the $n$th dimensional rotated quadratic cone is

$$\mathcal{K}^n = \left\{ x = (x_0; x_1; \hat{x}) \in \mathbb{R} \times \mathbb{R} \times \mathbb{R}^{n-2} : 2x_0 x_1 \geq \|\hat{x}\|^2,\ x_0 \geq 0, x_1 \geq 0 \right\}.$$

Recall also that the constraint on $x$ that satisfies the inequality $2x_0 x_1 \geq \|\hat{x}\|^2$ is called the hyperbolic constraint. As mentioned in Section 11.2, a rotated quadratic cone optimization problems, a linear objective function is minimized subject to linear constraints and hyperbolic constraints. Note that

$$
\begin{aligned}
(x_0; x_1; \hat{x}) \in \mathcal{K}^n \quad &\Longleftrightarrow \quad (2x_0 + x_1; 2x_0 - x_1; 2\hat{x}) \in \mathbb{E}_+^n \qquad\qquad\qquad\ \ \text{(By (11.6))} \\
&\Longleftrightarrow \quad \begin{bmatrix} (2x_0 + x_1)I & (2x_0 - x_1; 2\hat{x}) \\ (2x_0 - x_1; 2\hat{x})^\mathsf{T} & 2x_0 + x_1 \end{bmatrix} \in \mathbb{S}_+^n. \quad \text{(By (12.7))}
\end{aligned}
$$

This means that a rotated quadratic cone optimization problem can be expressed as an SDP problem because the hyperbolic constraint is equivalent to a linear matrix inequality.

## 12.3.  Applications in combinatorial optimization

In this section, we describe three combinatorial applications of SDP. For non-combinatorial applications, we point out that the applications that we described in Section 11.3 in the context of second-order cone programming were first formulated as applications of SDP (see [VB99]) The focus in this chapter is only on those applications in combinatorial optimization. The material of this section has appeared in some sections of [Goe98, VB99, Tod01]. For more applications of SDP, we refer the reader to [Goe98, KMS98, VB99, Ren99, Tod01, LP21]. In particular, the work in [LP21] presents an interesting application in Ramsey theory.

### Shannon capacity of graphs

In this problem, the objective is to bound the Shannon capacity, or the independence number of graphs. We need some definitions.

---

**Definition 12.1** *Let $G = (V, E)$ be an undirected graph. Then:*

*(a) A clique set (or simply, a clique) of $G$ is a set $S \subseteq V$ of mutually adjacent vertices.*

(b) *A clique cover of G is a collection C of cliques that together include all V. That is, $V \subseteq \cup_{S \in C} S$.*

(c) *The clique cover number of G, denoted by $\bar{\chi}(G)$, is the minimum cardinality of a clique cover of G. That is, $\bar{\chi}(G)$ is the smallest number of cliques of G whose union covers the set V.*

Note that the clique cover number of a graph and the chromatic number (see Definition 4.7) of its complement are identical, i.e., $\bar{\chi}(G) = \chi(\bar{G})$.

**Definition 12.2** *Let $G = (V, E)$ be an undirected graph. Then:*

(a) *An independent (or stable) set of G is a set $S \subseteq V$ of mutually nonadjacent vertices.*

(b) *The independence (or stability) number of G, denoted by $\alpha(G)$, is the maximum cardinality of an independent set G. In other words, $\alpha(G)$ is the largest number of vertices that can be colored with the same color in a graph coloring of G.*

Note that each node in a stable set must be in a different clique in a clique cover, hence we have

$$\alpha(G) \leq \bar{\chi}(G). \tag{12.8}$$

If the equality holds in (12.8), then we say that $G$ is a perfect graph.

It is known that $\alpha(G)$ and $\bar{\chi}(G)$ are both NP-hard to compute. In the Shannon capacity problem, the objective is to approximate a function that lies between $\alpha(G)$ and $\bar{\chi}(G)$. This function is called the Lovász's theta function (denoted by $\theta(G)$), and is the optimal solution of the SDP problem:

$$\begin{aligned} \max \quad & \mathbf{11}^T \bullet X \\ \text{s.t.} \quad & I \bullet X = 1, \\ & x_{ij} = 0, \quad for\ (i, j) \in E, \\ & X \geq 0, \end{aligned} \tag{12.9}$$

where $\mathbf{1}$ is a vector of ones with an appropriate dimension.

Note that the SDP problem (12.9) is in primal form, but in maximization form. The dual of Problem (12.9) is the SDP problem [Tod01]:

$$\begin{aligned} \min \quad & s \\ \text{s.t.} \quad & sI + \sum_{(i,j) \in E} y_{ij} M_{ij} \geq \mathbf{11}^T, \end{aligned}$$

where $M_{ij}$ is the symmetric matrix that is all zero except for ones in the $ij$th and $ji$th positions. Clearly, the above SDP models calculate the numbers $\alpha(G)$ and $\bar{\chi}(G)$ exactly if $G$ s perfect.

## Max-cut of graphs

In this problem, the objective is to find the cut of maximum weight in undirected graphs. This problem arises in finding the ground state of a spin glass; see [PT93]. We need the following definition.

---

**Definition 12.3** *Let $G = (V, E)$ be an undirected graph with a nonnegative weight vector $\mathbf{w} = (w_{ij})_{(i,j)\in E} \in \mathbb{R}_+^{|E|}$. The cut determined by a subset $S \subseteq V$ is the set $\delta(S) \triangleq \{(i, j) \in E : i \in S, j \notin S\}$, and its weight is defined as $w(\delta(S)) \triangleq \sum_{(i,j)\in\delta(S)} w_{ij}$.*

---

In light of Definition 12.3, we can assume that the underlying graph is complete by letting $w_{ij} = 0$ for all $(i, j) \notin E$ (we let also $w_{ii} = 0$ for all $i$). To represent the cut $\delta(S)$, we introduce the variable $\mathbf{x} \in \mathbb{R}^{|V|}$, which is defined as

$$x_i = \begin{cases} 1, & \text{if } i \in S, \\ -1, & \text{if } i \in V - S. \end{cases}$$

It follows that

$$1 - x_i x_j = \begin{cases} 2, & \text{if } (i, j) \in \delta(S), \\ 0, & \text{if } (i, j) \notin \delta(S). \end{cases}$$

We also define the matrix $C \in \mathbb{S}^{|V|}$ as

$$c_{ij} = \begin{cases} -w_{ij}/4, & \text{if } i \neq j, \\ \sum_j w_{ij}/4, & \text{if } i = j. \end{cases}$$

Note that

$$
\begin{aligned}
w(\delta(S)) &= \sum_{(i,j)\in\delta(S)} w_{ij} \\
&= \frac{1}{2} \sum_{i<j} w_{ij}(1 - x_i x_j) \\
&= \frac{1}{4} \sum_i \sum_j w_{ij}(1 - x_i x_j) \\
&= -\sum_i \left( x_i \left( \sum_j \frac{w_{ij}}{4} x_j \right) \right) \\
&= \sum_i \left( x_i \left( \sum_j c_{ij} x_j \right) \right) \\
&= \mathbf{x}^\mathsf{T} C \mathbf{x} = C \bullet \mathbf{x}\mathbf{x}^\mathsf{T}.
\end{aligned}
$$

Because every $(+1, -1)$-vector in $\mathbb{R}^{|V|}$ corresponds to a cut in the graph $G = (V, E)$, the max-cut problem can be formulated as the integer program:

$$\begin{aligned} \max \quad & C \bullet xx^{\mathsf{T}} \\ \text{s.t.} \quad & x \in \{+1, -1\}^{|V|}, \end{aligned} \tag{12.10}$$

or equivalently, as the quadratic program:

$$\begin{aligned} \max \quad & x^{\mathsf{T}} C x \\ \text{s.t.} \quad & x_i^2 = 1, \ i \in V. \end{aligned} \tag{12.11}$$

Observe that Problem (12.11) is linear in the products $x_i x_j$, and these are the entries of the rank one matrix $X \triangleq xx^{\mathsf{T}}$. Observe also that $X \in \mathbb{S}_+^{|V|}$, with $x_{ii} = 1$ for all $i \in V$. It follows that Problem (12.10) can be written as

$$\begin{aligned} \max \quad & C \bullet X \\ \text{s.t.} \quad & x_{ii} = 1, i \in V, \\ & X \geq 0, \\ & X \text{ is of rank one.} \end{aligned} \tag{12.12}$$

Relaxing the last constraint in Problem (12.12), we obtain the SDP problem [Tod01]:

$$\begin{aligned} \max \quad & C \bullet X \\ \text{s.t.} \quad & x_{ii} = 1, i \in V, \\ & X \geq 0. \end{aligned}$$

We also mention that Todd [Tod01] discusses two other different ways to arrive at an SDP relaxation of the max-cut problem.

## Combinatorial topology optimization

Ben-Tal and Bendsøe in [BTB93] consider the following problem from combinatorial topology optimization optimization of truss structures. A structure of $k$ linear elastic bars connects a set of $p$ nodes. They assume the geometry (topology and lengths of bars) and the material are fixed. The goal is to size the bars, i.e., determine appropriate cross-sectional areas of the bars.

For $i = 1, 2, \ldots, k$, and $j = 1, 2, \ldots, p$, we define the following decision variables and parameters:

- $f_j \triangleq$ the external force applied on the $j^{th}$ node,

- $d_j \triangleq$ the (small) displacement of the $j^{th}$ node resulting from the load force $f_j$,

- $x_i \triangleq$ the cross-sectional area of the $i^{th}$ bar,

- $\underline{x}_i \triangleq$ the lower bound on the cross-sectional area of the $i^{th}$ bar,

- $\overline{x}_i \triangleq$ the upper bound on the cross-sectional area of the $i^{th}$ bar,

- $l_i \triangleq$ the length of the $i^{th}$ bar,

- $v \triangleq$ the maximum allowed volume of the bars of the structure,

- $G(x) \triangleq \sum_{i=1}^{k} x_i G_i$ is the stiffness matrix, where the matrices $G_i \in \mathcal{S}^p$, $i = 1, 2, \ldots, k$, depend only on fixed parameters (such as length of bars and material).

In the simplest version of the problem they consider one fixed set of externally applied nodal forces $f_j$, $j = 1, 2, \ldots, p$. Given this, the elastic stored energy within the structure is given by

$$\varepsilon = f^\mathsf{T} d,$$

which is a measure of the inverse of the stiffness of the structure. In view of the definition of the stiffness matrix $G(x)$, we can also conclude that the following linear relationship between $f$ and $d$:

$$f = G(x)\, d.$$

The objective is to find the stiffest truss by minimizing $\varepsilon$ subject to the inequality $l^\mathsf{T} x \leq v$ as a constraint on the total volume (or equivalently, weight) and the constraint $\underline{x} \leq x \leq \bar{x}$ as upper and lower bounds on the cross-sectional areas.

For simplicity, we assume that $\underline{x} > 0$ and $G(x) > 0$, for all $x > 0$. In this case we can express the elastic stored energy in terms of the inverse of the stiffness matrix and the external applied nodal force as follows:

$$\varepsilon = f^\mathsf{T} G(x)^{-1} f.$$

In summary, they consider the problem [VB99]:

$$
\begin{aligned}
\min \quad & f^\mathsf{T} G(x)^{-1} f \\
\text{s.t.} \quad & \underline{x} \leq x \leq \bar{x}, \\
& l^\mathsf{T} x \leq v,
\end{aligned}
$$

which is equivalent to:

$$
\begin{aligned}
\min \quad & s \\
\text{s.t.} \quad & f^\mathsf{T} G(x)^{-1} f \leq s, \\
& \underline{x} \leq x \leq \bar{x}, \\
& l^\mathsf{T} x \leq v.
\end{aligned}
\tag{12.13}
$$

The first inequality constraint in (12.13) is just fractional quadratic function inequality constraint and it can be formulated as a positive semidefinite constraint. This problem can be cast as an SDP problem as follows:

$$\min \quad s$$

$$\text{s.t.} \quad \begin{bmatrix} G(x) & f \\ f^{\mathsf{T}} & s \end{bmatrix} \geq 0,$$

$$\underline{x} \leq x \leq \bar{x},$$

$$l^{\mathsf{T}} x \leq v.$$

## 12.4. Duality in semidefinite programming

In this section, we present a glimpse of the duality theory associated with SDP and introduce the complementarity condition as one of the optimality conditions of SDP. Much, but not all, of the duality theory for SDP is very similar to duality theory for linear programming (and for second-order cone programming). Recall that a regular cone is self-dual if it equals its dual cone (see Definition 3.22). Now we need the following lemma.

**Lemma 12.1** *The symmetric positive semidefinite cone* $\mathbb{S}^n_+$ *is self-dual under the Frobenius inner product.*

**Proof** We verify that the symmetric positive semidefinite cone $\mathbb{S}^n_+$ equals its dual cone, which is defined as

$$(\mathbb{S}^n_+)^{\star} \triangleq \{ X \in \mathbb{S}^n : X \bullet Y \geq 0 \text{ for all } Y \in \mathbb{S}^n_+ \}.$$

We first prove that $(\mathbb{S}^n_+)^{\star} \subseteq \mathbb{S}^n_+$. Assume that $X \notin \mathbb{S}^n_+$, then there exists a nonzero vector $y \in \mathbb{R}^n$ such that $X \bullet yy^{\mathsf{T}} = y^{\mathsf{T}} X y < 0$, which shows that $X \notin (\mathbb{S}^n_+)^{\star}$.

Now, we prove that $\mathbb{S}^n_+ \subseteq (\mathbb{S}^n_+)^{\star}$. Letting $X \in \mathbb{S}^n_+$, then for any $Y \in \mathbb{S}^n_+$, we have that

$$X \bullet Y = \text{trace}(XY) = \text{trace}\left( X^{\frac{1}{2}} Y X^{\frac{1}{2}} \right) \geq 0,$$

where we used the fact that $X^{\frac{1}{2}} Y X^{\frac{1}{2}} \in \mathbb{S}^n$ to obtain the inequality. In fact, if $U \in \mathbb{S}^n_+$ and $Q$ and $\Lambda$ are respectively its corresponding orthogonal and diagonal matrices obtained from the spectral decomposition (see Property 12.2), then we have

$$\text{trace}(U) = \text{trace}(Q\Lambda Q^{\mathsf{T}}) = \text{trace}(\Lambda Q Q^{\mathsf{T}}) = \text{trace}(\Lambda) = \sum_{i=1}^{n} \lambda_i \geq 0.$$

We have shown that $X \in (\mathbb{S}^n_+)^{\star}$ whenever $X \in \mathbb{S}^n_+$. The proof is complete. ∎

Theorem 12.1 presents the weak duality property in SDP, and states that the objective function values of primal feasible solutions always dominate those of dual feasible solutions.

> **Theorem 12.1 (Weak duality in SDP)**  *Let $X$ and $(y, S)$ be feasible for $(P|SDP)$ and $(D|SDP)$ respectively, then*
>
> $$C \bullet X - b^\mathsf{T} y = S \bullet X \geq 0.$$

**Proof**  Using (12.4), we have

$$
\begin{aligned}
C \bullet X - b^\mathsf{T} y &= (\mathcal{A}^\star y + S) \bullet X - (\mathcal{A}X)^\mathsf{T} y \\
&= \mathcal{A}^\star y \bullet X + S \bullet X - (\mathcal{A}X)^\mathsf{T} y \\
&= \mathcal{A}^\star y \bullet X + S \bullet X - \mathcal{A}^\star y \bullet X \\
&= S \bullet X \geq 0,
\end{aligned}
$$

where the last inequality follows from the self-duality of the cone of the symmetric positive semidefinite matrices. ∎

As we mentioned in Chapter 11, the strong duality property can fail in general conic optimization (see [NN94]). However, despite of this, a slightly weaker property can be always shown in conic optimization. Now, we give conditions for such a slightly weaker property to hold in SDP. We say that the primal problem is strictly feasible if there exists a primal feasible point $\hat{X}$ such that $\hat{X} > 0$. We make the following assumption for convenience.

> **Assumption 12.1**  *The $m$ matrices $A_1, A_2, \ldots, A_m$ are linearly independent in $\mathbb{S}^n$.*

Now we state and prove the following semi-strong duality result.

> **Lemma 12.2 (Semi-strong duality in SDP)**  *Consider the primal–dual pair $(P|SDP)$ and $(D|SDP)$. If the primal problem is strictly feasible and solvable, then the dual problem is solvable and their optimal values are equal.*

**Proof**  By the assumption of the lemma, the primal problem is strictly feasible and solvable. So, let $X$ be an optimal solution of the primal problem where we can apply the KKT conditions (see Theorem 11.1). This implies that there are Lagrange multipliers $y$ and $S$ such that $(X, y, S)$ satisfies

$$
\begin{aligned}
\mathcal{A}X &= b, \\
\mathcal{A}^\star y + S &= C, \\
X \bullet S &= 0, \\
X, S &\geq 0.
\end{aligned}
$$

This means that $(y, S)$ is feasible for the dual problem. Let $(v, Z)$ be any feasible solution of the dual problem, then we have that $b^\mathsf{T} v \leq C \bullet X = S \bullet X + b^\mathsf{T} y = b^\mathsf{T} y$, where we used the weak duality to obtain the inequality and the complementary slackness to obtain the last equality. Thus, $(y, S)$ is an optimal solution of the dual problem and $C \bullet X = b^\mathsf{T} y$ as desired. ∎

The following strong duality result can be obtained by applying the duality relations to our problem formulation (see also [Tod01, Theorem 4.1]).

---

**Theorem 12.2 (Strong duality in SDP)** *Consider the primal–dual pair (P|SDP) and (D|SDP). If both the primal and dual problems have strictly feasible solutions, then they both have optimal solutions $X^\star$ and $(y^\star, S^\star)$, respectively, and*

$$p^\star \triangleq C \bullet X^\star = d^\star \triangleq b^\mathsf{T} y^\star (i.e., X^\star \bullet S^\star = 0).$$

---

The following lemma describes the complementarity condition as one of the optimality conditions of SDP. This lemma is not hard to show using the spectral decomposition of $X$, and considering its positive and zero eigenvalues separately.

---

**Lemma 12.3 (Complementarity condition in SDP)** *If $X, S \geq 0$, then*

$$X \bullet S = 0 \iff XS = 0.$$

---

As a result of Lemma 11.6, the complementarity slackness condition for the primal and dual SDP problems (P|SDP) and (D|SDP) can be equivalently represented by the equation $XS = 0$. From the above results, we get the following corollary.

---

**Corollary 12.1 (Optimality conditions in SDP)** *Consider the primal–dual pair (P|SDP) and (D|SDP). Assume that both the primal and dual problems are strictly feasible, then $(X, (y, S)) \in \mathbb{S}^n \times \mathbb{R}^m \times \mathbb{S}^n$ is a pair of optimal solutions to the SDP (P|SDP) and (D|SDP) if and only if*

$$\begin{aligned}
\mathcal{A}X &= b, \\
\mathcal{A}^\star y + S &= C, \\
XS &= 0, \\
X, S &\geq 0.
\end{aligned} \tag{12.14}$$

---

We have established the duality relations in SDP. The focus in the remaining part of this chapter is to solve SDP algorithmically.

## 12.5. A primal-dual path-following algorithm

In Section 11.5, we presented a primal-dual path-following algorithm for solving SOCP problems. In this section, we present a primal-dual path-following algorithm for solving SDP problems. The material presented in this section is based on, and similar to, any primal-dual path-following algorithm proposed for SDP (see for instance [TBY17]). The general scheme of the path-following algorithms for SDP is as follows. We associate the perturbed problems to semidefinite programming problems (P|SDP) and (D|SDP), then we draw a path of the centers defined by the perturbed KKT optimality conditions. After that, Newton's method is applied to treat the corresponding perturbed equations in order to obtain a descent search direction.

Let $\mu > 0$ be a barrier parameter. The perturbed primal problem corresponding to the primal problem (P|SDP) is

$$
\begin{array}{rl}
\min & f_\mu(X) \triangleq C \bullet X - \mu \ln \det(X) + n\mu \ln \mu \\
(\text{P|SDP}_\mu) \quad \text{s.t.} & \mathcal{A}X = b, \\
& X > 0,
\end{array}
$$

and the perturbed dual problem corresponding to the dual problem (D|SDP) is

$$
\begin{array}{rl}
\max & g_\mu(y, S) \triangleq b^\top y + \mu \ln \det(S) - n\mu \ln \mu \\
(\text{D|SDP}_\mu) \quad \text{s.t.} & \mathcal{A}^\star y + S = C, \\
& S > 0.
\end{array}
$$

Now, we define the following feasibility sets:

$$
\begin{aligned}
\mathcal{F}_{\text{P|SDP}} &\triangleq \{X \in \mathbb{S}^n : \mathcal{A}X = b, \ X \geq 0\}, \\
\mathcal{F}_{\text{D|SDP}} &\triangleq \{(y, S) \in \mathbb{R}^m \times \mathbb{S}^n : \mathcal{A}^\star y + S = C, \ S \geq 0\}, \\
\mathcal{F}_{\text{P|SDP}}^\circ &\triangleq \{X \in \mathbb{S}^n : \mathcal{A}X = b, \ X > 0\}, \\
\mathcal{F}_{\text{D|SDP}}^\circ &\triangleq \{(y, S) \in \mathbb{R}^m \times \mathbb{S}^n : \mathcal{A}^\star y + S = C, \ S > 0\}, \\
\mathcal{F}_{\text{SDP}}^\circ &\triangleq \mathcal{F}_{\text{P|SDP}}^\circ \times \mathcal{F}_{\text{D|SDP}}^\circ.
\end{aligned}
$$

We also make the following assumption about the primal-dual pair (P|SDP) and (D|SDP).

**Assumption 12.2** *The set $\mathcal{F}_{\text{SDP}}^\circ$ is nonempty.*

Assumption 12.2 requires that Problem (P|SDP$_\mu$) and its dual (D|SDP$_\mu$) have strictly feasible solutions, which guarantees strong duality for the semidefinite programming problem. Note that the feasible region for Problems (P|SDP$_\mu$) and (D|SDP$_\mu$) is described implicitly by $\mathcal{F}_{\text{SDP}}^\circ$. Due to the coercivity of the function $f_\mu$ on the feasible set of (P|SDP$_\mu$), Problem (P|SDP$_\mu$) has an optimal solution.

The following lemma proves the convergence of the optimal solution of Problem (P|SDP$_\mu$) to the optimal solution of Problem (P|SDP) when $\mu$ approaches zero.

**Lemma 12.4** *Let $\bar{X}_\mu$ be an optimal primal solution of Problem (P|SDP$_\mu$), then $\bar{X} = \lim_{\mu \to 0} \bar{X}_\mu$ is an optimal solution of Problem (P|SDP).*

**Proof** Let $f_\mu(X) \triangleq f(X, \mu)$ and $f(X) \triangleq f(X, 0)$. Due to the coercivity of the function $f_\mu$ on the feasible set of (P|SDP$_\mu$), Problem (P|SDP$_\mu$) has an optimal solution, say $\bar{X}_\mu$, such that

$$
\nabla_X f_\mu(\bar{X}_\mu) = \nabla_X f(\bar{X}_\mu, \mu) = 0.
$$

Then, for all $X \in \mathcal{F}_{\text{PISDP}}^{\circ}$, we have that

$$
\begin{aligned}
f(X) &\geq f(\bar{X}_{\mu}, \mu) + (X - \bar{X}_{\mu}) \bullet \nabla_X f(\bar{X}_{\mu}, \mu) + (0 - \mu)\tfrac{\partial}{\partial \mu} f(\bar{X}_{\mu}, \mu) \\
&\geq f(\bar{X}_{\mu}, \mu) + \mu \ln \det \bar{X}_{\mu} - n\mu \ln \mu - n\mu \\
&\geq C \bullet \bar{X}_{\mu} - \mu \ln \det \bar{X}_{\mu} + n\mu \ln \mu + \mu \ln \det \bar{X}_{\mu} - n\mu \ln \mu - n\mu \\
&\geq C \bullet \bar{X}_{\mu} - n\mu.
\end{aligned}
$$

Since $X$ was arbitrary in $\mathcal{F}_{\text{PISDP}}^{\circ}$, this implies that

$$
\min_{X \in \mathcal{F}_{\text{PISDP}}^{\circ}} f(X) \geq C \bullet \bar{X}_{\mu} - n\mu \geq C \bullet \bar{X}_{\mu} = f(\bar{X}_{\mu}).
$$

On the other side, we have $f(\bar{X}_{\mu}) \geq \min_{X \in \mathcal{F}_{\text{PISDP}}^{\circ}} f(X)$. As $\mu$ goes to 0, it immediately follows that $f(\bar{X}) = \min_{X \in \mathcal{F}_{\text{PISDP}}^{\circ}} f(X)$. Thus, $\bar{X}$ is an optimal solution of Problem (PISDP). The proof is complete. ∎

### Newton's method and commutative directions

As we mentioned, the objective function of Problem (PISDP$_{\mu}$) is strictly convex, hence the KKT conditions are necessary and sufficient to characterize an optimal solution of Problem (PISDP$_{\mu}$). Consequently, and in light of Lemma 12.3, the points $\bar{X}_{\mu}$ and $(\bar{y}_{\mu}, \bar{S}_{\mu})$ are optimal solutions of (PISDP$_{\mu}$) and (DISDP$_{\mu}$) respectively if and only if they satisfy the perturbed nonlinear system

$$
\begin{aligned}
\mathcal{A}X &= b, & X &> 0, \\
\mathcal{A}^{\star}y + S &= C, & S &> 0, && (12.15) \\
XS &= \mu I, & \mu &> 0,
\end{aligned}
$$

where $I$ is the identity matrix of $\mathbb{S}^n$.

We call the set of all solutions of system (12.15), denoted by $(X_{\mu}, y_{\mu}, S_{\mu})$ with $\mu > 0$, the central path. We say that a point $(X, y, S)$ is near to the central path if it belongs to the set $\mathcal{N}_{\text{SDP}}(\mu)$, which is defined as

$$
\mathcal{N}_{\text{SDP}}(\mu) \triangleq \left\{ (X, y, S) \in \mathcal{F}_{\text{PISDP}}^{\circ} \times \mathcal{F}_{\text{DISDP}}^{\circ} : d_{\text{SDP}}(X, S) \leq \theta\mu, \theta \in (0, 1) \right\},
$$

where

$$
d_{\text{SDP}}(X, S) \triangleq \left\| X \circ S - \mu I \right\|_F.
$$

Since for $X, S \in \mathbb{S}^n$, the product $XS$ is generally not in $\mathbb{S}^n$, so the left-hand side of System (12.15) is a map from $\mathbb{R}^{n \times n} \times \mathbb{R}^m \times \mathbb{S}^n$ to $\mathbb{S}^n \times \mathbb{R}^m \times \mathbb{S}^n$. Thus, System (12.15) is not a square system when $X$ and $S$ are restricted to $\mathbb{S}^n$, which is needed for applying Newton's method. A remedy for this is to make System (12.15) square by modifying the left-hand side to a map from $\mathbb{S}^n \times \mathbb{R}^m \times \mathbb{S}^n$ to itself. To this end, we use the symmetrization operator $\mathcal{H}_P(\cdot)$ defined in (12.5). The following lemma is due to Zhang [Zha98a].

---

**Lemma 12.5** *For $M \in \mathbb{S}^n$, nonsingular $P \in \mathbb{R}^{n \times n}$, and a scalar $\tau$, we have*

$$\mathcal{H}_P(M) = \tau I \iff M = \tau I.$$

---

In light of Lemma 12.5, for any nonsingular matrix $P$, System (12.15) is equivalent to

$$
\begin{aligned}
\mathcal{A}X &= b, & X &> 0, \\
\mathcal{A}^{\star}y + S &= C, & S &> 0, \\
\mathcal{H}_P(XS) &= \mu I, & \mu &> 0.
\end{aligned}
\tag{12.16}
$$

Recall that $\mathcal{H}_P(XS) = \frac{1}{2}(PXSP^{-1} + P^{-1}SXP^{\mathsf{T}})$. So, an alternative way to view the above development is to scale (P|SDP) so that the variable $X$ is replaced by $\overline{X} \triangleq PXP^{\mathsf{T}}$ and to scale (D|SDP) so that the variable $S$ is replaced by $\underline{S} \triangleq P^{-1^{\mathsf{T}}}SP^{-1}$.

The need for the above symmetrization occurs because $X$ and $S$ do not commute(see Property 12.6). So, we choose $P$ so that the scaled matrices commute. Denote by $C(X, S)$ the set of all matrices so that the scaled matrices commute. That is,

$$C(X,S) \triangleq \left\{ P \in \mathbb{S}^n : P^{-1} \text{ exists, and } PXP^{\mathsf{T}} \& P^{-1^{\mathsf{T}}}SP^{-1} \text{ commute} \right\}. \tag{12.17}$$

Now, we can apply Newton's method to system (12.16) and obtain the following linear system

$$
\begin{aligned}
\mathcal{A}\Delta X &= 0, \\
\mathcal{A}^{\star}\Delta y + \Delta S &= 0, \\
\mathcal{H}_P(X\Delta S + \Delta XS) &= \sigma\mu I - \mathcal{H}_P(XS).
\end{aligned}
\tag{12.18}
$$

where $(\Delta X, \Delta y, \Delta S) \in \mathbb{S}^n \times \mathbb{R}^m \times \mathbb{S}^n$ is the search direction, $\sigma \in (0, 1)$ is the centering parameter, and $\mu = \frac{1}{n}X \bullet S = \frac{1}{n}\overline{X} \bullet \underline{S}$ is the normalized duality gap corresponding to $(X, y, S)$. In fact,

$$\frac{1}{n}\overline{X} \bullet \underline{S} = \frac{1}{n}\left(PXP^{\mathsf{T}}\right) \bullet \left(P^{-1^{\mathsf{T}}}SP^{-1}\right) = \frac{1}{n}X \bullet S.$$

Solving the scaled Newton system (12.18) yields the search direction $(\Delta X, \Delta y, \Delta S)$. Note that the search direction $(\Delta X, \Delta y, \Delta S)$ belongs to the so-called the *MZ family of directions* (due to Monteiro [Mon97] and Zhang [Zha98b]). In fact, such a way of scaling originally proposed for semidefinite programming by Monteiro [Mon97] and Zhang [Zha98b], and after that it was used for second-order cone programming by Schmieta and Alizadeh [SA03]. Clearly, the set $C(X, S)$ defined in (12.17) is a subclass of the MZ family of search directions. Our focus is in matrices $P \in C(X, S)$. We discuss the following three choices of $P$ (see [Tod01, Section 6]):

- The first one is to choose $P = X^{-1/2}$, which gives $\overline{X} = I$.

- The second one is to choose $P = S^{1/2}$, which gives $\underline{S} = I$.

- The third choice of $P$ is given by $P = (X^{1/2}(X^{1/2}SX^{1/2})^{-1/2}X^{1/2})^{-1/2}$, which yields $P^2XP^2 = S$, and therefore $\overline{X} = \underline{S}$.

As we mentioned in Chapter 11 in the case of second-order cone programming, the first two choices of directions are respectively called the HRVW/KSH/M direction and dual HRVW/KSH/M direction (due to Helmberg et al. [HRVW96], Monteiro [Mon97] and Kojima et al. [KSH97]). The third choice of direction is called the NT direction (due to Nesterov and Todd [NT98]).

## Path-following algorithm

We formally state the path-following algorithm for solving SDP problem in Algorithm 12.1.

---

**Algorithm 12.1:** Path-following algorithm for SDP

---

**Input:** Data in Problems (P|SDP) and (D|SDP), $k = 0$,
$\left(X^{(0)}, y^{(0)}, S^{(0)}\right) \in \mathcal{N}_{\mathrm{SDP}}\left(\mu^{(0)}\right), \epsilon > 0,\ \sigma^{(0)}, \theta \in (0, 1)$
**Output:** An $\epsilon$-optimal solution to Problem (P|SDP)

1: **while** $X^{(k)} \bullet S^{(k)} \geq \epsilon$ **do**

2:     choose $P^{(k)} \in C\left(X^{(k)}, S^{(k)}\right)$

3:     set $\mu^{(k)} \triangleq \frac{1}{n}X^{(k)} \bullet S^{(k)}$
      $H^{(k)} \triangleq \sigma^{(k)}\mu^{(k)}I - X^{(k)} \circ S^{(k)}$

4:     compute $\left(\Delta X^{(k)}; \Delta y^{(k)}; \Delta S^{(k)}\right)$ by solving the scaled system (12.18) to get

$$\Delta y^{(k)} = -\left(\mathcal{A}\left(S^{(k)^{-1}} \circ X^{(k)}\right)\mathcal{A}^\star\right)^{-1}\mathcal{A}\left(S^{(k)^{-1}} \circ H^{(k)}\right)$$
$$\Delta S^{(k)} = -\mathcal{A}^\star\Delta y^{(k)}$$
$$\Delta X^{(k)} = S^{(k)^{-1}} \circ \left(H^{(k)} - X^{(k)} \circ \Delta S^{(k)}\right)$$

5:     set the new iterate according to
$$X^{(k+1)} \triangleq X^{(k)} + \alpha^{(k)}\Delta X^{(k)}$$
$$y^{(k+1)} \triangleq y^{(k)} + \alpha^{(k)}\Delta y^{(k)}$$
$$S^{(k+1)} \triangleq S^{(k)} + \alpha^{(k)}\Delta S^{(k)}$$

6:     set $k = k + 1$

7: **end**

---

Algorithm 12.1 selects a sequence of displacement steps $\{\alpha^{(k)}\}$ and centrality parameters $\{\sigma^{(k)}\}$ according to the following rule: For all $k \geq 0$, we take $\sigma^{(k)} = 1 - \delta/\sqrt{n}$, where $\delta \in [0, \sqrt{n})$. The author in [TBY17] discusses in Section 3 various selections for calculating the displacement step $\alpha^{(k)}$.

In the rest of this section, we prove that the complementary gap and the function $f_\mu$ decrease for a given displacement step. The proof of this result depends essentially on the following lemma.

---

**Lemma 12.6** *Let $(X, y, S) \in int\, \mathbb{S}_+^n \times \mathbb{R}^m \times int\, \mathbb{S}_+^n$, $(X, y, S)$ be obtained by applying scaling to $(X, y, S)$, and $(\Delta X, \Delta y, \Delta S)$ be a solution of System (12.18). Then we have*

*(a)* $\Delta X \bullet \Delta S = 0.$

*(b)* $X \bullet \Delta S + \Delta X \bullet S = trace(H)$, where $H \triangleq \sigma \mu I - X \circ S$ such that $\sigma \in (0, 1)$ and $\mu = \frac{1}{n} X \bullet S$.

*(c)* $X^+ \bullet S^+ = \left(1 - \alpha\left(1 - \frac{\sigma}{2}\right)\right) X \bullet S$, $\forall \alpha \in \mathbb{R}$, where $X^+ \triangleq X + \alpha \Delta X$ and $S^+ \triangleq S + \alpha \Delta S$.

---

**Proof**   By the first two equations of System (12.18), we get

$$\Delta X \bullet \Delta S = -\Delta X \bullet \mathcal{A}^\star \Delta y = -\left(\mathcal{A}\Delta X\right)^\mathsf{T} \Delta y = 0.$$

This proves item (a).

We prove item (b) by noting that

$$
\begin{aligned}
trace(H) &= trace\left(\sigma \mu I - X \circ S\right) \\
&= trace\left(X \circ \Delta S + \Delta X \circ S\right) \\
&= trace\left(X \circ \Delta S\right) + trace\left(\Delta X \circ S\right) = X \bullet \Delta S + \Delta X \bullet S,
\end{aligned}
$$

where we used the last equation of System (12.18) to obtain the first equality. Item (c) is left as an exercise for the reader (see Exercise 12.1). The proof is complete.   ∎

The result in the following theorem is essentially those in [TBY17, Lemmas 4.2 and 4.3]. The result given here in SDP is the counterpart of that in Theorem 11.3 for second-order cone programming, and its proof is also the same as the proof of its counterpart.

---

**Theorem 12.3** *Let $(X, y, S)$ and $(X^+, y^+, S^+)$ be strictly feasible solutions of the pair of problems $(P|SDP_\mu)$ and $(D|SDP_\mu)$ with*

$$\left(X^+, y^+, S^+\right) = \left(X + \alpha \Delta X, \ y + \alpha \Delta y, \ S + \alpha \Delta S\right),$$

*where $\alpha$ is a displacement step and $(\Delta X, \Delta y, \Delta S)$ is the Newton direction. Then*

*(a)* $X^+ \bullet S^+ < X \bullet S.$       *(b)* $f_\mu(X^+) < f_\mu(X).$

---

**Proof**   Note that

$$X^+ \bullet S^+ = \left(1 - \alpha\left(1 - \frac{\sigma}{2}\right)\right) X \bullet S < X \bullet S,$$

where the equality follows from item (d) of Lemma 12.6 and the strict inequality follows from $(1 - \alpha(1 - \sigma/2)) < 1$ (as $\alpha > 0$ and $\sigma \in (0, 1)$). This proves item (a).

To prove item (b), note that

$$f_\mu(X^+) \simeq f_\mu(X) + \nabla_X f_\mu(X) \bullet (X^+ - X),$$

and hence

$$f_\mu(X^+) - f_\mu(X) \simeq \alpha \nabla_X f_\mu(X) \bullet \Delta X.$$

Since

$$\nabla_X f_\mu(X) = -\nabla^2_{XX} f_\mu(X) \Delta X,$$

we have

$$f_\mu(X^+) - f_\mu(X) \simeq -\alpha \, \Delta X \bullet \nabla^2_{XX} f_\mu(X) \Delta X < 0,$$

where the strict inequality follows from the positive definiteness of the Hessian matrix $\nabla^2_{XX} f_\mu(X)$ (as $f_\mu$ is strictly convex). Thus, $f_\mu(X^+) < f_\mu(X)$. The proof is complete. ∎

## Complexity estimates

In this part, we analyze the complexity of the proposed path-following algorithm for SDP. More specifically, we prove that the iteration-complexity of Algorithm 12.1 is bounded by

$$O\left( \sqrt{n} \ln \left[ \epsilon^{-1} X^{(0)} \bullet S^{(0)} \right] \right).$$

Our proof depends essentially on the following two lemmas.

---

**Lemma 12.7** *Let* $(X, y, S) \in \mathcal{F}^\circ_{P\!I\!S\!D\!P} \times \mathcal{F}^\circ_{D\!I\!S\!D\!P}$, *and* $(X, y, S)$ *be obtained by applying scaling to* $(X, y, S)$ *with* $H = \sigma\mu I - XS$, *and* $(\Delta X, \Delta y, \Delta S)$ *be a solution of System* (12.18). *For any* $\alpha \in \mathbb{R}$, *we set*

$$(X(\alpha), y(\alpha), S(\alpha)) \triangleq (X, y, S) + \alpha(\Delta X, \Delta y, \Delta S),$$
$$\mu(\alpha) \triangleq \tfrac{1}{n} X(\alpha) \bullet S(\alpha),$$
$$V(\alpha) \triangleq X(\alpha) \circ S(\alpha) - \mu(\alpha)I.$$

*Then*

$$V(\alpha) = (1 - \alpha)(X \circ S - \mu I) + \alpha^2 \, \Delta X \circ \Delta S. \tag{12.19}$$

---

**Proof**  See Exercise 12.2. ∎

**Lemma 12.8** *Let* $(X, y, S) \in \mathcal{F}^{\circ}_{P|SDP} \times \mathcal{F}^{\circ}_{D|SDP}$, *and* $(X, y, S)$ *be obtained by applying scaling to* $(X, y, S)$ *such that* $\|XS - \mu I\| \leq \theta\mu$, *for some* $\theta \in [0, 1)$ *and* $\mu > 0$. *Let also* $(\Delta X, \Delta y, \Delta S)$ *be a solution of System* (12.18), $H = \sigma\mu I - XS$, $\delta_X \triangleq \mu\|\Delta X X^{-1}\|_F, \delta_S \triangleq \|X\Delta S\|_F$. *Then, we have*

$$\delta_X \delta_S \leq \frac{1}{2}\left(\delta_X^2 + \delta_S^2\right) \leq \frac{\|H\|_F^2}{2(1 - \theta)^2}. \tag{12.20}$$

**Proof** See Exercise 12.3. ∎

The following theorem analyzes the behavior of one iteration of Algorithm 12.1. This theorem is due to [TBY17, Theorem 4.6]. The result here in SDP is the counterpart of that in Theorem 11.4 for second-order cone programming, and its proof is also the same as the proof of its counterpart.

**Theorem 12.4** *Let* $\theta \in (0, 1)$ *and* $\delta \in \left[0, \sqrt{n}\right)$ *be given such that*

$$\frac{\theta^2 + \delta^2}{2(1 - \theta)^2\left(1 - \frac{\delta}{\sqrt{n}}\right)} \leq \theta \leq \frac{1}{2}. \tag{12.21}$$

*Suppose that* $(X, y, S) \in N_{SDP}(\mu)$ *and let* $(\Delta X, \Delta y, \Delta S)$ *denote the solution of system* (12.18) *with* $H = \sigma\mu I - X \circ S$ *and* $\sigma = 1 - \delta/\sqrt{n}$. *Then, we have*

*(a)* $X^+ \bullet S^+ = (1 - \delta/\sqrt{n}) X \bullet S$.

*(b)* $(X^+, y^+, S^+) = (X, y, S) + (\Delta X, \Delta y, \Delta S) \in N_{SDP}(\mu)$.

**Proof** Item (a) follows directly from item (c) of Lemma 12.6 with $\alpha = 1$ and $\sigma = 1 - \delta/\sqrt{n}$. We now prove item (b). Define

$$\mu^+ \triangleq \frac{1}{n} X^+ \bullet S^+ = (1 - \delta/n)\,\mu, \tag{12.22}$$

and let $(X, y, S) \in N_{SDP}(\mu)$, we then have

$$\begin{aligned}\left\|\sigma\mu I - X \circ S\right\|_F^2 &\leq \left\|(\sigma - 1)\mu I\right\|_F^2 + \left\|\mu I - X \circ S\right\|_F^2 \\ &\leq \left((\sigma - 1)^2 r + \theta^2\right)\mu^2 = \left(\delta^2 + \theta^2\right)\mu^2.\end{aligned} \tag{12.23}$$

Since $\|X \circ S - \mu I\| \leq \theta\mu$ and $H = \sigma\mu I - X \circ S$, using Lemma 12.8 it follows that

$$\left\|\Delta x X^{-1}\right\|_F \|X\Delta S\|_F \leq \frac{\left\|\sigma\mu I - X \circ S\right\|_F^2}{2(1 - \theta)^2\mu}. \tag{12.24}$$

Defining $V^+ \triangleq V(1) = X^+ \circ S^+ - \mu^+ I$ and using (12.19) with $\alpha = 1$, (12.24), (12.23), (12.21) and (12.22), we get

$$
\begin{aligned}
\|V^+\|_F &= \|\Delta X \Delta S\|_F \\
&\leq \|\Delta X X^{-1}\|_F \|X \Delta S\|_F \\
&\leq \frac{\|\sigma \mu I - X \circ S\|_F^2}{2(1-\theta)^2 \mu} \\
&\leq \frac{(\delta^2 + \theta^2)\mu}{2(1-\theta)^2} \\
&\leq \theta \left(1 - \frac{\delta}{\sqrt{n}}\right)\mu = \theta \mu^+.
\end{aligned}
$$

Consequently,

$$
\left\|X^+ \circ S^+ - \mu^+ I\right\|_F \leq \theta \mu^+. \tag{12.25}
$$

By using the right-hand side inequality in (12.20), and using (12.23) and (12.21), we have

$$
\left\|\Delta X X^{-1}\right\|_F \leq \frac{\|\sigma \mu I - X \circ S\|_F}{(1-\theta)\mu} \leq \frac{\sqrt{\delta^2 + \theta^2}}{(1-\theta)} \leq \sqrt{2\theta\left(1 - \frac{\delta}{\sqrt{n}}\right)} < 1,
$$

where the strict inequality follows from $\theta \leq \frac{1}{2}$ and $0 < 1 - \frac{\delta}{\sqrt{n}} < 1$.

One can easily see that $\left\|\Delta X X^{-1}\right\|_F < 1$ implies that $I + \Delta X X^{-1} > 0$, and therefore

$$
X^+ = \Delta X + X = \left(I + \Delta X X^{-1}\right) X > 0.
$$

Note that, from (12.25), we have $\lambda_{\min}(X^+ \circ S^+) = \lambda_{\min}(X^+ S^+) \geq (1-\theta)\mu^+ > 0$, and therefore $X^+ S^+ > 0$. Since $X^+ > 0$ and $X^+$ and $S^+$ commute, we conclude that $S^+ > 0$. Using the first equation of system (12.18), we get

$$
\mathcal{A} X^+ = \mathcal{A}(X + \Delta X) = \mathcal{A}X + A\Delta X = b, \quad \text{and hence} \quad X^+ \in \mathcal{F}_{\text{PISDP}}^\circ.
$$

By using the second equation of system (12.18), we get

$$
\mathcal{A}^\star y^+ + S^+ = \mathcal{A}^\star(y + \Delta y) + (S + \Delta S) = \mathcal{A}^\star y + S + \mathcal{A}^\star \Delta y + \Delta S = C,
$$

and hence $(y^+, S^+ \in \mathcal{F}_{\text{DISDP}}^\circ$. Thus, in view of (12.25), we deduce that $(X^+, y^+, S^+) \in \mathcal{N}_{\text{SDP}}(\mu)$. Item (b) is therefore established. The proof is now complete. ∎

Now we present Corollary 12.2 in SDP to be the counterpart of Corollary 11.2 for second-order cone programming. The proof of Corollary 12.2 is also the same as that of its counterpart.

**Corollary 12.2** *Let $\theta$ and $\delta$ as given in Theorem 12.4 and $(X^0, y^0, S^{(0)}) \in \mathcal{N}_{SDP}(\mu)$. Then Algorithm 12.1 generates a sequence of points $\{(X^{(k)}, y^{(k)}, S^{(k)})\} \subset \mathcal{N}_{SDP}(\mu)$ such that*

$$X^{(k)} \bullet S^{(k)} = \left(1 - \frac{\delta}{\sqrt{n}}\right)^k X^{(0)} \bullet S^{(0)}, \quad \forall k \geq 0.$$

*Moreover, given a tolerance $\epsilon > 0$, Algorithm 12.1 computes an iterate $\{(X^{(k)}, y^{(k)}, S^{(k)})\}$ satisfying $X^{(k)} \bullet S^{(k)} \leq \epsilon$ in at most*

$$O\left(\sqrt{n} \ln\left(\frac{X^{(0)} \bullet S^{(0)}}{\epsilon}\right)\right)$$

*iterations.*

**Proof**  Looking recursively at item (a) of Theorem 12.4, for each $k$ we have that

$$X^{(k)} \bullet S^{(k)} = \left(1 - \frac{\delta}{\sqrt{n}}\right)^k X^{(0)} \bullet S^{(0)} \leq \epsilon.$$

By taking natural algorithm of both sides, we get

$$k \ln\left(1 - \frac{\delta}{\sqrt{n}}\right) \leq \ln\left(\frac{\epsilon}{X^{(0)} \bullet S^{(0)}}\right),$$

which holds only if

$$k\left(-\frac{\delta}{\sqrt{n}}\right) \leq \ln\left(\frac{\epsilon}{X^{(0)} \bullet S^{(0)}}\right),$$

or equivalently,

$$k \geq \frac{\sqrt{n}}{\delta} \ln\left(\frac{X^{(0)} \bullet S^{(0)}}{\epsilon}\right).$$

The proof is complete.                                                                        ∎

In this section, we have presented and analyzed a path-following algorithm for SDP by extending the path-following algorithm that was presented in Section 11.5 for second-order cone programming. Exercises 12.5-12.8 aim to derive a homogeneous self-dual algorithm for SDP by extending the homogeneous self-dual algorithms that were presented in Section 10.7 for linear programming, and in Section 11.6 for second-order cone programming. In fact, there are several interior-point methods that can be extended from linear programming and second-order cone programming to SDP. In this context, we end this chapter by presenting Table 12.1, which shows gradient and Hessian derivatives for the logarithmic barriers applied on the most three well-known examples of conic constraints: the polyhedral, second-order cone and semidefinite constraints.

| | Polyhedral cone [a] | Second-order cone [b] | Semidefinite cone [c] |
|---|---|---|---|
| Space | $\mathbb{R}^n$ | $\mathbb{E}^n = \left\{ s = \begin{pmatrix} s_0 \\ \bar{s} \end{pmatrix} : s \in \mathbb{R} \times \mathbb{R}^{n-1} \right\}$ | $\mathbb{S}^n = \{ S \in \mathcal{M}^n : S = S^T \}$ |
| Cone | $\mathbb{R}^n_+ = \{ s \in \mathbb{R}^n : s \geq 0 \}$ | $\mathbb{E}^n_+ = \{ s \in \mathbb{E}^n : s_0 \geq \|\bar{s}\|_2 \}$ | $\mathbb{S}^n_+ = \{ S \in \mathbb{S}^n : S \succeq 0 \}$ |
| Feasibility constraint | $s(x) = Ax - b \in \mathbb{R}^n_+$ | $s(x) = Ax - b \in \mathbb{E}^n_+$ | $S(x) = \sum_i x_i A_i - B \in \mathbb{S}^n_+$ |
| Logarithmic barrier $\ell(x)$ | $-\mathbf{1}^T \ln s(x)$ | $-\ln\det(s(x))$ | $-\ln\det(S(x))$ |
| Gradient $\nabla_x \ell(x)$ | $-A^T s^{-1}$ | $-A^T s^{-1}$ | $-\mathcal{A}^T \mathbf{vec}(S^{-1})$ |
| Hessian $\nabla^2_{xx} \ell(x)$ | $A^T S^{-2} A$ | $A^T Q_{s^{-1}} A$ | $\mathcal{A}^T (S^{-1} \otimes S^{-1}) \mathcal{A}$ |

Table 12.1: Comparison of gradient and Hessian derivatives of the logarithmic barrier for polyhedral, second-order cone, and semidefinite constraints.

[a][**Notations for specifying derivatives in** $\mathbb{R}^n_+$]: We use $\mathbf{1}$ to denote the vector of all ones of appropriate dimension. For any strictly positive vector $s \in \mathbb{R}^n$, we define $\ln s \triangleq (\ln s_1, \ldots, \ln s_n)^T$ and $s^{-1} \triangleq (s_1^{-1}, \ldots, s_n^{-1})^T$. We use $S \triangleq \mathrm{diag}(s)$ to denote the $n \times n$ diagonal matrix whose diagonal entries are $s_1, \ldots, s_n$. We define $\sigma_i \triangleq P_{ii}$ for $i = 1, 2, \ldots, n$, and $\Sigma \triangleq \mathrm{diag}(\sigma)$. We define $P \triangleq P(s) \triangleq S^{-1} A^T (AS^{-2} A^T)^{-1} AS^{-1}$ to act as the orthogonal projection onto the range of $AS^{-1}$.

[b][**Notations for specifying derivatives in** $\mathbb{E}^n_+$]: All these notations have been introduced in Table 11.1 including the arrow-shaped matrix and the quadratic representation. Having the quadratic representation specified for $\mathbb{E}^n$, we can now introduce the vector $\sigma$ in exactly the same way as we defined for any Euclidean Jordan algebra $\mathcal{J}$ (described early in this section) taking under consideration that the inner product "•" is the standard dot product.

[c][**Notations for specifying derivatives in** $\mathbb{S}^n_+$]: For two matrices $S, T \in \mathbb{S}^n$, the Kronecker product $S \otimes T$ is the $n^2 \times n^2$ block matrix whose $i,j$ block is $s_{ij}T$, $i,j = 1, \ldots, n$. The vectorization of a matrix $S$, denoted $\mathbf{vec}(S)$, is the vector obtained by stacking the columns of $S$ on top of one another. We use $\mathcal{A}$ to denote the matrix whose $i$th column is $\mathbf{vec}(A_i)$. We now define the matrix $\Sigma$ used for $\mathbb{S}^n$. By applying a Gram-Schmidt procedure to $\{S^{-1/2} A_i S^{-1/2}, i = 1, \ldots, n\}$, we obtain symmetric matrices $U_i, i = 1, \ldots, n$ having $\|U_i\| = 1$ for all $i$ and $\mathrm{trace}(U_i U_j) = 0, i \neq j$, such that the linear span of $\{U_i, i = 1, \ldots, n\}$ is equal to the span of $\{S^{-1/2} A_i S^{-1/2}, i = 1, \ldots, n\}$. We then define $\Sigma \triangleq \sum_{i=1}^n U_i^2$.

# EXERCISES

**12.1**   Prove item (c) in Lemma 12.6.

**12.2**   Prove Lemma 12.7.

**12.3**   Prove Lemma 12.8.

**12.4**   Implement Algorithm 12.1 and test it on the pair (P|SDP) and (D|SDP) with

$$C \triangleq \mathrm{Diag}(5; 8; 8; 5) \in \mathbb{S}^4, \quad b \triangleq (1; 1; 1; 2) \in \mathbb{R}^4, \quad A_4 \triangleq I \in \mathbb{S}^4,$$

and $A_i$'s, $i = 1, 2, 3$, are so that

$$(a_i)_{jk} \triangleq \begin{cases} 1, & \text{if } j = k = i, \text{ or } j = k = i+1, \\ -1, & \text{if } j = i, k = j+1, \text{ or } j = i+1, k = i, \\ 0, & \text{otherwise.} \end{cases}$$

Take $X^{(0)} = \frac{1}{2} I \in \mathbb{S}^4$, $y^{(0)} = (1.5; 1.5; 1.5; 1.5) \in \mathbb{R}^4$, and

$$S^{(0)} = \begin{bmatrix} 2 & 1.5 & 0 & 0 \\ 1.5 & 3.5 & 1.5 & 0 \\ 0 & 1.5 & 3.5 & 1.5 \\ 0 & 0 & 1.5 & 2 \end{bmatrix}$$

as your initial strictly feasible points. You may also take $\epsilon = 10^{-6}, \sigma = 0.1$ and $\rho = 0.99$.

**12.5**   In this and the following exercises, we practice extending interior-point methods from linear programming and second-order cone programming to SDP. In particular, this and the following exercises aim to derive homogeneous self-dual algorithm for SDP by extending the homogeneous self-dual algorithms that were presented in Section 10.7 for linear programming, and in Section 11.6 for second-order cone programming. In this exercise we ask the reader to write the homogeneous model for the pair (P|SDP) and (D|SDP).

**12.6**   Write the search direction system corresponding to the homogeneous model obtained in Exercise 12.5.

**12.7**   State the generic homogeneous algorithm for solving the pair (P|SDP) and (D|SDP) based on the search direction system obtained in Exercise 12.6.

**12.8**   Let $\epsilon_0 > 0$ be the residual error at a starting point, and $\epsilon > 0$ be a given tolerance. Estimate (without proof) the computational complexity (worst behavior) of the algorithm obtained in Exercise 12.7. This computational complexity must be in terms of the rank of the underlying positive semidefinite cone ($n$).

# APPENDIX A

# SOLUTIONS TO CHAPTER EXERCISES

1.1 (a) (i).     (e) (i).     (i) (iv).     (m) (i).

    (b) (iii).    (f) (ii).    (j) (iii).    (n) (i).

    (c) (i).    (g) (ii).    (k) (iii).    (o) (iv).

    (d) (i).    (h) (iv).    (l) (iv).    (p) (ii).

1.2 (a) A true proposition.    (f) A true proposition.    (k) Not a proposition.

    (b) Not a proposition.    (g) Not a proposition.    (l) A true proposition.

    (c) Not a proposition.    (h) A true proposition.

    (d) A true proposition.    (i) Not a proposition.    (m) Not a Proposition.

    (e) Not a proposition.    (j) Not a Proposition.    (n) Not a proposition.

© 2022 by Baha Alzalg | Kindle Direct Publishing, Washington, United States 2022/9
B. Alzalg, *Combinatorial and Algorithmic Mathematics: From Foundation to Optimization*,
DOI 10.5281/zenodo.7110553

1.3 (a) Today is not Thursday.

(b) There is pollution in New Jersey.

(c) $2 + 1 \neq 3$.

(d) Sara's first answer to item ($l$) in Exercise 1.1 was correct.

(e) The Summer in Santiago is hot.

(f) The summer in Santiago is not hot or not bearable.

(g) The summer in Santiago is hot and humid.

(h) The sun is shining in Santiago's sky and I am not going to the nearest beach.

(i) The sun is shining in Santiago's sky and I am not going to the nearest beach or not doing a little physical exercise.

1.4 (a) I did not buy a lottery ticket this week.

(b) I bought a lottery ticket this week or I won the million dollar jackpot on Friday.

(c) I bought a lottery ticket this week and I won the million dollar jackpot on Friday.

(d) I did not buy a lottery ticket this week and I did not win the million dollar jackpot on Friday.

1.5 (a) $P \wedge Q$.        (b) $P \wedge \neg Q$.        (c) $\neg P \wedge \neg Q$.        (d) $P \vee Q$.

1.6  ▪ If you have a passing score on the final exam, then you will receive a passing grade for the course.

▪ You will receive a passing grade for the course if you get a passing score on the final exam.

▪ Receiving a passing grade for the course is necessary for getting a passing score on the final exam.

▪ Getting a passing score on the final exam is sufficient to receive a passing score for the course.

▪ You will get a passing score on the final exam only if you receive a passing grade for the course.

1.7 (a) We are on the line TF, so it is false.

(b) We cannot be on the line TF, so it is true.

(c) We cannot be on the line TF, so it is true.

(d) We are on the line FF, so it is true.

(e) We are on the line TT or the line FF only, so it is true.

(f) We are on the line TF, so it is false.

(g) We are on the line TT, so it is true.

1.8 Restating the theorem gives us "If $x$ and $y$ are odd integers, then their sum $x + y$ is even." From here, all odd integers can be represented as $2k + 1$ for some integer $k$. That is,

$$x = 2n + 1 \text{ and } y = 2m + 1 \text{ for some integers } n \text{ and } m.$$

It follows that

$$x + y = (2n + 1) + (2m + 1) = 2(m + n + 1).$$

Given that $m$ and $n$ are integers, $(m + n + 1)$ is also an integer. Thus, $x$ is an integer multiplied by 2, hence it must be an even integer.

1.9 The implication statement is "If the Sun is shrunk to the size of your head, then the Earth will be the size of the pupil of your eye". So the contrapositive is "If the Earth will not be the size of the pupil of your eye, then the Sun is not shrunk to the size of your head", the converse is "If the Earth will be the size of the pupil of your eye, then the Sun is shrunk to the size of your head", and the is inverse "If the Sun is not shrunk to the size of your head, then the Earth will not be the size of the pupil of your eye".

1.10 (a) Any statement "If P then Q" for which P and Q are both true or both false. Take, for instance, the statement "If 1+1=2 then 2-1=1".

(b) Impossible.

1.11 (a) The contrapositive of this theorem is "If $x$ is odd, then $x^2$ is odd". An odd integer can be represented as $2n + 1$, where $n$ is some integer.

Now $x = 2n + 1$, which implies that $x^2 = (2n + 1)^2 = 2(2n^2 + 2n) + 1$.

Given that $n$ is an integer, $(2n^2 + 2n)$ must also be an integer. This means that $x^2$ must also be an odd integer, proving that the theorem is true.

(b) The theorem says that "$x^2$ is odd $\leftrightarrow x$ is odd".

($\rightarrow$) The part that "If $x$ is odd, then $x^2$ is odd" was already proved in item (a).

($\leftarrow$) To prove the part that "If $x^2$ is odd, then $x$ is odd", note that the contrapositive of this statement is "If $x$ is even, then $x^2$ is even". Given that $n$ is an integer, an even number can be represented as $2n$ where $n$ is some integer.

Now $x = 2n$, which implies that $x^2 = (2n)^2 = 4n^2 = 2(2n^2)$.

Given that $n$ is an integer, $2n^2$ must also be an integer. This means that $x^2$ must also be an even integer, proving that the theorem is true.

1.12 We construct the following truth table.

| $P$ | $Q$ | $P \leftrightarrow Q$ | $P \rightarrow Q$ | $Q \rightarrow P$ | $(P \rightarrow Q) \wedge (Q \rightarrow P)$ |
|---|---|---|---|---|---|
| F | F | T | T | T | T |
| F | T | F | T | F | F |
| T | F | F | F | T | F |
| T | T | T | T | T | T |

Because the truth values for $P \leftrightarrow Q$ and $(P \rightarrow Q) \wedge (Q \rightarrow P)$ are identical, this imply that they are logically equivalent.

**1.13** (a)

$$
\begin{array}{rll}
\neg[P \leftrightarrow Q] & \equiv & \neg[(P \rightarrow Q) \wedge (Q \rightarrow P)] \qquad \text{(The hint)} \\
& \equiv & \neg(P \rightarrow Q) \vee \neg(Q \rightarrow P) \qquad \text{(DeMorgan's law)} \\
& \equiv & \neg(\neg P \vee Q) \vee \neg(\neg Q \vee P) \qquad \text{(Implication law)} \\
& \equiv & (\neg\neg P \wedge \neg Q) \vee (\neg\neg Q \wedge \neg P) \qquad \text{(DeMorgan's law)} \\
& \equiv & (P \wedge \neg Q) \vee (Q \wedge \neg P). \qquad \text{(Double negation law)}
\end{array}
$$

(b)

$$
\begin{array}{rll}
\neg[P \oplus Q] & \equiv & \neg[(P \vee Q) \wedge \neg(P \wedge Q)] \qquad \text{(The hint)} \\
& \equiv & \neg(P \vee Q) \vee \neg\neg(P \wedge Q) \qquad \text{(DeMorgan's law)} \\
& \equiv & \neg(P \vee Q) \vee (P \wedge Q) \qquad \text{(Double negation law)} \\
& \equiv & (\neg P \wedge \neg Q) \vee (P \wedge Q). \qquad \text{(Double negation law)}
\end{array}
$$

**1.14** (a) Below is "$P \wedge \neg P$" truth table.

| $P$ | $\neg P$ | $P \wedge \neg P$ |
|---|---|---|
| T | F | F |
| F | T | F |

(b) Below is "$(P \vee \neg Q) \rightarrow Q$" truth table.

| $P$ | $Q$ | $\neg Q$ | $(P \vee \neg Q)$ | $(P \vee \neg Q) \rightarrow Q$ |
|---|---|---|---|---|
| T | T | F | T | T |
| T | F | T | T | F |
| F | T | F | F | T |
| F | F | T | T | F |

(c) Below is "$(P \vee Q) \rightarrow (P \wedge Q)$" truth table.

| $P$ | $Q$ | $P \vee Q$ | $P \wedge Q$ | $(P \vee Q) \rightarrow (P \wedge Q)$ |
|---|---|---|---|---|
| T | T | T | T | T |
| T | F | T | F | F |
| F | T | T | F | F |
| F | F | F | F | T |

(d) Below is "$(P \rightarrow Q) \leftrightarrow (\neg Q \rightarrow \neg P)$" truth table.

| $P$ | $Q$ | $\neg P$ | $\neg Q$ | $P \rightarrow Q$ | $\neg Q \rightarrow \neg P$ | $(P \rightarrow Q) \leftrightarrow (\neg Q \rightarrow \neg P)$ |
|---|---|---|---|---|---|---|
| T | T | F | F | T | T | T |
| T | F | F | T | F | F | T |
| F | T | T | F | T | T | T |
| F | F | T | T | T | T | T |

(e) Below is "$P \oplus P$" truth table.

| $P$ | $P \oplus P$ |
|---|---|
| T | F |
| F | F |

(f) Below is "$P \oplus \neg Q$" truth table.

| $P$ | $Q$ | $\neg Q$ | $P \oplus \neg Q$ |
|---|---|---|---|
| T | T | F | T |
| T | F | T | F |
| F | T | F | F |
| F | F | T | T |

(g) Below is "$\neg P \oplus \neg Q$" truth table.

| $P$ | $Q$ | $\neg P$ | $\neg Q$ | $\neg P \oplus \neg Q$ |
|---|---|---|---|---|
| T | T | F | F | F |
| T | F | F | T | T |
| F | T | T | F | T |
| F | F | T | T | F |

(h) Below is "$(P \oplus Q) \wedge (P \oplus \neg Q)$" truth table.

| $P$ | $Q$ | $\neg Q$ | $P \oplus Q$ | $P \oplus \neg Q$ | $(P \oplus Q) \wedge (P \oplus \neg Q)$ |
|---|---|---|---|---|---|
| T | T | F | F | T | F |
| T | F | T | T | F | F |
| F | T | F | T | F | F |
| F | F | T | F | T | F |

(i) Below is "$P \rightarrow \neg Q$" truth table.

| $P$ | $Q$ | $\neg Q$ | $P \rightarrow \neg Q$ |
|---|---|---|---|
| T | T | F | F |
| T | F | T | T |
| F | T | F | T |
| F | F | T | T |

(j) Below is "$\neg P \leftrightarrow Q$" truth table.

| $P$ | $Q$ | $\neg P$ | $\neg P \leftrightarrow Q$ |
|---|---|---|---|
| T | T | F | F |
| T | F | F | T |
| F | T | T | T |
| F | F | T | F |

(k) Below is "$(P \rightarrow Q) \vee (\neg P \rightarrow Q)$" truth table.

| $P$ | $Q$ | $\neg P$ | $\neg Q$ | $P \rightarrow Q$ | $\neg P \rightarrow Q$ | $(P \rightarrow Q) \vee (\neg P \rightarrow Q)$ |
|---|---|---|---|---|---|---|
| T | T | F | F | T | T | T |
| T | F | F | T | F | T | T |
| F | T | T | F | T | T | T |
| F | F | T | T | T | F | T |

(l) Below is "$(P \leftrightarrow Q) \vee (\neg P \leftrightarrow Q)$" truth table.

| $P$ | $Q$ | $\neg P$ | $P \leftrightarrow Q$ | $\neg P \leftrightarrow Q$ | $(P \leftrightarrow Q) \vee (\neg P \leftrightarrow Q)$ |
|---|---|---|---|---|---|
| T | T | F | T | F | T |
| T | F | F | F | T | T |
| F | T | T | F | T | T |
| F | F | T | T | F | T |

(m) Below is "$(P \wedge Q) \vee R$" truth table.

| $P$ | $Q$ | $R$ | $P \wedge Q$ | $(P \wedge Q) \vee R$ |
|---|---|---|---|---|
| T | T | T | T | T |
| T | T | F | T | T |
| T | F | T | F | T |
| T | F | F | F | F |
| F | T | T | F | T |
| F | T | F | F | F |
| F | F | T | F | T |
| F | F | F | F | F |

(n) "$(P \wedge Q) \wedge R$" truth table.

| $P$ | $Q$ | $R$ | $(P \wedge Q) \wedge R$ |
|---|---|---|---|
| T | T | T | T |
| T | T | F | F |
| T | F | T | F |
| T | F | F | F |
| F | T | T | F |
| F | T | F | F |
| F | F | T | F |
| F | F | F | F |

(o) Below is "$(P \vee Q) \vee R$" truth table.

| $P$ | $Q$ | $R$ | $P \vee Q$ | $(P \vee Q) \vee R$ |
|---|---|---|---|---|
| T | T | T | T | T |
| T | T | F | T | T |
| T | F | T | T | T |
| T | F | F | T | T |
| F | T | T | T | T |
| F | T | F | T | T |
| F | F | T | F | T |
| F | F | F | F | F |

(p)  Below is "$(P \land Q) \lor \neg R$" truth table.

| $P$ | $Q$ | $R$ | $\neg R$ | $P \land Q$ | $(P \land Q) \lor \neg R$ |
|---|---|---|---|---|---|
| T | T | T | F | T | T |
| T | T | F | T | T | T |
| T | F | T | F | F | F |
| T | F | F | T | F | T |
| F | T | T | F | F | F |
| F | T | F | T | F | T |
| F | F | T | F | F | F |
| F | F | F | T | F | T |

(q)  Below is "$P \to (\neg Q \lor R)$" truth table.

| $P$ | $Q$ | $R$ | $\neg Q$ | $(\neg Q \lor R)$ | $P \to (\neg Q \lor R)$ |
|---|---|---|---|---|---|
| T | T | T | F | T | T |
| T | T | F | F | F | F |
| T | F | T | T | T | T |
| T | F | F | T | T | T |
| F | T | T | F | T | T |
| F | T | F | F | F | T |
| F | F | T | T | T | T |
| F | F | F | T | T | T |

(r)  Below is "$(P \to Q) \lor (\neg P \to R)$" truth table.

| $P$ | $Q$ | $R$ | $\neg P$ | $P \to Q$ | $\neg P \to R$ | $(P \to Q) \lor (\neg P \to R)$ |
|---|---|---|---|---|---|---|
| T | T | T | F | T | T | T |
| T | T | F | F | T | T | T |
| T | F | T | F | F | T | T |
| T | F | F | F | F | T | T |
| F | T | T | T | T | T | T |
| F | T | F | T | T | F | T |
| F | F | T | T | T | T | T |
| F | F | F | T | T | F | T |

(s) Below is "$(P \to Q) \land (\neg P \to R)$" truth table.

| $P$ | $Q$ | $R$ | $\neg P$ | $P \to Q$ | $\neg P \to R$ | $(P \to Q) \land (\neg P \to R)$ |
|---|---|---|---|---|---|---|
| T | T | T | F | T | T | T |
| T | T | F | F | T | T | T |
| T | F | T | F | F | T | F |
| T | F | F | F | F | T | F |
| F | T | T | T | T | T | T |
| F | T | F | T | T | F | F |
| F | F | T | T | T | T | T |
| F | F | F | T | T | F | F |

(t) Below is "$(P \leftrightarrow Q) \lor (\neg Q \leftrightarrow R)$" truth table.

| $P$ | $Q$ | $R$ | $\neg Q$ | $P \leftrightarrow Q$ | $\neg Q \leftrightarrow R$ | $(P \leftrightarrow Q) \lor (\neg Q \leftrightarrow R)$ |
|---|---|---|---|---|---|---|
| T | T | T | F | T | F | T |
| T | T | F | F | T | T | T |
| T | F | T | T | F | T | T |
| T | F | F | T | F | F | F |
| F | T | T | F | F | F | F |
| F | T | F | F | F | T | T |
| F | F | T | T | T | T | T |
| F | F | F | T | T | F | T |

**1.15** Let $F$ be "Lillian is forceful", $G$ be "Lillian will be a good executive", $E$ be "Lillian is efficient", and $C$ be "Lillian is creative". Then the problem statement can be formulated in the following propositional logical model.

$$\text{Problem model:} \begin{cases} F \lor C, \\ F \to G, \\ \neg(E \land C), \\ \neg E \to (F \lor G). \end{cases}$$

The desired conclusion is to conclude that the following propositional formula

$$[(F \lor C) \land (F \to G) \land (\neg(E \land C)) \land (\neg E \to (F \lor G))] \longrightarrow G \qquad (\text{A.1})$$

is a tautology. Otherwise, we cannot conclude that Lillian will be a good executive. Using the DeMorgan's law and double negation law, the conditional statement (the fourth proposition) in (A.1) can be written as

$$[(F \lor C) \land (F \to G) \land \neg(E \land C) \land (E \lor F \lor G)] \longrightarrow G. \qquad (\text{A.2})$$

Now we prove whether the conditional statement in (A.1) is a tautology or not. The easiest way to do this is through a truth table. However, a lot of lines from this truth table can be eliminated since the only time this formula can be false is when $G$ is false and everything on the left side of the implication is true. So, if we can prove that for every combination of values where $G$ is false, the left side of the implication is also false, then this implication will always be true.

| | | | | (1) | (2) | (3) | (4) | |
|---|---|---|---|---|---|---|---|---|
| $F$ | $E$ | $C$ | $G$ | $F \vee C$ | $F \rightarrow G$ | $\neg(E \wedge C)$ | $(E \vee F \vee G)$ | $(1) \wedge (2) \wedge (3) \wedge (4)$ |
| F | F | F | F | F | T | T | F | F |
| F | F | T | F | T | T | T | F | F |
| F | T | F | F | F | T | T | T | F |
| F | T | T | F | T | T | F | T | F |
| T | F | F | F | T | F | T | T | F |
| T | F | T | F | T | F | T | T | F |
| T | T | F | F | T | F | T | T | F |
| T | T | T | F | T | F | F | T | F |

As you can see from the table, for every combination of values where $G$ is false, everything to the left of the implication is also false. This means the original implication in (A.1) can never be false, so $G$ must be true. Thus, Lillian will be a good executive.

**1.16** Implication law, associative law, commutative law, distributive law, contradiction law, domination law and idempotent law.

**1.17** All the given implications are tautologies as shown below.

(a)

$$
\begin{aligned}
[\neg P \wedge (P \vee Q)] \rightarrow Q &\equiv [(\neg P \wedge P) \vee (\neg P \wedge Q)] \rightarrow Q \\
&\equiv [F \vee (\neg P \wedge Q)] \rightarrow Q \\
&\equiv (\neg P \wedge Q) \rightarrow Q \\
&\equiv \neg(\neg P \wedge Q) \vee Q \\
&\equiv (P \vee \neg Q) \vee Q \\
&\equiv P \vee (\neg Q \vee Q) \\
&\equiv P \vee T \equiv T.
\end{aligned}
$$

(b)

$$
\begin{aligned}
[(P \to Q) \wedge (Q \to R)] \to (P \to R) \quad &\equiv \quad \neg[(\neg P \vee Q) \wedge (\neg Q \vee R)] \vee (\neg P \vee R) \\
&\equiv \quad [\neg(\neg P \vee Q) \vee \neg(\neg Q \vee R)] \vee (\neg P \vee R) \\
&\equiv \quad [(P \wedge \neg Q) \vee (Q \wedge \neg R)] \vee (\neg P \vee R) \\
&\equiv \quad (P \wedge \neg Q) \vee (Q \wedge \neg R) \vee \neg P \vee R \\
&\equiv \quad [(P \wedge \neg Q) \vee \neg P] \vee [(Q \wedge \neg R) \vee R] \\
&\equiv \quad [(P \vee \neg P) \wedge (\neg Q \vee \neg P)] \vee [(Q \vee R) \wedge (\neg R \vee R)] \\
&\equiv \quad [T \wedge (\neg Q \vee \neg P)] \vee [(Q \vee R) \wedge T] \\
&\equiv \quad (\neg Q \vee \neg P) \vee (Q \vee R) \\
&\equiv \quad (\neg Q \vee Q) \vee (\neg P \vee R) \\
&\equiv \quad T \vee (\neg P \vee R) \equiv T.
\end{aligned}
$$

(c)

$$
\begin{aligned}
[P \wedge (P \to Q)] \to Q \quad &\equiv \quad [P \wedge (\neg P \vee Q)] \to Q \\
&\equiv \quad [(P \wedge \neg P) \vee (P \wedge Q)] \to Q \\
&\equiv \quad [F \vee (P \wedge Q)] \to Q \\
&\equiv \quad (P \wedge Q) \to Q \\
&\equiv \quad \neg(P \wedge Q) \vee Q \\
&\equiv \quad (\neg P \vee \neg Q) \vee Q \\
&\equiv \quad \neg P \vee (\neg Q \vee Q) \\
&\equiv \quad \neg P \vee T \equiv T.
\end{aligned}
$$

(d)

$$
\begin{aligned}
[(P \vee Q) \wedge (P \to R) \wedge (Q \to R)] \to R \quad &\equiv \quad \neg[(P \vee Q) \wedge (\neg P \vee R) \wedge (\neg Q \vee R)] \vee R \\
&\equiv \quad [\neg(P \vee Q) \vee \neg(\neg P \vee R) \vee \neg(\neg Q \vee R)] \vee R \\
&\equiv \quad (\neg P \wedge \neg Q) \vee (P \wedge \neg R) \vee (Q \wedge \neg R) \vee R \\
&\equiv \quad (\neg P \wedge \neg Q) \vee (P \wedge \neg R) \vee [(Q \vee R) \wedge (\neg R \vee R)] \\
&\equiv \quad (\neg P \wedge \neg Q) \vee (P \wedge \neg R) \vee [(Q \vee R) \wedge T] \\
&\equiv \quad (\neg P \wedge \neg Q) \vee (P \wedge \neg R) \vee Q \vee R \\
&\equiv \quad [(\neg P \wedge \neg Q) \vee Q] \vee [(P \wedge \neg R) \vee R] \\
&\equiv \quad [(\neg P \vee Q) \wedge (\neg Q \vee Q)] \vee [(P \vee R) \wedge (\neg R \vee R)] \\
&\equiv \quad [(\neg P \vee Q) \wedge T] \vee [(P \vee R) \wedge T] \\
&\equiv \quad (\neg P \vee Q) \vee (P \vee R) \\
&\equiv \quad (\neg P \vee P) \vee (Q \vee R) \\
&\equiv \quad T \vee (Q \vee R) \equiv T.
\end{aligned}
$$

**1.18** Starting with item (a), we have

$$
\begin{aligned}
P \leftrightarrow Q &\equiv (P \rightarrow Q) \wedge (Q \rightarrow P) \\
&\equiv (\neg P \vee Q) \wedge (\neg Q \vee P) \\
&\equiv [(\neg P \vee Q) \wedge \neg Q] \vee [(\neg P \vee Q) \wedge P] \\
&\equiv [(\neg P \wedge \neg Q) \vee (Q \wedge \neg Q)] \vee [(\neg P \wedge P) \vee (Q \wedge P)] \\
&\equiv [(\neg P \wedge \neg Q) \vee F] \vee [F \vee (Q \wedge P)] \\
&\equiv (\neg P \wedge \neg Q) \vee (Q \wedge P) \quad \text{(This item (b))} \\
&\equiv \neg(P \vee Q) \vee (Q \wedge P) \\
&\equiv \neg\neg[\neg(P \vee Q) \vee (Q \wedge P)] \\
&\equiv \neg[(P \vee Q) \wedge \neg(Q \wedge P)] \equiv \neg[P \oplus Q]. \quad \text{(This item (c))}.
\end{aligned}
$$

**1.19** To construct a DNF having the given truth table, note that

| $P$ | $Q$ | $R$ | Statement | Equivalent To |
|-----|-----|-----|-----------|---------------|
| T | T | F | T | $P \wedge Q \wedge \neg R$ |
| T | F | T | T | $P \wedge \neg Q \wedge R$ |
| F | T | F | T | $\neg P \wedge Q \wedge \neg R$ |

A DNF would be $(P \wedge Q \wedge \neg R) \vee (P \wedge \neg Q \wedge R) \vee (\neg P \wedge Q \wedge \neg R)$.

**1.20** (a) $\neg(A \vee B) \equiv \neg A \wedge \neg B$ (CNF).

(b) $\neg(A \wedge B) \equiv \neg A \vee \neg B \equiv (\neg A \vee \neg B) \wedge (A \vee \neg A)$ (CNF).

(c) $A \vee (B \wedge C) \equiv (A \vee B) \wedge (A \vee C)$ (CNF).

**1.21** $(P \wedge \neg P \wedge Q) \vee (Q \wedge \neg Q) \vee (R \wedge \neg R) \vee (P \wedge \neg Q \wedge \neg R \wedge \neg P)$.

**1.22** Note that $P \wedge Q \wedge (\neg R \vee S \vee \neg T) \equiv (P \wedge Q \wedge \neg R) \vee (P \wedge Q \wedge S) \vee (P \wedge Q \wedge \neg T)$, which is in DNF. According to Remark 1.5, the given proposition is satisfiable.

**1.23** (a) (iii).   (b) (iv).   (c) (ii).   (d) (iv).   (e) (ii).   (f) (iii).

**1.24** (a) False. The value $x = -1$ can make this proposition false.

(b) False. It is true only if we take $x = 0$, but this is not in the domain.

(c) True. The value $x = 2$ can make this proposition true.

(d) False. The value $x = -1$ can make this proposition false.

(e) True. The value $x = 1$ can make this proposition true.

**1.25** Note that $\mathbb{N}$ is defined as $\mathbb{N} \triangleq \{1, 2, 3, \ldots\}$.

    (a) Take $P(x, y) = $ "$x + y = x + y$".    (c) Take $P(x, y) = $ "$xy = y$".

    (b) Take $P(x, y) = $ "$xy = \frac{1}{2}$".    (d) Take $P(x, y) = $ "$xy = x$".

**1.26** The negation of the given proposition is $\forall x \in \mathbb{R}, \exists y \in \mathbb{R}, x + y \neq 0$ as it is seen below.

$$
\begin{aligned}
\neg(\exists x \in \mathbb{R}, \forall y \in \mathbb{R}, x + y = 0) &\equiv \forall x \in \mathbb{R}, \neg(\forall y \in \mathbb{R}, x + y = 0) \\
&\equiv \forall x \in \mathbb{R}, \exists y \in \mathbb{R}, \neg(x + y = 0) \\
&\equiv \forall x \in \mathbb{R}, \exists y \in \mathbb{R}, x + y \neq 0.
\end{aligned}
$$

**1.27** Denote by $D$ the set of all people.

    (a) $\exists x \in D, \neg L(x)$.

    (b) $\exists x \in D, (I(x) \wedge \neg C(x) \wedge \neg A(x))$.

    (c) $\forall x \in D, ([A(x) \vee C(x)] \wedge I(x) \to L(x))$.

**1.28** The negation of the sentence "Some children do not like mimes" is the sentence "Every child has some mime that he/she likes", which can be symbolized as $\forall x \in D, [\text{Child}(x) \to (\exists y \in D, [\text{Mime}(y) \wedge \text{Like}(x, y)])]$. As another approach, the quantified statement "Some children do not like mimes" is symbolized as $\exists x \in D, [\text{Child}(x) \wedge (\forall y \in D, [\text{Mime}(y) \to \neg\text{Like}(x, y)])]$, which can be negated as $\forall x \in D, [\text{Child}(x) \to \neg(\forall y \in D, [\text{Mime}(y) \to \neg\text{Like}(x, y)])]$.

**1.29** We let $D_1$ be the set of all houses and $D_2$ be the set of all owners. We also define the following two predicates: $\text{OWN}(v, x) = $ "The house $v$ has owner $x$", and $\text{ADJ}(v, u) = $ "The house $v$ is adjacent to house $u$".

    (a) $\forall x \in D_2, \exists v \in D_1, \text{OWN}(v, x)$.

    (b) $\forall v \in D_1, \forall u \in D_1, \forall x \in D_2, (\text{ADJ}(v, u) \wedge \text{OWN}(v, x) \longrightarrow \neg\text{OWN}(u, x))$.

    (c) $\forall v \in D_1, \forall x \in D_2, (\text{OWN}(v, x)) \longrightarrow [\neg\exists y \in D_2, ((y \neq x) \wedge \text{OWN}(v, y))])$.

**1.30** Denote by $D$ the set of all people. We also define the following three predicates: $\text{Fr}(x, y) = $ "The two persons $x$ and $y$ are friends", $\text{Sm}(x) = $ "The person $x$ smokes", and $\text{Ca}(x) = $ "The person $x$ has a cancer".

    (a) $\forall x \in D, \forall y \in D, \forall z \in D, (\text{Fr}(x, y) \wedge \text{Fr}(y, z) \longrightarrow \text{Fr}(x, z))$.

    (b) $\forall x \in D, (\neg[\exists y \in D, \text{Fr}(x, y)] \longrightarrow \text{Sm}(x))$.

    (c) $\forall x \in D, (\text{Sm}(x) \longrightarrow \text{Ca}(x))$.

    (d) $\forall x \in D, \forall y \in D, (\text{Fr}(x, y) \longrightarrow [\text{Sm}(x) \longleftrightarrow \text{Sm}(y)])$.

**2.1** (a) (iii).      (c) (ii).      (e) (iv).      (g) (ii).      (i) (ii).

(b) (iv).      (d) (ii).      (f) (iv).      (h) (iii).      (j) (iv).

**2.2** Suppose the principle is true when $n = k$, for some $k \in \mathbb{N}$. Then there is some natural number $s$ such that $m < sk$. Choosing the same $s$, we have $m < sk < s(k+1)$, so the principle is true when $n = k+1$. The proof is complete.

**2.3** By induction, let $P(n)$ be

$$"1 + 2^2 + 3^2 + ... + n^2 = \frac{n(n+1)(2n+1)}{6}". \qquad (A.3)$$

First, we look at the base case for $n = 1$ to see whether or not $P(1)$ is true. Substituting $n = 1$ into both sides of the equation, we see

$$\sum_{i=1}^{1} i^2 = 1^2 = \frac{(1)(1+1)(2*1+1)}{6} = 1$$

is true. Now, we look at the inductive step for $n = k$ to see whether or not $P(k+1)$ is true. By substituting all $n$'s with $(n+1)$ on the right side of (A.3), we get

$$\frac{(n+1)((n+1)+1)(2(n+1)+1)}{6} = \frac{(n+1)(n+2)(2n+3)}{6}$$
$$= \frac{(n^2+3n+2)(2n+3)}{6} \qquad (A.4)$$
$$= \frac{2n^3+9n^2+13n+6}{6}.$$

If the inductive hypothesis $P(k)$, which is

$$"1 + 2^2 + 3^2 + ... + k^2 = \frac{k(k+1)(2k+1)}{6}"$$

is true, then $P(k+1)$, which is

$$1 + 2^2 + 3^2 + ... + k^2 + (k+1)^2 = \frac{k(k+1)(2k+1)}{6} + (k+1)^2$$
$$= \frac{k(k+1)(2k+1)}{6} + \frac{6(k+1)^2}{6}$$
$$= \frac{k(k+1)(2k+1) + 6(k+1)^2}{6}$$
$$= \frac{(k^2+k)(2k+1) + 6(k+1)^2}{6}$$
$$= \frac{2k^3+9k^2+13k+6}{6},$$

is also true because this matches in (A.4) that we found by substituting $(n + 1)$ into the right-hand side of (A.3). We conclude if $P(k)$ is true, then $P(k + 1)$ is true for any $k \geq 1$ which proves the inductive conclusion.

2.4 Define the following predicate:

$P(n) = $ "The cardinality of the powerset of a finite set $A$ is equal to $2^n$ if the cardinality of $A$ is $n$."

Base case (for $n = 0$): The set $A$ with cardinality 0 is the empty set $\emptyset$. Its powerset (set of all subsets) is $\emptyset$. Since $\mathcal{P}(A) = \{\emptyset\}$, we have $|\mathcal{P}(A)| = 1 = 2^0$. Therefore $P(0)$ is true.

Inductive step: Assume that $P(k)$ is true. To prove that $P(k+1)$ is also true, let $B$ be any set of cardinality $k + 1$. Enumerating the elements of $B$ as $a_1, a_2, a_3, ..., a_k, a_{k+1}$. That is, $B = \{a_1, a_2, a_3, ..., a_k, a_{k+1}\}$ and $|B| = k + 1$. Define

$$A = \{a_1, a_2, a_3, ..., a_k\},$$

then $|A| = k$ and $|\mathcal{P}(A)| = 2^k$ (by the inductive hypothesis).

Note that $B = A \cup \{a_{k+1}\}$ and that every subset of $A$ is also a subset of $B$.

Now any subset of $B$ either contains $a_{k+1}$ or it does not. Every subset $C$ (of $B$) that does not contain $a_{k+1}$ is also a subset of $A$, of which there are $2^k$ (by the inductive hypothesis of $P(k)$). For every subset $C$ not containing $a_{k+1}$, there is another subset of the form $C \cup \{a_{k+1}\}$ containing it. Since there are $2^k$ such possible subsets $C$, there are $2^k$ subsets of $B$ containing $a_{k+1}$. We now have that $B$ has $2^k$ subsets not containing $a_{k+1}$ and $2^k$ containing it. Therefore, $|\mathcal{P}(B)| = 2^k + 2^k = 2^{k+1}$, and hence $P(k + 1)$ is true.

Thus, by induction on $n$, $P(n)$ is true for all $n \in \mathbb{N}$.

2.5 The base case for $n = 1$ is seen to be true as $T(1) = 1 = 3^1 - 2 = 1$, where the first equality follows from the piecewise function. Now, assume that the inductive hypothesis holds for $n = k$, i.e., $T(k) = 3^k - 2$. Then

$$T(k + 1) = 3T(k) + 4 = 3(3^k - 2) + 4 = 3(3^k) - 2 = 3^{k+1} - 2. \qquad \text{(A.5)}$$

Therefore, if $T(k) = 3^k - 2$, then $T(k + 1) = 3^{k+1} - 2$ for an integer $k \geq 0$. This proves the inductive conclusion. Thus, by induction on $n$, the proof is complete.

2.6 (a) False. To see this, let $a$ be any real number, and consider $S_1 = \{a\}$, $S_2 = \{\{a\}\}$, and $S_3 = \{\{\{a\}\}\}$. Then $S_1 \in S_2$ and $S_2 \in S_3$, but $S_1 \notin S_3$ because $S_3$ only contains one element which is not $S_1$.

Note that the statement becomes true if the symbol membership symbol "$\in$" is replaced with inclusion symbol "$\subseteq$". That is, the following statement is true.

$$(S_1 \subseteq S_2) \wedge (S_2 \subseteq S_3) \longrightarrow (S_1 \subseteq S_3).$$

(b) False. Take, for example, $A = \{1, 2\}$, $B = \{1, 3\}$ and $C = \{2, 3, 4\}$.

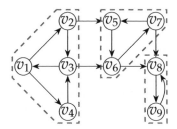

Figure A.1: The strongly connected components of the digraph of Exercise 4.10.

2.7 (a) Since $|S| = 5$, the number of subsets of $S$ is $2^{|S|} = 2^5 = 32$.

(b) Note that $T = \{22, 24, 26, \ldots, 38, 40\}$ and that $|T| = 10$. So, the number of subsets of $T$ is $2^{|T|} = 2^{10} = 1024$.

2.8 $(A \cup C') \cap B' = \{u, w, r, t\}$.

2.9 We have to show that the relation $\mathcal{R}$ is reflexive, symmetric, and transitive. For reflexivity, each person $x$ has the same birthday of him/herself. For symmetry, if $x$ and $y$ have the same birthday, then $y$ and $x$ have the same birthday. For transitivity, if $x$ and $y$ have the same birthday and $y$ and $z$ have the same birthday, then $x$ and $z$ have the same birthday. Thus, the relation $\mathcal{R}$ is an equivalence relation. Since there are 365 ways two people can have the same birthday, $\mathcal{R}$ has 365 equivalence classes.

2.10 The exponential function $f : \mathbb{R} \to [0, \infty)$ defined by $f(x) = e^x$ is an injection. To see this, let $x, y \in \mathbb{R}$ such that $e^x = e^y$, then $x = \ln e^x = \ln e^y = y$. Now, we show that the function $f$ is a surjection. To see this, note that any $y \in [0, \infty)$, we can choose $x = \ln y$ and have $f(x) = e^x = e^{\ln y} = y$. Therefore $f$ is both an injective and surjective function. Thus, $f$ is a bijection.

The exponential function $g : \mathbb{R} \to \mathbb{R}$ defined by $g(x) = e^x$ is not a bijection because it is not a surjection. For instance, because $e^x > 0$ for any $x \in \mathbb{R}$, there is no $x \in \mathbb{R}$ such that $g(x) = -1$.

3.1 (a) (iii).     (c) (ii).     (e) (i).     (g) (ii).

(b) (iv).     (d) (i).     (f) (i).     (h) (i).

3.2 Let $S$ be the set of positive integers for which the statement is true. Since $\lim_{n \to \infty} \frac{\ln n}{n} = 0, 1 \in S$. Assume that $k \in S$, i.e., $\lim_{n \to \infty} \frac{(\ln n)^k}{n}$. Using Hospital's rule, it follows that

$$\lim_{n \to \infty} \frac{(\ln n)^{k+1}}{n} = \lim_{n \to \infty} \frac{(k+1)(\ln n)^k}{n} = (k+1) \lim_{n \to \infty} \frac{(\ln n)^k}{n} = 0.$$

Thus, $k + 1 \in S$. This proves the result by induction.

**3.3** Using Table 3.1 (see also the result in Exercise 2.3), we have

$$\begin{aligned}
\sum_{k=30}^{60} k^2 &= \sum_{k=1}^{60} k^2 - \sum_{k=1}^{29} k^2 \\
&= \frac{60 \cdot 61 \cdot 62}{6} - \frac{29 \cdot 30 \cdot 31}{6} \\
&= 10 \cdot 61 \cdot 62 - 29 \cdot 5 \cdot 31 \\
&= 37,820 - 4,495 = 33,325.
\end{aligned}$$

**3.4** Note that

$$\frac{1}{(2k+1)(2k-1)} = \frac{(2k+1)/2 - (2k-1)/2}{(2k+1)(2k-1)} = \frac{1/2}{2k-1} - \frac{1/2}{2k+1}.$$

The sequence of partial sums is

$$\begin{aligned}
S_n &= \sum_{k=1}^{n} \frac{1}{(2k+1)(2k-1)} \\
&= \sum_{k=1}^{n} \left( \frac{1/2}{2k-1} - \frac{1/2}{2k+1} \right) \\
&= \left( \frac{1/2}{1} - \frac{1/2}{3} \right) + \left( \frac{1/2}{3} - \frac{1/2}{5} \right) + \cdots + \left( \frac{1/2}{2n-1} - \frac{1/2}{2n+1} \right) \\
&= \frac{1}{2} - \frac{1/2}{2n+1} \longrightarrow \frac{1}{2}.
\end{aligned}$$

Thus, the series converges to 1/2.

**3.5**    (*i*) Note that $\lim_{k \to \infty} 3^{-k} = \lim_{k \to \infty} 1/3^k = 0$. It follows that

$$\lim_{k \to \infty} \frac{1}{2 + 3^{-k}} = \frac{1}{2+0} = \frac{1}{2}.$$

Using the divergence test, the series $\displaystyle\sum_{k=1}^{\infty} \frac{1}{2 + 3^{-k}}$ diverges.

(*ii*) Using Theorem 3.3 (*iv*), we have $\lim_{k \to \infty}(0.3)^k = 0$. It follows that

$$\lim_{k \to \infty} \frac{1}{2 + (0.3)^k} = \frac{1}{2+0} = \frac{1}{2}.$$

Using the divergence test, the series $\sum_{k=1}^{\infty} \dfrac{1}{2 + (0.3)^k}$ diverges.

**3.6** (*i*) Since $2/3 < 1$, using the geometric series we have

$$\sum_{k=1}^{\infty} \left(\frac{2}{3}\right)^k = \frac{\frac{2}{3}}{1 - \frac{2}{3}} = 2.$$

The series converges to 2.

(*ii*) Note that

$$\sum_{k=1}^{\infty} \left(\frac{4}{5}\right)^{3-k} = \left(\frac{4}{5}\right)^3 \sum_{k=1}^{\infty} \left(\frac{5}{4}\right)^k.$$

Since $5/4 > 1$, using the geometric series, we conclude that the series diverges.

(*iii*) Using the geometric series, we have

$$\sum_{k=2}^{\infty} x^k = \frac{x^2}{1 - x}, \quad \text{if } |x| < 1.$$

Replacing $x$ with $-x$, we get

$$\sum_{k=2}^{\infty} (-1)^k x^k = \sum_{k=2}^{\infty} (-x)^k = \frac{(-x)^2}{1 - (-x)} = \frac{x^2}{1 + x}.$$

When $x = 2/5$, we have

$$\sum_{k=2}^{\infty} (-1)^k \left(\frac{2}{5}\right)^{k-2} = \left(\frac{5}{2}\right)^2 \sum_{k=2}^{\infty} (-1)^k \left(\frac{2}{5}\right)^k = \left(\frac{5}{2}\right)^2 \frac{(\frac{2}{5})^2}{1 + \frac{2}{5}} = \frac{1}{1 + \frac{2}{5}} = \frac{5}{7}.$$

The series converges to 5/7.

(*iv*) Using the geometric series, we have

$$\sum_{k=1}^{\infty} x^k = \frac{x}{1 - x}, \quad \text{if } |x| < 1.$$

Differentiating both sides with respect to $x$, we get

$$\sum_{k=1}^{\infty} k x^k = \frac{d}{dx}\left(\frac{x}{1 - x}\right) = \frac{1}{(1 - x)^2}, \quad \text{if } |x| < 1.$$

When $x = 2/3$, we have

$$\sum_{k=1}^{\infty} k \left( \frac{2}{3} \right)^k = \frac{1}{\left( 1 - \frac{2}{3} \right)^2} = \frac{1}{\left( \frac{1}{3} \right)^2} = 9.$$

The series converges to 9.

3.7 If $M \in \mathbb{R}^{n \times n}$ is a symmetric positive definite matrix, and $x \in \mathbb{R}^n$ is a nonzero vector, then

$$x^\mathsf{T} M^{-1} x = x^\mathsf{T} (M M^{-1} M)^{-1} x = (M^{-1} x)^\mathsf{T} M M^{-1} x = y^\mathsf{T} M y > 0,$$

where $y = M^{-1} x$. Thus, $M^{-1}$ is also positive definite.

3.8 Using Proposition 3.1, we have

$$
\begin{aligned}
H^\perp &= \left\{ x \in \mathbb{R}^2 : x_1 - 2x_2 = 0 \right\}^\perp \\
&= \left\{ \begin{bmatrix} x_1 \\ x_2 \end{bmatrix} \in \mathbb{R}^2 : \begin{bmatrix} 1 & -2 \end{bmatrix} \begin{bmatrix} x_1 \\ x_2 \end{bmatrix} = 0 \right\}^\perp \\
&= \left\{ \begin{bmatrix} x_1 \\ x_2 \end{bmatrix} \in \mathbb{R}^2 : x = \begin{bmatrix} 1 \\ -2 \end{bmatrix} u, u \in \mathbb{R} \right\} = \left\{ x \in \mathbb{R}^2 : 2x_1 = -x_2 \right\}.
\end{aligned}
$$

3.9 Let $x, y \in \mathbb{R}^n$ and $\lambda \in [0, 1]$. We have

$$
\begin{aligned}
f(\lambda x + (1 - \lambda) y) &= \max_{i=1,\ldots,m} f_i(\lambda x + (1-)y) \\
&\leq \max_{i=1,\ldots,m} (\lambda f_i(x) + (1 - \lambda) f_i(y)) \\
&\leq \lambda \max_{i=1,\ldots,m} f_i(x) + (1 - \lambda) \max_{i=1,\ldots,m} f_i(y) \\
&= \lambda f(x) + (1 - \lambda) f(y),
\end{aligned}
$$

where the first inequality follows from the convexity of $f_i, i = 1, 2, \ldots, m$.

3.10 Assume that $S_1, S_2, \ldots, S_m$ are convex sets. Let $x, y \in \cap_{i=1}^m S_i$ and $\lambda \in [0, 1]$. By the convexity of $S_i, i = 1, 2, \ldots, m$, we have $\lambda x + (1 - \lambda) y \in S_i, i = 1, 2, \ldots, m$. It follows that $\lambda x + (1 - \lambda) y \in \cap_{i=1}^m S_i$. Therefore, $\cap_{i=1}^m S_i$ is convex. If an infinite number of convex sets intersect, the result is also correct and a proof similar to the above one can be used. In addition, if the number of convex sets is infinite but countable, a proof by induction can be also thought of.

3.11 Under the assumption in Farkas' lemma (Version II). It is clear that if (2') holds, then so does (2). Now, assume that (2) is true, i.e., $Ax = 0, c^\mathsf{T} x < 0$ and $x \geq 0$, and let $\bar{x} = -x/(c^\mathsf{T} x)$. Then, the nonnegativity of $\bar{x}$ follows from the nonnegativity of $x$ and the negativity of $c^\mathsf{T}x$. Moreover,

$$c^\mathsf{T}\bar{x} = c^\mathsf{T}\left(\frac{-1}{c^\mathsf{T}x}x\right) = -\frac{1}{c^\mathsf{T}x}c^\mathsf{T}x = -1, \text{ and } A\bar{x} = A\left(\frac{-1}{c^\mathsf{T}x}x\right) = -\frac{1}{c^\mathsf{T}x}Ax = 0.$$

This proves that if (2) holds, then so does (2′).

**4.1** (a) (iv).     (c) (i).     (e) (ii).     (g) (i).     (i) (iv).

(b) (ii).     (d) (iv).     (f) (iii).     (h) (iii).     (j) (iii).

**4.2** A spanning tree so that no vertex has a degree of 4 is shown to the right. (Other answers might be possible).

**4.3** Yes. To see this, labeling the vertices of $G_1$ and that of $G_2$ creates a matching between them in a way that preserves adjacency as shown below.

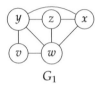                              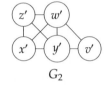

$G_1$                                                     $G_2$

**4.4** (*a*) The answer is no, because we have an odd number of odd degree vertices.

(*b*) No. The justification for this answer is that there are two 4-degree vertices. The minimum number of vertices (outside of themselves) that those two vertices must be connected to is 6. Therefore, all 8 vertices are connected. However, the degree sequence has 2 vertices of degree 0, contradicting this.

(*c*) Yes. To justify this answer, a graph that represents the given degree sequence is given in shown to the right.

**4.5** Assume that $G$ is acyclic but adding $(x, y)$ to $G$ creates a cycle for every $x, y \in V$ with $(x, y) \notin E$. To prove that $G$ is a tree we must show that $G$ is connected. Let $u$ and $v$ be arbitrary vertices in $G$. If $u$ and $v$ are not already adjacent, adding the edge $(u, v)$ creates a cycle in which all edges but $(u, v)$ belong to $G$. Thus, there is a path from $u$ to $v$. Since $u$ and $v$ were chosen arbitrarily, $G$ is connected. The proof is complete.

**4.6** Only one graph is possible, which is the complete graph $K_5$.

**4.7** The graph $K_{5,7}$ has no a Hamiltonian cycle because the sizes of its partition subsets are not identical. The graph $K_{5,7}$ has no a Hamiltonian path because the sizes of its partition subsets differ by more than one.

**4.8** (*a*) The tight upper bound is $k + 1$. We may need to add a single color, i.e., change the color of one endpoint to a new color.

(*b*) While $k$ is still an upper bound, the tight upper bound would be $k - 1$. Let us consider worst case which is the complete graph $K_n$. The graph $K_n$ is not $(n - 1)$-colorable (Theorem 4.12). Now, removing any edge from $K_n$ reduces the number of colors required to properly color it to $n - 1$.

**4.9** The statement is false. From Theorem 4.12, the complete graph $K_{d+2}$ is not $(d + 1)$-colorable. Let $v_1, v_2, \ldots, v_d, v_{d+1}, v_{d+2}$ be the vertices of $K_{d+2}$. Let also $G$ be the graph formed by adding a new vertex $u$ of degree $d$ and adjacent to each of $v_1, v_2, \ldots, v_d$. Then $G$ has a vertex of degree $d$ but it is not $(d + 1)$-colorable.

**4.10** The strongly connected components of the given digraph are surrounded by blue dashed polygons in Figure A.1.

**5.1** (a) (iv).    (b) (ii).    (c) (iii).    (d) (iii).    (e) (iv).

**5.2** (a) The recurrence relation $T(n) = 2T(n/2) + n$ is of the merge sort algorithm. In Example 5.5, we used the iteration method to solve this recurrence. Now, we use the guess-and-confirm method to obtain the same answer. In this recurrence, each $n$ should be a power of 2 (i.e., $n = 1, 2, 4, 8, \ldots$), otherwise we have either non-integer inputs or inputs with undefined references. It is given that $T(1) = 1$. Note that

$$
\begin{aligned}
T(2) &= 2T(1) + 2 = 4 = 2\log 2 + 2, \\
T(4) &= 2T(2) + 4 = 12 = 4\log 4 + 4, \\
T(8) &= 2T(4) + 8 = 32 = 8\log 8 + 8, \\
T(16) &= 2T(8) + 16 = 80 = 16\log 16 + 16, \\
&\vdots \\
T(k) &= k\log k + k.
\end{aligned}
$$

It seems that $T(n) = n\log n + n$ for every $n = 2^i$ and $i = 0, 1, 2, 3, \ldots$. We prove this by induction on $n$. The base case is trivial: $T(1) = 1 = 1\log 1 + 1$. Assume that the statement is true for all $m < k$, i.e., $T(m) = m\log m + m$ for any $m < k$. Now, we prove that $T(k) = k\log k + k$. Since $k$ is a power of 2, we have $k = 2^i$ (and hence $i = \log k$) for some $i \in \mathbb{N}$. Since $2^{i-1} < k$, we particularly have $T(2^{i-1}) = 2^{i-1}\log 2^{i-1} + 2^{i-1} = 2^{i-1}(i - 1) + 2^{i-1} = i(2^{i-1})$ by the inductive step. It follows that

$$
T(k) = 2T\left(\frac{k}{2}\right) + k = 2T\left(2^{i-1}\right) + k = 2i\left(2^{i-1}\right) + k = i\left(2^i\right) + k = k\log k + k.
$$

This confirms that $T(n) = n\log n + n$.

(b) The recurrence relation $T(n) = 3T(n-1) + 4$ is the one given in (5.1). We use the guess-and-confirm method to solve this recurrence. It is given that $T(1) = 1$. Note that

$$
\begin{aligned}
T(2) &= 3T(1) + 4 = 7 = 3^2 - 2, \\
T(3) &= 3T(2) + 4 = 25 = 3^3 - 2, \\
T(4) &= 3T(3) + 4 = 79 = 3^4 - 2, \\
&\vdots \\
T(k) &= 3^k - 2.
\end{aligned}
$$

It seems that $T(n) = 3^n - 2$ for $n = 1, 2, 3, \ldots$. We prove this by induction on $n$. The base case is trivial: $T(1) = 1 = 3^1 - 2$. Assume that the statement is true for $n = k$, i.e., $T(k) = 3^k - 2$ for some $k \in \mathbb{N}$. Now, we prove that $T(k+1) = 3^{(k+1)} - 2$. Note that

$$T(k+1) = 3T(k) + 4 = 3\left(3^k - 2\right) + 4 = 3\left(3^k\right) - 6 + 4 = 3^{k+1} - 2.$$

Thus, $T(n) = 3^n - 2$ for any $n = 1, 2, 3, \ldots$. This confirms the solution given in (7.10).

**5.3** Using repeated substitutions, we have

$$
\begin{aligned}
T(1) &= 1 = (0-3)2^0 + 4, \\
T(2) &= 0 = (1-3)2^1 + 4, \\
T(3) &= 0 = (2-3)2^2 + 4, \\
T(3) &= 5T(2) - 8T(1) + 4T(0) = 4 = (3-3)2^3 + 4, \\
&\vdots \\
T(n) &= 5T(n-1) - 8T(n-2) + 4T(n-3) = (n-3)2^n + 4, \text{ for all } n \geq 1.
\end{aligned}
$$

We prove the $n$th-term guess by mathematical induction. The base case is trivial: $T(0) = 1 = (0-3)2^0 + 4$, $T(1) = 0 = (1-3)2^1 + 4$, and $T(2) = 0 = (2-3)2^2 + 4$. Assume that the statement is true for all $m < k$, i.e., $T(m) = (m-3)2^m + 4$ for any $m < k$. Now, we prove that $T(k) = (k-3)2^k + 4$. This can be seen from the following:

$$
\begin{aligned}
T(k) &= 5T(k-1) - 8T(k-2) + 4T(k-3) \\
&= 5((k-4)2^{(k-1)} + 4) - 8((k-5)2^{(k-2)} + 4) + 4((k-6)2^{(k-3)} + 4) \\
&= (5k)2^{(k-1)} - (20)2^{(k-1)} + 20 - (8k)2^{(k-2)} + (40)2^{(k-2)} - 32 \\
&\quad + (4k)2^{(k-3)} - (24)2^{(k-3)} + 16 \\
&= (2.5)(k)2^k - (10)2^k + 20 - (2k)2^k + (10)2^k - 32 \\
&\quad + (0.5)(k)2^k - (3)2^k + 16 = (k-3)2^k + 4.
\end{aligned}
$$

5.4 (a) Note that

$$T(n) = 5T(n-1) = 5(5T(n-2)) = 5(5(5T(n-3))) = \cdots = 5^k T(n-k).$$

Let $k = n$, then $T(n) = 5^n T(0) = 3(5^n)$.

(b) Note that

$$
\begin{aligned}
T(n) &= 2T(\tfrac{n}{2}) + n \log n \\
&= 2(2T(\tfrac{n}{2^2}) + \tfrac{n}{2} \log(\tfrac{n}{2})) + n \log n \\
&= 2^2 T(\tfrac{n}{2^2}) + n(\log n - \log 2) + n \log n \\
&= 2^2 T(\tfrac{n}{2^2}) + 2n \log n - n \\
&= 2^3 T(\tfrac{n}{2^3}) + 3n \log n - n - 2n \\
&\ \ \vdots \\
&= 2^k T(\tfrac{n}{2^k}) + kn \log n - n \sum_{i=0}^{k-1} i \\
&= 2^k T(\tfrac{n}{2^k}) + kn \log n - \tfrac{n}{2}(k-1)k.
\end{aligned}
$$

Let $n = 2^k$, then

$$T(n) = nT(1) + n \log^2 n - \frac{n}{2}(\log n - 1)\log n = cn + \frac{1}{2}n \log^2 n + \frac{1}{2}n \log n.$$

5.5 Let $g(x)$ be the generating function for the sequence $\{a_n\} = \{T(n)\}$. Then

$$g(x) = \sum_{k=0}^{\infty} T(k)\, x^k.$$

Using the recurrence relation, we have

$$
\begin{aligned}
g(x) &= T(0) + T(1)x + \sum_{k=2}^{\infty} T(k)x^k \\
&= 1 + 3x + \sum_{k=2}^{\infty} (-T(k-1) + 6T(k-2))x^k \\
&= 1 + 3x - x\sum_{k=2}^{\infty} T(k-1)x^{k-1} + 6x^2 \sum_{k=2}^{\infty} T(k-2)x^{k-2} \\
&= 1 + 3x + xT(0) - x\sum_{k=1}^{\infty} T(k-1)x^{k-1} + 6x^2 \sum_{k=2}^{\infty} T(k-2)x^{k-2} \\
&= 1 + 4x - x\sum_{k=0}^{\infty} T(k)x^k + 6x^2 \sum_{k=0}^{\infty} T(k)x^k = 1 + 4x - xg(x) + 6x^2 g(x).
\end{aligned}
$$

It follows that

$$g(x) = 1 + 4x - xg(x) + 6x^2 g(x).$$

Solving for $g(x)$, we get

$$
\begin{aligned}
g(x) &= \frac{1+4x}{1+x-6x^2} \\
&= \frac{1+4x}{(1-2x)(1+3x)} \\
&= \frac{\frac{-1}{5}(1-2x)+\frac{6}{5}(1+3x)}{(1-2x)(1+3x)} = \frac{6}{5}\frac{1}{1-2x} - \frac{1}{5}\frac{1}{1+3x}.
\end{aligned}
$$

Using the geometric series (5.7), we get

$$
g(x) = \frac{6}{5}\sum_{k=0}^{\infty} 2^k x^k - \frac{1}{5}\sum_{k=0}^{\infty}(-3)^k x^k = \sum_{k=0}^{\infty}\left(\underbrace{\frac{6}{5}2^k - \frac{1}{5}(-3)^k}_{T(k)}\right)x^k.
$$

Therefore, $T(n) = (1/5)\left((6)\,2^n + (-1)^{n+1}3^n\right)$ for any $n \ge 0$.

**5.6** Let $g(x)$ be the generating function for the sequence $\{a_n\} = \{T(n)\}$. Then

$$
g(x) = \sum_{k=0}^{\infty} T(k)\,x^k.
$$

Using the recurrence relation, we have

$$
\begin{aligned}
g(x) &= T(0) + T(1)x + \sum_{k=2}^{\infty} T(k)x^k \\
&= T(0) + T(1)x + \sum_{k=2}^{\infty}(T(k-1) + T(k-2))x^k \\
&= T(0) + T(1)x + \sum_{k=2}^{\infty} T(k-1)x^k + \sum_{k=2}^{\infty} T(k-2)x^k \\
&= T(0) + T(1)x + x\sum_{k=2}^{\infty} T(k-1)x^{k-1} + x^2\sum_{k=2}^{\infty} T(k-2)x^{k-2} \\
&= T(0) + T(1)x - xT(0) + x\sum_{k=1}^{\infty} T(k-1)x^{k-1} + x^2\sum_{k=2}^{\infty} T(k-2)x^{k-2} \\
&= x + x\sum_{k=0}^{\infty} T(k)x^k + x^2\sum_{k=0}^{\infty} T(k)x^k = x + xg(x) + x^2 g(x).
\end{aligned}
$$

It follows that
$$
g(x) = x + xg(x) + x^2 g(x).
$$
Solving for $g(x)$, we get $g(x) = x/(x^2 + x - 1)$.

It can be shown that

$$1 - x - x^2 = -(x + \alpha)(x + \beta), \quad \text{and hence} \quad g(x) = \frac{1}{\sqrt{5}} \left( \frac{\beta}{\beta + x} - \frac{\alpha}{\alpha + x} \right),$$

where $\alpha = (1 + \sqrt{5})/2$ and $\beta = (1 - \sqrt{5})/2$.

Note that $\alpha = -1/\beta$. It follows that

$$\frac{\beta}{\beta + x} = \frac{1}{1 + x/\beta} = \frac{1}{1 - \alpha x} = \sum_{k=0}^{\infty} \alpha^k x^k,$$

where the last equality was obtained by the geometric series (5.7). Similarly, we also have

$$\frac{\alpha}{\alpha + x} = \sum_{k=0}^{\infty} \beta^k x^k.$$

Thus, we have

$$g(x) = \frac{1}{\sqrt{5}} \left( \sum_{k=0}^{\infty} \alpha^k x^k - \sum_{k=0}^{\infty} \beta^k x^k \right) = \sum_{k=0}^{\infty} \left( \underbrace{\frac{1}{\sqrt{5}} \left( \alpha^k - \beta^k \right)}_{T(k)} x^k \right).$$

Therefore, $T(n) = \frac{1}{\sqrt{5}} \left( \alpha^n - \beta^n \right)$ for any $n \geq 0$.

**5.7** The recursion tree for the recurrence $T(n) = 2T(n - 1) + 1$ is shown below.

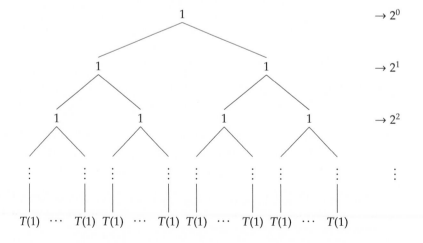

From the tree, the total cost at level $i$ is $(3/2)^i n$. The height of the tree, $h$, is obtained when $n - h = 1$. So, we have $h = n - 1$. It follows that

$$T(n) = \sum_{i=0}^{n-1} 2^i = \frac{1 - 2^n}{1 - 2} = 2^n - 1 = \Theta(2^n).$$

**5.8** (a) The recursion tree for the recurrence $T(n) = T(n/2) + n^2$ is shown below.

From the recursion tree shown to the right, the total cost at level $i$ is $n/2^{2i}$. The height of the tree, $h$, is obtained when $n/2^h = 1$. So, we have $h = \log n$. It follows that

$$
\begin{aligned}
T(n) &= \sum_{i=0}^{\log n} \left(\frac{1}{2}\right)^{2i} n^2 \\
&\leq n^2 \sum_{i=0}^{\infty} \left(\frac{1}{4}\right)^i \\
&= \frac{1}{1 - \frac{1}{4}} n^2 = \frac{4}{3} n^2.
\end{aligned}
$$

Thus, using the asymptotic notation introduced in Chapter 7, we have $T(n) = O(n^2)$.

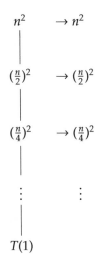

(b) The recursion tree for the recurrence $T(n) = T(n-1) + T(n/2) + n$ is shown below.

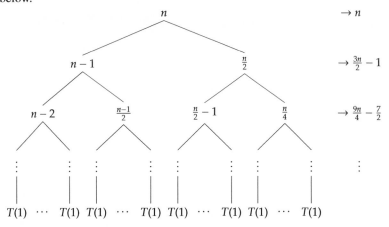

Since we are finding an upper bound, the height of the tree is decided by the *longest* path, which is the leftmost path. So, the height of the tree, $h$, is obtained when $n - h = 1$. Hence, $h = n - 1$.

Also, because we are finding an upper bound, we can drop the subtracted constant at each level. Then the total cost at level $i$ can be taken to be $(3/2)^i n$. It follows that

$$T(n) \leq \sum_{i=1}^{n-1} \left(\frac{3}{2}\right)^i n = \frac{1 - \left(\frac{3}{2}\right)^n}{1 - \frac{3}{2}} n = 2n\left(\left(\frac{3}{2}\right)^n - 1\right),$$

where we used the finite geometric series formula (3.3) to obtain the first equality. Thus, using the asymptotic notation introduced in Chapter 7, we have $T(n) = O(n(3/2)^n)$.

5.9 (a) The recursion tree for the recurrence $T(n) = 4T(\frac{n}{2} + 2) + n$ is shown below.

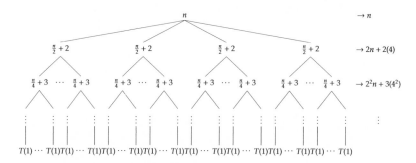

Although growing, the growth rate of the constant in front of $4^i$ is much smaller, so we can ignore it. So, we can take $2^i n + 4^i$ to be the total cost at level $i$. Also, the height of the tree, $h$, can obtained when $n/2^h = 1$. So, we have $h = \log n$. It follows that

$$T(n) \geq \sum_{i=0}^{\log n} \left(2^i n + 4^i\right) = n \sum_{i=0}^{\log n} 2^i + \sum_{i=0}^{\log n} 4^i.$$

Using the finite geometric series formula, we have

$$T(n) \geq \frac{1 - 2^{(1+\log n)}}{1 - 2} n + \frac{1 - 4^{(1+\log n)}}{1 - 4} = (2n-1)n + \frac{4n^2 - 1}{3} = \frac{10}{3}n^2 - n - \frac{1}{3}.$$

Thus, using the asymptotic notation introduced in Chapter 7, we have $T(n) = \Omega(n^2)$.

(b) The recursion tree for the recurrence $T(n) = T(n/3) + T(2n/3) + cn$ is shown in what follows.

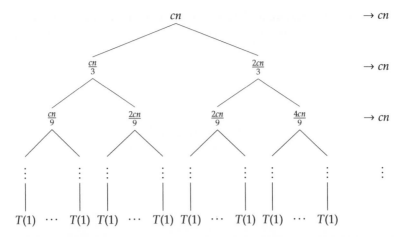

Since we are finding a lower bound, the height of the tree is decided by the *shortest* path, which is the leftmost path. So, the height of the tree, $h$, is obtained when $n/3^h = 1$. Hence, $h = \log_3 n$. From the tree, the total cost at level $i$ is $cn$. It follows that

$$T(n) \geq \sum_{i=0}^{\log_3 n} cn = cn(1 + \log_3 n) = cn\left(1 + \frac{\log n}{\log 3}\right).$$

Thus, using the asymptotic notation introduced in Chapter 7, we have $T(n) = \Omega(n \log n)$.

**6.1** (a) (ii).     (d) (ii).     (g) (iii).     (j) (i).     (m) (i).     (p) (iv).

     (b) (i).     (e) (iii).     (h) (ii).     (k) (i).     (n) (ii).

     (c) (iv).     (f) (iv).     (i) (i).     (l) (iv).     (o) (iii).

**6.2** From the binomial theorem it follows that

$$
\begin{aligned}
(x+y)^5 &= \sum_{k=0}^{5} \binom{5}{k} x^{5-k} y^k \\
&= \binom{5}{0}x^5 + \binom{5}{1}x^4 y + \binom{5}{2}x^3 y^2 + \binom{5}{3}x^2 y^3 + \binom{5}{4}xy^4 + \binom{5}{5}y^5 \\
&= x^5 + 5x^4 y + 10x^3 y^2 + 10x^2 y^3 + 5xy^4 + y^5.
\end{aligned}
$$

**6.3** A set with $n$ elements has a total of $2^n$ different subsets. Each subset has zero elements, one element, two elements, ..., or $n$ elements in it. There are $\binom{n}{0}$ subsets with zero elements, $\binom{n}{1}$ subsets with one elements, $\binom{n}{2}$ subsets with two elements, ..., and $\binom{n}{n}$ subsets with $n$ elements. Therefore, $\sum_{k=0}^{n} \binom{n}{k}$ counts the total number of subsets of a set with $n$ elements. By equating the two formulas we have for the number of subsets of a set with $n$ elements, we see that

$$\sum_{k=0}^{n} \binom{n}{k} = 2^n.$$

**6.4** Using the binomial theorem with $x = 2$ and $y = 1$, we have

$$3^n = (2 + 1)^n = \sum_{k=0}^{n} \binom{n}{k} 2^k 1^{n-k} = \sum_{k=0}^{n} \binom{n}{k} 2^k.$$

**6.5** Suppose that $T$ is a set containing $n$ elements. Let $a$ be an element in $T$, and let $S = T - \{a\}$. Note that there are $\binom{n}{k}$ subsets of $T$ containing $k$ elements. However, a subset of $T$ with $k$ elements either contains $a$ together with $k - 1$ elements of $S$, or contains $k$ elements of $S$ and does not contain $a$. Because there are $\binom{n-1}{k-1}$ subsets of $k - 1$ elements of $S$, there are $\binom{n-1}{k-1}$ subsets of $k$ elements of $T$ that contain $a$. And there are $\binom{n-1}{k}$ subsets of $k$ elements of $T$ that do not contain $a$, because there are $\binom{n-1}{k}$ subsets of $k$ elements of $S$. Consequently,

$$\binom{n}{k} = \binom{n-1}{k-1} + \binom{n-1}{k}.$$

**6.6** Suppose that there are $m$ items in one set and $n$ items in a second set. Then the total number of ways to pick $r$ elements from the union of these sets is $\binom{m+n}{r}$.

Another way to pick $r$ elements from the union is to pick $k$ elements from the second set and $r - k$ elements from the first set, where $k$ is an integer with $0 \le k \le r$. Because there are $\binom{n}{k}$ ways to choose $k$ elements from the second set and $\binom{m}{r-k}$ ways to choose $r - k$ elements from the first set, the product principle of counting tells us that this can be done in $\binom{m}{r-k}\binom{n}{k}$ ways. Hence, the total number of ways to pick $r$ elements from the union also equals $\sum_{k=0}^{r} \binom{m}{r-k}\binom{n}{k}$.

We have found two expressions for the number of ways to pick $r$ elements from the union of a set with $m$ items and a set of $n$ items. Equating them gives us Vandermonde's identity.

**6.7** Define the following predicate:

$P(m) = $ "If there are $n_k$ ways to perform task $T_k$ for $k = 1, \ldots, m$, then there are $n_1 \times n_2 \times \cdots \times n_m$ ways to perform all $m$ tasks."

Base case (for $m = 1$): $P(1)$ is trivially true.

Inductive step: Assume that $P(k)$ is true for all $k \le m$. To prove that $P(m + 1)$ is also true, assume that we have $m + 1$ tasks, and that task $T_k$ can be performed in $n_k$ ways, for $k = 1, 2, \ldots, m + 1$. We would like to show that the entire set of $m + 1$ tasks can be performed in $n_1 \times n_2 \times \cdots \times n_m \times n_{m+1}$ ways.

Let $T$ be a (large) task that comprises of the combination of tasks $T_1$ through $T_m$. Now, to do the tasks $T_1$ through $T_{m+1}$, we must perform the pair of tasks $T$ and $T_{m+1}$. From the inductive hypothesis, we conclude that the task $T$ can be done in $n$ ways where $n = n_1 \times n_2 \times \cdots \times n_m$ ways. Now, the inductive hypothesis can be also used to conclude that the number of ways to perform the pair of tasks $T$ and $Tm + 1$ is $n \times n_{m+1}$ ways. Thus, the entire set of $m + 1$ tasks can be performed in $n \times n_{m+1} = n_1 \times n_2 \times \cdots \times n_m \times n_{m+1}$ ways.

6.8 Define the following predicate:

$P(m) = $ "If a procedure can be done in one of $n_1$ ways, in one of $n_2$ ways, …, or in one of $n_m$ ways, then the total number of ways to do the procedure is $n_1 + n_2 + \cdots + n_m$ ways, provided that none of the set of $n_i$ ways of doing the procedure is the same as any of the set of $n_j$ ways, for all pairs $i$ and $j$ with $1 \le i < j \le m$."

Base case (for $m = 1$): $P(1)$ is trivially true.

Inductive step: Assume that $P(k)$ is true for all $k \le m$. To prove that $P(m + 1)$ is also true, assume that a procedure can be done in one of $n_1$ ways, in one of $n_2$ ways, …, in one of $n_m$ ways, or in one of $n_{m+1}$ ways, where none of the set of $n_i$ ways of doing the task is the same as any of the set of $n_j$ ways, for all pairs $i$ and $j$ with $1 \le i < j \le m + 1$. We would like to show that the total number of ways to do the procedure is $n_1 + n_2 + \cdots + n_{m+1}$ ways.

Note that it is assumed that the procedure can be done in one of $n_1$ ways, in one of $n_2$ ways, …, in one of $n_m$ ways. So, by the inductive hypothesis, we conclude that the procedure can also be done in one of $n$ ways where $n = n_1 + n_2 + \cdots + n_m$ ways. Note that it is also assumed that the procedure can be done in one of $n_{m+1}$ ways. Because none of the set of $n_i$ ways of doing the procedure is the same as any of the set of $n_j$ ways, for all pairs $i$ and $j$ with $1 \le i < j \le m + 1$, it follows that none of the set of $n$ ways of doing the procedure is the same as any of the set of $n_{m+1}$ ways. Now, the inductive hypothesis can be also used to conclude that the procedure can be performed in $n + n_{m+1} = n_1 + n_2 + \cdots + n_m + n_{m+1}$ ways.

6.9 Suppose, in the contrary, that none of the $k$ boxes contains more than $\lceil N/k \rceil - 1$ objects. Then the total number of objects would be at most

$$k\left(\left\lceil \frac{N}{k} \right\rceil - 1\right) < k\left(\left(\frac{N}{k} + 1\right) - 1\right) = N,$$

where we have used the inequality $\lceil N/k \rceil < (N/k) + 1$. This contradicts the fact that there are a total of $N$ objects.

6.10 Note that $A' \cap B \cap C = B \cap C \cap A' = B \cap C - A$. Then $|A' \cap B \cap C| = |B \cap C| - |A \cap B \cap C| = 19 - 11 = 8$.

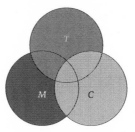

Figure A.2: A Venn diagram for Exercise 6.11.

6.11 Let $S$ be the set of all people who took the survey, $T$ be the set of all people who like tea, $C$ be the set of all people who like coffee, and $M$ be the set of all people who like milk. Then $|S| = 240, |T| = 91, |C| = 70, |T \cap C| = 31, |T' \cap C' \cap M'| = 91$, and $|T \cap C \cap M| = 7$. We are looking for $|T' \cap C' \cap M|$.

Note that $|T \cup C \cup M| = |S| - |(T \cup C \cup M)'| = |S| - |T' \cap C' \cap M'| = 240 - 91 = 149$. Note also that the subset $M \cap T' \cap C'$ is the dark-colored area in the Venn diagram in Figure A.2. It follows that

$$|M \cap T' \cap C'| = |M \cup T \cup C| - |T| - |C| + |T \cap C| = 149 - 91 - 70 + 31 = 19.$$

6.12 Note that the "for" loops in lines (5) and (8) are independent, but this pair of loops and "for" loops in lines (2) and (3) are nested. Let $T_s$ be the number of times of executing the statement given in line $(s)$ for each $s = 1, 2, 3, 4, 5, 6, 8, 9$. The initial value of "sum" is one. Each time the "sum" statement in each of lines (4), (6) and (9) is executed, 1 is added to "sum". Therefore, by the sum principle of counting, the value of "sum" after the fragment given in Algorithm 6.5 has been executed is equal to the number of ways to carry out the task $T_1$, plus the number of ways to carry out the task $T_4$, plus the number of ways to carry out the task $T_6$, plus the number of ways to carry out the task $T_9$. By using the product principle of counting, we can find the number of ways to carry out the task $T_s, s = 1, 2, 3, 4, 5, 6, 8, 9$. We added a column for these numbers in Algorithm A.1 (see the comments in gray in Algorithm A.1). Thus, the value of "sum" after the fragment given in Algorithm 6.5 has been executed is equal to

$$1 + nm + nmp + nmq = 1 + nm(1 + p + q).$$

6.13 Suppose that for each element $y$ in the codomain of $f$ we have a box that contains all elements $x$ of the domain of $f$ such that $f(x) = y$. Because the domain contains $k + 1$ or more elements and the codomain contains only $k$ elements, the pigeonhole principle tells us that one of these boxes contains two or more elements $x$ of the domain. This means that $f$ cannot be an injection.

---

**Algorithm A.1:** Algorithm 6.5 revisited

| | |
|---|---|
| 1:  sum = 1 | // Task $T_1$ is done in 1 way |
| 2:  **for** $(i = 1; i \leq n; i + +)$ **do** | // $T_2$ is done in $n + 1$ ways |
| 3:      **for** $(j = 0; j < m; j + +)$ **do** | // $T_3$ is done in $n \times (m + 1)$ ways |
| 4:          sum = sum + 1 | // $T_4$ is done in $n \times m$ ways |
| 5:          **for** $(k = p; k \geq 1; k - -)$ **do** | // $T_5$ is done in $n \times m \times (p + 1)$ ways |
| 6:              sum = sum + 1 | // $T_6$ is done in $n \times m \times p$ ways |
| 7:          **end** | |
| 8:          **for** $(r = q; r > 0; r - -)$ **do** | // $T_8$ is done in $n \times m \times (q + 1)$ ways |
| 9:              sum = sum + 1 | // $T_9$ is done in $n \times m \times q$ ways |
| 10:         **end** | |
| 11:     **end** | |
| 12: **end** | |

---

**6.14** Note that the order in which the shirts can be selected does not matter, and the shirts can be repeated. Total number of different colored shirts is $n = 6$. The number of shirts to be selected is $r = 4$. Here, the desired number is equal to the number of 4-combinations with repetition allowed from a set with six elements. From Theorem 6.13, the shirts can be displayed in

$$C(6 + 4 - 1, 4) = C(9, 4) = C(9, 5) = \frac{9 \cdot 8 \cdot 7 \cdot 6}{1 \cdot 2 \cdot 3 \cdot 4} = 126$$

different ways.

**6.15** (a) We count permutations of 2 O's, 1 N, 1 E, 1 W, 1 R, and 1 D, a total of 7 symbols. By Theorem 6.14, the number of these is

$$\frac{7!}{2! \, 1! \, 1! \, 1! \, 1! \, 1!} = 7 \cdot 6 \cdot 5 \cdot 4 \cdot 3 = 2,520.$$

(b) Under the given condition, we should treat the O-block as a single unit "OO". We then have to count permutations of 1 N, 1 E, 1 W, 1 R, 1 D, and 1 "OO", a total of 6 symbols. The number of these is

$$\frac{6!}{1! \, 1! \, 1! \, 1! \, 1! \, 1!} = 6! = 720.$$

**6.16** (a) We count permutations of 3 P's, 2 E's, 1 I, 1 N, 1 A, and 1 L, a total of 9 symbols. By Theorem 6.14, the number of these is

$$\frac{9!}{3! \, 2! \, 1! \, 1! \, 1! \, 1!} = 30,240.$$

(b) Under the given condition, we should treat the P-block as a single unit "PPP". We then have to count permutations of 2 E's, 1 I, 1 N, 1 A, 1 L, and 1 "PPP", a total of 8 symbols. The number of these is

$$\frac{8!}{2!\,1!\,1!\,1!\,1!\,1!} = 20,160.$$

**7.1** (a) (iii).   (e) (i).   (i) (iii).   (m) (ii).   (q) (iii).   (u) (iv).

(b) (i).   (f) (i).   (j) (iv).   (n) (iii).   (r) (iv).   (v) (iv).

(c) (i).   (g) (ii).   (k) (iv).   (o) (iv).   (s) (ii).

(d) (i).   (h) (iv).   (l) (iv).   (p) (iii).   (t) (ii).

**7.2** An algorithmic code that solves the given problem is stated in Algorithm A.2.

---

**Algorithm A.2:** The algorithm of Exercise 7.2

---

**Input:** An array $A[0 : n - 1]$ of positive integers and length $n$
**Output:** The largest number of the array
1: find-maxi$(A, n)$
2: **if** $(n = 1)$ **then**
3:    |  **return** $A[0]$
4: **end**
5: $v1 = A[0]$
6: $v2 = $ find-maxi$(A, n - 1)$
7: **if** $(v1 > v2)$ **then**
8:    |  **return** $v_1$
9: **end**
10: **else**
11:    |  **return** $v_2$
12: **end**

---

**7.3** (a) Yes. The input for Algorithm A.2 is a finite array of positive numbers whose length is n.

(b) Yes. The output for Algorithm A.2 is the maximum in the array.

(c) Yes. Algorithm A.2 terminates when n is equal to 1.

(d) Yes. One can prove the following equality by mathematical induction.

find-maxi$(A[i : n-1], n-i) = \max\{A[i], \text{find-maxi}(A[i+1 : n-1], n-i-1)\}.$

**7.4** Let $T(n)$ be the running time of Algorithm 7.12. Carrying out a line-by-line analysis for Algorithm 7.12, we have

| Line | One-time Cost | Number of Times |
|------|---------------|-----------------|
| 2 | $c_1$ | 1 |
| 3 | $c_2$ | $n + 1$ |
| 4 | $c_3$ | $n$ |
| 5 | $c_4$ | $n$ [worst-case] |
| 8 | $c_5$ | 1 |

The total runtime is

$$T(n) = c_1 + c_2(n + 1) + c_3 n + c_4 n + c_5 = (c_2 + c_3 + c_4)n + (c_1 + c_2 + c_5).$$

**7.5** Let $T(n)$ be the running time of Algorithm 7.33. Carrying out a line-by-line analysis for Algorithm 7.33, we have

| Line | One-time Cost | Number of Times |
|------|---------------|-----------------|
| 1 | $c_1$ | 1 |
| 2 | $c_2$ | $1 + \log n$ |
| 3 | $c_3$ | $\log n(1 + \log n)$ |
| 4 | $c_4$ | $\log^2 n$ |

The total runtime is

$$\begin{aligned} T(n) &= c_1 + c_2(1 + \log n) + c_3 \log n(1 + \log n) + c_4 \log^2 n \\ &= (c_3 + c_4)\log^2 n + (c_2 + c_3)\log n + c_1 + c_2. \end{aligned}$$

**7.6** (a) Since the loop goes from $p$ to $q$ in increments of 1, and $p \leq q$, the solution would be $q - p$. Since this loop includes $q$ as part of the end condition ($\leq q$), add one to the solution, making it $q - p + 1$.

(b) Since the loop goes from $q$ to $p$ in decrements of 1, and $p \leq q$, the solution would be $q - p$. Since this loop includes $p$ as part of the end condition ($\geq p$), add one to the solution, making it $q - p + 1$.

(c) Since the loop goes from $p$ to $q$ in increments of $k$, and $p \leq q$, the solution would be $\lfloor (q-p)/k \rfloor + 1$. Since the loop includes $q$ as part of the end condition ($\leq q$), add one to the solution, making it $\lfloor (q - p)/k \rfloor + 2$.

(d) Since the loop goes from $q$ to $p$ in decrements of $k$, and $p \leq q$, the solution would be $\lfloor (q-p)/k \rfloor + 1$. Since the loop includes $p$ as part of the end condition ($\geq p$), add one to the solution, making it $\lfloor (q - p)/k \rfloor + 2$.

(e) The loop goes from $p$ to $q$ as follows: $p, kp, k^2p, k^3p, \ldots, q - 1, q$. Note that $k^h p = q$ when $h = \log_k(q/p)$ So, the solution would be $\log_k(p/q) + 1$. Since the loop includes $q$ as part of the end condition ($\leq q$), add one to the solution, making it $\log_k(p/q) + 2$.

(f) The loop goes from $q$ to $p$ as follows: $q, q/k, q/k^2, q/k^3, \ldots, p + 1, p$. Note that $q/k^h = p$ when $h = \log_k(q/p)$. So, the solution would be $\log_k(p/q) + 1$. Since the loop includes $p$ as part of the end condition ($\geq p$), add one to the solution, making it $\log_k(p/q) + 2$.

**7.7** Let $T(n)$ be the running time of Algorithm 7.34. Writing the summations that represent the running time of the code in Algorithm 7.34 and solving the summations, we get

$$T(n) = \sum_{i=n^2}^{n^2+5} \sum_{j=4}^{n} c_4 = \sum_{i=n^2}^{n^2+5} c_4(n-3) = c_4(n^2+5-n^2+1)(n-3) = 6c_4(n-3).$$

**7.8** Let $T(n)$ be the running time of Algorithm 7.35. Writing the summations that represent the running time of Algorithm 7.35 and solving them by bounding, we get

$$T(n) = \sum_{i=1}^{n} \sum_{j=1}^{3i^3} c = \sum_{i=1}^{n} 3i^3 c = 3c \sum_{i=1}^{n} i^3 \leq 3c \sum_{i=1}^{n} n^3 = 3cn^4.$$

Thus, $T(n) = O(n^4)$.

**7.9** For an upper bound, we have

$$f(n) = \sum_{i=n}^{4n^3} \sum_{j=i}^{8n^3} c = \sum_{i=n}^{4n^3} (8n^3 - i + 1)c \leq \sum_{i=n}^{4n^3} (8n^3)c \leq (4n^3 - n + 1)(8n^3)c.$$

It follows that $f(n) \leq (32n^6 - 8n^4 + 8n^3)c$, and therefore $f(n) \in O(n^6)$.

For a lower bound, we have

$$f(n) = \sum_{i=n}^{4n^3} \sum_{j=i}^{8n^3} c \geq \sum_{i=n}^{4n^3} (8n^3 - i)c \geq \sum_{i=n}^{4n^3} (8n^3 - 4n^3)c \geq (4n^3 - n + 1)(4n^3)c.$$

It follows that $f(n) \geq (16n^6 - 4n^4 + 4n^3)c$, and therefore $f(n) \in \Omega(n^6)$.

Thus, from Property 7.1, we have $f(n) \in \Theta(n^6)$.

**7.10** From Algorithm 7.30, we have

$$f(n) = \sum_{i=1}^{n-1} \sum_{j=i+1}^{n} \sum_{k=i}^{j} c = \sum_{i=1}^{n-1} \sum_{j=i+1}^{n} c(j-i+1).$$

For an upper bound, we have

$$f(n) \le c \sum_{i=1}^{n-1} \sum_{j=i+1}^{n} j \le c \sum_{i=1}^{n-1} \sum_{j=i+1}^{n} n \le c \sum_{i=1}^{n-1} (n-i)n \le c \sum_{i=1}^{n-1} n^2 = cn^2(n-1) \le cn^3.$$

For a lower bound, we have

$$
\begin{aligned}
f(n) \;\ge\;& c \sum_{i=1}^{n-1} \sum_{j=i+1}^{n} (j-i) \\
\ge\;& c \sum_{i=1}^{n-1} \sum_{j=(n-i)/2}^{n} (j-i) \\
\ge\;& c \sum_{i=1}^{n-1} \sum_{j=(n-i)/2}^{n} \left(\frac{n-i}{2} - i\right) \\
=\;& c \sum_{i=1}^{n-1} \sum_{j=(n-i)/2}^{n} \left(\frac{n}{2} - \frac{3i}{2}\right) \\
\ge\;& c \sum_{i=1}^{n-1} \left(\frac{n}{2} - \frac{3i}{2}\right)\left(n - \frac{n-i}{2}\right) \\
=\;& c \sum_{i=1}^{n-1} \left(\frac{n}{2} - \frac{3i}{2}\right)\left(\frac{n}{2} + \frac{i}{2}\right) \\
\ge\;& c \sum_{i=1}^{n-1} \left(\frac{n}{2} - \frac{i}{2}\right)\left(\frac{n}{2} + \frac{i}{2}\right) \\
=\;& c \sum_{i=1}^{n-1} \left(\frac{n^2}{4} - \frac{i^2}{4}\right) \\
\ge\;& c \sum_{i=1}^{n/2} \left(\frac{n^2}{4} - \frac{i^2}{4}\right) \\
\ge\;& c \sum_{i=1}^{n/2} \left(\frac{n^2}{4} - \frac{1}{4}\left(\frac{n}{2}\right)^2\right) \\
=\;& c \left(\frac{3n^2}{16}\right)\left(\frac{n}{2}\right) = \frac{3c}{32} n^3.
\end{aligned}
$$

Thus $\frac{3}{32}cn^3 \le f(n) \le cn^3$, and therefore $f(n) = \Theta(n^3)$.

**7.11** (a) If $n \ge 1$, then

$$5n^2 - 3n + 20 \le 5n^2 + 20 \le 5n^2 + 20n^2 = 25n^2.$$

Therefore, $5n^2 - 3n + 20 \in O(n^2)$ with $c = 25$ and $n_0 = 1$.

(b) If $n \geq 1$, then

$$4n^2 - 12n + 10 \leq 4n^2 + 10 \leq 4n^2 + 10n^2 = 14n^2.$$

Therefore, $4n^2 - 12n + 10 \in O(n^2)$ with $c = 14$ and $n_0 = 1$.

(c) If $n \geq 1$, then

$$5n^5 - 4n^4 - 2n^2 + n \leq 5n^5 + n \leq 5n^5 + n^5 = 6n^5.$$

Therefore, $5n^5 - 4n^4 - 2n^2 + n \in O(n^5)$ with $c = 6$ and $n_0 = 1$.

(d) If $n \geq 2$, then

$$n^{\frac{3}{2}} + \sqrt{n}\sin(n) + n\log(n) \leq n^2 + (n)(n) + (n)(n) \leq 3n^2.$$

Therefore, $n^{\frac{3}{2}} + \sqrt{n}\sin(n) + n\log(n) \in O(n^2)$ with $c = 3$ and $n_0 = 2$.

7.12 (a) If $n \geq 1$, then $4n^2 + n + 1 \geq 4n^2$. Therefore, $4n^2 + n + 1 \in \Omega(n^2)$ with $c = 4$ and $n_0 = 1$.

(b) If $n \geq 1$, then $n\log(n) - 2n + 13 \geq n\log(n) - 2n$. It is enough to find positive constants $c$ and $n_0$ such that $n\log(n) - 2n \geq cn\log(n)$, or equivalently $1 - 2/\log(n) \geq c$, for all $n \geq n_0$. Note that, for $n \geq 8$, we have $1 - 2/\log(n) \geq 1/3$. Therefore, $n\log(n) - 2n + 13 = \Omega(n\log(n))$ with $c = 1/3$ and $n_0 = 8$.

7.13 (a) If $n \geq 1$, then

$$n^5 \leq n^5 + n^3 + 7n + 1 \leq n^5 + n^5 + 7n^5 + n^5 = 10n^5.$$

Therefore, $n^5 + n^3 + 7n + 1 \in \Theta(n^5)$ with $c_1 = 1, c_2 = 10$ and $n_0 = 1$.

(b) If $n \geq 7$, then

$$\frac{1}{14}n^2 \leq \frac{1}{2}n^2 - 3n \leq \frac{1}{2}n^2,$$

where the first inequality follows by noting that

$$\frac{1}{2}n^2 - 3n \geq c_1 n^2 \iff \frac{1}{2} - \frac{3}{n} \geq c_1. \text{ Hence, for } n \geq 7, \frac{1}{2} - \frac{3}{n} \geq \frac{1}{14} = c_1.$$

Therefore, $\frac{1}{2}n^2 - 3n = \Theta(n^2)$ with $c_1 = 1/14, c_2 = 1/2$ and $n_0 = 7$.

7.14 Given that $f_i(n) = O(g_i(n))$ for each $i = 1, 2, \ldots, k$, then by definition there are positive constants $c_i$ and $n_i$ such that $f_i(n) \leq c_i g_i(n)$ for $n \geq n_i$.

(a) Let $n_0 = \max_{1 \leq i \leq k} n_i$ and $c_0 = c_1 + c_2 + \cdots + c_k$, then for all $n \geq n_0$, we have

$$\sum_{i=1}^{k} f_i(n) \leq \sum_{i=1}^{k} (c_i g_i(n)) \leq \left(\sum_{i=1}^{k} c_i\right) \max_{1 \leq i \leq k} g_i(n) \leq c_0 \max_{1 \leq i \leq k} g_i(n).$$

Thus, $\sum_{i=1}^{k} f_i(n) = O(\max_{1 \le i \le k} g_i(n))$.

(b) Let $n_0 = \max_{1 \le i \le k} n_i$ and $c_0 = c_1 c_2 \cdots c_k$, then for all $n \ge n_0$, we have

$$\prod_{i=1}^{k} f_i(n) \le \prod_{i=1}^{k} (c_i g_i(n)) = \left( \prod_{i=1}^{k} c_i \right) \prod_{i=1}^{k} g_i(n) = c_0 \prod_{i=1}^{k} g_i(n).$$

Thus, $\prod_{i=1}^{k} f_i(n) = O(\prod_{i=1}^{k} g_i(n))$.

**7.15** We prove Properties 7.1 & 7.6, and prove also Properties 7.2 & 7.7 for Big-O.

**Pr 7.1 Pf.** By definitions, $f(n) = O(g(n))$ and $f(n) = \Omega(g(n))$ iff there are positive constants $c_1, c_2, n_1$ and $n_2$ such that $f(n) \le c_1 g(n)$ for $n \ge n_1$, and that $f(n) \ge c_2 g(n)$ for $n \ge n_2$. This is equivalent to $c_2 g(n) \le f(n) \le c_1 g(n)$ for all $n \ge \max\{n_1, n_2\}$, which is also equivalent to $f(n) = \theta(g(n))$, with $n_0 = \max\{n_1, n_2\}$, by definition. Thus, we conclude $f(n) = \theta(g(n))$ iff $f(n) = O(g(n))$ and $f(n) = \Omega(g(n))$.

**Pr 7.2 Pf.** Given that $f(n) = O(g(n))$, then by definition there are positive constants $c_1$ and $n_1$ such that $f(n) \le c_1 g(n)$ for $n \ge n_1$. Similarly, given that $g(n) = O(h(n))$, then by definition there are positive constants $c_2$ and $n_2$ such that $g(n) \le c_2 h(n)$ for $n \ge n_2$. Let $n_0 = \max\{n_1, n_2\}$, then $f(n) \le c_1 g(n) \le c_1 c_2 h(n)$ for all $n \ge n_0$. Let $c_0 = c_1 c_2$, then $f(n) \le c_0 h(n)$ for all $n \ge n_0$, which means we can conclude that $f(n) = O(h(n))$.

**Pr 7.6 Pf.** By definition of Big-Oh, $f(n) = O(g(n))$ iff there are positive constants $c$ and $n_0$ such that $f(n) \le c_1 g(n)$, or equivalently $g(n) \ge \frac{1}{c} f(n)$, for all $n \ge n_0$. By definition of Big-Omega, this equivalent to $g(n) = \Omega(f(n))$ with positive constants $1/c$ and $n_0$.

**Pr 7.7 Pf.** Note that $f(n) \le f(n)$ for $n \ge 1$. Thus, $f(n) = O(f(n))$ with $c = n_0 = 1$.

**7.16** (a) We have $\sqrt{4n^2 + 1} = \Theta(n)$ because

$$\lim_{n \to \infty} \frac{\sqrt{4n^2 + 1}}{n} = \lim_{n \to \infty} \sqrt{\frac{4n^2 + 1}{n^2}} = \lim_{n \to \infty} \sqrt{4 + \frac{1}{n^2}} = \sqrt{4 + \lim_{n \to \infty} \frac{1}{n^2}} = 2.$$

(b) We have $n^n = \Omega(n!)$ because

$$\lim_{n \to \infty} \frac{n^n}{n!} = \lim_{n \to \infty} \frac{\overbrace{n \cdot n \cdots n \cdot n}^{n-\text{times}}}{n \cdot (n - 1) \cdots 2 \cdot 1} = \lim_{n \to \infty} \left( 1 \cdot \frac{n}{n - 1} \cdot \frac{n}{n - 2} \cdots \frac{n}{2} \cdot n \right) = \infty.$$

(c)  We have $\sqrt{n+4} - \sqrt{n} = O(1)$ because

$$
\begin{aligned}
\lim_{n \to \infty} \frac{\sqrt{n+4} - \sqrt{n}}{1} &= \lim_{n \to \infty} \left( \left( \sqrt{n+4} - \sqrt{n} \right) \cdot \frac{\sqrt{n+4} + \sqrt{n}}{\sqrt{n+4} + \sqrt{n}} \right) \\
&= \lim_{n \to \infty} \frac{(n+4) - n}{\sqrt{n+4} + \sqrt{n}} \\
&= \lim_{n \to \infty} \frac{4}{\sqrt{n+4} + \sqrt{n}} = 0.
\end{aligned}
$$

(d)  We have $n \log(n^2) + (n-1)^2 \log(\frac{n}{2}) = \Theta(n^2 \log(n))$ because

$$
\begin{aligned}
\lim_{n \to \infty} \frac{n \log(n^2) + (n-1)^2 \log(\frac{n}{2})}{n^2 \log n} &= \lim_{n \to \infty} \frac{2n \log n + (n-1)^2 (\log n - \log 2)}{n^2 \log n} \\
&= \lim_{n \to \infty} \left( \frac{2}{n} + \frac{(n-1)^2}{n^2} \frac{(\log n - \log 2)}{\log n} \right) \\
&= \lim_{n \to \infty} \left( \frac{2}{n} + \left( 1 - \frac{1}{n} \right)^2 \left( 1 - \frac{\log 2}{\log n} \right) \right) = 1.
\end{aligned}
$$

**7.17**  Yes, it is true that $2^{(n+10)} = O(2^n)$. To prove this, note that, if $n \geq 1$, then $2^{(n+10)} = 2^{10} 2^n \leq (1024)(2^n)$. Thus, $2^{(n+10)} = O(2^n)$ with $c = 1024$ and $n_0 = 1$.

**7.18**  (a) (iii).          (b) (i).          (c) (ii).          (d) (i).

**7.19**  The first statement is an assignment statement with constant time. The inside of the for loop is also a simple statement with constant time. So, the part that will affect overall runtime is the runtime and value of $f(n)$. The answers are (a) $O(n^2)$ and (b) $O(n(n!))$. Below is the reasoning behind these answers, which is similar to that reasoning of Example 7.36 (b).

(a)  The"test" runtime in the for loop is $O(n)$, since that is how long it takes $f(n)$ to run. The "body" of the for loop takes $O(1)$, as it is just a simple statement. The "reinitialization" runtime still takes $O(1)$. We perform the loop $O(n!)$ times, since the value of $f(n)$ is $n$. Thus, putting it all together, we have $O(1 + (n + 1 + 1)n) = O(n^2)$.

(b)  The"test" runtime in the for loop is $O(n)$, since that is how long it takes $f(n)$ to run. The "body" of the for loop takes $O(1)$, as it is just a simple statement. The "reinitialization" runtime still takes $O(1)$. We perform the loop $O(n!)$ times, since the value of $f(n)$ is $n!$. Thus, putting it all together, we have $O(1 + (n + 1 + 1)n!) = O(nn!)$.

**7.20**  The answer is $O(n^3)$. Below is the reasoning behind the answer.

In Figure 7.15, we see how the runtime of for loops is calculated. We add the initialization runtime (usually $O(1)$) plus the cost of going around the loop once multiplied by the number of times we go around the loop, represented

$O(1 + (f(n) + 1)g(n))$. Keep in mind "the cost of going around the loop once" is represented in Figure 7.15 as the runtimes of "test" plus "body" plus "reinitialize", with "test" being the condition of the for loop and "body" obviously being the body. Now, we apply this to our problem.

The "test" runtime in the for loop is $O(n)$, since that is how long it takes the function cat(.) to run and, as previously found, cat(.) has a runtime of $O(n)$. The "body" of the for loop takes $O(n)$, since, again, that is how long it takes cat(.) to run. The "reinitialization" runtime takes $O(1)$.

The "for" loop in line (10) goes from 1 to the value of the function cat(n,n). Thus, we must find what is the value of cat(n,n). The function cat(.) takes a number and adds values from 1 to $n$ to the original number, in this case $n$. So, the value of cat(n,n) equals $n + 1 + 2 + 3 + \cdots + n$. In other words,

$$cat(n,n) = n + \sum_{i=1}^{n} i = n + \frac{n(n+1)}{2} = \frac{n^2 + 3n}{2} \in O(n^2).$$

Thus, the for loop is iterated $O(n^2)$ times (from $i = 1$ to $i = (n^2 + 3n)/2$). Now, putting it all together we have $O(1 + (n + n + 1)n^2) = O(n^3)$.

The first two statements of the function main(.) are $O(1)$, as they are simple statements. The statement in line (3) calls the function cow(.), which takes $O(n^3)$ as we found above. The runtime of the statement in line (4), which has a call to cat(.), is $O(n)$ (as previously stated, cat(.) has a runtime of $O(n)$). So, the total runtime for main(.) is $O(1) + O(1) + O(n^3) + O(n) = O(n^3)$.

7.21 The graph structure of Algorithm 7.38 is shown in Figure A.3.

7.22 The running time of the block in lines (17)-(19) is $O(m)$, so the running time of the for-statement in lines (16)-(20) is $O(Km)$. Since the running time of the block in lines (21)-(24) is $O(n)$ and those in the statements in lines (14) and (15) are $O(1)$, the running time of the block in lines (14)-(25) is $O(n + Km)$, and hence the running time of the while-statement in lines (13)-(25) is $O((n + Km)N_{in})$. Here, $N_{in}$ is the number of times we go around the inner while loop of line (13). The running time of the block in lines (5)-(7) is $O(m)$, so the running time of the for-statement in lines (4)-(8) is $O(Km)$. Since the running time of the block in lines (9)-(12) is $O(n)$, the running time of the block in lines (4)-(12) is $O(n + Km)$, and hence the running time of the block in lines (4)-(26) is $O((n + Km) + (n + Km)N_{in}) = O((n + Km)N_{in})$, it follows that the while-statement in lines (3)-(27) is $O((n + Km)N_{in}N_{out})$, where $N_{out}$ is the number of times we go around the outer while loop of line (3). As a result, based on the information obtained and because the running time of the for-statement in lines (28)-(30) is $O(Km)$, and all other statements are of constant times $O(1)$, the running time of the algorithm is

$$O(1 + Km + (n + Km)N_{in}N_{out}) = O((n + Km)N_{in}N_{out}).$$

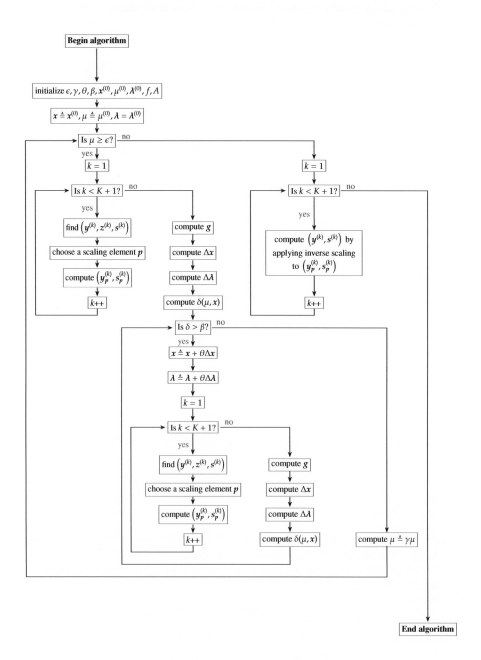

Figure A.3: The graph structure of Algorithm 7.38.

Bounding $N_{\text{out}}$ is exactly similar to bounding $N_{\text{out}}$ in Example 7.35. Using a similar argument to that in Example 7.35, we can show that

$$N_{\text{out}} \le \frac{\log(\mu^{(0)}/\epsilon)}{-\log\gamma}.$$

(a) It is given that $N_{\text{in}} = O(n + Km)$. Now, if $\gamma \in (0, 1)$ is an arbitrarily chosen constant, then $\gamma = O(1)$, and hence

$$N_{\text{out}} \le \log\left(\frac{\mu^{(0)}}{\epsilon}\right) O(1).$$

Thus, the running time of the algorithm as a whole is

$$O((n + Km)N_{\text{in}}N_{\text{out}}) = O\left((n + Km)^2 \log\left(\frac{\mu^{(0)}}{\epsilon}\right)\right).$$

(b) It is given that $N_{\text{in}} = O(1)$. Now, if $\gamma = 1 - \sigma/\sqrt{n + Km}$ ($\sigma > 0$), then[1]

$$\log\gamma = \log\left(1 - \sigma/\sqrt{n + Km}\right) \approx -\sigma/\sqrt{n + Km},$$

and hence

$$N_{\text{out}} \le \frac{\log(\mu^{(0)}/\epsilon)}{-\log\gamma} \approx \frac{\log(\mu^{(0)}/\epsilon)}{\sigma/\sqrt{n + Km}} = \sqrt{n + Km}\log\left(\frac{\mu^{(0)}}{\epsilon}\right) O(1).$$

Thus, the running time of the algorithm as a whole is

$$O((n + Km)N_{\text{in}}N_{\text{out}}) = O\left((n + Km)^{3/2} \log\left(\frac{\mu^{(0)}}{\epsilon}\right)\right).$$

**8.1** (a) (iii).    (c) (iii).    (e) (iii).    (g) (iv).    (i) (ii).
    (b) (iv).    (d) (i).    (f) (ii).    (h) (iii).

**8.2** When we write $O(n^{2.376})$, we mean that there is a positive constant $c$ such that the Coppersmith and Winograd's algorithm takes no more than $cn^{2.376}$ flops. For this algorithm, the constant $c$ is so large that it does not beat Strassen's method until $n$ is really enormous.

**8.3** The program in Algorithm 8.14 is nonrecursive. Figure A.4 shows a basic scheme of this program. From Algorithm 8.5, the function linear-search is $O(n)$. In view of Figure A.4, we analyze the function karger before analyzing the function main because the later calls the function karger. Now, if the function call in the body of a for loop, we add its cost to the bound on the time for each

---

[1] For small positive values of $x$, we have $\log(1 + x) \approx (1 + x) - 1 = x$.

The function `main` calls the function `karger`, and the function `karger` calls the function `linear-search`. We already analyzed the function `linear-search` in Algorithm 8.5. Thus, we analyze the function `karger`, and then analyze the function `main`.

Figure A.4: A basic scheme of the program shown in Algorithm 8.14.

iteration. It follows that the running time of a call to `karger` is $O(mn)$. Next, when the function call is within a simple statement, we add its cost to the cost of that statement. Thus, the function `main` takes $O(mn)$ times. Therefore, the running time of this program is $O(mn)$.

**8.4** The program in Algorithm 8.15 is nonrecursive. Figure A.5 shows a basic scheme of this program. From Algorithms 8.6 and 8.9, the functions `linear-search` and `merge-sort` are $O(\log n)$ and $O(n \log n)$ respectively. In view of Figure A.5, we analyze the function `dinic` before analyzing the function `main` because the later calls the first. Now, if the function call in the body of a for loop, we add its cost to the bound on the time for each iteration. It follows that the running time of a call to `dinic` is $O(\log m \log n)$. Next, when the function call is within a simple statement, we add its cost to the cost of that statement. Thus, the function `main` takes $O(\log m \log n + n \log n)$ times. Therefore, the running time of this program is $O((n + \log m) \log n)$.

The function `main` calls the functions `dinic` and `merge-sort`, and the function `dinic` calls the function `binary-search`. We already analyzed the functions `binary-search` and `merge-sort` in Algorithms 8.6 and 8.9, respectively. So, we analyze the function `dinic`, and then analyze the function `main`.

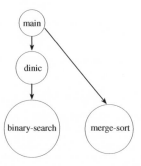

Figure A.5: A basic scheme of the program shown in Algorithm 8.15.

**8.5** If $x$ divides both $a$ and $b$, then it clearly also divides $a - b\lfloor a/b \rfloor = a \bmod b$. Conversely, if $x$ divides $a \bmod b$ and $b$, then it also divides $(a \bmod b) + b\lfloor a/b \rfloor = a$.

**8.6** We need to show that if $a > b > 0$ and Algorithm 8.11 performs $n \geq 1$ iterations of the while loop, then $a \geq f_{n+2}$ and $b \geq f_{n+1}$.

The proof is by induction on $n$. The statement holds for $n = 1$ because $b \geq 1 = f_2$ and $a > b$ imply that $a \geq 2 = f_3$. Now assume that $n \geq 2$. Because $a > b \geq 1$ and $n \geq 2$, the next iteration will use the number $b$ and $a \bmod b$. To these we can apply the induction hypothesis, because $n - 1$ iterations remain and $b > a \bmod b > 0$ (as $n - 1 > 0$). Thus, $b \geq f_{n+1}$ and $a \bmod b \geq f_n$. Thus, $a = \lfloor a/b \rfloor \cdot b + (a \bmod b) \geq b + (a \bmod b) \geq f_{n+1} + f_n = f_{n+2}$.

**8.7** This is an implementation exercise. Applying Newton's method, we obtain the root 3.15145 if we start with $x_0 = 1$.

**9.1** (a) (iv).    (c) (iii).    (e) (iii).    (g) (iv).    (i) (iii).
(b) (ii).    (d) (ii).    (f) (i).    (h) (iv).    (j) (iii).

**9.2** (a) True.    (c) True.    (e) True.    (g) True.    (i) False.
(b) True.    (d) False.    (f) False.    (h) True.    (j) False.

**9.3** (a) The adjacency list representation for the graph is:

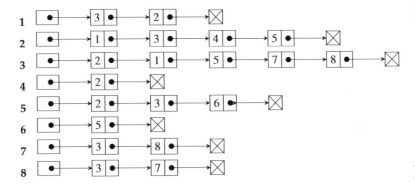

(b) The adjacency matrix representation for the graph is:

|   | 1 | 2 | 3 | 4 | 5 | 6 | 7 | 8 |
|---|---|---|---|---|---|---|---|---|
| 1 | 0 | 1 | 1 | 0 | 0 | 0 | 0 | 0 |
| 2 | 1 | 0 | 1 | 1 | 1 | 0 | 0 | 0 |
| 3 | 1 | 1 | 0 | 0 | 1 | 0 | 1 | 1 |
| 4 | 0 | 1 | 0 | 0 | 0 | 0 | 0 | 0 |
| 5 | 0 | 1 | 1 | 0 | 0 | 1 | 0 | 0 |
| 6 | 0 | 0 | 0 | 0 | 1 | 0 | 0 | 0 |
| 7 | 0 | 0 | 1 | 0 | 0 | 0 | 0 | 1 |
| 8 | 0 | 0 | 1 | 0 | 0 | 0 | 1 | 0 |

**9.4** (*a*) Running a breadth-first search, we obtain the breadth-first tree shown to the right. It follows that $(v_1, v_2, v_5, v_6, v_8)$ is the shortest path from vertex $v_1$ to vertex $v_8$. This shortest path is clearly unique.

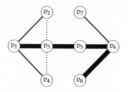

(*b*) As shown below, running a breadth-first search, we find that there is a conflicting assignment which occurs when we color the vertex $v_3$. Hence this is not a bipartite graph.

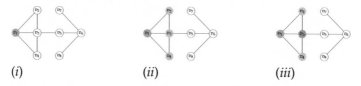

(*i*)            (*ii*)            (*iii*)

(*c*) As shown below, running a depth-first search, we get the following depth-first tree.

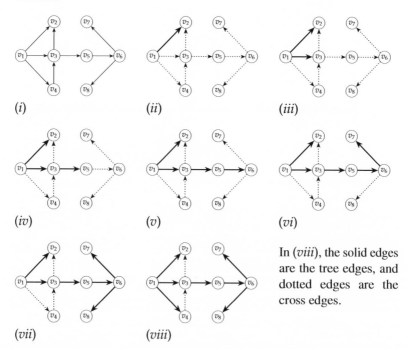

In (*viii*), the solid edges are the tree edges, and dotted edges are the cross edges.

(*d*) Because we did not find a back edge we performed a depth-first search in item (*c*), the directed graph has no a directed cycle.

(*e*) A topological ordering is computed and shown below.

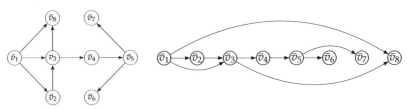

**9.5** (*a*) Running a breadth-first search on the graph, we obtain the following breadth-first tree.

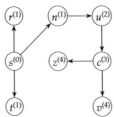

(*b*) Running a breadth-first search on the given graph, we find that the vertices $s$ and $r$ have the same color as shown below. Thus, this is not a bipartite graph.

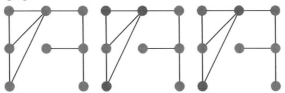

(*c*) Running a depth-first search on the graph, we obtain the following depth-first tree.

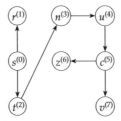

(*d*) Because we did not find a back edge we performed a depth-first search in item (*c*), the directed graph has no a directed cycle.

(*e*) See the solution of item (*e*) in Exercise 9.4. This graph and that in Exercise 9.4 are isomorphic, and they have the same topological ordering.

**9.6** (*a*) Running a breadth-first search on the graph, we obtain the following breadth-first tree.

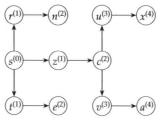

(*b*) The undirected version of the graph is bipartite as shown below.

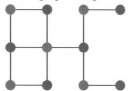

(*c*) Running a depth-first search on the graph, we obtain the following depth-first tree.

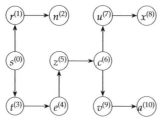

(*d*) Because we did not find a back edge we performed a depth-first search in item (*c*), the directed graph has no a directed cycle.

(*e*) A topological ordering is computed and shown below.

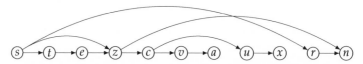

**9.7** This is a false statement. The depth-first search may produce different depth-first trees with different numbers of tree edges depending on the starting vertex and upon the order in which vertices are searched.

As a counterexample, consider the graph shown to the right. If the depth-first search starts at $u$, then it will visit $v$ next, and $(u, v)$ will become a tree edge. But if the depth-first search starts at $v$, then $u$ and $v$ become separate trees in the depth-first forest, and $(u, v)$ becomes a cross edge.

**10.1** (*a*) (ii).     (*d*) (ii).     (*g*) (v).     (*j*) (iii).     (*m*) (i).     (*p*) (i).     (*s*) (v).

(*b*) (ii).     (*e*) (v).     (*h*) (i).     (*k*) (v).     (*n*) (v).     (*q*) (i).     (*t*) (ii).

(*c*) (ii).     (*f*) (iv).     (*i*) (i).     (*l*) (ii).     (*o*) (ii).     (*r*) (iii).     (*u*) (ii).

**10.2** The decision variables are:

$$x_1 = \text{ kg of food } F_1;$$
$$x_2 = \text{ kg of food } F_2.$$

Minimizing the cost of the mixtures, we have the following LP problem.

$$\begin{aligned} \min \quad & 60x_1 + 80x_2 \\ \text{s.t.} \quad & 5x_1 + 2x_2 \geq 11, \\ & 3x_1 + 4x_2 \geq 8. \end{aligned}$$

**10.3** The decision variables are:

$x_1$: The number of loaves of bread baked;

$x_2$: The change in the supply of flour through financial transactions (in ounces).

Measuring profits in cents, we have the following LP problem.

$$\begin{aligned} \max \quad & 30x_1 - 4x_2 \\ \text{s.t.} \quad & 5x_1 - x_2 \leq 30, \\ & x_1 \qquad \leq 5, \\ & x_1 \qquad \geq 0. \end{aligned}$$

**10.4** The decision variables are:

$x_1$: Acres of radishes produced;

$x_2$: Acres of onions produced;

$x_3$: Acres of potatoes produced.

We wish to maximize the agriculturist's profit (in $). Subtracting labor and water costs, we obtain the following LP problem.

$$\begin{aligned} \max \quad & 7072.5x_1 + 6775x_2 + 4812.5x_3 \\ \text{s.t.} \quad & x_1 + x_2 + x_3 \leq 126, \\ & 6x_1 + 5x_2 + 6x_3 \leq 500, \\ & x_1, x_2, x_3 \geq 0. \end{aligned}$$

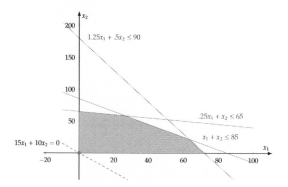

Figure A.6: Graphical solution of the optimization problem in Exercise 10.5.

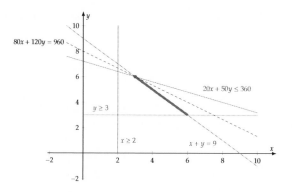

Figure A.7: Graphical solution of the optimization problem in Exercise 10.6.

**10.5** The graphical representation of the given LP problem is shown in Figure A.6, with the feasible region shaded in gray. From the graph, we find that the minimum value for $z$ is 0 at $x = (0, 0)$.

**10.6** (*a*) The LP problem that can be used to maximize the profit is seen below:

$$\begin{array}{rlcl}
\max & 80x + 120y & & \\
\text{s.t.} & x & \geq & 2, \\
& y & \geq & 3, \\
& x + y & = & 9, \\
& 20x + 50y & \leq & 360.
\end{array}$$

(*b*) The feasible region is the thick black line segment shown in the graph in Figure A.7 (b). We find that the maximum value for the objective function is 960 at $x = (3, 6)$.

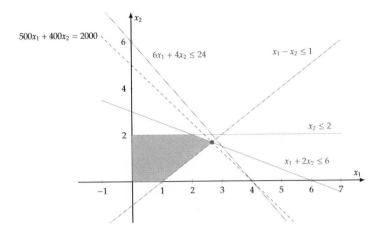

Figure A.8: Graphical solution of the optimization problem in Exercise 10.7 (b).

**10.7** (*a*) Define decision variables: Letting the objective function be the maximum total daily profit (in $), we obtain the following LP:

$$
\begin{aligned}
\max \quad & 500x_1 + 400x_2 \\
\text{s.t.} \quad & 6x_1 + 4x_2 \leq 24, \\
& x_1 + 2x_2 \leq 6, \\
& x_1 - x_2 \leq 1, \\
& x_2 \leq 2, \\
& x_1, \quad x_2 \geq 0.
\end{aligned}
$$

(*b*) The graphical representation of the given LP problem is shown in Figure A.8, with the feasible region shaded in gray. From the graph, we find that the maximum value for the objective function is 2000 at $x = (8/3, 5/3)$.

(*c*) Restricting $x_2$ to be integer-valued changes the feasible region, we find that the feasible region is the thick black line segments shown in the graph in Figure A.9. We also find that the optimal solution is now 1800 at $x = (2, 2)$.

(*d*) Restricting both $x_1$ and $x_2$ to be integer-valued again changes the feasible region to become the black bullets shown in the graph in Figure A.10. The optimal solution remains the same.

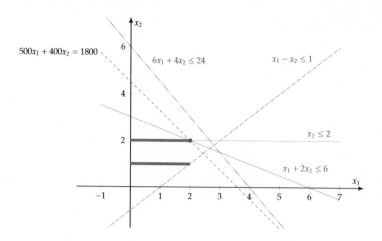

Figure A.9: Graphical solution of the optimization problem in Exercise 10.7 (c).

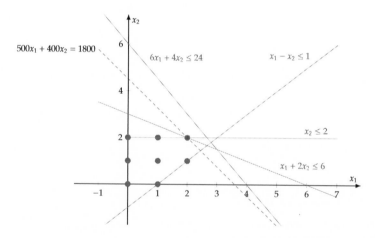

Figure A.10: Graphical solution of the optimization problem in Exercise 10.7 (d).

10.8 (*a*) The graphical representation of the given LP problem is shown in Figure A.11, with the feasible region shaded in gray. From the graph, we find that the minimum value for the objective function is $230/13$ at $x = (18/13, 20/13)$.

(*b*) The feasible region is the thick black line segment shown in the graph in Figure A.12. From the graph, we find that the maximum value for the objective function is 18 at $x = (2, 2)$.

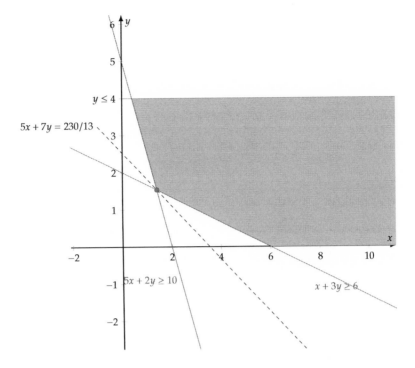

Figure A.11: Graphical solution of the optimization problem in Exercise 10.8 (a).

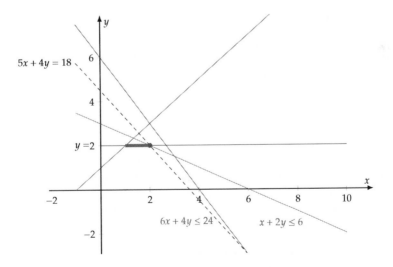

Figure A.12: Graphical solution of the optimization problem in Exercise 10.8 (b).

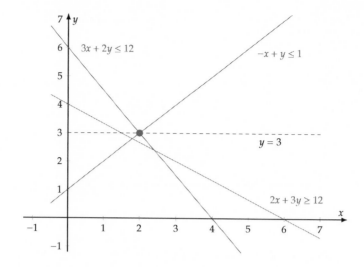

Figure A.13: Graphical solution of the optimization problem in Exercise 10.8 (c).

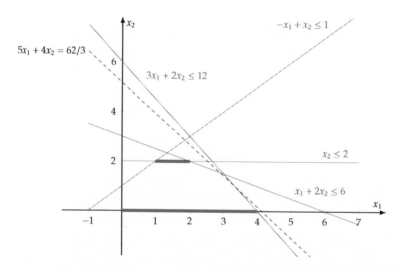

Figure A.14: Graphical solution of the optimization problem in Exercise 10.9 (a).

(c) The feasible region is the black bullet shown in the graph in Figure A.13. From the graph, we find that the maximum value for the objective function is 3 at $x = (2, 3)$.

**10.9** (a) Restricting $x_2$ to be integer-valued changes the feasible region to become the thick black line segments shown in the graph in Figure A.14. We find that the optimal solution is $62/3$ at $x = (10/3, 1)$.

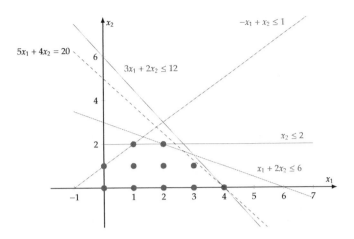

Figure A.15: Graphical solution of the optimization problem in Exercise 10.9 (b).

(b) Restricting both $x_1$ and $x_2$ to be integer-valued again changes to become the black bullets shown in the graph in Figure A.15. We find that the optimal solution is 20 at $x = (4, 0)$.

10.10 Letting $x_i = x_i^+ - x_i^-$ for $i = 1, 2, 3$, and introducing the excess variable $x_4$ and the slack variables $x_5$ and $x_6$, the given LP problem is equivalent to the standard form problem:

$$
\begin{aligned}
\min z = \ & 2x_1^+ - 2x_1^- - 4x_2^+ + 4x_2^- + 5x_3^+ - 5x_3^- - 30 \\
\text{s.t.} \quad & 3x_1^+ - 3x_1^- + 2x_2^+ - 2x_2^- - x_3^+ + x_3^- - x_4 && = 10, \\
& -2x_1^+ + 2x_1^- && + 4x_3^+ - 4x_3^- && + x_5 && = 35, \\
& 4x_1^+ - 4x_1^- - x_2^+ + x_2^- && && + x_6 && = 20, \\
& x_1^+ - x_1^- && && \leq 6, \\
& \qquad\quad x_2^+ - x_2^- && && \leq 8, \\
& \qquad\qquad\qquad x_3^+ - x_3^- && \leq 10, \\
& x_1^+, \ x_1^-, \ x_2^+, \ x_2^-, \ x_3^+, \ x_3^-, \ x_4, \ x_5, \ x_6 \geq 0.
\end{aligned}
$$

10.11 We can easily see that $x^\star$ and $y^\star$ are feasible in the primal and dual problems, respectively. One can also easily see that $b^\top y^\star = 19$ and $c^\top x^\star = 19$. Based on the strong duality property (Theorem 10.2), since $b^\top y^\star = c^\top x^\star$, we conclude that $x^\star$ and $y^\star$ are are optimal in the primal and dual problems, respectively, and their optimal value is 19.

10.12 The problem $\max y_1 + y_2$, subject to $y_1 - y_2 \leq -1$, $-y_1 + y_2 \leq -1$, and $y_1, y_2 \geq 0$, and its dual problem are a pair with such a property.

10.13 We need to show that if (DILP) is infeasible, then (PILP) is either infeasible or unbounded. The possibility that Problems (PILP) and (DILP) could be both infeasible has been grounded in Exercise 10.12. So, to prove the desired result, it remains to show that if (D|LP) is infeasible and (P|LP) is feasible, then (P|LP) must be unbounded. Assume that (D|LP) is infeasible and let $\bar{x}$ be a feasible solution for (P|LP). Due to the infeasibility of (D|LP), there does not exist $y$ satisfying $A^\mathsf{T} y \le c$. Using Farkas' lemma (Version II); see Theorem 3.16, there is a vector $\hat{x}$ satisfying $A\hat{x} = \mathbf{0}$, $c^\mathsf{T}\hat{x} < 0$, and $\hat{x} \ge \mathbf{0}$. Due to the feasibility of $\bar{x}$ in (P|LP), we have $A\bar{x} = b$ and $\bar{x} \ge \mathbf{0}$. Define $x_\alpha \triangleq \bar{x} + \alpha\hat{x}$ for $\alpha \ge 0$. Then

$$A x_\alpha = A(\bar{x} + \alpha\hat{x}) = A\bar{x} + \alpha A\hat{x} = b + \alpha A\hat{x} = b, \text{ and } x_\alpha = \bar{x} + \alpha\hat{x} \ge \mathbf{0}.$$

This means that $x_\alpha$ is feasible in (P|LP). Note that, because $c^\mathsf{T}\hat{x} < 0$, we have

$$c^\mathsf{T} x_\alpha = c^\mathsf{T}(\bar{x} + \alpha\hat{x}) = c^\mathsf{T}\bar{x} + \alpha c^\mathsf{T}\hat{x} \longrightarrow c^\mathsf{T}\bar{x} - \infty = -\infty,$$

as $\alpha \longrightarrow \infty$, which implies that Problem (P|LP) is unbounded.

**10.14** $(x_1, x_2) = (4, -1/3)$.

**10.15** (*a*) (ii).    (*b*) (i).    (*c*) (iv).    (*d*) (ii).    (*e*) (iii).    (*f*) (ii).    (*g*) (iii).

**10.16** (*a*) After introducing slack variables, say $x_3$ and $x_4$, we obtain the following standard form problem:

$$\begin{aligned}
\max \quad & z = x_1 + 1.5 x_2 \\
\text{s.t.} \quad & 2x_1 + 4x_2 + x_3 = 12, \\
& 3x_1 + 2x_2 + x_4 = 10, \\
& x_1, x_2, x_3, x_4 \ge 0.
\end{aligned}$$

Note that $x = (0, 0, 12, 10)$ is a basic feasible solution. Hence, we have the following initial tableau:

|        | rhs | $x_1$ | $x_2$ | $x_3$ | $x_4$ |
|--------|-----|-------|-------|-------|-------|
|        | 0   | 1     | 1.5   | 0     | 0     |
| $x_3 =$ | 12  | 2     | ④     | 1     | 0     |
| $x_4 =$ | 10  | 3     | 2     | 0     | 1     |

Since we are maximizing the objective function, we select a nonbasic variable with the greatest positive reduced cost to be the one that enters the basis. Indicating the pivot element with a circled number, we obtain the following two tableaux:

|        | rhs  | $x_1$ | $x_2$ | $x_3$ | $x_4$ |
|--------|------|-------|-------|-------|-------|
|        | -9/2 | 1/4   | 0     | -1/4  | 0     |
| $x_2 =$ | 3    | 1/2   | 1     | 1/4   | 0     |
| $x_4 =$ | 4    | ②     | 0     | -1/2  | 1     |

|        | rhs | $x_1$ | $x_2$ | $x_3$ | $x_4$ |
|--------|-----|-------|-------|-------|-------|
|        | -20 | 0     | 0     | -3/4  | 0     |
| $x_2 =$ | 2   | 0     | 1     | 3/8   | -1/2  |
| $x_1 =$ | 2   | 1     | 0     | -1/4  | 1     |

The reduced costs in the zeroth row of the tableau are all nonpositive, so the current basic feasible solution is optimal. In terms of the original variables $x_1$ and $x_2$, this solution is $x = (2, 2)$.

(b) After introducing the slack variables, say $s_1$ and $s_2$, we obtain the following standard form problem:

$$
\begin{aligned}
\max \quad & z = 3x_1 + 5x_2 + 4x_3 \\
\text{s.t.} \quad & 2x_1 + 3x_2 + s_1 = 8, \\
& 2x_2 + 5x_3 + s_2 = 10, \\
& 3x_1 + 2x_2 + 4x_3 + s_3 = 15, \\
& x_1, x_2, x_3 \geq 0.
\end{aligned}
$$

We then obtain the following sequence of tableaux when using the circled pivot point.

| $x_1$ | $x_2$ | $x_3$ | $s_1$ | $s_2$ | $s_3$ | rhs |
|---|---|---|---|---|---|---|
| 2 | ③ | 0 | 1 | 0 | 0 | 8 |
| 0 | 2 | 5 | 0 | 1 | 0 | 10 |
| 3 | 2 | 4 | 0 | 0 | 1 | 15 |
| −3 | −5 | −4 | 0 | 0 | 0 | 0 |

| $x_1$ | $x_2$ | $x_3$ | $s_1$ | $s_2$ | $s_3$ | rhs |
|---|---|---|---|---|---|---|
| 2/3 | 1 | 0 | 1/3 | 0 | 0 | 8/3 |
| −4/3 | 0 | ⑤ | −2/3 | 1 | 0 | 14/3 |
| 5/3 | 0 | 4 | −2/3 | 0 | 1 | 29/3 |
| 1/3 | 0 | −4 | 5/3 | 0 | 0 | 40/3 |

| $x_1$ | $x_2$ | $x_3$ | $s_1$ | $s_2$ | $s_3$ | rhs |
|---|---|---|---|---|---|---|
| 2/3 | 1 | 0 | 1/3 | 0 | 0 | 8/3 |
| −4/15 | 0 | 1 | −2/15 | 1/5 | 0 | 14/15 |
| ⑪/15 | 0 | 0 | −2/15 | −4/5 | 1 | 98/15 |
| −11/15 | 0 | 0 | 17/15 | 4/5 | 0 | 256/15 |

| $x_1$ | $x_2$ | $x_3$ | $s_1$ | $s_2$ | $s_3$ | rhs |
|---|---|---|---|---|---|---|
| 0 | 1 | 0 | 15/41 | 8/41 | −10/41 | 50/41 |
| 0 | 0 | 1 | −6/41 | 5/41 | 4/41 | 62/41 |
| 1 | 0 | 0 | −2/41 | −12/41 | −15/41 | 89/41 |
| 0 | 0 | 0 | 45/41 | 24/41 | 11/41 | 765/41 |

The reduced costs in the zeroth row of the tableau are all positive, so the basic feasible solution is optimal. In terms of the original variables, $x_1, x_2$, and $x_3$ this solutions is $x = (89/41, 50/41, 62/41)$.

(c) After introducing the slack variables, say $x_4, x_5$ and $x_6$, we obtain the following standard form problem:

$$
\begin{aligned}
\max \quad & 2x_1 - x_2 + x_3 \\
\text{s.t.} \quad & 3x_1 + x_2 + x_3 + x_4 &= 6, \\
& x_1 - x_2 + 2x_3 + x_5 &= 1, \\
& x_1 + x_2 - x_3 + x_6 &= 2, \\
& x_1, \quad x_2, \quad x_3, \quad x_4, \quad x_5, \quad x_6 \geq 0.
\end{aligned}
$$

Note that $x = (0, 0, 0, 6, 1, 2)$ is a basic feasible solution. Hence, we have the following initial tableau:

|         | rhs | $x_1$ | $x_2$ | $x_3$ | $x_4$ | $x_5$ | $x_6$ |
|---------|-----|-------|-------|-------|-------|-------|-------|
|         | 0   | 2     | −1    | 1     | 0     | 0     | 0     |
| $x_4 =$ | 6   | 3     | 1     | 1     | 1     | 0     | 0     |
| $x_5 =$ | 1   | ①     | −1    | 2     | 0     | 1     | 0     |
| $x_6 =$ | 2   | 1     | 1     | −1    | 0     | 0     | 1     |

Since we are maximizing the objective function, we choose a nonbasic variable with the greatest positive reduced cost to be the one that enters the basis. Indicating the pivot variable with a circled number, we obtain the following tableaux:

|         | rhs | $x_1$ | $x_2$ | $x_3$ | $x_4$ | $x_5$ | $x_6$ |
|---------|-----|-------|-------|-------|-------|-------|-------|
|         | −2  | 0     | 1     | −3    | 0     | −2    | 0     |
| $x_4 =$ | 3   | 0     | 4     | −5    | 1     | −3    | 0     |
| $x_1 =$ | 1   | 1     | −1    | 2     | 0     | 1     | 0     |
| $x_6 =$ | 1   | 0     | ②     | −3    | 0     | −1    | 1     |

|         | rhs  | $x_1$ | $x_2$ | $x_3$ | $x_4$ | $x_5$ | $x_6$ |
|---------|------|-------|-------|-------|-------|-------|-------|
|         | −2.5 | 0     | 0     | −1.5  | 0     | −1.5  | −0.5  |
| $x_4 =$ | 1    | 0     | 0     | 1     | 1     | −1    | −2    |
| $x_1 =$ | 1.5  | 1     | 0     | 0.5   | 0     | 0.5   | 0.5   |
| $x_2 =$ | 0.5  | 0     | 1     | −1.5  | 0     | −0.5  | 0.5   |

The reduced costs in the zeroth row of the tableau are all nonpositive, so the current basic feasible solution is optimal. In terms of the original variables $x_1, x_2$, and $x_3$, this solution is $x = (1.5, 0.5, 0)$.

(*d*) After introducing slack variables, say $x_4, x_5, x_6$ and $x_7$, we obtain the following standard form problem:

$$\max z = 60x_1 + 30x_2 + 20x_3$$

$$\begin{aligned}
\text{s.t.} \quad 8x_1 + 6x_2 + x_3 + x_4 &= 48, \\
4x_1 + 2x_2 + 1.5x_3 + x_5 &= 20, \\
2x_1 + 1.5x_2 + 0.5x_3 + x_6 &= 8, \\
x_2 + x_7 &= 5, \\
x_1, \quad x_2, \quad x_3, \quad x_4, \quad x_5, \quad x_6, \quad x_7 &\geq 0.
\end{aligned}$$

Note that $x = (0,0,0,48,20,8,5)$ is a basic feasible solution. Hence, we have the following initial tableau:

|         | rhs | $x_1$ | $x_2$ | $x_3$ | $x_4$ | $x_5$ | $x_6$ | $x_7$ |
|---------|-----|-------|-------|-------|-------|-------|-------|-------|
|         | 0   | 60    | 30    | 20    | 0     | 0     | 0     | 0     |
| $x_4 =$ | 48  | 8     | 6     | 1     | 1     | 0     | 0     | 0     |
| $x_5 =$ | 20  | 4     | 2     | 1.5   | 0     | 1     | 0     | 0     |
| $x_6 =$ | 8   | (2)   | 1.5   | 0.5   | 0     | 0     | 1     | 0     |
| $x_7 =$ | 5   | 0     | 1     | 0     | 0     | 0     | 0     | 1     |

Again indicating the pivot element with a circled number, we obtain the following tableaux:

|         | rhs  | $x_1$ | $x_2$ | $x_3$ | $x_4$ | $x_5$ | $x_6$ | $x_7$ |
|---------|------|-------|-------|-------|-------|-------|-------|-------|
|         | −240 | 0     | −15   | 5     | 0     | 0     | −30   | 0     |
| $x_4 =$ | 16   | 0     | 0     | −1    | 1     | 0     | −4    | 0     |
| $x_5 =$ | 4    | 0     | −1    | (0.5) | 0     | 1     | −2    | 0     |
| $x_1 =$ | 4    | 1     | 0.75  | 0.25  | 0     | 0     | 0.5   | 0     |
| $x_7 =$ | 5    | 0     | 1     | 0     | 0     | 0     | 0     | 1     |

|         | rhs  | $x_1$ | $x_2$ | $x_3$ | $x_4$ | $x_5$ | $x_6$ | $x_7$ |
|---------|------|-------|-------|-------|-------|-------|-------|-------|
|         | −280 | 0     | −5    | 0     | 0     | −10   | −10   | 0     |
| $x_4 =$ | 24   | 0     | −2    | 0     | 1     | 2     | −8    | 0     |
| $x_3 =$ | 8    | 0     | −2    | 1     | 0     | 2     | −4    | 0     |
| $x_1 =$ | 2    | 1     | 1.25  | 0     | 0     | −0.5  | 1.5   | 0     |
| $x_7 =$ | 5    | 0     | 1     | 0     | 0     | 0     | 0     | 1     |

The reduced costs in the zeroth row of the tableau are all nonpositive, so the current basic feasible solution is optimal. In terms of the original variables $x_1, x_2,$ and $x_3$, this solution is $x = (2, 0, 8)$.

**10.17** (*a*) We have the following tableau:

| $x_1$ | $x_2$ | $s_1$ | $s_2$ | $s_3$ | rhs |
|---|---|---|---|---|---|
| 0 | 0 | 17 | −7 | 0 | 10 |
| 1 | 0 | 3 | −1 | 0 | 2 |
| 0 | 1 | 4 | −2 | 0 | 2 |
| 0 | 0 | 1 | 0 | 1 | 6 |

Since every element of the $s_2$ column is less than or equal to zero, this LP is unbounded. The LP does not have any optimal solution.

(*b*) We have the following tableau:

| $x_1$ | $x_2$ | $s_1$ | $s_2$ | $s_3$ | rhs |
|---|---|---|---|---|---|
| 0 | 0 | 17 | 1 | 0 | 10 |
| 1 | 0 | 3 | −1 | 0 | 2 |
| 0 | 1 | 4 | 2 | 0 | 2 |
| 0 | 0 | 1 | −1 | 1 | 6 |

The LP has only one optimal solution.

(*c*) We have the following tableau:

| $x_1$ | $x_2$ | $s_1$ | $s_2$ | $s_3$ | rhs |
|---|---|---|---|---|---|
| 0 | 0 | 17 | 0 | 0 | 10 |
| 1 | 0 | 3 | −1 | 0 | 2 |
| 0 | 1 | 4 | 3/2 | 0 | 2 |
| 0 | 0 | 1 | 1 | 1 | 6 |

There is a nonbasic column we can pivot to, so this LP has many optimal solutions.

**10.18** (*a*) The basic variables are $x_1$, $x_3$, and $x_6$. For $x_3$ to be basic, $a_3 = 1$ and $a_4 = 0$. For $x_6$ to be basic, $c_3 = 0$, $a_7 = 0$, and $a_8 = 1$.

(*b*) Setting $b \geq 0$ makes the LP feasible. Setting $c_1 \geq 0$, $c_2 \geq 0$, and $c_3 \geq 0$ makes the LP optimal.

(*c*) Setting $b \geq 0$, there are three variables we can introduce into the basis to obtain alternative optimal solutions:

(*i*) Variable $x_2$: To do this, at least one of $a_1, a_2$ must be greater than zero to pivot into Column 2.

(*ii*) Variable $x_4$: To do this, $c_1 = 0$ and $a_5 \geq 0$ to pivot into Column 4.

(*iii*) Variable $x_5$: To do this, $c_2 = 0$ and $a_6 \geq 0$ to pivot into Column 5.

**10.19** After introducing slack variables, we obtain the following standard form problem:

$$\max \quad z = 5x_1 - x_2$$
$$\text{s.t.} \quad x_1 - 3x_2 + x_3 \qquad\qquad = 1,$$
$$x_1 - 4x_2 \qquad\quad + x_4 = 3,$$
$$x_1, \quad x_2, \quad x_3, \quad x_4 \geq 0.$$

We form the initial tableau:

|       | rhs | $x_1$ | $x_2$ | $x_3$ | $x_4$ |
|-------|-----|-------|-------|-------|-------|
|       | 0   | 5     | −1    | 0     | 0     |
| $x_3 =$ | 1   | −3    | 1     | 1     | 0     |
| $x_4 =$ | 1   | (-4)  | −1    | 0     | 1     |

The tableau has a nonbasic variable $x_1$ that could enter and improve the value of $z$, but there are no candidates for the minimum ratio test. Suppose $x_1$ entered the basis. The equations representing the constraints are

$$x_3 = 1 + 3x_1$$
$$x_4 = 1 + 4x_1.$$

As $x_1$ increases, both $x_3$ and $x_4$ stay positive. Thus we could keep increasing $x_1$ (improving $z$) without ever encountering infeasibility. Hence the LP is unbounded.

**10.20** *(a)* The dual problem is:

$$\max \quad w = 4y_1 + 9y_2 + 5y_3$$
$$\text{s.t.} \quad y_1 + 2y_2 \qquad\quad \geq 5,$$
$$y_1 + 3y_2 - y_3 \geq 3,$$
$$y_1 - y_2 + 3y_3 \leq -2,$$
$$y_1, \quad y_2, \quad y_3 \geq 0.$$

*(b)* *(i)* The optimal solution to the primal problem is $x = (0; 3.5; 1.5)$. The optimal value to the primal problem is 7.5.

*(ii)* The optimal solution to the dual problem is $y = (0; 3.5; 1.5)$. The optimal value to the dual problem is 7.5.

**10.21** *(a)*

```
1  function [x, optimal_cost, iters, Running_time] = rsm (A,
       b, c, N, M)
2
3  t1=cputime;
4
5  c=c';
```

```
 6  [m n] = size(A);
 7  bfs=[zeros(1,n-m)';b];
 8  B_indices = find(bfs);
 9  N_indices = find(ones(1,n) - abs(sign(bfs))');
10
11  rsm_nnz = zeros(5000,2);
12
13  % disp('Please determine the way of choosing the entering
         varialble as follows: ');
14  % fprintf(1,'Input 1 if you want to choose the variable
         with the smallest reduced cost to enter the basis.\n')
         ;
15  % fprintf(1,'Or input 0 if you want to choose the variable
          that first gives a negative reduced cost to enter the
          basis.\n');
16  % N = input(' ');
17  %
18  % disp('Please determine the way of choosing the leaving
         varialble as follows: ');
19  % fprintf(1,'Input "1" if you want to choose the smallest
         index rule.\n');
20  % fprintf(1,'Or input "0" if you want to choose the
         lexicographic.\n');
21  % M = input(' ');
22
23  iters=0;
24  while 1==1
25  iters=iters+1;
26
27  Binv = inv(A(:,B_indices));
28
29  rsm_nnz(iters,1) = nnz(A(:,B_indices));
30  rsm_nnz(iters,2) = nnz(Binv);
31
32  d = Binv * b;
33
34  if N == 1
35      c_tilde = zeros(1,n);
36      c_tilde(:,N_indices) = c(:,N_indices) - c(:,B_indices)
             * Binv * A(:,N_indices);
37      [cj j]=min(c_tilde);
38
39      if cj >= 0
40          x = zeros(n,1);
41          x(B_indices,:) = d;
```

```
42          optimal_cost = c*x;
43          x=x';
44          break;
45      end;
46  end
47
48  if N == 0
49      c_tilde = zeros(1,n);
50      for k=1:length(N_indices)
51          c_tilde(:,N_indices(k)) = c(:,N_indices(k)) - c(:,
                B_indices) * Binv * A(:,N_indices(k));
52          cj = c_tilde(:,N_indices(k));
53          if cj < 0
54              j = N_indices(k);
55              break
56          end;
57      end;
58      if cj >= 0
59          x = zeros(n,1);
60          x(B_indices,:) = d;
61          optimal_cost = c*x;
62          x=x';
63          break;
64      end;
65  end
66
67  u = Binv * A(:,j);
68  mn = inf;
69  i=0;
70  zz = find (u > 0)';
71
72  if (length(zz) == 0)
73      x='The LP is unbounded';
74      optimal_cost='The LP is unbounded';
75      break
76  else
77      [yy, ii] = min (d(zz) ./ u (zz)) ;
78      i = zz(ii(1));
79      k = B_indices(i);
80      B_indices(i) = j;
81      N_indices(j == N_indices) = k;
82  end;
83
84  end;
85
```

```
86  t2=cputime;
87  Running_time=t2-t1;
```

**10.21** (*b*)

```
1   function [x, optimal_cost, iters, Running_time] = tsm (A,
        b, c,N,M)
2
3   t1=cputime;
4
5   c=c';
6   [m n] = size(A);
7   bfs=[zeros(1,n-m)';b];
8   B_indices = find(bfs);
9
10  % disp('Please determine the way of choosing the entering
        varialble as follows: ');
11  % fprintf(1,'Input "1" if you want to choose the variable
        with the smallest reduced cost to enter the basis.\n')
        ;
12  % fprintf(1,'Or input "0" if you want to choose the
        variable with the smallest index with a negative
        reduced cost to enter the basis.\n');
13  % N = input(' ');
14  %
15  % disp('Please determine the way of choosing the leaving
        varialble as follows: ');
16  % fprintf(1,'Input "1" if you want to choose the smallest
        index rule.\n');
17  % fprintf(1,'Or input "0" if you want to choose the
        lexicographic.\n');
18  % M = input(' ');
19
20  Binv = inv(A(:,B_indices));
21  x_B=Binv * b;
22  c_tilde = c - c(:,B_indices) * Binv * A;
23  T_1=[- c(:,B_indices)*x_B, c_tilde; x_B, Binv * A];
24
25  T=T_1;
26
27  [m,n] = size(T);
28
29  m=m-1;
30  n=n-1;
31
32  iters=0;
```

```
33  while 1==1
34  iters=iters+1;
35
36
37  if N==0
38      y=find(T(1,:)<0);
39      if length(y)>0
40          j=y(1)-1;
41      else
42          x=zeros(1,n);
43          for k=2:n+1
44              z=find(T(:,k));
45              if length(z)==1
46                  x(k-1)=T(z(1),1);
47              end
48          end
49          optimal_cost = -T(1,1);
50          break
51      end
52  end
53
54
55  if N==1
56      [x_j j]=min(T(1,:));
57      if x_j<0
58          j=j-1;
59      else
60          x=zeros(1,n);
61          for k=2:n+1
62              z=find(T(:,k));
63              if length(z)==1
64                  x(k-1)=T(z(1),1);
65              end
66          end
67          optimal_cost = -T(1,1);
68          break
69      end
70  end
71
72  u = Binv * A(:,j);
73  zz = find (u > 0)' ;
74  if (length(zz) == 0)
75      x='The LP is unbounded';
76      optimal_cost='The LP is unbounded';
77      break
```

```
78   end
79
80   dind = find(T(2:end,j+1)>0);
81
82   if M== 1
83   [thetast,l] = min( T(1+dind,1)./ T(1+dind,j+1) );
84   l=dind(l);
85   end
86
87   if M==0
88   T(dind+1,:)=T(dind+1,:)./repmat(T(dind+1,j+1),1,n+1);
89   [Ts,sind]=sortrows(T(dind+1,:));
90   l=dind(sind(1));
91   thetast = T(l+1,1)/T(l+1,j+1);
92   end
93
94
95   T(l+1,:)=T(l+1,:)/T(l+1,j+1);
96   for i=setdiff( (1:m+1), l+1 )
97       T(i,:) = T(i,:) − T(i,j+1)*T(l+1,:);
98   end
99
100  for k=2:n+1
101      z=find(T(:,k));
102      if length(z)==1 && T(z(1),k) ~= 1
103          T(z(1),:)=T(z(1),:)/T(z(1),k);
104      end
105  end
106
107  if norm(T − T_1)<1.0e−2
108      disp('The simplex method cycles'),break
109  end
110
111  end
112
113  t2=cputime;
114  Running_time=t2−t1;
```

**10.22** Note that from the first three equations in (10.22) we have that

$$
\begin{aligned}
(Ax)^{\mathsf{T}}y - b^{\mathsf{T}}y\tau &= \mathbf{0}, \\
-x^{\mathsf{T}}A^{\mathsf{T}}y - x^{\mathsf{T}}s + x^{\mathsf{T}}c\tau &= \mathbf{0}, \\
-c^{\mathsf{T}}x\tau + b^{\mathsf{T}}y\tau - \kappa\tau &= 0.
\end{aligned}
$$

Now we get the desired equation by adding the above three equations.

**11.1** For each $i = 1, 2, \ldots, k$, it is known that the matrix $\bar{M}_i \succeq 0$ if and only if every principle minor of $\bar{M}_i$ is nonnegative. Since

$$
\det(\tau_i \Lambda_i - I) = \begin{vmatrix} \tau_1 \lambda_{i1} - 1 & 0 & \cdots & 0 \\ 0 & \tau_2 \lambda_{i2} - 1 & \cdots & 0 \\ \vdots & \vdots & \ddots & \vdots \\ 0 & 0 & \cdots & \tau_k \lambda_{ik} - 1 \end{vmatrix} = \prod_{j=1}^{k} (\tau_j \lambda_{ij} - 1).
$$

It follows that $\bar{M}_i \succeq 0$ if and only if $\Pi_{j=1}^{s}(\tau_j \lambda_{ij} - 1) \geq 0$, for all $s = 1, 2, \ldots, k$ and $\det(\bar{M}_i) \geq 0$. Thus, $\bar{M}_i \succeq 0$ if and only if $\tau_j \lambda_{ij} \geq 1$, for all $j = 1, 2, \ldots, k$ and $\det(\bar{M}_i) \geq 0$.

Notice that

$$
\det(\bar{M}_i) = \left( \prod_{j=1}^{k} (\tau_i \lambda_{ij} - 1) \right) (\tau_i v_i + d_2 - \bar{x}^\mathsf{T} \bar{x}) - \sum_{j=1}^{k} u_{ij}^2.
$$

This means the inequality $\det(\bar{M}_i) \geq 0$ strictly holds for each $i \leq j \leq k$ such that $\tau_j \lambda_{ij} = 1$. Hence, $\det(\bar{M}_i) \geq 0$ if and only if

$$
(\tau_i v_i + d_2 - \bar{x}^\mathsf{T} \bar{x}) - \mathbf{1}^\mathsf{T} s_i = (\tau_i v_i + d_2 - \bar{x}^\mathsf{T} \bar{x}) - \sum_{\tau_i \lambda_{ij} > 1} (u_{ij}^2 / (\tau_j \lambda_{ij} - 1)) \geq 0.
$$

Therefore, $\bar{M}_i \succeq 0$ if and only if $\tau_i \lambda_{\min}(H_i) \geq 1$ and $d_2 \geq \bar{x}^\mathsf{T} \bar{x} - \tau_i v_i + \mathbf{1}^\mathsf{T} s_i$.

**11.2** Let $p \in [1, \infty]$. We first prove that $\mathcal{P}_q^n \subseteq \mathcal{P}_p^{n\star}$. Let $x = (x_0; \widetilde{x}) \in \mathcal{P}_q^n$, we show that $x \in \mathcal{P}_p^{n\star}$ by verifying that $x^\mathsf{T} y \geq 0$ for any $y \in \mathcal{P}_p^n$. So let $y = (y_0; \widetilde{y}) \in \mathcal{P}_p^n$, then

$$
x^\mathsf{T} y = x_0 y_0 + \widetilde{x}^\mathsf{T} \widetilde{y} \geq \|\widetilde{x}\|_q \|\widetilde{y}\|_p + \widetilde{x}^\mathsf{T} \widetilde{y} \geq |\widetilde{x}^\mathsf{T} \widetilde{y}| + \widetilde{x}^\mathsf{T} \widetilde{y} \geq 0,
$$

where the first inequality follows from the fact that $x \in \mathcal{P}_q^n$ and $y \in \mathcal{P}_p^n$, and the second one from Hölder's inequality. Thus, $\mathcal{P}_q^n \subseteq \mathcal{P}_p^{n\star}$.

Now we show $\mathcal{P}_p^{n\star} \subseteq \mathcal{P}_q^n$. Let $y = (y_0; \widetilde{y}) \in \mathcal{P}_p^{n\star}$, we need to show that $y \in \mathcal{P}_q^n$. This is trivial if $\widetilde{y} = 0$ or $p = \infty$. If $\widetilde{y} \neq 0$ and $1 \leq p < \infty$, let $u \triangleq (y_1^{p/q}; y_2^{p/q}; \ldots; y_{n-1}^{p/q})$ and consider $x \triangleq (\|u\|_p; -u) \in \mathcal{P}_p^n$. Then by using Hölder's inequality, where the equality is attained, we obtain

$$
0 \leq x^\mathsf{T} y = \|u\|_p y_0 - u^\mathsf{T} \widetilde{y} = \|u\|_p y_0 - \|u\|_p \|\widetilde{y}\|_q = \|u\|_p (y_0 - \|\widetilde{y}\|_q).
$$

This implies that $y_0 \geq \|\widetilde{y}\|_q$, and therefore means that $y \in \mathcal{P}_q^n$. Thus, $\mathcal{P}_p^{n\star} \subseteq \mathcal{P}_q^n$. The proof is complete.

**11.3** We first show that $\mathcal{K}^n_{(M^{-1})^\mathsf{T}} \subseteq (\mathcal{K}^n_M)^\star$. Let $x = (x_0; \widetilde{x}) \in \mathcal{K}^n_{(M^{-1})^\mathsf{T}}$, we need to show that $x \in (\mathcal{K}^n_M)^\star$. For any $y = (y_0; \widetilde{y}) \in \mathcal{K}^n_M$, we have

$$
\begin{aligned}
x^\mathsf{T} y &= x_0 y_0 + \widetilde{x}^\mathsf{T} \widetilde{y} \\
&\geq \left\| \left( M^{-1} \right)^\mathsf{T} \widetilde{x} \right\| \left\| M \widetilde{y} \right\| + \widetilde{x}^\mathsf{T} \widetilde{y} \\
&\geq \left| \widetilde{x}^\mathsf{T} M^{-1} M \widetilde{y} \right| + \widetilde{x}^\mathsf{T} \widetilde{y} \\
&= \left| \widetilde{x}^\mathsf{T} \widetilde{y} \right| + \widetilde{x}^\mathsf{T} \widetilde{y} \geq 0,
\end{aligned}
$$

where we used the assumptions that $x \in \mathcal{K}^n_{(M^{-1})^\mathsf{T}}$ and $y \in \mathcal{K}^n_M$ to obtain the first inequality, and we used Cauchy-Schwartz inequality to obtain the second inequality. This means that $x \in (\mathcal{K}^n_M)^\star$ and hence $\mathcal{K}^n_{(M^{-1})^\mathsf{T}} \subseteq (\mathcal{K}^n_M)^\star$.

To prove the reverse inclusion, let $y \in (\mathcal{K}^n_M)^\star$, we need to show that $y \in \mathcal{K}^n_{(M^{-1})^\mathsf{T}}$, which is trivial if $\widetilde{y} = 0$. If $\widetilde{y} \neq 0$, let $x \triangleq \left( \left\| M \widetilde{y} \right\| ; -\widetilde{y} \right) \in \mathcal{K}^n_M$. Then, we have

$$
\begin{aligned}
x^\mathsf{T} y &= x_0 y_0 + \widetilde{x}^\mathsf{T} \widetilde{y} \\
&= y_0 \left\| M \widetilde{y} \right\| - \widetilde{y}^\mathsf{T} \widetilde{y} \\
&= y_0 \left\| M \widetilde{y} \right\| - \widetilde{y}^\mathsf{T} M^\mathsf{T} \left( M^{-1} \right)^\mathsf{T} \widetilde{y} = y_0 \left\| M \widetilde{y} \right\| - \left\| M \widetilde{y} \right\| \left\| \left( M^{-1} \right)^\mathsf{T} \widetilde{y} \right\|,
\end{aligned}
$$

where we used Cauchy-Schwartz inequality, where the equality is attained, to obtain the last equality. Since $x$ belongs to the elliptic cone $\mathcal{K}^n_M$ and $y$ belongs to its dual, it follows that

$$
0 \leq x^\mathsf{T} y = \left\| M \widetilde{y} \right\| \left( y_0 - \left\| \left( M^{-1} \right)^\mathsf{T} \widetilde{y} \right\| \right).
$$

As $\widetilde{y} \neq 0$, this implies that $y_0 \geq \left\| \left( M^{-1} \right)^\mathsf{T} \widetilde{y} \right\|$. That is, $y \in \mathcal{K}^n_{(M^{-1})^\mathsf{T}}$ and hence $(\mathcal{K}^n_M)^\star \subseteq \mathcal{K}^n_{(M^{-1})^\mathsf{T}}$. The proof is complete.

**11.4** To prove item (c) of Lemma 11.8, note that

$$
\begin{aligned}
\overline{x}^{+\mathsf{T}} \underline{s}^+ &= \left( \overline{x} + \alpha \overline{\Delta x} \right)^\mathsf{T} \left( \underline{s} + \alpha \underline{\Delta s} \right) \\
&= \overline{x}^\mathsf{T} \underline{s} + \alpha \left( \overline{\Delta x}^\mathsf{T} \underline{s} + \overline{x}^\mathsf{T} \underline{\Delta s} \right) + \alpha^2 \overline{\Delta x}^\mathsf{T} \underline{\Delta s} \\
&= \overline{x}^\mathsf{T} \underline{s} + \tfrac{1}{2} \alpha \, \text{trace} \left( \sigma \mu e - \overline{x} \circ \underline{s} \right) \\
&= \overline{x}^\mathsf{T} \underline{s} + \tfrac{1}{2} \alpha \sigma \mu \, \text{trace} \left( e \right) - \tfrac{1}{2} \alpha \, \text{trace} \left( \overline{x} \circ \underline{s} \right) \\
&= \overline{x}^\mathsf{T} \underline{s} + \alpha \sigma \mu r - \alpha \overline{x}^\mathsf{T} \underline{s} \\
&= \overline{x}^\mathsf{T} \underline{s} + \tfrac{1}{2} \alpha \sigma \overline{x}^\mathsf{T} \underline{s} - \alpha \overline{x}^\mathsf{T} \underline{s} = \left( 1 - \alpha \left( 1 - \tfrac{\sigma}{2} \right) \right) \overline{x}^\mathsf{T} \underline{s},
\end{aligned}
$$

where the third equality follows from items (a) and (b) of Lemma 11.8.

**11.5** Given $\alpha \in \mathbb{R}$, using item (c) of Lemma 11.8, we have

$$x(\alpha)^\mathsf{T} s(\alpha) = (1 - \alpha + \alpha\sigma)\overline{x}^\mathsf{T}\underline{s}, \quad \text{and hence} \quad \mu(\alpha) = (1 - \alpha + \alpha\sigma)\mu.$$

Thus, we get

$$
\begin{aligned}
V(\alpha) &= x(\alpha) \circ s(\alpha) - \mu(\alpha)e \\
&= \left(\overline{x} + \alpha\overline{\Delta x}\right) \circ \left(\underline{s} + \alpha\underline{\Delta s}\right) - (1 - \alpha + \alpha\sigma)\mu e \\
&= (1 - \alpha)(\overline{x} \circ \underline{s} - \mu e) + \overbrace{\alpha(\overline{x} \circ \underline{s} - \sigma\mu e)}^{-h} + \overbrace{\alpha\left(\overline{x} \circ \underline{\Delta s} + \overline{\Delta x} \circ \underline{s}\right)}^{h} \\
&\quad + \alpha^2\, \overline{\Delta x} \circ \underline{\Delta s} \\
&= (1 - \alpha)(\overline{x} \circ \underline{s} - \mu e) + \alpha^2\, \overline{\Delta x} \circ \underline{\Delta s}.
\end{aligned}
$$

This completes the proof.

**11.6** By the last equation of system (11.15) and from the operator commutativity, we have

$$h = \overline{x} \circ \underline{\Delta s} + \overline{\Delta x} \circ \underline{s} = \overline{x} \circ \underline{\Delta s} + \mu\overline{\Delta x} \circ \overline{x}^{-1} + \left(\overline{\Delta x} \circ \overline{x}^{-1}\right) \circ \left(\overline{x} \circ \underline{s} - \mu e\right).$$

It immediately follows that

$$
\begin{aligned}
\|h\|_F &\geq \left\|\overline{x} \circ \underline{\Delta s} + \mu\,\overline{\Delta x} \circ \overline{x}^{-1}\right\|_F - \left\|\overline{\Delta x} \circ \overline{x}^{-1}\right\|_F \left\|\overline{x} \circ \underline{s} - \mu e\right\| \\
&\geq \left\|\overline{x} \circ \underline{\Delta s} + \mu\,\overline{\Delta x} \circ \overline{x}^{-1}\right\|_F - \theta\delta_x \\
&= \sqrt{\left\|\overline{x} \circ \underline{\Delta s}\right\|_F^2 + \left\|\mu\,\overline{\Delta x} \circ \overline{x}^{-1}\right\|_F^2} - \theta\delta_x \\
&= \sqrt{\delta_x^2 + \delta_s^2} - \theta\delta_x \geq (1 - \theta)\sqrt{\delta_x^2 + \delta_s^2},
\end{aligned}
\tag{A.6}
$$

where the second inequality follows from the assumption that $\left\|\overline{x} \circ \underline{s} - \mu e\right\| \leq \theta\mu$, and the first equality follows from (11.2) and the fact that

$$
\begin{aligned}
\left(\overline{x} \circ \underline{\Delta s}\right)^\mathsf{T} \left(\overline{\Delta x} \circ \overline{x}^{-1}\right) &= \operatorname{trace}\left(\left(\overline{x} \circ \underline{\Delta s}\right) \circ \left(\overline{\Delta x} \circ \overline{x}^{-1}\right)\right) \\
&= \operatorname{trace}\left(\underline{\Delta s} \circ \overline{\Delta x}\right) = \overline{\Delta x}^\mathsf{T}\underline{\Delta s},
\end{aligned}
$$

which is essentially zero due to item (a) of Lemma 11.8.

The right-hand side inequality in (11.18) follows by noting that $(\delta_x - \delta_s)^2 \geq 0$, and the left-hand side inequality in (11.18) follows from the last inequality in (A.6). The proof is complete.

**11.7** This is an implementation exercise. As a sample answer, see [Alz17, Example 7.1].

**11.8** The homogeneous model for the pair (11.29) and (11.30) is as follows:

$$W_0 x_0 - h_0 \tau = \mathbf{0},$$
$$B_k x_0 + W_k x_k - h_k \tau = \mathbf{0}, k = 1, 2, \dots, K,$$
$$-W_0^\mathsf{T} y_0 - \sum_{k=1}^{K} B_k^\mathsf{T} y_k + \tau c_0 - s_0 = \mathbf{0},$$
$$-W_k^\mathsf{T} y_k + \tau c_k - s_k = \mathbf{0}, k = 1, 2, \dots, K, \qquad (A.7)$$
$$\sum_{k=0}^{K} h_k^\mathsf{T} y_k - \sum_{k=0}^{K} c_k^\mathsf{T} x_k - \kappa = 0,$$
$$x_k \geq \mathbf{0}, s_k \geq \mathbf{0}, k = 0, 1, \dots, K,$$
$$\tau \geq 0, \kappa \geq 0.$$

**11.9** The search direction system corresponding to (A.7) is defined by the following system:

$$W_0 \Delta x_0 - h_0 \Delta \tau = \eta r_{p0},$$
$$B_k \Delta x_0 + W_k \Delta x_k - h_k \Delta \tau = \eta r_{pk}, k = 1, 2, \dots, K,$$
$$-W_0^\mathsf{T} \Delta y_0 - \sum_{k=1}^{K} B_k^\mathsf{T} \Delta y_k + \Delta \tau c_0 - \Delta s_0 = \eta r_{d0},$$
$$-W_k^\mathsf{T} \Delta y_k + \Delta \tau \, c_k - \Delta s_k = \eta r_{dk}, k = 1, 2, \dots, K,$$
$$\sum_{k=0}^{K} h_k^\mathsf{T} \Delta y_k - \sum_{k=0}^{K} c_k^\mathsf{T} \Delta x_k - \Delta \kappa = \eta r_g,$$
$$\kappa \Delta \tau + \tau \Delta \kappa = \gamma \mu - \tau \kappa,$$
$$\Delta x_0 \circ s_0 + x_0 \circ \Delta s_0 = \gamma \mu e_0 - x_0 \circ s_0,$$
$$\Delta x_k \circ s_k + x_k \circ \Delta s_k = \gamma \mu e_k - x_k \circ s_k, k = 1, 2, \dots, K,$$

where

$$r_{p0} \triangleq h_0 \tau - W_0 x_0;$$
$$r_{pk} \triangleq h_k \tau - B_k x_0 - W_k x_k;$$
$$r_{d0} \triangleq W_0^\mathsf{T} y_0 + \sum_{k=1}^{K} B_k^\mathsf{T} y_k + s_0 - \tau c_0;$$
$$r_{dk} \triangleq W_k^\mathsf{T} y_k + s_k - \tau c_k;$$
$$r_g \triangleq \kappa - \sum_{k=0}^{K} h_k^\mathsf{T} y_k + \sum_{k=0}^{K} c_k^\mathsf{T} x_k;$$
$$\mu \triangleq \frac{1}{2r(K+1)+1} \left( \sum_{k=0}^{K} x_k^\mathsf{T} s_k + \tau \kappa \right),$$

and $\eta$ and $\gamma$ are two parameters.

**12.1** To prove item (c) of Lemma 12.6, note that

$$
\begin{aligned}
X^+ \bullet S^+ &= (X + \alpha \Delta X) \bullet (S + \alpha \Delta S) \\
&= X \bullet S + \alpha \left( \Delta X \bullet S + X \bullet \Delta S \right) + \alpha^2 \Delta X \bullet \Delta S \\
&= X \bullet S + \tfrac{1}{2}\alpha \, \mathrm{trace} \left( \sigma \mu I - X \circ S \right) \\
&= X \bullet S + \tfrac{1}{2}\alpha \sigma \mu \, \mathrm{trace}\,(I) - \tfrac{1}{2}\alpha \, \mathrm{trace}\,(X \circ S) \\
&= X \bullet S + \alpha \sigma \mu r - \alpha X \bullet S \\
&= X \bullet S + \tfrac{1}{2}\alpha \sigma X \bullet S - \alpha X \bullet S = \left( 1 - \alpha \left( 1 - \tfrac{\sigma}{2} \right) \right) X \bullet S,
\end{aligned}
$$

where the third equality follows from items (a) and (b) of Lemma 12.6.

**12.2** Given $\alpha \in \mathbb{R}$, using item (c) of Lemma 12.6, we have

$$
X(\alpha) \bullet S(\alpha) = (1 - \alpha + \alpha \sigma) \, X \bullet S, \quad \text{and hence} \quad \mu(\alpha) = (1 - \alpha + \alpha \sigma)\mu.
$$

Thus, we get

$$
\begin{aligned}
V(\alpha) &= X(\alpha) \circ s(\alpha) - \mu(\alpha)I \\
&= (X + \alpha \Delta X) \circ (S + \alpha \Delta S) - (1 - \alpha + \alpha \sigma)\mu I \\
&= (1 - \alpha)(X \circ S - \mu I) + \alpha \overbrace{(X \circ S - \sigma \mu I)}^{-H} + \alpha \overbrace{(X \circ \Delta S + \Delta X \circ S)}^{H} \\
&\quad + \alpha^2 \, \Delta X \circ \Delta S \\
&= (1 - \alpha)(X \circ S - \mu I) + \alpha^2 \, \Delta X \circ \Delta S.
\end{aligned}
$$

This completes the proof.

**12.3** Using the last equation of system (12.18) and the commutativity property, we have

$$
H = X \circ \Delta S + \Delta X \circ S = X \circ \Delta S + \mu \Delta X \circ X^{-1} + \left( \Delta X \circ X^{-1} \right) \circ (XS - \mu I).
$$

It immediately follows that

$$
\begin{aligned}
\|H\|_F &\geq \left\| X \circ \Delta S + \mu \, \Delta X \circ X^{-1} \right\|_F - \left\| \Delta X X^{-1} \right\|_F \left\| XS - \mu I \right\|_2 \\
&\geq \left\| X \circ \Delta S + \mu \, \Delta X \circ X^{-1} \right\|_F - \theta \delta_X \\
&= \sqrt{ \| X \Delta S \|_F^2 + \left\| \mu \, \Delta X X^{-1} \right\|_F^2 } - \theta \delta_X \\
&= \sqrt{ \delta_X^2 + \delta_S^2 } - \theta \delta_X \geq (1 - \theta) \sqrt{ \delta_X^2 + \delta_S^2 },
\end{aligned}
\tag{A.8}
$$

where the second inequality follows from the assumption that $\left\|XS - \mu I\right\|_2 \leq \theta \mu$, and the first equality follows from (12.2) and the fact that

$$
\begin{aligned}
(X \circ \Delta S) \bullet \left(\Delta X \circ X^{-1}\right) &= \operatorname{trace}\left((X \circ \Delta S) \circ \left(\Delta X \circ X^{-1}\right)\right) \\
&= \operatorname{trace}\left(\Delta S \circ \Delta X\right) = \Delta X \bullet \Delta S,
\end{aligned}
$$

which is essentially zero due to item (a) of Lemma 12.6.

The right-hand side inequality in (12.20) follows by noting that $(\delta_X - \delta_S)^2 \geq 0$, and the left-hand side inequality in (12.20) follows from the last inequality in (A.8). The proof is complete.

**12.4** This is an implementation exercise. As a sample answer, see [TBY17, Example 2].

**12.5** The homogeneous model for the pair (P|SDP) and (D|SDP) is as follows (see also [PS98, JAZ12]):

$$
\begin{aligned}
\mathcal{A}X &\quad &\quad -b\tau &\quad &= \mathbf{0}, \\
&-\mathcal{A}^\star y \quad -S \quad +C\tau &\quad &= \mathbf{0}, \\
-C \bullet X \quad +b^\mathsf{T} y &\quad &\quad -\kappa &= 0, \\
X &\quad &\quad &\geq \mathbf{0}, \\
&\quad S &\quad &\geq \mathbf{0}, \\
&\quad &\quad \tau &\geq 0, \\
&\quad &\quad \kappa &\geq 0.
\end{aligned} \tag{A.9}
$$

**12.6** The search direction system corresponding to (A.9) is defined by the following system (see also [PS98, JAZ12]):

$$
\begin{aligned}
\mathcal{A}\,\Delta X &\quad &\quad -b\,\Delta\tau &\quad &= \eta r_p, \\
&-\mathcal{A}^\star\,\Delta y \quad -\Delta S \quad +C\,\Delta\tau &\quad &= \eta R_d, \\
-C \bullet \Delta X \quad +b^\mathsf{T}\,\Delta y &\quad &\quad -\Delta\kappa &= \eta r_g, \\
&\quad \kappa\,\Delta\tau \quad + \quad \tau\,\Delta\kappa &\quad &= \gamma\mu - \tau\kappa, \\
\mathcal{H}_P(\Delta X S) &\quad + \mathcal{H}_P(X \Delta S) &\quad &= \gamma\mu I - \mathcal{H}_P(XS),
\end{aligned} \tag{A.10}
$$

where $\mathcal{H}_P(\cdot)$ is the symmetrization operator $\mathcal{H}_P : \mathbb{R}^{n \times n} \longrightarrow \mathbb{S}^n$ defined in (12.5), $\eta$ and $\gamma$ are two parameters, and

$$
\begin{aligned}
r_p &\triangleq b\tau - \mathcal{A}X, \\
R_d &\triangleq \mathcal{A}^\star y + S - \tau C, \\
r_g &\triangleq C \bullet X - b^\mathsf{T} y + \kappa, \\
\mu &\triangleq \tfrac{1}{n+1}(X \bullet S + \tau\kappa).
\end{aligned}
$$

---

**Algorithm A.3:** Generic homogeneous self-dual algorithm for SDP

---

**Input:** Data in Problems (P|SDP) and (D|SDP) $(X, y, S, \tau, \kappa) \triangleq (I, 0, I, 1, 1)$
**Output:** An approximate optimal solution to Problem (P|SDP)

1: **while** a stopping criterion is not satisfied **do**
2:     choose $\eta, \gamma$
3:     compute the solution $(\Delta X, \Delta y, \Delta S, \Delta \tau, \Delta \kappa)$ of the linear system (A.10)
4:     compute a step length $\theta$ so that
    $X + \theta \Delta X \succ 0$
    $S + \theta \Delta S \succ 0$
    $\tau + \theta \Delta \tau > 0$
    $\kappa + \theta \Delta \kappa > 0$
5:     set the new iterate according to
    $(X, y, S, \tau, \kappa) \triangleq (X, y, S, \tau, \kappa) + \theta(\Delta X, \Delta y, \Delta S, \Delta \tau, \Delta \kappa)$
6: **end**

---

**12.7** We state the generic homogeneous algorithm for solving the pair (P|SDP) and (D|SDP) in Algorithm A.3 (see also [JAZ12, Algorithm 1]).

**12.8** Under Assumptions 12.2 and 12.1, if the pair (P|SDP) and (D|SDP) has a solution $(X^\star, y^\star, S^\star)$, then Algorithm A.3 finds an $\epsilon$-approximate solution in at most

$$O\left(\sqrt{n} \ln \left(\text{trace}\, (X^\star + S^\star) \left(\frac{\epsilon_0}{\epsilon}\right)\right)\right)$$

iteration (see also [PS98, Theorems 5.2 and 6.2]).

# REFERENCES

AA22.    Baha Alzalg and Hadjer Alioui. Applications of stochastic mixed-integer second-order cone optimization. *IEEE Access*, 10:3522–3547, 2022.

ABA19.    Baha Alzalg, Khaled Badarneh, and Ayat Ababneh. An infeasible interior-point algorithm for stochastic second-order cone optimization. *J. Optim. Theory Appl.*, 181(1):324–346, apr 2019.

AG03.    Farid Alizadeh and Donald Goldfarb. Second-order cone programming. *Mathematical programming*, 95(1):3–51, 2003.

Alz11.    Baha Alzalg. A class of polynomial volumetric barrier decomposition algorithms for stochastic symmetric programming. Technical report, WSU Technical Report, 2011.

Alz12.    Baha M. Alzalg. Stochastic second-order cone programming: Applications models. *Applied Mathematical Modelling*, 36(10):5122–5134, 2012.

Alz14a.    Baha Alzalg. Decomposition-based interior point methods for stochastic quadratic second-order cone programming. *Applied Mathematics and Computation*, 249:1–18, 2014.

Alz14b.    Baha Alzalg. Homogeneous self-dual algorithms for stochastic second-order cone programming. *J. Optim. Theory Appl.*, 163(1):148–164, oct 2014.

Alz17.    Baha Alzalg. A primal-dual interior-point method based on various selections of displacement step for second-order cone programming. 2017.

Alz20.      Baha Alzalg. A logarithmic barrier interior-point method based on majorant functions for second-order cone programming. *Optim. Lett.*, 14(3):729–746, 2020.

AP17a.      Baha Alzalg and Mohammad Pirhaji. Elliptic cone optimization and primal–dual path-following algorithms. *Optimization*, 66(12):2245–2274, 2017.

AP17b.      Baha Alzalg and Mohammad Pirhaji. Primal-dual path-following algorithms for circular programming. 2017.

AR14.       Howard Anton and Chris Rorres. *Elementary Linear Algebra: Applications Version.* Wiley, eleventh edition, 2014.

AU94.       Alfred V. Aho and Jeffrey D. Ullman. *Foundations of Computer Science.* W. H. Freeman Co., USA, 1st edition, 1994.

BF10.       Richard L. Burden and J. Douglas Faires. *Numerical Analysis: 9th Ed.* Cengage Learning Inc., USA, 2010.

BT97.       Dimitris Bertsimas and John Tsitsiklis. *Introduction to Linear Optimization.* Athena Scientific, 1st edition, 1997.

BTB93.      Aharon Ben-Tal and Martin P. Bendsøe. A new method for optimal truss topology design. *SIAM J. Optim.*, 3:322–358, 1993.

BUS13.      Hande Y. Benson and Ümit Seğlam. *Mixed-Integer Second-Order Cone Programming: A Survey*, chapter 2, pages 13–36. 2013.

CK07.       E. Ward Cheney and David R. Kincaid. *Numerical Mathematics and Computing.* Brooks/Cole Publishing Co., USA, 6th edition, 2007.

CLRS01.     Thomas H. Cormen, Charles E. Leiserson, Ronald L. Rivest, and Clifford Stein. *Introduction to Algorithms.* The MIT Press, 2nd edition, 2001.

CS21.       Charles A. Cusack and David A. Santos, editors. *An Active Introduction to Discrete Mathematics and Algorithms.* USA, 2021.

CZ13.       E.K.P. Chong and S.H. Zak. *An Introduction to Optimization.* Wiley Series in Discrete Mathematics and Optimization. Wiley, 2013.

Far94.      A Faraut, J Korányi. Analysis on symmetric cones. *The Clarendon Press Oxford University Press, New York*, 1994.

Fre04.      Robert M. Freund. Newton's method for unconstrained optimization. 2004.

Goe98.      Michel X. Goemans. Semidefinite programming and combinatorial optimization. 1998.

Hig02.      Nicholas J. Higham. *Accuracy and Stability of Numerical Algorithms.* Society for Industrial and Applied Mathematics, USA, 2nd edition, 2002.

HJ90.       Roger A. Horn and Charles R. Johnson. *Matrix Analysis.* Cambridge University Press, 1990.

HRVW96.    Christoph Helmberg, Franz Rendl, Robert J. Vanderbei, and Henry Wolkowicz. An interior-point method for semidefinite programming. *SIAM Journal on Optimization*, 6(2):342–361, 1996.

HV16.       Stefan Hougardy and Jens Vygen. *Algorithmic Mathematics.* Springer Publishing Company, Incorporated, 1st edition, 2016.

JAZ12.      Siqiao Jin, K. A. Ariyawansa, and Yuntao Zhu. Homogeneous self-dual algorithms for stochastic semidefinite programming. *Journal of Optimization Theory and Applications*, 155:1073–1083, 2012.

JM07.    B. Jaradat M., ALzalg. The cycle-complete graph ramsey number r(c-8;k-8). *SUT J. Math.*, 43, 2007.

JM08.    B. Jaradat M., ALzalg. The cycle-complete graph ramsey number r(c-6;k-8) <= 38. *SUT J. Math.*, 44, 2008.

Jos89.   K.D. Joshi. *Foundations of Discrete Mathematics*. Wiley, 1989.

KMS98.   David Karger, Rajeev Motwani, and Madhu Sudan. Approximate graph coloring by semidefinite programming. *J. ACM*, 45(2):246–265, mar 1998.

KSH97.   Masakazu Kojima, Susumu Shindoh, and Shinji Hara. Interior-point methods for the monotone semidefinite linear complementarity problem in symmetric matrices. *SIAM Journal on Optimization*, 7(1):86–125, 1997.

KVL63.   M.M. Khuwārizmī, K. Vogel, and Cambridge University Library. *Mohammed ibn Musa Alchwarizmi's Algorismus: das früheste Lehrbuch zum Rechnen mit indischen Ziffern : nach der einzigen (lateinischen) Handschrift (Cambridge Un. Lib. Ms. Ii. 6.5.) in Faksimile mit Transkription und Kommentar*. Milliaria. Faksimiledrucke zur Dokumentation der Geistesentwicklung. O. Zeller, 1963.

LP21.    Bernard Lidický and Florian Pfender. Semidefinite programming and ramsey numbers. *SIAM Journal on Discrete Mathematics*, 35(4):2328–2344, 2021.

LVBL98.  Miguel Sousa Lobo, Lieven Vandenberghe, Stephen Boyd, and Hervé Lebret. Applications of second-order cone programming. *Linear Algebra and its Applications*, 284(1):193–228, 1998. International Linear Algebra Society (ILAS) Symposium on Fast Algorithms for Control, Signals and Image Processing.

MKB86.   Joe L. Mott, Abraham Kandel, and Theodore P. Baker, editors. *Discrete Mathematics for Computer Scientists  Mathematicians (2nd Ed.)*. Prentice-Hall, Inc., USA, 1986.

Mon97.   Renato D. C. Monteiro. Primal–dual path-following algorithms for semidefinite programming. *SIAM Journal on Optimization*, 7(3):663–678, 1997.

MR04.    Stephen B. Maurer and Anthony Ralston. *Discrete Algorithmic Mathematics*. AK Peters Ltd, 2004.

NN94.    Yurii Nesterov and Arkadii Nemirovskii. *Interior-Point Polynomial Algorithms in Convex Programming*. Society for Industrial and Applied Mathematics, 1994.

NT98.    Yu. E. Nesterov and M. J. Todd. Primal-dual interior-point methods for self-scaled cones. *SIAM Journal on Optimization*, 8(2):324–364, 1998.

NW88.    George L. Nemhauser and Laurence A. Wolsey. *Integer and Combinatorial Optimization*. Wiley-Interscience, USA, 1988.

PS98.    Florian A. Potra and Rongqin Sheng. On homogeneous interrior-point algorithms for semidefinite programming. *Optimization Methods and Software*, 9(1-3):161–184, 1998.

PT93.    Svatopluk Poljak and Zsolt Tuza. Maximum cuts and largest bipartite subgraphs. In *Combinatorial Optimization*, 1993.

Rad11.   Stanislaw P. Radziszowski. Small ramsey numbers. *Electronic Journal of Combinatorics*, 1000, 2011.

Ren96.   Michael Renardy. Singular value decomposition in minkowski space. *Linear Algebra and its Applications*, 236:53–58, 1996.

Ren99.      Franz Rendl. Semidefinite programming and combinatorial optimization. *Applied Numerical Mathematics*, 29(3):255–281, 1999. Proceedings of the Stieltjes Workshop on High Performance Optimization Techniques.

Ros02.      Kenneth H. Rosen. *Discrete Mathematics and Its Applications*. McGraw-Hill Higher Education, 5th edition, 2002.

SA03.       SH Schmieta and Farid Alizadeh. Extension of primal-dual interior point algorithms to symmetric cones. *Mathematical Programming*, 96(3):409–438, 2003.

SEA14.      D. Smith, M. Eggen, and R.S. Andre. *A Transition to Advanced Mathematics*. Cengage Learning, 2014.

SF04.       Peng Sun and Robert M. Freund. Computation of minimum-volume covering ellipsoids. *Operations Research*, 52(5):690–706, 2004.

SH71.       Saturnino L. Salas and Einar Hille. Calculus: One and several variables. 1971.

Str69.      Volker Strassen. Gaussian elimination is not optimal. *Numer. Math.*, 13(4):354–356, aug 1969.

TBY17.      Imene Touil, Djamel Benterki, and Adnan Yassine. A feasible primal–dual interior point method for linear semidefinite programming. *Journal of Computational and Applied Mathematics*, 312:216–230, 2017. ICMCMST 2015.

Tod01.      M. J. Todd. Semidefinite optimization. *Acta Numerica*, 10:515–560, 2001.

Tuc57.      Albert W. Tucker. 1 . dual systems of homogeneous linear relations. 1957.

VB99.       Lieven Vandenberghe and Stephen Boyd. Applications of semidefinite programming. *Applied Numerical Mathematics*, 29(3):283–299, 1999. Proceedings of the Stieltjes Workshop on High Performance Optimization Techniques.

Wat91.      David S. Watkins. *Fundamentals of Matrix Computations*. John Wiley Sons, Inc., USA, 1991.

YTM94.      Yinyu Ye, Michael J. Todd, and Shinji Mizuno. An $o(\sqrt{n}\,l)$-iteration homogeneous and self-dual linear programming algorithm. *Mathematics of Operations Research*, 19(1):53–67, 1994.

Zha98a.     Yin Zhang. On extending some primal–dual interior-point algorithms from linear programming to semidefinite programming. *SIAM Journal on Optimization*, 8(2):365–386, 1998.

Zha98b.     Yin Zhang. On extending some primal-dual interior-point algorithms from linear programming to semidefinite programming. *SIAM Journal on Optimization*, 8:365–386, 1998.

# Index

**523**

Made in the USA
Las Vegas, NV
30 November 2022

60765491R00314